LAY—Convex Sets and Their Applications

LINZ—Theoretical Numerical Analysis: An Introduction to Advanced Techniques

LOVELOCK and RUND—Tensors, Differential Forms, and Variational Principles

MARTIN—Nonlinear Operators and Differential Equations in Banach Spaces

MELZAK—Companion to Concrete Mathematics

MELZAK—Invitation to Geometry

NAYFEH—Perturbation Methods

NAYFEH and MOOK—Nonlinear Oscillations

ODEN and REDDY—An Introduction to the Mathematical Theory of Finite Elements

PASSMAN—The Algebraic Structure of Group Rings

PETRICH—Inverse Semigroups

PRENTER—Splines and Variational Methods

RIBENBOIM—Algebraic Numbers

RICHTMYER and MORTON—Difference Methods for Initial-Value Problems, 2nd Edition

RIVLIN—The Chebyshev Polynomials

ROCKAFELLAR—Network Flows and Monotropic Optimization

RUDIN—Fourier Analysis on Groups

SAMELSON—An Introduction to Linear Algebra

SCHUMAKER—Spline Functions: Basic Theory

SHAPIRO—Introduction to the Theory of Numbers

SIEGEL—Topics in Complex Function Theory

Volume 1—Elliptic Functions and Uniformization Theory

Volume 2—Automorphic Functions and Abelian Integrals

Volume 3—Abelian Functions and Modular Functions of Several Variables

STAKGOLD—Green's Functions and Boundary Value Problems

STOKER—Differential Geometry

STOKER—Nonlinear Vibrations in Mechanical and Electrical Systems

STOKER—Water Waves

TURÁN—On A New Method of Analysis and Its Applications

WHITHAM—Linear and Nonlinear Waves

WOUK—A Course of Applied Functional Analysis

ZAUDERER—Partial Differential Equations of Applied Mathematics

APPLIED NONLINEAR ANALYSIS

APPLIED NONLINEAR ANALYSIS

JEAN-PIERRE AUBIN and IVAR EKELAND

A WILEY-INTERSCIENCE PUBLICATION

JOHN WILEY & SONS

New York Chichester Brisbane Toronto Singapore

Library of Congress Cataloging in Publication Data:

Aubin, Jean Pierre.
Applied nonlinear analysis.

(Pure and applied mathematics)
"A Wiley-Interscience publication."
Bibliography: p.
Includes index.
1. Nonlinear functional analysis. I. Ekeland, I. (Ivar), 1944- . II. Title. III. Series: Pure and applied mathematics (John Wiley & Sons)
QA321.5.A93 1984 515.7 83-26011
ISBN 0-471-05998-6

Printed in the United States of America
10 9 8 7 6 5 4 3 2 1

This book is dedicated to

LAURENT SCHWARTZ

PREFACE

For a long time now, functional analysis has been linear. Since the beginning of this century, thanks mainly to David Hilbert and Stefan Banach, the theory of infinite-dimensional vector spaces has become the appropriate framework for studying linear equations. The tremendous success of this approach is illustrated by Laurent Schwartz's theory of distributions, the gateway to partial differential equations. The methods of linear analysis are now fixed by tradition in a standard package, which can be found in numerous textbooks, and belong to the common background of all mathematicians.

On the other hand, the same period also saw the growth of fixed-point theory, from Brouwer's finite-dimensional result to the Leray–Schauder theory in Banach spaces, with its application to nonlinear partial differential equations.

In more recent years, the interest in nonlinear problems and methods has dramatically increased. Scholars in many fields, from partial differential equations to economic theory, have developed methods of coping with nonlinear problems, such as solving $f(x)=0$ with x an infinite-dimensional variable or even $F(x) \ni 0$ wtih F a set-valued map.

In addition, the success of linear programming has set a pattern for the whole of optimization theory to follow: Find necessary conditions for optimality, discover dual formulations of the problem, and use them to evolve efficient numerical methods for computing the solution. This program has been completed for convex optimization and is now in the process of being carried over to some nonconvex problems. On the way, new tools have been developed for analyzing nonsmooth functions, which are constantly encountered in optimization theory.

Nonlinear analysis now stands on its own; it is no longer a subsidiary of linear analysis, but has its own methods and its own applications. Better still, it is now possible to introduce students to this very active field without going through the whole of linear analysis as a preliminary. The ideas in nonlinear analysis are simple, their proofs direct, and their applications clear. No more prerequisites are needed than the elementary theory of Hilbert spaces; indeed, many of the results are most interesting in Euclidian spaces.

In order to remain at an introductory level, we have chosen not to delve into technical difficulties or sophisticated results that are not of current use. Applications are given as soon as possible, and the theoretical parts are written with that purpose in mind. We feel that the added benefit from the availability of applications more than compensates for scattering topics.

Several themes run throughout the book. The first one is the resolution of nonlinear equations $f(x)=0$ and inclusions $F(x) \ni 0$. Chapter 6 gives a unified treatment of these problems in close relationship with game theory and mathematical economics, which provide insight and applications.

In the particular case when the equation to be solved is $V'(x)=0$ or $\partial V(x) \ni 0$, where V is a real-valued function and ∂ denotes a generalized gradient, we are dealing with a variational problem. This is the second theme of the book. In Chapter 2, these problems are studied geometrically by means of the inverse function theorem. The framework is finite dimensional or becomes so after a suitable reduction. In Chapter 4, we investigate the special case when V is convex: Any solution of $\partial V(x) \ni 0$ must then minimize V, and we enter the realm of convex optimization. We give, and sometimes improve, the classical results of convex analysis and duality theory.

In Chapter 5, we state a general variational principle, which is applicable to infinite-dimensional nonconvex situations, and give several applications to nonconvex variational problems. The tools developed in Chapter 5 are applied in Chapter 8 to a specific example of great practical importance: finding periodic solutions to differential equations of Hamiltonian type, in other words, describing periodic oscillations of conservative nonlinear systems.

Another theme is nonsmooth analysis, which is developed for the purposes of optimization and stability theory in dynamical systems. This is done in Chapter 7: Two generalized derivatives are introduced and compared and the calculus rules extended accordingly up to the inverse function theorem, even for set-valued maps.

Of course, this is but a glimpse into the vast and expanding field of nonlinear analysis. Much has been left unsaid—degree theory, inverse function theorems of Nash–Moser type, Morse theory, Liusternik–Schnirelman theory, obstacle and free boundary problems in partial differential equations, problems from differential geometry or nonlinear elasticity—and the list is depressingly long. Ours is but a personal selection, strongly motivated by our background in optimization theory. It ranges from very smooth functions to nonsmooth ones, from convex variational problems to nonconvex ones, from economics to mechanics. It is our hope that this will be enough to lead students into the field, to make them feel how lively and exciting it is, and to stimulate them to learn more.

JEAN-PIERRE AUBIN
IVAR EKELAND

Paris, France
February 1984

CONTENTS

APPLIED NONLINEAR ANALYSIS

CHAPTER 1

Background Notes

In this preliminary chapter, we group together some basic definitions, notations, and facts that will be used later in the book.

In the first section, we define set-valued maps and fix the vocabulary employed. Baire's theorem and some of its fundamental consequences are recalled in the second section. Examples of lower semicontinuous functionals on Banach spaces of functions are provided in the third section. The fourth section deals with different classical notions of differentiability for functionals and presents several examples.

These background notes end with the presentation of support functions of closed convex sets and a quite useful *closed image* theorem.

1. SET-VALUED MAPS

We shall not escape using set-valued maps in this book, for the good reason that we deal with them in a very natural way, as shown in the list of examples mentioned in the first section. Nor do we choose to regard a set-valued map from a set X to a set Y as a single-valued map from X to the set $\mathscr{P}(Y)$ of the subsets of Y: By doing so, we would lose a lot of information, since, in most cases, the structures on $\mathscr{P}(Y)$ are much poorer than the original structures on Y.

Let X and Y be two sets. A *set-valued map*, or a *correspondence* F from X to Y, is a map that associates with any $x \in X$ a subset $F(x)$ of Y, called the *image* or the *value* of F at x.

We say that a set-valued map is *proper* if there exists at least an element $x \in X$ such that $F(x) \neq \varnothing$, that is, if F is not the constant map $\varnothing$. In this case, we say that the subset

$$\text{Dom}\,(F) := \{x \in X \mid F(x) \neq \varnothing\} \tag{1}$$

is the *domain* of F.

Actually, a set-valued map F is characterized by its *graph*, the subset of $X \times Y$ defined by

$$\text{Graph}\,(F) := \{(x, y) \mid y \in F(x)\} \tag{2}$$

Indeed, if G is a nonempty subset of the product space $X \times Y$, it is the graph of the set-valued map F defined by

$$y \in F(x) \quad \text{if and only if } (x, y) \in G \tag{3}$$

The domain of F is the projection on X of Graph (F) and the *image* of F, the subset of Y defined by

$$\operatorname{Im}(F) := \bigcup_{x \in X} F(x) = \bigcup_{x \in \operatorname{Dom}(F)} F(x) \tag{4}$$

is the projection on Y of Graph (F).

The *inverse* F^{-1} of F is the set-valued map F from Y to X defined by

$$x \in F^{-1}(y) \quad \text{if and only if } y \in F(x) \tag{5}$$

or, equivalently,

$$x \in F^{-1}(y) \quad \text{if and only if } (x, y) \in \operatorname{Graph}(F) \tag{6}$$

Therefore, we obtain the formulas

$$\operatorname{Dom}(F^{-1}) = \operatorname{Im}(F), \qquad \operatorname{Im}(F^{-1}) = \operatorname{Dom}(F) \tag{7}$$

and

$$\operatorname{Graph}(F^{-1}) = \{(y, x) \in Y \times X \mid (x, y) \in \operatorname{Graph}(F)\} \tag{8}$$

We shall say that a set-valued map F from X to Y is *strict* if Dom $(F) = X$, that is, if the images $F(x)$ are nonempty for all $x \in X$. When $K \subset X$ is a nonempty subset and when F is a *strict set-valued map* from K to Y, it may be useful to "extend it" to the set-valued map F_K from X to Y defined by

$$F_K(x) := \begin{cases} F(x) & \text{when } x \in X \\ \varnothing & \text{when } x \notin K \end{cases} \tag{9}$$

whose domain Dom (F_K) is K.

When F is a set-valued map from X to Y and $K \subset X$, we denote by $F|_K$ its restriction to K.

Let ($\mathscr{P}$) be a property of a subset (for instance, closed, convex, monotone, maximal monotone, etc.). As a general rule, we shall say that a set-valued map F satisfies the property ($\mathscr{P}$) if the graph of F satisfies this property.

For instance, we shall speak of a closed, convex, monotone, maximal monotone map, which is a set-valued map whose graph is closed, convex, monotone, maximal monotone, and so on.

If the images of a set-valued map F are closed, convex, bounded, compact, and so on, we say that F is a closed-valued map, convex-valued map, bounded-valued map, compact-valued map, and so on.

When $\star$ denotes an operation on the subsets, we use the same notation for the operation on set-valued maps, which is defined by

$$F_1 \star F_2 : x \to F_1(x) \star F_2(x) \tag{10}$$

We define in that way $F_1 \cap F_2$, $F_1 \cup F_2$, $F_1 \setminus F_2$, and $F_1 + F_2$ (in vector spaces).

Similarly, if α is a map from the subsets of Y to the subsets of Y, we define

$$\alpha(F) : x \to \alpha(F(x)) \tag{11}$$

For instance $\bar{F} := \mathrm{cl}(F) : x \to \overline{F(x)}$, $\mathrm{Int}\, F : x \to \mathrm{Int}\, F(x)$, $\mathrm{co}(F) : x \to \mathrm{co}(F(x))$, $\overline{\mathrm{co}}(F) : x \to \overline{\mathrm{co}}\, F(x)$, and so on.

Let us mention the following elementary properties:

$$\begin{cases} \text{i.} & F(K_1 \cup K_2) = F(K_1) \cup F(K_2) \\ \text{ii.} & F(K_1 \cap K_2) \subset F(K_1) \cap F(K_2) \\ \text{iii.} & F(X \setminus K) \supset F(x) \setminus F(K) \\ \text{iv.} & K_1 \subset K_2 \Rightarrow F(K_1) \subset F(K_2) \end{cases} \tag{12}$$

Examples

a. The first natural instance where set-valued maps occur is the inverse f^{-1} of a single-valued map from X to Y. We always can define f^{-1} as a set-valued map whose domain is $\mathrm{Im}(f)$, which is strict when f is surjective and single-valued when f is injective.

This map plays an important role when we study equations $f(x) = y$ and are interested in the behavior of the set of solutions $f^{-1}(y)$ as y ranges over Y.

b. We shall associate with a function V from a set X to $\mathbb{R} \cup \{+\infty\}$ the set-valued map

$$\mathbf{V}_+(x) = \begin{cases} V(x) + \mathbb{R}_+ & \text{when } V(x) < +\infty \\ \varnothing & \text{when } V(x) = +\infty \end{cases} \tag{13}$$

The domain of $\mathbf{V}_+$ is the subset of elements x satisfying $V(x) < +\infty$, and the graph of $\mathbf{V}_+$ is the *epigraph* of the function V. We say that V is *proper* if $\mathbf{V}_+$ is proper and that $\mathrm{Dom}\, V := \mathrm{Dom}\, \mathbf{V}_+$ is the *domain* of V

$$\begin{cases} \text{i.} & V \text{ proper} \Leftrightarrow \{\exists x \in X \mid V(x) < +\infty\} \\ \text{ii.} & \mathrm{Dom}\, V := \{x \in X \mid V(x) < +\infty\} \end{cases} \tag{14}$$

c. Another instance of set-valued maps is associated with a family of functions

$f(\cdot, u)$ from X to Y when u ranges over a set U of parameters. In this case, we set

(15) $$F(x) = \{f(x, u)\}_{u \in U}$$

Control theory provides examples of such maps, called *parametrized maps.*

d. We shall associate with a lower semicontinuous convex function V from a Hilbert space X to $\mathbb{R} \cup \{+\infty\}$ its subdifferential $\partial V(x)$, defined by

(16) $$\partial V(x) := \left\{ p \in X^\star \mid \langle p, x \rangle - V(x) = \max_{y \in X} [\langle p, y \rangle - V(y)] \right\}$$

It is a closed convex subset of $X^\star$, which may be empty. It "generalizes" the concept of gradient in the sense that if V has a gradient $\nabla V(x) \in X^\star$ at x, then $\partial V(x) = \{\nabla V(x)\}$.

The set-valued map $x \to \partial V(x)$ will play a crucial role in many applications described in this book.

Another important fact is that the inverse of ∂V is also the subdifferential of a lower semicontinuous convex function $V^\star$, called the *conjugate* of V, defined on the dual $X^\star$ of X by

(17) $$V^\star(p) = \sup_{y \in X} [\langle p, y \rangle - V(y)]$$

More generally, when V is a locally Lipschitz function from an open subset $\mathscr{U}$ of a Hilbert space X to $\mathbb{R}$, we shall define its *generalized gradient*, $\partial V(x)$, which is a bounded closed convex subset that reduces to $\{\nabla V(x)\}$ whenever V is continuously differentiable, and which was introduced by Clarke.

e. Let W be a function from $X \times Y$ to $\mathbb{R}$. We consider the minimization problems

(18) $$\forall y \in X, \qquad V(y) = \inf_{x \in X} W(x, y)$$

The function V is called the *marginal* (or *performance* or *value*) function.

Let

$$G(y) := \{x \in X \mid W(x, y) = V(y)\}$$

be the subset of solutions to the minimization problem $V(y)$. One of the main purposes of optimization theory is to study the set-valued map G (continuity and differentiability in a suitable sense, and so on). We shall call it the *marginal map.*

It is no wonder that game theory and mathematical economics use set-valued maps in a natural way.

2. COMPLETE METRIC SPACES

A *distance* d on a set X is a real-valued function on $X \times X$ that satisfies the following for all x, y, and $z \in X$:

$$d(x, y) \geqslant 0 \tag{1}$$

$$d(x, y) = 0 \Leftrightarrow x = y \tag{2}$$

$$d(x, z) \leqslant d(x, y) + d(y, z) \tag{3}$$

The third property is the triangle inequality. A set X, together with a distance d, is a *metric space*. On such a space, we can define convergent sequences and Cauchy sequences.

A sequence x_n, $n \in \mathbb{N}$, in X is *convergent* if there exists some $\bar{x} \in X$ such that for every $\varepsilon > 0$, some $N \in \mathbb{N}$ can be found so large that

$$\forall n \geqslant N, \qquad d(x_n, \bar{x}) \leqslant \varepsilon \tag{4}$$

The point $\bar{x}$ is uniquely defined and called the limit of the sequence x_n.

A sequence x_n, $n \in \mathbb{N}$, in X is *Cauchy* if for every $\varepsilon > 0$, some $N \in \mathbb{N}$ can be found so large that

$$\forall n \geqslant N, \qquad \forall m \geqslant N, \qquad d(x_n, x_m) \leqslant \varepsilon \tag{5}$$

Any convergent sequence is Cauchy. The converse is not true without further assumptions on the sequence or on the space. For instance, if a Cauchy sequence contains a convergent subsequence, then it is itself convergent. Assumptions about the space are much more fruitful; we are led to the following definition.

DEFINITION 1
A metric space (X, d) is complete if every Cauchy sequence is convergent. ▲

Most familiar metric spaces, and certainly all those we shall deal with in this book, are complete. This includes $\mathbb{R}$, finite-dimensional vector spaces like $\mathbb{R}^n$, function spaces like L^p or $\mathscr{C}^k$, and all closed subsets thereof. It is clear from the definition that if (X, d) is complete and $F \subset X$ is closed, then (F, d) is complete.

Of course, the most important property of complete metric spaces lies in definition 1 itself: there is practically no way to prove that a sequence x_n, $n \in \mathbb{N}$, is convergent, except by proving that it is Cauchy and lies in a complete metric space. The major difficulty, of course, is to prove completeness: It may be easy for us, knowing that $\mathbb{R}$ is complete, but it certainly was a major achievement of nineteenth-century mathematics to build up $\mathbb{R}$ from the set of all rationals (noncomplete) in such a way that it *was* complete.

However, working from the definition, we obtain other properties of complete metric spaces. First an easy one; recall that the *diameter* of a subset in a metric space is the upper bound of all distances measured within that subset

(6) $$F \subset X, \qquad \text{diam } F = \sup \{d(x, y) | x \in F, y \in F\}$$

PROPOSITION 2

Let (X, d) be a metric space and F_n, $n \in \mathbb{N}$, a decreasing sequence of closed subsets whose diameters go to zero

(7) $$F_n \subset F_{n+1} \text{ and } \text{diam } F_n \to 0$$

Then there is a single point $\bar{x}$ belonging to all of the F_n

(8) $$\bigcap_{n=1}^{\infty} F_n = \{\bar{x}\}$$ ▲

Proof. Choose one point x_n in each of the F_n. We claim the sequence x_n, $n \in \mathbb{N}$, thus obtained is Cauchy. Indeed, let $\varepsilon > 0$ be given and choose N so large that:

$$\text{diam } F_N \leqslant \varepsilon$$

If both n and m are greater than N, then both x_n and x_m belong to F_N, since the sequence is decreasing, and we get the Cauchy property

$$\forall n \geqslant N, \qquad \forall m \geqslant N, \qquad d(x_n, x_m) \leqslant \text{diam } F_N \leqslant \varepsilon$$

Since the space (X, d) is complete, the Cauchy sequence x_n has to converge to some $\bar{x}$. Any one subset F_k contains $\bar{x}$, since it is closed and contains the sequence x_n except for the starting terms $x_1, \ldots, x_{k-1}$. Since $\bar{x}$ belongs to all of the F_n, it belongs to their intersection, as in equation (8).

Let $\tilde{x}$ be another point with the same property. Since both $\bar{x}$ and $\tilde{x}$ belong to every F_n, we must have

$$\forall n, \qquad d(\bar{x}, \tilde{x}) \leqslant \text{diam } F_n$$

Since the right-hand side goes to zero, we get $d(\bar{x}, \tilde{x}) = 0$; hence, $\bar{x} = \tilde{x}$, and uniqueness is proved. ■

Example

Equation (7), particularly the fact that the diameters must go to zero, is essential. Let us give two counterexamples, one in finite dimension and the other infinite.

Take $X = \mathbb{R}$ with the usual distance, and set $F_n := [n, +\infty]$. This is a decreasing sequence of closed subsets with empty intersection (note that diam $F_n = +\infty$).

Take $X = \ell^2$, the Hilbert space of square-summable sequences, $x = (\xi_i, i \in \mathbb{N})$, with $\sum \xi_i^2 < \infty$. Set

$$F_n := \left\{ x = (0, \ldots, 0, \xi_n, \xi_{n+1}, \ldots) \in \ell^2 \,\middle|\, \sum_{i=n}^{\infty} \xi_i^2 = 1 \right\}$$

This is a decreasing sequence of closed subsets with empty intersection (note that diam $F_n = 1$).

One step further away from the definition, we find a very important result: the Baire theorem. ■

THEOREM 3

Let (X, d) be a complete metric space and U_n, $n \in \mathbb{N}$, a sequence of open sets, each of which is dense in X. Then so is their intersection

$$X = \operatorname{cl} \bigcap_{n=1}^{\infty} U_n \tag{9}$$ ▲

Proof. Call this intersection G. We want to prove that $G \cap B_0 \neq \varnothing$, where B_0 is any prescribed open ball in X.

Since U_1 is dense and B_0 is open, $U_1 \cap B_0$ is nonempty. Since U_1 is open, so is $U_1 \cap B_0$. It follows that $U_1 \cap B_0$ contains some closed ball $\bar{B}_1$, the diameter of which can be taken less than 1.

We now proceed by induction. Assume that we have found a closed ball

$$\bar{B}_n \subset B_0 \cap \bigcap_{k=1}^{n} U_k \tag{10}$$

with diam $\bar{B}_n \leq 1/n$. We then call B_n the interior of $\bar{B}_n$: It is an open ball. Since U_{n+1} is open and dense, the intersection $U_{n+1} \cap B_n$ is open and nonempty. Thus, we are able to pick another closed ball

$$\bar{B}_{n+1} \subset B_n \cap U_{n+1} \subset B_0 \cap \bigcap_{k=1}^{n+1} U_k$$

$$\operatorname{diam} \bar{B}_{n+1} < \frac{1}{n+1}$$

We now apply theorem 2 to the sequence $\bar{B}_n$, $n \in \mathbb{N}$ and obtain a point $\bar{x}$ that belongs to all of the $\bar{B}_n$. It follows from equation (10) that

$$\bar{x} \in B_0 \quad \text{and} \quad \bar{x} \in U_n, \qquad \text{for all } n$$

Hence, $\bar{x} \in G \cap \bar{B}_0$, and the result is proved. ■

An alternative version is

THEOREM 4

Let (X, d) be a complete metric space and F_n, $n \in \mathbb{N}$, a sequence of closed sets. If their union covers X, then one of them has nonempty interior

$$\bigcup_{n=1}^{\infty} F_n = X \Rightarrow \exists n\colon \mathring{F}_n \neq \varnothing \tag{11}$$ ▲

Proof. Set $U_n = X \setminus F_n$. The U_n are open sets with empty intersection. If they were all dense in X, so would their intersection be by the Baire theorem. Since the empty set cannot be dense, it follows that one of the U_n at least, say U_N, is not dense in X. This means that there is an open ball B in X that does not intersect U_N.

$$B \cap U_N = \varnothing \Leftrightarrow B \subset F_N$$

It follows that the interior of F_N contains B. ■

In both these theorems, the fact that we are dealing with *sequences*, that is, countable families, is very important. For instance, with each real number x associate the open set

$$U_x := \{y \in \mathbb{R} \mid y \neq x\}$$

which is obviously dense in $\mathbb{R}$. Now

$$G := \cap\, U_x \qquad \text{for } x \in Q$$

is dense (it is just the set of irrational numbers), whereas

$$G' := \cap\, U_x \qquad \text{for } x \in \mathbb{R}$$

is empty! The reason the Baire theorem applies in one case and not in the other is because the first intersection is countable (recall that the set Q of rational numbers is countable), whereas the second is not.

Note also that not all dense subsets can be obtained as countable intersections of open dense sets. In fact, we introduce

DEFINITION 5

Let (X, d) be a complete metric space. A subset $G \subset X$ is residual *if it contains a countable intersection of open dense subsets.* ▲

It follows from the definition that the intersection of two residual subsets, or a countable number of residual subsets, is still residual. Indeed

$$G_1 \cap G_2 \supset \left(\bigcap_n U_n \right) \cap \left(\bigcap_k V_k \right) = \bigcap_{(n,k)} (U_n \cap V_k)$$

$$\bigcap_n G_n \supset \bigcap_n \left(\bigcap_k U_{n,k} \right) = \bigcap_{(n,k)} U_{n,k}$$

where n and k range over $\mathbb{N}$ and (n, k) over $\mathbb{N} \times \mathbb{N}$, which is countable.

The Baire theorem states that in a complete metric space, any residual subset is dense. For instance, two residual subsets have a nonempty intersection: It is residual, hence dense, hence nonempty. In the same way, any countable family of residual subsets has a nonempty intersection. In the real line R, for instance, the set $G := \mathbb{R} \setminus Q$ of irrational numbers is residual, whereas Q itself is not (although it is dense!). For if G and Q were both residual, they would have to intersect, which is not the case.

We see that residual subsets behave much better than ordinary dense subsets: Two dense subsets need not intersect (e.g., Q and $\mathbb{R} \setminus Q$), whereas two residual subsets always do. In this way, we can think of residual subsets G as being "full" subsets of X, as $\mathbb{R} \setminus Q$ is a full subset of $\mathbb{R}$. The following definition emphasizes this and has become very popular in various areas of mathematics.

DEFINITION 6

A property $P(x)$, where x runs through a complete metric space (X, d) is called generic *if the set of points $G \subset X$ where it holds true is residual.* ▲

For instance, the property that a real number be irrational is generic. The interesting thing is that if two properties $P_1(x)$ and $P_2(x)$ are generic, then so is the property "$P_1(x)$ *and* $P_2(x)$." For that matter, if the properties $P_n(x)$, for $n \in \mathbb{N}$, are all generic, then so is the property that all $P_n(x)$ hold simultaneously.

It happens very often in mathematics, and it will happen to us, that to prove that there is one point $\bar{x} \in X$ such that $P(\bar{x})$ holds, we have to prove that $P(x)$ is generic. In other words, we find a large number of points where $P(x)$ is satisfied, just because we need *one*.

3. BANACH SPACES

A *norm* on a vector space X is a function $x \rightarrow \|x\|$ from X to $\mathbb{R}$ that satisfies the following for all x, y in X and all $\lambda \in \mathbb{R}$:

(1) $$\|x\| \geqslant 0$$

(2) $$\|x\| = 0 \Leftrightarrow \|x\| = 0$$

(3) $$\|\lambda x\| = |\lambda| \, \|x\|$$

(4) $$\|x + y\| \leqslant \|x\| + \|y\|$$

It follows that $d(x, y) = \|x-y\|$ is a distance on X. Unless otherwise specified, X will be endowed with this distance, which turns it into a metric space: It is the so-called *norm topology* on X. If the space (X, d) is complete, it will be called a *Banach space*.

In the normed linear space X, the (closed) unit ball B is defined by

$$B := \{x \in X | \; \|x\| \leqslant 1\} \tag{5}$$

The corresponding sphere S will be the set of points in X at a distance 1 from the origin.

The closed ball with center $x \in X$ and radius $\rho = 0$ is the subset $x + \rho B$. The open ball, denoted by $x + \rho \mathring{B}$, is obtained by removing the boundary

$$x + \rho B = \langle y \in X | \; \|x-y\| \leqslant \rho\}$$

$$x + \rho \mathring{B} = \{y \in X | \; \|x-y\| < p\}$$

It should be noted that balls in a normed linear space are *never* compact unless the space is finite dimensional (a theorem of Riesz). Finite-dimensional spaces are just $\mathbb{R}^n$, for some suitable n, albeit perhaps with some non-Euclidian norm, and all closed balls will certainly be compact. Now this cannot happen in infinite-dimensional spaces, which puts us on the lookout for another, weaker, topology.

If X is a normed linear space, denote by $X^\star$ the set of all *continuous* linear functionals p on X. It is a normed linear space, and we call it the (topological) dual of X; its norm is given by

$$\|p\|_\star = \sup \{|\langle x, p\rangle| \; | \; \|x\| \leqslant 1\}. \tag{6}$$

The *weak topology* (as opposed to the norm topology) on X will be defined as follows: x converges weakly to $\bar{x}$ in X if for every *fixed* p in $X^\star$

$$\langle p, x\rangle \rightarrow \langle p, \bar{x}\rangle \tag{7}$$

Of course, if the point x converges to $\bar{x}$ in the norm topology, then it converges in the weak topology. The converse is not true, except again in the finite-dimensional case. For most practical cases, that is, for $L^p(\Omega)$ and $W^{k,p}(\Omega)$, with $1 \leqslant p < \infty$, the weak topology is metrizable, so that we can be content with considering sequences x_n.

The main usefulness of weak topologies lies in the following result.

THEOREM 1

Assume X is reflexive, that is, $X = (X^\star)^\star$. Then all norm-closed balls in X are weakly compact. ▲

The most important example of reflexive spaces are *Hilbert spaces*, such as the space $\ell^2(\mathbb{N})$ of square-summable sequences or the space $L^2(\Omega)$ of square-integrable functions over some subset $\Omega \subset \mathbb{R}^n$ or the Sobolev spaces $H^k(\Omega)$ and $H_0^k(\Omega)$.

But there are other examples, such as the spaces $\ell^p(\mathbb{N})$ and $L^p(\Omega)$, for $1 < p < \infty$, the duals of which are $\ell^q(\mathbb{N})$ and $L^q(\Omega)$, $p^{-1} + q^{-1} = 1$, or the Sobolev spaces $W^{p,k}(\Omega)$ and $W_0^{p,k}(\Omega)$, always with $1 < p < \infty$ and $k < \infty$.

Now that we have indeed found some compact subsets in some infinite-dimensional Banach spaces, we shall try to exploit this for minimization purposes. For this, we need yet another notion, lower semicontinuity of functionals. The following definition is stated in terms of a *topological* space, that is, it can apply to either the norm topology or the weak topology of a Banach space. Note also that we shall allow $+\infty$ as a possible value for the functions we consider. This device will be very useful in minimization problems.

DEFINITION 2

Let X be a topological space. A function $U: X \to \mathbb{R} \cup \{+\infty\}$ is lower semicontinuous (sometimes shortened to l.s.c.) *if for each point $\bar{x} \in X$, we have*

$$\liminf_{x \to \bar{x}} U(x) \geq U(\bar{x}) \tag{8}$$

▲

This means that when x goes to $\bar{x}$ in X, all cluster points of $U(x)$ in $\mathbb{R} \cup \{+\infty\}$ are to be above $U(\bar{x})$. In other words, jumps are allowed [inequality (8) may be strict], but they must occur downward. The most useful characterization of such functions is given by the following classical result.

PROPOSITION 3

Let X be a topological space, $U: X \to \mathbb{R} \cup \{+\infty\}$ a function, epi U *its* epigraph

$$\text{epi } U = \{(x, a) \in X \times \mathbb{R} \mid a \geq U(x)\}$$

Then U is lower semicontinuous if and only if epi U *is a closed subset of $X \times \mathbb{R}$.* ▲

Of course, a function $U: X \to \mathbb{R} \cup \{-\infty\}$ will be called upper semicontinuous if for each point $\bar{x} \in X$, we have

$$\limsup_{x \to \bar{x}} U(x) \leq U(\bar{x})$$

In other words, U is upper semicontinuous if $(-U)$ is lower semicontinuous. If a finite function U is both upper and lower semicontinuous, it is continuous.

Lower semicontinuity is a much weaker property than continuity, and it will be much easier to check in practical situations. On the other hand, it is also all that is required for the purposes of minimization, as the following result shows.

PROPOSITION 4

Let the topological space X be compact. Then any lower semicontinuous function $U: X \to \mathbb{R} \cup \{+\infty\}$ attains its minimum

$$\exists \bar{x}: \forall x \in X, \qquad U(\bar{x}) \leqslant U(x) \qquad \blacktriangle$$

Let us now go back to Banach spaces. We have stressed the fact that there are two possible topologies on such a space, the norm (strong) topology or the weak one. We speak accordingly of strongly or weakly lower semicontinuous functions, and they are not the same. Indeed, we note the following important result.

PROPOSITION 5

Let X be a Banach. Any weakly l.s.c. *function is strongly* l.s.c. *The converse is true in the convex case: Any convex function that is strongly* l.s.c. *is also weakly* l.s.c. ▲

Proof. Let $U: X \to \mathbb{R} \cup \{+\infty\}$ be some function and consider its epigraph. Any weakly closed subset of $X \to \mathbb{R}$ is also strongly closed, so by proposition 3, if U is weakly l.s.c., it also is strongly l.s.c.

Conversely, by the Hahn–Banach theorem, any strongly closed subset of $X \times \mathbb{R}$ that is convex is also weakly closed. The result follows by proposition 3. ■

This leads to the following consequence, which is about the only way to find minimizers in infinite-dimensional problems. Note that it is stated in terms of the norm topology, although its proof requires the weak topology.

PROPOSITION 6

Let X be a reflexive Banach space and $U: X \to \mathbb{R} \cup \{+\infty\}$ a convex and lower semicontinuous function. Let K be a bounded subset of X, convex and closed. Then U attains its minimum on K.

$$\exists \bar{x} \in K: U(\bar{x}) \leqslant U(x), \qquad \textit{for all } x \in K \qquad \blacktriangle$$

Proof. Switch to the weak topology. The function U is still lower semicontinuous by proposition 5. The subset K is still closed by the Hahn–Banach theorem (see Section 5). Being bounded, K is contained in some closed ball, which, by theorem 1, is weakly compact. Now any closed subset of a compact set is compact, so K is (weakly) compact. By proposition 4, U attains its minimum on K. ■

Of course, convexity is a very stringent assumption. In Chapters 2 and 5, we shall introduce ways of doing without it and still obtaining some information about the existence of minimizers.

A typical way of constructing functionals on function spaces is by using integrals. We conclude by studying such functionals on L^2.

Example 1

Let Ω be an open subset of $\mathbb{R}^n$, and let $u: \Omega \times \mathbb{R}^k \to \mathbb{R} \cup \{+\infty\}$ be a borelian function such that

a. $$\forall \omega \in \Omega, \quad y \to u(\omega, y) \quad \text{is lower semicontinuous}$$

b. $$\forall (\omega, y) \in \Omega \times \mathbb{R}^k, \quad u(\omega, y) \geqslant 0$$

Define a functional U on $L^2(\Omega)$ by

$$\forall x \in L^2(\Omega), \qquad U(x) = \int_\Omega u(\omega, x(\omega)) d\omega$$

The right-hand side is always well defined, since the function $\omega \to u(\omega, x(\omega))$ is measurable and nonnegative: The integral is either finite or $+\infty$. It follows that U is well-defined, with values in $\mathbb{R} \cup \{+\infty\}$. (This could be the case even if u itself were restricted to values in $\mathbb{R}$.)

We claim that $U: L^2(\Omega) \to \mathbb{R} \cup \{+\infty\}$ is lower semicontinuous. ▲

Indeed, let x_n, $n \in \mathbb{N}$, be a sequence converging to some $\bar{x}$ in $L^2(\Omega)$. We first extract a subsequence $x_{n'}$ such that

$$\liminf_{n \to \infty} U(x_n) = \lim_{n' \to \infty} U(x_{n'})$$

and then we extract from $x_{n'}$ a second subsequence $x_{n''}$ that converges almost everywhere

$$x_{n''}(\omega) \xrightarrow{\text{a.e.}} \bar{x}(\omega) \quad \text{in } \mathbb{R}^n$$

From assumption (a) it follows that for almost every ω in Ω

$$\liminf_{n'' \to \infty} u(\omega, x_{n''}(\omega)) \geqslant u(\omega, \bar{x}(\omega))$$

Integrating both sides yields

$$\int_\Omega \liminf_{n'' \to \infty} u(\omega, x_{n''}(\omega)) d\omega \geqslant \int_\Omega u(\omega, \bar{x}(\omega)) d\omega$$

Since the integrands are nonnegative, we can apply Fatou's lemma

$$\liminf_{n'' \to \infty} \int_\Omega u(\omega, x_{n''}(\omega)) d\omega \geqslant \int_\Omega \liminf_{n'' \to \infty} u(\omega, x_{n''}(\omega)) d\omega$$

Adding the two last inequalities

$$\liminf_{n''\to\infty} \int_\Omega u(\omega, x_{n''}(\omega))d\omega \geq \int_\Omega u(\omega, \bar{x}(\omega))d\omega$$

But the integral on the left-hand side is $U(x_{n''})$, and the right-hand side is $U(\bar{x})$. Finally, we obtain the desired result

$$\liminf_{n\to\infty} U(x_n) = \lim_{n''\to\infty} U(x_{n''}) \geq U(\bar{x})$$ ■

Example 2
Let Ω be a borelian subset of $\mathbb{R}^n$, and let $u: \Omega \times \mathbb{R}^k \to \mathbb{R} \cup \{+\infty\}$ be a borelian function such that for some $a \in L^1(\Omega)$ and $c \in \mathbb{R}$

(a) $\forall\omega, \quad y \to u(\omega, y)$ is lower semicontinuous

(b) $\forall(\omega, y), \quad u(\omega, y) \geq -a(\omega) - cy^2$

Then the functional $U: L^2(\Omega) \to \mathbb{R} \cup \{+\infty\}$, defined as before, is lower semicontinuous. ▲

Indeed, consider the functional V on $L^2(\Omega)$ defined by

$$V(x) = \int_\Omega [u(\omega, x(\omega)) + a(\omega) + cx^2(\omega)]d\omega$$

$$= U(x) + \int_\Omega a(\omega)d\omega + c\|x\|_2^2$$

The integrand, inside the brackets, is nonnegative, so V is lower semicontinuous by example 1. But V differs from U by a constant [the integral of (a)] plus a continuous function (the square of the norm in L^2). So U itself must be lower semicontinuous.

Example 3
Let Ω be a borelian subset of $\mathbb{R}^n$, and let $u: \Omega \times \mathbb{R}^k \to \mathbb{R}$ satisfy the following:

(a) $\forall y \in \mathbb{R}^k, \quad \omega \to u(\omega, y)$ is measurable

(b) $\forall \omega \in \Omega, \quad y \to u(\omega, y)$ is continuous

(c) $\forall(\omega, y), \quad |u(\omega, y)| \leq a(\omega) + cy^2$

for some $a \in L^1(\Omega)$ and $c \in \mathbb{R}$. Then the functional $U: L^2(\Omega) \to \mathbb{R}$ is continuous. ▲

Indeed, we break inequality (c) into two parts

$$u(\omega, y) \geqslant -a(\omega) - cy^2$$

$$-u(\omega, y) \geqslant -a(\omega) - cy^2$$

The first one with example 2 tells us that the functional U is lower semicontinuous with values in $\mathbb{R} \cup \{+\infty\}$, and the second one tells us that the functional $-U$ is also lower semicontinuous with values in $\mathbb{R} \cup \{+\infty\}$. The result follows immediately. ■

Example 4: A Minimization Problem Without Solutions
Let Ω be the interval (0, 1) in $\mathbb{R}$, and consider the functional U: $H_0^1(0, 1) \rightarrow \mathbb{R} \cup \{+\infty\}$ defined by

$$U(x) = \int_0^1 [x^2(\omega) + (1 - \dot{x}(\omega)^2)^2] \, d\omega$$

with $\dot{x} := dx/d\omega$. Recall that the functions x in $H_0^1(0, 1)$ are all continuous and vanish on the boundary

$$x(0) = 0 = x(1)$$

We claim that there is no point $\bar{x}$ that minimizes U on $X = H_0^1(0, 1)$. In other words, the problem of minimizing U on X has no solution

$$\forall x \in X, \qquad U(x) > \inf_X U \qquad \blacktriangle$$

To see this, we note that the infimum of U is zero. On the one hand, U is clearly nonnegative, that is, inf $U \geqslant 0$. A particular sequence x_n can be built in X such that $U(x_n) \rightarrow 0$. We define it by cutting up the interval (0, 1) into $2n$ equal intervals, $I_k = ((k-1)/2n, k/2n)$, and setting

$$\dot{x}_n(\omega) = +1 \quad \text{if} \quad \omega \in I_k \quad \text{with } k \text{ odd}$$

$$\dot{x}_n(\omega) = -1 \quad \text{if} \quad \omega \in I_k \quad \text{with } k \text{ even}$$

This, together with $x_n(0) = 0$, defines a function $x_n \in H_0^1(0, 1)$. It is easily checked that $|x_n(\omega)| \leqslant 1/2n$ and $\dot{x}_n(\omega)^2 = 1$ for almost every ω in (0, 1). Substituting this into the definition of U, we have

$$U(x_n) = (2n)^{-2} \rightarrow 0$$

Finally, inf $U = 0$ as claimed.

On the other hand, there could never be an $\bar{x}$ with $U(\bar{x})$ actually zero. Indeed, the integrand is the sum of two squares and could not be zero without each of them being zero. The condition

$$U(\bar{x})=0$$

breaks down into two pointwise conditions

$$\bar{x}^2(\omega)=0 \qquad \text{for almost every } \omega$$

$$\frac{d\bar{x}}{d\omega}(\omega)=\pm 1 \quad \text{for almost every } \omega$$

These conditions are clearly incompatible: There is no $\bar{x}$ minimizing U. ■

Note that the functional U itself is lower semicontinuous on H_0^1, by example 1. Changing $(1-\dot{x}^2)^2$ to $(1-\dot{x}^2)^2(1+\dot{x}^2)^{-1}$ makes it continuous, and even C^1, but the corresponding minimization problem will still have no solution (argument unchanged). In other words, there is nothing pathological about the function U that causes this situation to occur.

4. DIFFERENTIABLE FUNCTIONALS

In this section, X will be a normed linear space and $X^\star$ its dual. Let U be a real-valued function defined on an open subset, x some point in X, and p some continuous linear functional. There are several possible meanings to the statement that p is the derivative of U at x. We list four of them:

(a) *Gâteaux Differentiability.* For any $v \in X$

$$\lim_{h\to 0} \frac{1}{h}\left[U(x+hv)-U(x)-\langle p, hv\rangle\right]=0$$

(b) *Frêchet Differentiability.*

$$\lim_{v\to 0} \frac{1}{\|v\|}\left[U(x+v)-U(x)-\langle p, v\rangle\right]=0$$

(c) *Strict Differentiability.*

$$\lim_{\substack{y\to x\\ v\to 0}} \frac{1}{\|v\|}\left[U(y+v)-U(y)-\langle p, v\rangle\right]=0$$

Any one of these definitions will give a single possible value for $p \in X^\star$, henceforth denoted by $U'(x)$ or $\nabla U(x)$, to show dependence on the functional U and the point x, and called the *gradient*, or the *derivative*, of U at x.

We give a fourth definition

(d) *The C^1 Property.* U is Gâteaux differentiable on a neighborhood of x in X, and $U'(y) \to U'(x)$ in $X^\star$ when $y \to x$ in X.

LEMMA 1

(d) ⇒ (c) ⇒ (b) ⇒ (a), with the same p. ▲

Proof. The last two, **(c) ⇒ (b)** and **(b) ⇒ (a)**, are obvious. As for the first one, assume **(d)** is satisfied, choose $\varepsilon > 0$ so small that U is Gâteaux differentiable on the ball with center x and radius ε, and pick any y and v with $\|y-x\| \leqslant \varepsilon/3$ and $\|v\| \leqslant \varepsilon/3$. Consider the function $\phi = [0, 1] \to \mathbb{R}$ defined by

$$\phi(t) = U(y+tv) - U(y)$$

It is derivable, and we can use the mean value theorem

$$\exists \theta \in [0, 1]: \qquad \phi(1) - \phi(0) = \phi'(\theta)$$

$$U(y+v) - U(y) = \langle U'(y+\theta v), v \rangle$$

When $y \to x$ and $v \to 0$, we have

$$[U(y+v) - U(y) - \langle p, v \rangle] \|v\|^{-1} = \langle U'(y+\theta v) - U'(y), v\|v\|^{-1} \rangle \to 0$$

which is exactly property **(c)**. ■

Some comments on these definitions may be useful. Gâteaux differentiability is the weakest notion—and therefore the easiest to check; for purposes of minimization, it is often enough. For instance, it is easily seen that if a Gâteaux-differentiable function U attains its minimum over X at some point $\bar{x}$, then its derivative at $\bar{x}$ is zero

$$U(\bar{x}) = \inf_X U \Rightarrow U'(\bar{x}) = 0 \in X^\star \tag{1}$$

On the other hand, for matters relying on the inverse function theorem, Gâteaux differentiability is not enough: We require strict differentiability at least.

To understand Gâteaux differentiability, we write it as follows.

$$\lim_{h \to 0+} \frac{1}{h} [U(x+hv) - U(x)] = \langle U'(x), v \rangle \tag{2}$$

The left-hand side is the directional derivative toward v, and Gâteaux differentiability simply expresses that it is a continuous linear functional of v. On the other hand, the function U need not even be continuous at x! The reader will check that the function $U\colon \mathbb{R}^2 \to \mathbb{R}$ defined by

$$\begin{cases} U(x, y) = \left(\dfrac{y}{x}\right)(x^2 + y^2) & \text{for } x \neq 0 \\ U(0, y) = 0 \end{cases}$$

is Gâteaux differentiable, but not continuous, at the origin [with $U'(0)=0$].

Fréchet differentiability, on the other hand, implies continuity. It simply means that the first-order Taylor expansion is valid

$$U(x+v) = U(x) + \langle U'(x), v\rangle + \varepsilon(v)\|v\| \tag{3}$$

with $\varepsilon(v) \to 0$ when $\|v\| \to 0$. Strict differentiability will mean that the same expansion is going to be valid in a uniform way for points y close to x

$$U(y+v) = U(y) + \langle U'(x), v\rangle + \varepsilon(v, y)\|v\| \tag{4}$$

with $\varepsilon(v, y) \to 0$ when $v \to 0$ uniformly for all y in a neighbourhood of x. As for the C^1-property, it means that the derivative $U'(x)$ depends continuously on the point x, that is, U' is continuous as a (nonlinear) map from X to $X^\star$.

Observe that the definition of Fréchet differentiability involves a priori knowledge of the gradient! Therefore, to compute a Fréchet derivative, we must begin by computing the limit [equation (2)] of the differential quotients by means of the usual calculus of functions of one variable, then check whether the dependence on v is linear and continuous, and, finally, verify whether property (b) holds true.

Even the requirement of Gâteaux differentiability is too stringent, for the limit of the differential quotients may not exist, and if it does exist, it may be either nonlinear or discontinuous. We shall relax the concept of limit of differential quotients in several ways to obtain even weaker notions of directional derivatives, retaining enough properties, however, to be useful. We shall investigate these properties for convex functions in Chapter 4 and for more general functions in Chapter 7.

Now for some examples.

Example 1

Let Ω be an open subset of $\mathbb{R}^n$, with the usual Lebesgue measure $d\omega$. Let there be given a function $u\colon \Omega \times \mathbb{R}^k \to \mathbb{R}$, borelian with respect to both variables (ω, x). Assume that

(5) for any fixed $\omega \in \Omega$, the function $x \to u(\omega, x)$ is C^1 over $\mathbb{R}^k$

(6) there is some $a \in L^2(\Omega, \mathbb{R})$ and some constant b such that

$$|u'_y(\omega, y)| \leqslant a(\omega) + b|y| \qquad \text{for all } y \in \mathbb{R}^k$$

(7)
$$\int_\Omega |u(\omega, 0)| d\omega < \infty$$

Then the functional $U: L^2(\Omega, \mathbb{R}^k) \to \mathbb{R}$ given by

$$U(x) = \int_\Omega u(\omega, x(\omega)) d\omega$$

is well defined and Gâteaux differentiable. ▲

To see this, we must make a computation. Let x and v be given in $L^2(\Omega, \mathbb{R}^k)$. For each fixed $\omega \in \Omega$, consider the derivative

(8)
$$\frac{\partial}{\partial t} u(\omega, x(\omega) + tv(\omega)) = u'_y(\omega, x(\omega) + tv(\omega)) \cdot v(\omega)$$

Using estimate (6), we obtain

(9)
$$\left| \frac{\partial}{\partial t} u(\omega, x(\omega) + tv(\omega)) \right| \leqslant |v(\omega)|(a(\omega) + b|x(\omega) + tv(\omega)|)$$
$$\leqslant |v(\omega)|(a(\omega) + b|x(\omega)| + b|v(\omega)|)$$

provided $0 \leqslant t \leqslant 1$. We note that the right-hand side is an integrable function of ω

(10)
$$\left| \frac{\partial}{\partial t} u(\omega, x(\omega) + tv(\omega)) \right| \leqslant g(\omega), \qquad \text{with } g \in L^1(\Omega, \mathbb{R})$$

The consequences are twofold. First, for any $v \in L^2$, we have, setting $x = 0$ in the preceding formula,

$$u(\omega, v(\omega)) = u(\omega, 0) + \int_0^1 \frac{\partial}{\partial t} u(\omega, tv(\omega)) dt$$
$$|u(\omega, v(\omega))| \leqslant |u(\omega, 0)| + g(\omega)$$
$$|U(v)| \leqslant \int |u(\omega, 0)| d\omega + \int g(\omega) d\omega < \infty$$

So U is well defined. We now take any $x \in L^2$ and write the directional derivative toward $v \in L^2$

$$\frac{d}{dt} U(x+tv)\bigg|_{t=0} = \frac{\partial}{\partial t} \int_\Omega u(\omega, x(\omega)+tv(\omega))d\omega$$

Thanks to the estimate in inequality (10), we can use the Lebesgue theorem on the differentiation of integrals with respect to a parameter

$$\frac{\partial}{\partial t} \int_\Omega u(\omega, x(\omega)+tv(\omega))d\omega = \int_\Omega \frac{\partial}{\partial t} u(\omega, x(\omega)+tv(\omega))d\omega$$

Using equation (8), we finally obtain

$$\frac{d}{dt} U(x+tv)\bigg|_{t=0} = \int_\Omega u_y'(\omega, x(\omega))v(\omega)d\omega \tag{11}$$

Note that because of condition (6), the function $\omega \to u_y'(\omega, x(\omega))$ is square integrable, just as x itself

$$u_y' \circ x \in L^2(\Omega, \mathbb{R}^k)$$

so that the right-hand side of equation (11) is indeed a continuous linear functional on $X = L^2(\Omega, \mathbb{R}^k)$. We write it as

$$\frac{d}{dt} U(x+tv)\bigg|_{t=0} = \langle u_y' \circ x, v \rangle \tag{12}$$

$$U'(x) = u_y' \circ x \in X^\star \tag{13}$$

■

Example 2

Same assumptions as example 1. We now claim that U has the C^1 property.

We have already shown that U is Gâteaux differentiable and that its derivative U' is given by equation (13). We now have to show that U' is a continuous map from X to $X^\star$, that is, from $L^2(\Omega, \mathbb{R}^k)$ into itself. This is a consequence of a theorem by Krasnosel'skii, which we now state as theorem 2.

THEOREM 2

Let X and Y be two Banach spaces, Ω a Borel subset of $\mathbb{R}^n$ and $u: \Omega \times X \to Y$ a mapping, measurable with respect to ω in Ω, continuous with respect to x in X. For all x in $L^p(\Omega, X)$, let $\phi(x)$ be the (measurable) function from Ω to Y defined by

$$\phi(x): \omega \to u(\omega, x(\omega)) \tag{14}$$

If ϕ maps $L^p(\Omega, X)$ into $L^q(\Omega, Y)$, then it is continuous $(1 \leq p, q < \infty)$. ▲

Proof. We shall prove that ϕ is continuous at the origin 0; continuity at any point $\bar{x}$ follows by replacing $\phi(x)$ with $\phi(x-\bar{x})$.

Let x_n, $n \in \mathbb{N}$, be any sequence of functions in $L^p(\Omega, X)$ such that $\|x_n-\bar{x}\|_p \to 0$. We are going to show that there is a subsequence x_{n_k}, $k \in \mathbb{N}$, such that $\phi(x_{n_k}) \to \phi(0)$. This is enough to ensure continuity at the origin. We choose the subsequence x_{n_k} to satisfy $\|x_{n_k}\|_p^p \leqslant 2^{-k}$. It is well known that this implies that $x_{n_k}(\omega) \to 0$ almost everywhere. It follows that

$$u(\omega, x_{n_k}(\omega)) \to u(\omega, 0) \quad \text{a.e.} \tag{15}$$

In other words, $\phi(x_{n_k}) \to \phi(0)$ almost everywhere.

To show convergence in L^q, we have to use Lebesgue's theorem. For this, we note that for almost every ω in Ω, the sequence $u(\omega, x_{n_k}(\omega))$ converges to $u(\omega, 0)$ in Y, and, hence, there must be some $k(\omega) \in \mathbb{N}$ such that the distance from $u(\omega, x_{n_k}(\omega))$ to $u(\omega, 0)$ is maximum. We choose the $k(\omega)$ in such a way that the function defined by

$$\bar{x}(\omega) = x_{k(\omega)}(\omega)$$

is measurable (here, we need a measurable selection theorem). By definition, we have

$$\|u(\omega, x_{n_k}(\omega)) - u(\omega, 0)\| \leqslant \|u(\omega, \bar{x}(\omega)) - u(\omega, 0)\| \quad \text{a.e.} \tag{16}$$

We claim that $\bar{x}$ belongs to L^p, indeed

$$\begin{aligned}\int_\Omega \|\bar{x}(\omega)\|^p d\omega &\leqslant \int_\Omega \operatorname{Sup}_k \|x_{n_k}(\omega)\|^p d\omega \\ &\leqslant \int_\Omega \sum_k \|x_{n_k}(\omega)\|^p d\omega \\ &\leqslant \sum_k \|x_{n_k}\|_p^p \leqslant \sum_k 2^{-k} < \infty\end{aligned}$$

By the assumption on ϕ, it follows that $\phi(\bar{x})$ has to belong to L^q. The right-hand side of inequality (16) thus belongs to L^q. Putting (15) and (16) together and using Lebesgue's theorem, we obtain the desired result

$$\phi(x_{n_k}) \to \phi(0) \quad \text{in} \quad L^q(\Omega, Y)$$ ■

This applies immediately to the map $x \to u'_y \circ x$ in example 1. Indeed, because of estimate (6), this map sends $L^2(\Omega, \mathbb{R}^k)$ into itself

$$\|u'_y \circ x\|_2 \leqslant \|a + b|x|\,\|_2 \leqslant \|a\|_2 + b\|x\|_2$$

It follows that this map is continuous. Because of equation (13), the function U has the C^1 property. ■

Example 3
Again take a function u satisfying condition (5), (6), and (7) in example 1, with $k=n$. This time, we are going to define a functional V on the Sobolev space $H_0^1(\Omega, \mathbb{R})$. We set

$$V(x)=\int_\Omega u(\omega, \nabla x(\omega))d\omega$$

It is easily seen that V is finite everywhere and has the C^1 property over H_0^1. Equation (8) for the derivative here becomes

$$\frac{d}{dt}V(x+tv)\bigg|_{t=0}=\int_\Omega u_u'(\omega, \nabla x(\omega))\nabla v(\omega)d\omega$$

The right-hand side is certainly a continuous linear functional of v. We want to write it as $\langle V'(x), v\rangle$, with $V'(x)\in H^{-1}$. This is done by integrating by parts

$$\frac{d}{dt}V(x+tv)\bigg|_{t=0}=-\int_\Omega v(\omega)\sum_{i=1}^{n}\frac{\partial}{\partial\omega_i}u_y'(\omega, \nabla x(\omega))d\omega$$

There is no boundary term because of the condition $v\in H_0^1$ (and not H^1). The sum on the right-hand side is the *divergence* operator. We finally write

$$V'(x)=-\text{div}\,(u_y'\circ\nabla x)$$ ■

Example 4
We can now combine examples 1, 2, and 3 to get the usual functionals of the calculus of variations. For instance, let u_1 and u_2 be two functions satisfying conditions (5), (6), and (7), with $k_1=k$ and $k_2=nk$, and consider the functional U: $H_0^1(\Omega, \mathbb{R}^k)\to\mathbb{R}$ defined by

$$U(x)=\int_\Omega [u_1(\omega, x(\omega))+u_2(\omega, \nabla x(\omega))]d\omega$$

This is a C^1 functional, and its derivative $U'(x)$ is the distribution in H^{-1} which is defined as

$$u_1'(\omega, x(\omega))-\text{div}\,u_2'(\omega, \nabla u(\omega))$$

If U attains its minimum over H_0^1 at some point $\bar{x}$, then the function $\bar{x}$

satisfies the *Euler–Lagrange equation*

$$u'_1(\omega, x(\omega)) - \operatorname{div} u'_2(\omega, \nabla u(\omega)) = 0 \qquad \text{in } H^{-1}(\Omega)$$

For instance, if $\bar{x}$ minimizes Dirichlet's integral (with φ given in L^2)

$$U(x) = \int_\Omega \left[\varphi(\omega)x(\omega) + \frac{1}{2}(\nabla x(\omega))^2\right] d\omega$$

then $\bar{x}$ solves Laplace's equation

$$\varphi(\omega) = \operatorname{div} \nabla x(\omega) = \sum_{i=1}^{n} \frac{\partial^2 x}{\partial \omega_i^2}(\omega)$$

$$\varphi = \Delta u$$

Of course, in most cases the growth conditions on u_1 will be too restrictive. They can be considerably relaxed by using the Sobolev embedding theorems and inequalities. For instance, if $n=1$ and Ω is a bounded domain, the H^1 norm is stronger than the C^0 norm, and it will be sufficient that $u'_1(\omega, y)$ be bounded for $(\omega, y) \in \Omega \times B$, whenever $B \subset \mathbb{R}^k$ is bounded. ■

These various definitions can be extended to maps between Banach spaces; there are two different ways of doing this. Given a map $\phi: X \to Y$, a point $x \in X$ and a map $A \in \mathscr{L}(X, Y)$, we shall say that

(a)′ A is the *Gâteaux derivative* of ϕ at x if

$$\lim_{h \to 0} \frac{1}{h} \|\phi(x+hy) - \phi(x) - hAy\| = 0$$

(b)′ A is the *Fréchet derivative* of ϕ at x if

$$\lim_{y \to 0} \frac{1}{\|y\|} \|\phi(x+y) - \phi(x) - Ay\| = 0$$

(a)″ A is the *weak Gâteaux derivative* of ϕ at x if

$$\forall p \in Y^\star, \quad \lim_{h \to 0} \frac{1}{h} \langle p, \phi(x+hy) - \phi(x) - hAy \rangle = 0$$

(b)″ A is the *weak Fréchet derivative* of ϕ at x if

$$\forall p \in Y^\star, \quad \lim_{\|y\| \to 0} \frac{1}{\|y\|} \langle p, \phi(x+y) - \phi(x) - Ay \rangle = 0$$

Clearly, **(b)′** $\Rightarrow$ **(b)″** $\Rightarrow$ **(a)″**, and **(b)′** $\Rightarrow$ **(a)′** $\Rightarrow$ **(a)″** with the same A, henceforth denoted by $\phi'(x)$, and called the derivative of ϕ at x.

This relationship enables us to define—again, in several different ways—the second derivative of a C^1 function $U: X \to \mathbb{R}$ as the derivative of $U': X \to \mathscr{L}(X, X^\star)$. We single out the most and the least restrictive definitions to single out two important classes.

A C^1 function $U: X \to \mathbb{R}$ has *the C^2 property* if $U': X \to X^\star$ is Fréchet differentiable everywhere and $U''(x) \in \mathscr{L}(X, X^\star)$ depends continuously on X.

A C^1 function $U: X \to \mathbb{R}$ *is twice weakly differentiable Gâteaux* if $U': X \to X^\star$ is weakly Gâteaux differentiable everywhere.

Example 5

Let Ω be an open subset on $\mathbb{R}^n$, with the usual Lebesgue measure $d\omega$. Let there be given a function $u: \Omega \times \mathbb{R}^k \to \mathbb{R}$, borelian with respect to both variables (ω, y). Assume that

(17) for any fixed $\omega \in \Omega$, the function $y \to u'(\omega, y)$ is C^2 over $\mathbb{R}^k$

(18) there is some constant c such that

$$|u''_{yy}(\omega, y)| \leqslant c \quad \text{all } (\omega, y) \in \Omega \times \mathbb{R}^k$$

(19) $$\int_\Omega |u(\omega, 0)| dx < \infty \quad \text{and} \quad \int_\Omega |u'_y(\omega, 0)| d\omega < \infty$$

Then the functional $U: L^2(\Omega, \mathbb{R}^k) \to \mathbb{R}$ given by

$$U(x) = \int_\Omega u(\omega, x(\omega)) d\omega$$

is twice weakly differentiable Gâteaux. ■

It is a simple matter to check that the growth condition in example 1 is satisfied, so that by example 2, U is a C^1 functional, and

$$U'(x) = u'_y \circ x \in L^2$$

For any $z \in L^2$, we have

$$\langle U'(x), z\rangle = \int_{\Omega} (z(\omega), u'_y(\omega, x(\omega)))d\omega$$

Again, the integrand satisfies all the assumptions in example 1, so

$$\lim_{h\to 0} \frac{1}{h} \langle U'(x+hy) - U'(x), z\rangle = \int_{\Omega} (z(\omega), u''_{yy}(\omega, x(\omega))y(\omega))d\omega$$

So U' is Gâteaux differentiable everywhere, and

$$\langle z, U''x)y\rangle = \int_{\Omega} (z(\omega), u''_{yy}(\omega, x(\omega)y(\omega))d\omega \tag{20}$$

Example 6

Same assumptions as example 5. We now claim that U *cannot* have the C^2 property unless u is precisely quadratic, that is,

$$u(\omega, y) = \tfrac{1}{2}(y, A(\omega)y) \quad \text{with} \quad A(\omega) \in \mathscr{L}(\mathbb{R}^k, \mathbb{R}^k) \qquad \blacksquare$$

This fact was noticed by Skrypniak, who used it to point out a few gaps in the original work of Palais and Smale, which fortunately turned out to be of little consequence. However, it shows how prudent one must be in making differentiability assumptions in infinite-dimensional spaces.

What we have to prove is that if $u''_{yy}(\omega, \cdot)$ is not a constant matrix for almost every $\omega \in \Omega$, then the map $U'': L^2 \to \mathscr{L}(L^2, L^2)$ cannot be continuous. This map is defined by the integral in equation (20), from which it follows that U'' will be continuous if and only if the map

$$\phi(x) = u''_{yy} \circ x$$

is continuous from L^2 to L^∞.

Now, it follows from condition (18) on u''_{yy} that ϕ maps L^2 into L^∞. Unfortunately, we have the case $q=\infty$ where Krasnosel'skii's theorem does not hold (see example 2): We cannot conclude that ϕ is continuous.

To avoid technicalities, we assume from now on that u''_{yy} is continuous in (ω, y) jointly. Assume that for some $\omega_0 \in \Omega$, $u''_{yy}(\omega_0, \cdot)$ is not constant, say

$$u''_{yy}(\omega_0, \xi_1) \neq u''_{yy}(\omega_0, \xi_2)$$

Let x_0 be any continuous function such that $x_0(\omega_0) = \xi_1$. With any $\varepsilon > 0$, we associate the function

$$x_\varepsilon(\omega) = \begin{cases} \xi_2 & \text{if } |\omega - \omega_0| < \varepsilon \\ x_0(\omega) & \text{if } |\omega - \omega_0| \geq \varepsilon \end{cases}$$

For small enough ε, we shall have

$$u''_{yy}(\omega, x_\varepsilon(\omega)) \neq u''_{yy}(\omega, x_0(\omega)) \text{ if } |\omega - \omega_0| < \varepsilon$$

When $\varepsilon \to 0$, we have $x_\varepsilon \to x_0$ in L^2. However, $u''_{yy} \circ x_\varepsilon$ does not converge to $u''_{yy} \circ x_0$ in L^∞, since the set of points where they differ always has positive measure.

So ϕ is not continuous, and U is not C^2. ■

5. SUPPORT FUNCTIONS AND BARRIER CONES OF CONVEX SUBSETS

Linear and convex analysis are based to a great extent on the Hahn–Banach Theorem, which, as is well known, has many different—but equivalent—formulations. We shall use the formulation that links the geometrical and analytical approaches, that is, the formulation that allows a characterization of closed convex subsets in terms of convex functions.

With this tool at hand, we can pass at will from the sometimes cumbersome handling of closed convex sets to the more flexible and traditional handling of convex functions.

We shall prove that any nonempty *closed convex* subset K of a Banach space X is characterized by its *support function* σ_K defined on the dual $X^\star$ of X by

$$\forall p \in X^\star, \quad \sigma_K(p) := \sup_{x \in K} \langle p, x \rangle \in]-\infty, +\infty]$$

Indeed, the Hahn–Banach Theorem states that

$$K = \{x \in X \mid \forall p \in X^\star, \ \langle p, x \rangle \leqslant \sigma_K(p)\}$$

In this formula, we can restrict p to range in the convex cone

$$\mathrm{b}(K) := \{p \in X^\star \mid \sigma_K(p) < +\infty\}$$

which is called the *barrier cone* of K. It measures the "boundedness" of K: The larger the barrier cone is, the smaller is K at infinity.

The negative polar cone of the barrier cone is the *recession cone* of K, since the following formula holds true

$$\forall x \in K, \quad \mathrm{b}(K)^- = \bigcap_{\lambda > 0} \lambda(K - x)$$

But the importance of the role played by the barrier cone $\mathrm{b}(K)$ of K lies in the following result: Let X and Y be Banach spaces, A a continuous linear operator from X to Y, and K a weakly closed subset of $Y^\star$.

If zero belongs to the interior of the set $\operatorname{Im} A + \mathrm{b}(K)$, *then* $A^\star(K)$ *is strongly closed in* $X^\star$. This "closed image" theorem will obviously be a precious tool.

We conclude Section 5 with a "calculus" of support functions and barrier cones, which allows us to "compute" support functions and characterize barrier cones of images, inverse images, sums, products, and intersections of closed convex sets.

DEFINITION 1

Let K *be a nonempty subset of a Hausdorff locally convex space* X. *The function* $\sigma_K: X^\star \to R \cup \{+\infty\}$ *defined by*

$$\text{(1)} \qquad \forall p \in X^\star, \qquad \sigma_K(p) := \sigma(K, p) := \sup_{x \in K} \langle p, x \rangle$$

is called the support function *of* K. *Its domain*

$$\text{(2)} \qquad \mathrm{b}(K) := \{p \in X^\star | \sigma_K(p) < +\infty\}$$

is called the barrier cone *of* K. ▲

It is readily seen that

$$\text{(3)} \qquad \begin{cases} \sigma_K \text{ is a proper lower semicontinuous convex positively} \\ \text{homogeneous function that is equal to } \sigma_{\overline{\mathrm{co}}(K)}\text{, the support} \\ \text{function of the closed convex hull } \overline{\mathrm{co}}(K) \text{ of } K \end{cases}$$

and

$$\text{(4)} \qquad b(K) \text{ is a convex cone (not necessarily closed).}$$

Before giving examples and formulas, we state and prove the version of the Hahn–Banach theorem that motivates the introduction of support functions.

THEOREM 2 (HAHN–BANACH)

Let K *be a subset of a Hausdorff locally convex vector space* X. *The closed convex hull* $\overline{\mathrm{co}}(K)$ *of* K *is given by*

$$\text{(5)} \qquad \overline{\mathrm{co}}(K) = \{x \in X | \forall p \in X^\star, \langle p, x \rangle \leq \sigma_K(p)\}$$ ▲

Proof. Set $\tilde{K} := \{x \in X | \forall p \in X^\star, \langle p, x \rangle \leq \sigma_K(p)\}$, which is a closed convex subset containing K. If $\overline{\mathrm{co}}(K) \neq \tilde{K}$, there exists some point $x \in \tilde{K}$ that does not belong to $\overline{\mathrm{co}}(K)$. The separation theorem implies the existence of $p_0 \in X^\star$ such that $\sigma_K(p_0) < \langle p_0, x \rangle$, which is a contradiction to the fact that x belongs to $\tilde{K}$. ■

We now introduce the negative polar cone of the barrier cone. It is the cone $b(K)^-$ defined by

$$b(K)^- := \{v \in X | \forall p \in b(K), \langle p, v\rangle \leq 0\}$$

PROPOSITION 3

Let K be a closed convex subset. Then for every $x_0 \in K$,

$$b(K)^- = \bigcap_{\lambda > 0} \lambda(K - x_0) \tag{6}$$ ▲

Proof. Take any point x_0 in K and let us set $L_0 := \cap_{\lambda>0} \lambda(K - x_0)$.

a. Take $x \in L_0$. For any $\lambda > 0$, there exists $y_\lambda \in K$ such that $x = \lambda(y_\lambda - x_0)$. Hence, $\langle p, x\rangle = \lambda(\langle p, y_\lambda\rangle - \langle p, x_0\rangle) \leq \lambda(\sigma_K(p) - \langle p, x_0\rangle)$. By letting λ converge to 0, we find that $\langle p, x\rangle \leq 0$ whenever $p \in b(K)$. Hence, $x \in b(K)^-$.

b. *Conversely,* let $x \in b(K)^-$ and $\lambda > 0$. Since x/λ belongs to $b(K)^-$, for all $p \in b(K)$, we have

$$\left\langle p, \frac{x}{\lambda} + x_0 \right\rangle = \left\langle p, \frac{x}{\lambda} \right\rangle + \langle p, x_0\rangle \leq \langle p, x_0\rangle \leq \sigma_K(p)$$

Consequently, since K is closed and convex, theorem 2 implies that $x/\lambda + x_0$ belongs to K, that is, $x \in L_0$. ■

DEFINITION 4

The negative polar cone $b(K)^-$ of the barrier cone is called the recession cone of K. ▲

The first use of the barrier cone that we mention is the closed image theorem, which illustrates the merits of this concept,

THEOREM 5 (CLOSED IMAGE)

Let X and Y be Banach spaces, A a continuous linear operator from X to Y, and $K \subset Y^\star$ a weakly closed subset of $Y^\star$. Assume that

$$0 \in \operatorname{Int}(\operatorname{Im} A + b(K)) \tag{7}$$

Let $A^\star \in \mathscr{L}(Y^\star, X^\star)$ denote the transpose of A. Then

$$A^\star(K) \text{ is strongly closed in } X^\star \tag{8}$$

More generally, $A^\star$ is proper in the sense that

(9)
- **i.** *$A^\star$ maps weakly closed subsets of K to closed subsets of $X^\star$.*
- **ii.** *For all strongly compact $M \subset X^\star$, the set $A^{\star-1}(M)$ is weakly compact.*

▲

Proof. Let us consider a sequence of elements $q_n \in K$ such that $A^\star q_n$ converges strongly to some p in $X^\star$. Let $\gamma > 0$ be such that by assumption (7), γB is contained in Im $A + \mathrm{b}(K)$. Then for all $y \in Y$, there exist points $x \in X$ and $z \in \mathrm{b}(K)$ such that $(\gamma/\|y\|)y = Ax + z$. Therefore,

$$\langle q_n, y\rangle = \left\langle q_n, \frac{\|y\|}{\gamma}(Ax+z)\right\rangle = \frac{\|y\|}{\gamma}(\langle A^\star q_n, x\rangle + \langle q_n, z\rangle)$$

$$\leqslant \frac{\|y\|}{\gamma}(\|x\|\,\|A^\star q_n\| + \sigma_K(z)) < +\infty \tag{10}$$

because the sequence of elements $A^\star q_n$ is bounded. It follows that the sequence of elements q_n is also weakly bounded and thus relatively weakly compact in $Y^\star$. Therefore, a generalized subsequence of elements $q_{n'} \in K$ converges to some q that belongs to K, because K is weakly closed. Hence, $A^\star q_{n'}$ converges weakly to $A^\star q$. Since $A^\star q_{n'}$ converges strongly to p, we deduce that $p = A^\star q$. ■

Since we used the weak compactness of weakly bounded subsets, we have to be careful of the dual statement.

THEOREM 6

Let X be a reflexive Banach space, Y a Banach space, A a continuous linear operator from X to Y, and K a weakly closed subset of X. Assume that

$$0 \in \mathrm{Int}\,(\mathrm{Im}\, A^\star + \mathrm{b}(K)) \tag{11}$$

Then

$$A(K) \text{ is strongly closed in } Y. \tag{12}$$

▲

Examples and Elementary Properties

We shall agree to set

$$\sigma_\varnothing(p) = -\infty, \qquad \forall p \in X^\star$$

We observe that

$$\sigma_X(p) = \begin{cases} +\infty & \text{when } p \neq 0 \\ 0 & \text{when } p = 0 \end{cases} \tag{13}$$

More generally, let $P \subset X$ be a *cone* of X. Then the barrier cone of P is the negative polar cone P^- of P

$$\mathrm{b}(P) = P^- := \{p \in X^\star \mid \sigma_K(p) \leqslant 0\} \tag{14}$$

and the support function of P is defined by

(15) $$\sigma_K(p) = \begin{cases} 0 & \text{when } p \in P^- \\ +\infty & \text{when } p \notin P^- \end{cases}$$ ■

As a consequence, we deduce from theorem 2 the following important result.

PROPOSITION 7 (THE BIPOLAR LEMMA)
If P is a closed convex cone of X, then

(16) $$P = (P^-)^-$$ ▲

When M is a vector subspace of X, the preceding formulas become

(17) $b(M) = M^\perp$ is the orthogonal subspace (annihilator) of M

and if M is a closed subspace of X, then

(18) $$M = M^{\perp\perp}$$

Support functions of points $\{x\} \subset M$ are the weakly continuous linear functionals on $X^\star$ because

(19) $$\sigma_{\{x\}}(p) = \langle p, x\rangle$$

It is obvious that if B denotes the unit ball of X, then

(20) $\sigma_B(p) = \|p\|_\star$ is the dual norm of $X^\star$.

Let K be a closed convex subset of X; then

- **(a)** $0 \in K$ if and only if $\sigma_K \geqslant 0$.
- **(b)** K is symmetric if and only if σ_K is even.
- **(c)** K is weakly bounded if and only if σ_K is finite everywhere, that is, if and only if $b(K) = X^\star$.

In Chapter 4, we shall prove the converse statement in theorem 8.

THEOREM 8
Let $\sigma: X^\star \to R \cup \{+\infty\}$ be a proper positively homogeneous lower semicontinuous convex function. It is the support of the closed convex subset K defined by

(21) $$K := \{x \in X \mid \forall p \in X^\star, \langle p, x\rangle \leqslant \sigma(p)\}$$ ▲

Formulas for Support Functions and Barrier Cones

(22) If $K \subset L$, then $\mathrm{b}(L) \subset \mathrm{b}(K)$ and $\sigma_K \leqslant \sigma_L$

(23) If $K := \prod_{i=1}^{n} K_i$ where $K_i \subset X_i$ $(i = 1, \ldots, n)$, then

$$\mathrm{b}(K) = \prod_{i=1}^{n} \mathrm{b}(K_i), \quad \text{and} \quad \sigma_K(p) = \sum_{i=1}^{n} \sigma_{K_i}(p_i)$$

(24) If $K := \overline{\mathrm{co}}\left(\bigcup_{i \in I} K_i\right)$, then $\mathrm{b}(K) = \bigcap_{i \in I} \mathrm{b}(K_i)$ and $\sigma_K(p) = \sup_{i \in I} \sigma_{K_i}(p)$

(25) If $A \subset \mathscr{L}(X, Y)$, then $\mathrm{b}(\overline{A(K)}) = A^{\star -1}\mathrm{b}(K)$, and $\sigma_{\overline{A(K)}}(p) = \sigma_K(A^\star p)$

In particular,

(26) If $K_1, K_2 \subset X$, then $\mathrm{b}(K_1 + K_2) = \mathrm{b}(K_1) \cap \mathrm{b}(K_2)$ and

$$\sigma_{K_1 + K_2}(p) = \sigma_{K_1}(p) + \sigma_{K_2}(p)$$

(27) If P is a convex cone, then $\mathrm{b}(K + P) = \mathrm{b}(K) \cap P^-$ and

$$\sigma_{K+P}(p) = \begin{cases} \sigma_K(p) & \text{if } p \in P^- \\ +\infty & \text{if } p \notin P^- \end{cases}$$

(28) If $x_0 \in X$, then

$$\sigma_{K + x_0}(p) = \sigma_K(p) + \langle p, x_0 \rangle$$

(29) If $A \in \mathscr{L}(X, Y)$, $L \subset X$ and $M \subset Y$ are closed and convex, and if

$$0 \in \mathrm{Int}\,(A(L) - M)$$

then $\mathrm{b}(\mathrm{L} \cap A^{-1}(M)) = \mathrm{b}(L) + A^* \mathrm{b}(M)$, and for all $p \in \mathrm{b}(K)$, there exists $\bar{q} \in \mathrm{b}(M)$ such that

$$\sigma_{L \cap A^{-1}(M)}(p) = \sigma_L(p - A^\star \bar{q}) + \sigma_M(\bar{q}) = \inf_{q \in Y^*} [\sigma_L(p - A^\star q) + \sigma_M(q)]$$

In particular,

(30) If $A \in \mathscr{L}(X, Y)$, if M is a closed convex subset of Y, and if

$$0 \in \mathrm{Int}\,(\mathrm{Im}\, A - M),$$

then

$$\mathrm{b}(A^{-1}(M)) = A^\star \mathrm{b}(M)$$

and for all $p \in \mathrm{b}(A^{-1}(M))$, there exists $\bar{q} \in \mathrm{b}(M)$ such that

$$A^\star \bar{q} = p \quad \text{and } \sigma_{A^{-1}(M)}(p) = \sigma_M(\bar{q}) = \inf_{A^* q = p} \sigma_M(q)$$

(31) If K_1 and K_2 are two closed convex subsets of X such that

$$0 \in \mathrm{Int}\,(K_1 - K_2)$$

then

$$\mathrm{b}(K_1 \cap K_2) = \mathrm{b}(K_1) + \mathrm{b}(K_2)$$

and for all $p \in \mathrm{b}(K_1 \cap K_2)$, there exists $\bar{p}_1 \in \mathrm{b}(K_1)$ and $\bar{p}_2 \in \mathrm{b}(K_2)$ such that $p = \bar{p}_1 + \bar{p}_2$ and

$$\sigma_{K_1 \cap K_2}(p) = \sigma_{K_1}(\bar{p}_1) + \sigma_{K_2}(\bar{p}_2) = \inf_{p = p_1 + p_2} (\sigma_{K_1}(p_1) + \sigma_{K_2}(p_2))$$

Remark

Formulas (22)–(28) are straightforward. Formula (29) is not obvious at all: It follows from Chapter 5.

Formulas (30) and (31) are obvious consequences of formula (29).

CHAPTER 2

Smooth Analysis

The inverse function theorem and the (closely related) implicit function theorem are two of the very few general methods available for studying nonlinear problems. The importance of these theorems in modern analysis, whether finite or infinite dimensional, can hardly be overrated. This explains why the inverse function theorem appears at both the beginning and the end of this book. In Chapter 7, the theorem is proved under very weak hypothesis for set-valued, nonsmooth maps, whereas in the present chapter, we prove it with strong differentiability assumptions.

The reader may wonder why we give two proofs, and two statements, of the inverse function theorem, since the later statement encompasses the first one. The reason is twofold. We do not believe that the more general statements are the clearer ones, and we prefer to give the theorem, explain it, and use it, in the differentiable case, before turning to more general, and perhaps less common, cases. Moreover, in the differentiable case, it is possible to give a constructive proof, that is, to devise an iterative procedure that actually converges toward a solution of the nonlinear problem under consideration. In other words, the solution is not only shown to exist, but can also be computed.

The structure of this chapter now becomes clear. We state the inverse function theorem and deduce the implicit function theorem, fully aware that better statements will be available later on. We emphasize the constructive aspects of the proof by giving two different algorithms to compute the solution. The second one (Newton's method) is much more efficient than the first, classical, one and has recently received a great deal of attention.

Later sections are devoted to applications. The depth of the inverse function theorem is revealed in that we are able to derive from it two very deep finite-dimensional results: Brouwer's fixed-point theorem and Morse's lemma on the singularities of a function.

We then find we have all the tools needed to investigate equations $f(\lambda, x)=0$, depending on a parameter λ. In Section 4, our equation will be $\phi'_x(\lambda, x)=0$, and x will be a finite-dimensional variable, so that, in fact, we are investigating the critical points of a real-valued function $\phi(\lambda, \cdot)$ depending on a parameter λ: We shall show that they appear (or disappear) in pairs. In Section 5, x will be infinite dimensional, and the equation $f(\lambda, x)=0$ will have the trivial solution $x=0$ for all values of λ: We look for another set of solutions, "bifurcating" from the trivial one at some value of λ. Because of the practical importance of the problem, we also investigate the stability of the solutions we find, trivial or not.

We conclude by stating and proving Thom's transversality theorem, probably one of the most brilliant achievements of modern mathematics. It is an everyday tool in topology and geometry and is making its way into analysis: We cannot do it complete justice, but we do provide a full proof and immediate applications. This enables us to end the chapter as we began it, by exploring Newton's method.

1. ITERATIVE PROCEDURES FOR INVERTING A MAP

We denote by X and Y Banach spaces and by $\mathscr{U}$ an open subset of X. Let $f:\mathscr{U}\to Y$ be a (nonlinear) map. We are interested in solving the equation

$$f(x)=y \tag{1}$$

Assume that equation (1) has been solved for some particular case, that is, values $(x_0, y_0)\in\mathscr{U}\times Y$ have been found such that

$$f(x_0)=y_0 \tag{2}$$

The inverse function theorem then gives us conditions under which equation (1) can be uniquely solved in x for all y, provided we consider only values of x and y close to x_0 and y_0. In its classical version, the theorem reads as follows.

INVERSE FUNCTION THEOREM

Assume $f:\mathscr{U}\to Y$ is a C^1 mapping. Let $x_0\in\mathscr{U}$ and $y_0\in Y$ be given such that

$$f(x_0)=y_0 \tag{2}$$

$$f'(x_0)\in\mathscr{L}(X, Y) \text{ has an inverse } f'(x_0)^{-1}\in\mathscr{L}(Y, X) \tag{3}$$

Then there is an open neighborhood $\mathscr{N}\subset\mathscr{U}$ of x_0 such that $f(\mathscr{N})=\mathscr{V}$ is an open neighborhood of y_0 and the map $f:\mathscr{N}\to\mathscr{V}$ has a C^1 inverse $f_{\mathscr{N}}^{-1}:\mathscr{V}\to\mathscr{N}$. The derivative is given by

$$[f_{\mathscr{N}}^{-1}]'(y)=[f'\circ f_{\mathscr{N}}^{-1}(y)]^{-1}, \qquad \text{for all } y\in\mathscr{V} \tag{4}$$ ▲

Before proceeding with the proof, let us spell out what the theorem says. Note first that the derivative $f'(x_0)$ is required to be invertible. In the case where X is finite dimensional, this is tantamount to requiring $f'(x_0)$ to be injective or surjective. This means that Y has the same dimension as X and that in any system of coordinates for X and Y, the jacobian of f at x_0 does not vanish. Conditions (2) and (3) now read

(5)
$$y_0=(f_1(x_0),\ldots,f_n(x_0))$$

(6)
$$\begin{vmatrix} \frac{\partial f_1}{\partial x_1}(x_0) & \cdots & \frac{\partial f_1}{\partial x_n}(x_0) \\ \vdots & & \vdots \\ \frac{\partial f_n}{\partial x_1}(x_0) & \cdots & \frac{\partial f_n}{\partial x_n}(x_0) \end{vmatrix} \neq 0$$

The conclusion of the inverse function theorem is that for any $y \in \mathscr{V}$, there is a unique $x \in \mathscr{U}$ solving the equation $f(x)=y$ (there might be many more outside), and it depends smoothly (C^1) on x.

But this is not all. As pointed out in the introduction, there are also iterative procedures to compute x from y; here is the simplest one

(7)
$$\begin{cases} \text{start from } x_0 \\ \text{run } x_{n+1}:=x_n+f'(x_0)^{-1}(y-f(x_n)) \end{cases}$$

Here is another one, known as *Newton's method*

(8)
$$\begin{cases} \text{start from } x_0 \\ \text{run } x_{n+1}:=x_n+f'(x_n)^{-1}(y-f(x_n)) \end{cases}$$

These procedures are illustrated in the following figure; we have taken $X=R=Y$, $y=0$, and $f(x)=(x^3-1)/4$.

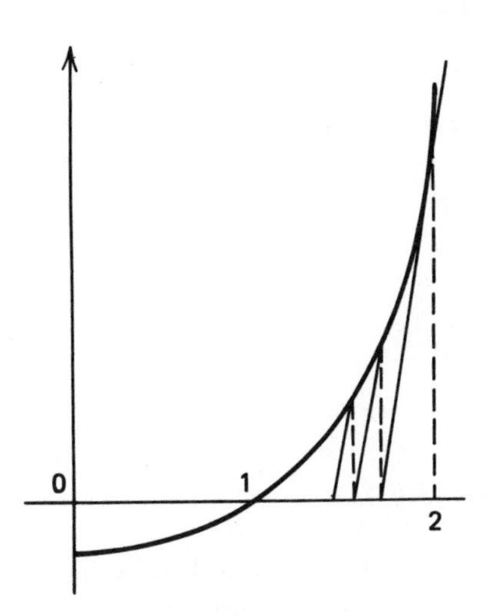

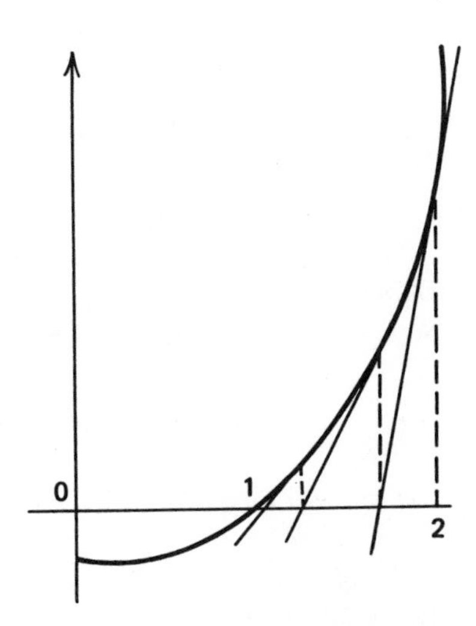

Figure 1.

Simple method	Newton's method
$x_0=2,\ x_{n+1}=x_n-f(x_n)/6$	$x_0=2,\ x_{n+1}=x_n-4f(x_n)/3x_n^2$
$x_1=1.708333333$	$x_1=1.416666667$
$x_2=1.542266469$	$x_2=1.11053441$
$x_3=1.431082585$	$x_3=1.010636768$
$x_4=1.350630361$	$x_4=1.000111557$
$x_5=1.289637732$	$x_5=1.000000012$

Both methods are seen to converge to 1, the second one much faster than the first. Let us confirm these results.

LEMMA 1

Choose any $k \in]0, 1[$. *For any* $\alpha \in]0, 1[$, *there is some* $\varepsilon > 0$ *such that whenever* $\|x-x_0\| \leqslant \varepsilon$, *we have*

$$\|I - f'(x_0)^{-1}f'(x)\| \leqslant k \tag{9}$$

$$\|f'(x_0)^{-1}(f(x)-f(x_0)-f'(x_0)(x-x_0))\| \leqslant \alpha\|x-x_0\| \tag{10}$$

▲

Proof. Consider the map from $\mathscr{U}$ to $\mathscr{L}(X, X)$ defined by

$$x \to I - f'(x_0)^{-1}f'(x) \tag{11}$$

This map is continuous, since f is a C^1 map and it is zero for $x=x_0$. Inequality (9) then follows by continuity.

Since f is Fréchet differentiable at x_0, for any $\beta > 0$ some $\varepsilon > 0$ can be found such that $\|x-x_0\| \leqslant \varepsilon$ implies

$$\|f(x)-f(x_0)-f'(x_0)(x-x_0)\| \leqslant \beta\|x-x_0\| \tag{12}$$

Taking $\beta = \alpha/\|f'(x_0)^{-1}\|$ yields inequality (10). ■

LEMMA 2

Set $\eta = (1-\alpha)\varepsilon/\|f'(x_0)^{-1}\|$. *For any* y *such that* $\|y-f(x_0)\| \leqslant \eta$, *the map*

$$g(x) = x + f'(x_0)^{-1}(y - f(x)) \tag{13}$$

is a contraction of the closed ball $x_0 + \varepsilon B$ *into itself.* ▲

Proof. We have $g'(x) = I - f'(x_0)^{-1}f'(x)$. It follows from lemma 1 that $\|g'(x)\| \leqslant k$ on $x_0 + \varepsilon B$, so that g is k Lipschitz on that ball, with $k < 1$. All we have to prove is that g sends $x_0 + \varepsilon B$ into itself.

Take any x such that $\|x-x_0\| \leqslant \varepsilon$. Let us check that $\|g(x)-x_0\| < \varepsilon$. We have

$$g(x)-x_0 = x-x_0+f'(x_0)^{-1}(y-f(x)) \tag{14}$$
$$= f'(x_0)^{-1}(f(x_0)+f'(x_0)(x-x_0)-f(x))+f'(x_0)^{-1}(y-f(x_0))$$

Hence, by inequality (10)

$$\|g(x)-x_0\| \leqslant \alpha\|x-x_0\|+(1-\alpha)\varepsilon \leqslant \varepsilon \tag{15}$$

■

LEMMA 3

For any y such that $\|y-f(x_0)\| \leqslant \eta$, the sequence x_n generated by algorithm (7) converges geometrically to the one and only solution of $f(x)=y$ in the ball $x_0+\varepsilon B$

$$\|x_n-x\| \leqslant constant \cdot k^n \tag{16}$$

Proof. The first part, and estimate (16), follow immediately from Banach's contraction principle applied to the contraction g and the closed ball $x_0+\varepsilon B$. ■

LEMMA 4

Take any two points $\hat{y}$ and $\bar{y}$ within distance η of $f(x_0)$. Let $\hat{x}$ and $\bar{x}$ solve equations $f(\hat{x})=\hat{y}$ and $f(\bar{x})=\bar{y}$ in the ball $x_0+\varepsilon B$. We then have

$$\|x-\bar{x}\| \leqslant \frac{\|f'(x_0)^{-1}\|}{(1-k)}\|y-\bar{y}\| \tag{17}$$

▲

Proof. Set

$$g(x)=x+f'(x_0)^{-1}(\hat{y}-f(x)) \tag{18}$$

$$\bar{g}(x)=x+f'(x_0)^{-1}(\bar{y}-f(x)) \tag{19}$$

We have $g(\hat{x})=\hat{x}$ *and* $\bar{g}(\bar{x})=\bar{x}$. It follows that

$$\|\hat{x}-\bar{x}\| = \|g(\hat{x})-\bar{g}(\bar{x})\| \tag{20}$$
$$\leqslant \|g(\hat{x})-\bar{g}(\hat{x})\|+\|\bar{g}(\hat{x})-\bar{g}(\bar{x})\|$$

Using equations (18) and (19), we get

$$\|g(\hat{x})-\bar{g}(\hat{x})\| \leqslant \|f'(x_0)^{-1}\|\,\|\hat{y}-\bar{y}\| \tag{21}$$

Lemma 2 tells us that $\bar{g}$ is k Lipschitz, so that

$$\|\bar{g}(\hat{x})-\bar{g}(\bar{x})\| \leqslant k\|\hat{x}-\bar{x}\| \tag{22}$$

Substituting this into inequality (20) and remembering that $0<k<1$, we get the desired result. ■

This almost proves the inverse function theorem. If we define

$$\mathscr{V}:=\{y \mid \|y-f(x_0)\|<\eta\} \tag{23}$$

$$\mathscr{N}:=f^{-1}(\mathscr{V})\cap\{x \mid \|x-x_0\|<\varepsilon\} \tag{24}$$

and we denote by $f_{\mathscr{N}}$ the restriction of f to $\mathscr{N}$, lemma 3 tells us that $f_{\mathscr{N}}$ is bijective. It is known to be continuous, and lemma 4 tells us that $f_{\mathscr{N}}^{-1}$ also is continuous (even Lipschitz). All that remains to be proved is that it is C^1.

To begin with, note that $f'(x)$ is invertible for all x in $\mathscr{N}$. Indeed, because of inequality (9), the series

$$\sum_{n=0}^{\infty}(I-f'(x_0)^{-1}f'(x))^n=S(x) \tag{25}$$

is convergent in $\mathscr{L}(X, X)$. It is easily checked that

$$S(x)-(I-f'(x_0)^{-1}f'(x))S(x)=I \tag{26}$$

$$S(x)-S(x)(I-f'(x_0)^{-1}f'(x))=I \tag{27}$$

In other words,

$$f'(x_0)^{-1}f'(x)S(x)=I=S(x)f'(x_0)^{-1}f'(x) \tag{28}$$

So $f'(x)^{-1}$ is well defined and equal to $S(x)f'(x_0)^{-1}$.

Now fix $\bar{y}$ in $\mathscr{V}$, and take some other point y in $\mathscr{V}$. Set $\bar{x}=f_{\mathscr{N}}^{-1}(\bar{y})$ and $x=f_{\mathscr{N}}^{-1}(y)$. Since f is Fréchet differentiable at $\bar{x}$, we have

$$f(x)-f(\bar{x})=f'(\bar{x})(x-\bar{x})+\|x-\bar{x}\|\varepsilon(x) \tag{29}$$

with $\varepsilon(x)\to 0$ in Y when $x\to\bar{x}$ in X.

We rewrite this as

$$y-\bar{y}=f'(\bar{x})(f_{\mathscr{N}}^{-1}(y)-f_{\mathscr{N}}^{-1}(\bar{y}))+\varepsilon\circ f_{\mathscr{N}}^{-1}(y)\|f_{\mathscr{N}}^{-1}(y)-f_{\mathscr{N}}^{-1}(\bar{y})\| \tag{30}$$

We note that $\varepsilon\circ f_{\mathscr{N}}^{-1}(y)\to 0$ when $y\to\bar{y}$ and that $f_{\mathscr{N}}^{-1}$ is Lipschitz. It follows that the last term can be rewritten as $\eta(y)\|y-\bar{y}\|$, with $\eta(y)\to 0$ when $y\to\bar{y}$

$$y-\bar{y}-\eta(y)\|y-\bar{y}\|=f'(\bar{x})(f_{\mathcal{N}}^{-1}(y)-f_{\mathcal{N}}^{-1}(\bar{y})) \tag{31}$$

$$f_{\mathcal{N}}^{-1}(y)-f_{\mathcal{N}}^{-1}(\bar{y})=f'(\bar{x})^{-1}(y-\bar{y})-f'(\bar{x})^{-1}\eta(y)\|y-\bar{y}\| \tag{32}$$

The latter relationship means precisely that $f_{\mathcal{N}}^{-1}$ is Fréchet differentiable at $\bar{y}$, with derivative

$$[f_{\mathcal{N}}^{-1}]'(\bar{y})=f'(\bar{x})^{-1}=[f'\circ f_{\mathcal{N}}^{-1}(\bar{y})]^{-1} \tag{33}$$

The last term depends continuously on $\bar{y}$, since $f_{\mathcal{N}}^{-1}$ is continuous and f is C^1, so $f_{\mathcal{N}}^{-1}$ is C^1 on $\mathcal{V}$, and the inverse function theorem is completely proved. ■

We detailed the proof in this way because we wanted to show that the neighborhoods $\mathcal{N}$ of x_0 and $\mathcal{V}$ of $y_0=f(x_0)$ could be evaluated. This is important for some applications. Of course, we want $\mathcal{V}$ and $\mathcal{N}$ to be as large as possible, and formulas (23) and (24) tell us that ε and η should then be chosen as large as possible. Looking up lemmas 1 and 2, we find that k should be chosen close to 1 and that α expresses a trade-off between ε and η.

Using Newton's method (8) instead of the preceding one will result in much faster convergence, albeit possibly on smaller domains $\mathcal{N}$ and $\mathcal{V}$. However, Newton's method requires the function f to be at least C^2. To be precise,

LEMMA 5

Assume f is C^2, and pick any δ with $0<\delta<1$. Then there exists some $\alpha>0$ and $\varepsilon>0$ such that for all x and $\bar{x}$ belonging to $x_0+\varepsilon B$, we have

$$\|f'(\bar{x})^{-1}(f(x)-f(\bar{x})-f'(x)(x-\bar{x}))\|\leqslant\alpha\|x-\bar{x}\|^2 \tag{34}$$

$$\|I-f'(\bar{x})^{-1}f'(x)\|\leqslant\delta \tag{35}$$

Set $\gamma=\alpha(1-k)^{-1}$ and $\eta=\varepsilon[2\|f'(x_0)^{-1}\|\ \max\ (1,\ \gamma)]^{-1}$. For any y with $\|y-f(x_0)\|\leqslant\eta$, the sequence x_n generated by Newton's procedure (8) *converges to a solution x of the equation $f'(x)=y$ and satisfies*

$$\|x_{n+1}-x_n\|\leqslant\gamma^{2^n-1}\|x_1-x_0\|^{2^n} \tag{36}$$ ▲

Proof. Since $f'(x_0)$ is invertible and f' is a continuous map, $f'(x)$ will be invertible for all x in some neighborhood of x_0. The existence of ε such that inequalities (35) and (34) are satisfied follows from the fact that f is C^2.

Assume for the time being that $x_1, x_2, \ldots, x_n$ all belong to $x_0+\varepsilon B$. Calculating

x_{n+1} by Newton's procedure, we have

$$\begin{aligned}(37)\qquad x_{n+1}-x_n&=x_n-x_{n-1}+f'(x_n)^{-1}(y-f(x_n))-f'(x_{n-1})^{-1}(y-f(x_{n-1}))\\&=f'(x_{n-1})^{-1}(f(x_{n-1})+f'(x_{n-1})(x_n-x_{n-1})-f(x_n))\\&\quad+(f'(x_n)^{-1}-f'(x_{n-1})^{-1})(y-f(x_n))\\&=f'(x_{n-1})^{-1}(f(x_{n-1})+f'(x_{n-1})(x_n-x_{n-1})-f(x_n))\\&\quad+(I-f'(x_{n-1})^{-1}f'(x_n))(x_{n+1}-x_n)\end{aligned}$$

Using estimates (35) and (34) with $\bar{x}=x_{n-1}$ and $x=x_n$ yields

$$(38)\qquad \|x_{n+1}-x_n\|\leqslant\alpha\|x_n-x_{n-1}\|^2+k\|x_{n+1}-x_n\|$$

Finally,

$$(39)\qquad \|x_{n+1}-x_n\|\leqslant\frac{\alpha}{1-k}\|x_n-x_{n-1}\|^2$$

From this, we easily derive relationship (36). Of course, all these calculations hold true only if the successive points x_n belong to $x_0+\varepsilon B$. We check this by induction. We have

$$(40)\qquad \|x_1-x_0\|\leqslant\|f'(x_0)^{-1}\|\eta\leqslant\frac{\varepsilon}{2}$$

We claim $\|x_n-x_0\|\leqslant\varepsilon(1-2^{-n})$. It is true for $n=1$. If it is true for n, it is true for $(n+1)$

$$\begin{aligned}(41)\qquad \|x_{n+1}-x_0\|&\leqslant\|x_n-x_0\|+\gamma^{2^n-1}\|x_1-x_0\|^{2^n}\\&\leqslant\varepsilon(1-2^{-n})+\left(\frac{\varepsilon}{2}\right)^{2^n}\leqslant\varepsilon(1-2^{-n-1})\end{aligned}$$

The result follows. ∎

It should be noted that the very rapid convergence we obtain by Newton's methods exacts its price, since it requires us to invert $f'(x_n)$ at every step. It has therefore been attempted to improve the procedure by doing away with operator inversions while retaining the quadratic convergence. Here is one such scheme

$$(42)\qquad x_{n+1}=x_n+A_n(y-f(x_n))$$

$$(43)\qquad A_{n+1}=A_n+A_n(I-f'(x_{n+1})A_n)\quad\text{and}\quad A_0=I$$

Here $A_n\in\mathscr{L}(Y,X)$ can be thought of as an approximation of $f'(x_n)^{-1}$ that is close enough to give quadratic convergence. Such schemes, however, do not in principle require that $f'(x)$ be invertible: Proceeding in that direction, we are able to prove very sophisticated "hard" inverse function theorems, which happily fall beyond the scope of this book.

Another remark about Newton's method. We can imagine a continuous procedure instead of a discrete one, thus transforming the induction formula (8) into a differential equation

$$\frac{dx}{dt}=f'(x)^{-1}(y-f(x)) \tag{44}$$

This can be rewritten as

$$f'(x)\frac{dx}{dt}=y-f(x) \tag{45}$$

Any solution $x(t)$ of this equation, starting at a point x_0 where $f'(x_0)$ is invertible, has the property that

$$\frac{d}{dt}[y-f(x(t))]=-[y-f(x(t))] \tag{46}$$

This integrates to

$$y-f(x(t))=(y-f(x_0))e^{-t} \tag{47}$$

as long as the solution $x(t)$ is defined. Now certainly, if $f'(x_0)$ is invertible and y is close enough to $f(x_0)$, the solution $x(t)$ is going to stay in some neighborhood of x_0 where $f'(x)$ is invertible, so that $x(t)$ will be defined for all $t>0$, and $f(x(t)) \to y$ exponentially when $t\to\infty$.

But the really interesting things occur when we start at a point x_0 where $f(x_0)$ and y are far apart. Then, either the solution of the differential equation (45) runs into a point x where $f'(x)$ is not invertible, or it converges to a solution x of $f(x)=y$. This situation lends itself to further analysis, and we shall come back to it in later sections. ■

2. MILNOR'S PROOF OF BROUWER'S FIXED POINT THEOREM

Nothing seems more appropriate to illustrate the depth of the inverse function theorem than to derive Brouwer's fixed point theorem from it. The proof we give is due to John Milnor (1978). As a matter of fact, he derives a stronger geometrical statement: There is no way to "comb" an even-dimensional sphere.

THEOREM 1

Let S^{2n} be the unit sphere in R^{2n+1}. Then any continuous tangent vector field on S^{2n} must have a zero. ▲

Note that this is obviously false for S^1 and (less obviously) for all odd-dimensional spheres. The way dimensionality comes into Milnor's proof is perhaps its most pleasing feature. We do not assume anything about dimensionality to begin with: We simply work on an m-dimensional sphere S^m.

We start by restricting ourselves to C^1 vector fields. Indeed, let $\xi: S^m \to R^{m+1}$ be a continuous tangent vector field. Using, for instance, the Stone–Weierstrass theorem, we can find for each $\varepsilon > 0$ some C^1 vector field $\xi_\varepsilon: S^m \to R^{m+1}$ such that $\|\xi - \xi_\varepsilon\| \leqslant \varepsilon$ in the uniform norm. Projecting $\xi_\varepsilon(x)$ on the tangent space to S^m at x, we get a tangent vector field $\xi_\varepsilon^t(x)$ that is still C^1 and still satisfies $\|\xi - \xi_\varepsilon^t\| \leqslant \varepsilon$. Now if the theorem has been proved for the C^1 case, then for each $\varepsilon > 0$, there will be a point $x_\varepsilon \in S^m$ where $\xi_\varepsilon^t(x_\varepsilon) = 0$. Letting $\varepsilon \to 0$ and using the compactness of S^m, we obtain a cluster point $\bar{x}$ where $\xi(\bar{x}) = 0$, as desired.

We now proceed with the proof. Assume $\xi: S^m \to \mathbb{R}^{m+1}$ is a nonvanishing C^1 vector field. We shall prove that m is odd.

Since $\xi(x) \neq 0$ for all x, we can associate with every $t \in R$ a map $\phi_t: S^m \to \sqrt{1+t^2}S^m$, as follows:

$$\phi_t(x) = x + t\xi(x)\|\xi(x)\|^{-1}$$

Since $\xi(x)$ is tangent to S^m at x, we have $\|\phi_t(x)\| = \sqrt{1+t^2}$, so that ϕ_t does send the unit sphere S^m into the sphere $\sqrt{1+t^2}S^m$. Certainly, ϕ_t is C^1 for all t.

LEMMA 2

There is some $\varepsilon > 0$ such that ϕ_t is a C^1 diffeomorphism for all $|t| < \varepsilon$. ▲

Proof.

Define a map $\psi_t: S^m \to S^m$ by

$$\psi_t(x) = (1+t^2)^{-1/2}\phi_t(x)$$

Clearly, ϕ_t is a diffeomorphism from S^m to $\sqrt{1+t^2}S^m$ if and only if ψ_t is a diffeomorphism from S^m to S^m. We shall work with ψ_t instead of ϕ_t so that the range does not change with t.

Take any point $\bar{x} \in S^m$, and choose a local coordinate system $(\theta_1, \ldots, \theta_m)$ for the sphere, valid in some neighborhood of $\bar{x}$. Since ψ_0 is the identity, so that $\psi_0'(\bar{x}) = I$, and ψ_t' depends continuously on (t, x), there is some $\varepsilon > 0$ and some neighborhood $\mathscr{U}$ of $\bar{x}$ such that whenever $|t| < \varepsilon$ and $x \in \mathscr{U}$, we have

$$\|I - \psi_t'(\bar{x})^{-1}\psi_t'(x)\| \leqslant \frac{1}{2}$$

$$\|\psi_t'(\bar{x})^{-1}(\psi_t(x) - \psi_t(\bar{x}) - \psi_t'(\bar{x})(x - \bar{x}))\| \leqslant \frac{\|x - \bar{x}\|}{2}$$

From the results of the preceding section, there will be an open neighborhood $\mathscr{V}$ of $\bar{x}=\psi_0(\bar{x})$, and for every $|t|<\varepsilon$, a C^1 map ω_t: $\mathscr{V}\to\mathscr{U}$ such that $\psi_t\circ\omega_t=I_{\mathscr{V}}$.

If we carry out this construction for all $\bar{x}$ in S^m, we define two coverings of the sphere by open subsets. By compactness, there will be some $\varepsilon_0>0$ and finite subcoverings $(\mathscr{U}_1, \ldots, \mathscr{U}_n)$ and $(\mathscr{V}_1, \ldots, \mathscr{V}_n)$ with the property that for all $|t|<\varepsilon_0$, there is a C^1 map ω_t^k: $\mathscr{V}_k\to\mathscr{U}_k$ such that $\psi_t\circ\omega_t^k=I_{\mathscr{V}_k}$.

Since $(\mathscr{V}_1, \ldots, \mathscr{V}_n)$ is a covering, it follows that ψ_t is surjective for $|t|<\varepsilon_0$. We claim there is an $\varepsilon_1>0$ such that ψ_t is one to one for $|t|<\varepsilon_1$. If it were not so, there would be a sequence $t_n\to 0$ and points $y_n\neq y'_n$ such that $\psi_{t_n}(y_n)=\psi_{t_n}(y'_n)$. By compactness, these sequences would have cluster points y and y', with $\psi_0(y)=\psi_0(y')$, so that $y=y'$. For n large enough, y_n and y'_n would belong to some common $\mathscr{V}_k$, and this would contradict the existence of ω_{t_n}.

We have proved that, for $|t|<\min(\varepsilon_0, \varepsilon_1)$, the map ψ_t is bijective. Its inverse must be ω_t^k on $\mathscr{V}_k$; hence the result. ■

We now extend ϕ_t to the unit ball B^{m+1} by setting

$$\bar{\phi}_t(x):=\|x\|\phi_t(x\|x\|^{-1}) \qquad \text{for } \|x\|\leqslant 1 \tag{1}$$

The map $\bar{\phi}_t$: $B^{m+1}\to\sqrt{1+t^2}B^{m+1}$ defined in this way is clearly a diffeomorphism for $|t|<\varepsilon$. We use it to compute the volume of $\sqrt{1+t^2}B^{m+1}$. Namely, we use the well-known formula for changing variables in a multiple integral.

$$\text{Vol}\,(\sqrt{1+t^2}B^{m+1})=\int_{B^{m+1}}|\text{Det}\,\bar{\phi}'_t(x)|dx$$

There is no need to compute precisely the right-hand side. Note only that Det $\bar{\phi}'_t(x)$ does not vanish, so that it has the sign of Det $\bar{\phi}'_0(x)$, which is positive. Note also that all derivatives are taken with respect to x, so that all the terms of the determinant are affine functions of t and the final result may be a polynomial in t

$$\text{Vol}\,(\sqrt{1+t^2}B^{m+1})=\text{polynomial}\,(t)$$

Now this is where we have a contradiction, for the left-hand side is homogeneous of degree $(m+1)$:

$$\sqrt{1+t^2}^{(m+1)}\,\text{Vol}\,(B^{m+1})=\text{polynomial}\,(t)$$

This obviously implies that $(m+1)$ is even, hence m is odd, as stated. ■

We now derive a similar result for vector fields on the unit ball, which will now hold true in all dimensions.

THEOREM 3

Let $\xi\colon B^n \to R^n$ *be a continuous vector field on the n-dimensional unit ball. Assume that on the boundary* S^{n-1} *of* B^n, *this vector field points outward*

$$\|x\|=1 \Rightarrow \langle x, \xi(x)\rangle \geqslant 0 \tag{2}$$

Then ξ *has a zero*

$$\exists \bar{x} \in B^n \quad \text{such that} \quad \xi(\bar{x})=0 \tag{3}$$ ▲

Proof. We study successively the cases when n is even and odd.

a. Assume first that n is even. Extend ξ to a vector field $\bar{\xi}\colon 2B^n \to R^n$ by setting setting

$$\bar{\xi}(x) := \xi(x) \qquad \text{if } \|x\|<1$$

$$\bar{\xi}(x) := (\|x\|-1)x + (2-\|x\|)\xi\left(\frac{x}{\|x\|}\right) \qquad \text{if } 1 \leqslant \|x\| \leqslant 2$$

Clearly, $\bar{\xi}$ is continuous and coincides with ξ on B^n. For $\|x\|>1$, that is, for all x outside B^n, we have

$$\langle \bar{\xi}(x), x\rangle = (\|x\|-1)\|x\|^2 + (2-\|x\|)\|x\| \left\langle \xi\left(\frac{x}{\|x\|}\right), \frac{x}{\|x\|}\right\rangle$$

$$\geqslant (\|x\|-1)\|x\|^2 > 0$$

So $\bar{\xi}$ does not vanish on $2B^n \setminus B^n$. On the boundary $2S^{n-1}$ of $2B^n$, we have $\bar{\xi}(x)=x$.

We now identify $2B^n$ with the southern hemisphere of the sphere $2S^n$ and extend $\bar{\xi}$ by symmetry to a continuous tangent vector field $\bar{\bar{\xi}}$ on $2S^n$. This is best explained by Figure 1; the continuity across the equator is ensured by the fact that $\bar{\xi}$ is normal to the boundary.

According to theorem 1, since n is even, $\bar{\bar{\xi}}$ must have a zero on $2S^n$. By symmetry, there must be a zero in the southern hemisphere, so that $\bar{\xi}$ has a zero on $2B^n$. Since $\bar{\xi}$ does not vanish on $2B^n \setminus B^n$, this zero must belong to B^n, and the result is proved.

b. Assume now that n is odd. Then B^n can be identified with a subset of B^{n+1}, and any continuous vector field $\xi\colon B^n \to R^n$ can be extended to a continuous vector field $\bar{\xi}\colon B^{n+1} \to \mathbb{R}^{n+1}$ as follows:

$$B^{n+1} = \left\{ (x_1, \ldots, x_{n+1}) \middle| \sum_{i=1}^{n+1} x_i^2 \leq 1 \right\}$$

$$B^n = \{(x_1, \ldots, x_{n+1}) \in B^{n+1} | x_{n+1} = 0\}$$

$$\bar{\xi}(x_1, \ldots, x_{n+1}) = (\xi(x_1, \ldots, x_n), ax_{n+1})$$

Here a is a positive number, to be determined presently. For any point $(x_1, \ldots, x_{n+1})$ on the boundary of B^{n+1}, we have

$$\sum_{i=1}^{n+1} x_i \bar{\xi}_i(x_1, \ldots, x_{n+1}) = \sum_{i=1}^{n} x_i \xi_i(x_1, \ldots, x_n) + ax_{n+1}^2$$
$$= \sum_{i=1}^{n} x_i \xi_i(x_1, \ldots, x_n) + a\left(1 - \sum_{i=1}^{n} x_i^2\right)$$

Now if the vector field ξ on B^n points outward on the boundary, we will be able to choose $a > 0$ so large that the vector field $\bar{\xi}$ on B^{n+1} also points outward on the boundary. Since n is odd, $n+1$ is even, and so $\bar{\xi}$ must have a zero: $\bar{\xi}(\bar{x}_1, \ldots, \bar{x}_{n+1}) = 0$. Looking up the definition of $\bar{\xi}$, we see this means that $\bar{x}_{n+1} = 0$ and $\xi(\bar{x}_1, \ldots, \bar{x}_n) = 0$, as desired. ■

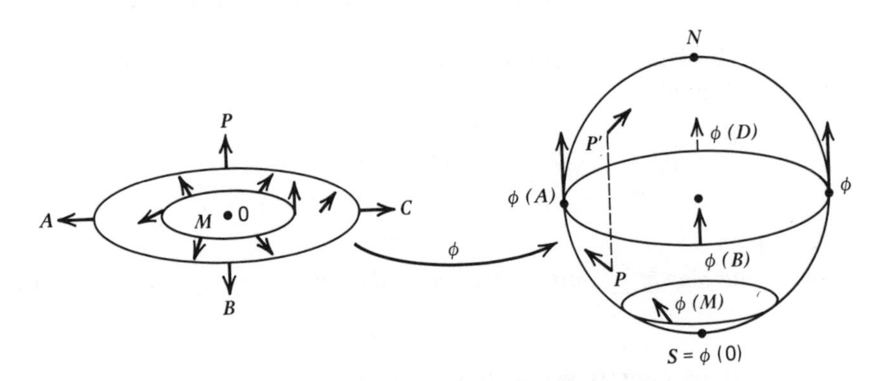

$$2B^n = \left\{ (x_1, \ldots, x_n) \middle| \sum_{i=1}^{n} x_i^2 \leq 4 \right\}$$

$$2S^n = \left\{ (y_1, \ldots, y_n, y_{n+1}) \middle| \sum_{i=1}^{n+1} y_i^2 = 4 \right\}$$

We have represented a diffeomorphism of $2B^n$ onto the southern hemisphere of $2S^n$, taking the center of $2B^n$ to the south pole of $2S^n$ and the boundary to the equator. The vector field $\bar{\xi}$ on $2B^n$ is transformed into a tangent vector field on the northern hemisphere by the formula

$$\bar{\bar{\xi}}_i(y_1, \ldots, y_n, y_{n+1}) = -\bar{\xi}_i(y_1, \ldots, y_n, -y_{n+1}) \qquad \text{for } 1 \leq i \leq n$$

$$\bar{\bar{\xi}}_{n+1}(y_1, \ldots, y_n, y_{n+1}) = \bar{\bar{\xi}}_{n+1}(y_1, \ldots, y_n, -y_{n+1})$$

Of course, if a continuous vector field ξ on B^n points inward on the boundary, $\langle x, \xi(x)\rangle \leqslant 0$ for $\|x\|=1$, then it also has a zero, because $-\xi$ will point outward and ξ has the same zeroes as $-\xi$. As an immediate corollary, we have Brouwer's theorem.

THEOREM 4 (BROUWER)

Any continuous map of the unit ball B^n into itself has a fixed point. ▲

Proof. Let $f\colon B^n \to B^n$ be a continuous map. Then $\xi(x)=f(x)-x$ is a continuous vector field on B^n that points inward on the boundary. By theorem 3, there is some point $\bar{x}$ where $0=\xi(\bar{x})=f(\bar{x})-\bar{x}$. ■

Of course, all these results extend to a wider class of objects. We shall later see that theorem 3 extends to closed convex subsets of infinite-dimensional vector spaces and continuous compact maps ξ.

We note now that Brouwer's theorem extends to any topological space X that is homeomorphic to some n-dimensional ball B^n. Indeed, let $g\colon X \to B^n$ be this homeomorphism, and let $f\colon X \to X$ be a continuous map. Then $g \circ f \circ g^{-1}$ is a continuous map of B^n into itself, which, therefore, has a fixed point $\bar{y}= g \circ f \circ g^{-1}(\bar{y})$. Setting $\bar{x}=g^{-1}(\bar{y}) \in X$, we get $\bar{x}=f(\bar{x})$, so that $\bar{x}$ is a fixed point of f.

Among such spaces X homeomorphic to B^n, we single out for later use the n-dimensional simplex

$$\sum\nolimits^n = \left\{(x_1, \ldots, x_{n+1}) \,\middle|\, \sum_{i=1}^{n+1} x_i = 1 \text{ and } x_i \geqslant 0 \text{ for all } i\right\}$$

COROLLARY 5

Any continuous map of $\sum^n$ into itself has a fixed point. ▲

3. LOCAL STUDY OF THE EQUATION $f(x)=0$

Let X and Y be Banach spaces and $f\colon X \to Y$ a smooth map. We are interested in the subset E of X defined by the equation $f(x)=0$

$$E := f^{-1}(0) = \{x \mid f(x)=0\} \tag{1}$$

More precisely, we wish to know if the subset E is in some sense "smooth." Of course, if f is continuous, then E is closed. But this is meager information, since a closed subset can be very bad indeed, such as a Cantor subset of the real line or even worse. Perhaps we can do better if f is smoother, C^r, for instance, with $r \geqslant 1$? The following result gives us the answer, and it is negative.

THEOREM 1
Let E be any closed subset of $\mathbb{R}^n$. Then there is some C^∞ function $f: \mathbb{R}^n \to \mathbb{R}$ such that $E=f^{-1}(0)$. ▲

Proof. If F is closed, its complement $\Omega=\mathbb{R}^n \setminus F$ is open. It is a standard property of $\mathbb{R}^n$ that any open subset is the reunion of a countable family of open balls. There is, therefore, a sequence B_k, $k \in \mathbb{N}$, of balls in $\mathbb{R}^n$, the center of B_k being x_k and its radius ρ_k, such that

$$\Omega = \bigcup_{k \in \mathbb{N}} B_k$$

Define for each k a C^∞ function f_k: $\mathbb{R}^n \to \mathbb{R}$ by

$$\text{(2)} \qquad \begin{cases} \textbf{i.} \quad f_k(x)=\exp\left[-(\rho_k-\|x-x_k\|)^{-2}\right] & \text{if } x \in B_k \\ \textbf{ii.} \quad f_k(x)=0 \qquad \text{if } x \notin B_k \end{cases}$$

It is easily checked that f_k is C^∞ and vanishes with all its derivatives outside B_k. Denote by $p=(p_1, \ldots, p_n)$ an element of $\mathbb{N}^n$, by $|p|$ the sum $\sum_{i=1}^n p_i$ and by $D^p f$ the partial derivative $\partial^{|p|}f/\partial x_1^{p_1} \cdots \partial x_n^{p_n}$. For each $k \in \mathbb{N}$, choose $\varepsilon_k > 0$ so small that

$$\text{(3)} \qquad \forall x \in B_k, \ |p| \leqslant k \Rightarrow |D^p f_k(x)| \leqslant (\varepsilon_k 2^k)^{-1}$$

Now set

$$f(x) = \sum_{k=0}^{\infty} \varepsilon_k f_k(x)$$

We claim this defines a C^∞ function f: $\mathbb{R}^n \to \mathbb{R}$. It will obviously be non-negative everywhere and zero only if all the $f_k(x)$ are zero; that is, if x does not belong to $\bigcup_{k=0}^\infty B_k = \Omega = \mathbb{R}^n \setminus F$. The result then follows.

Now to prove our claim. Take any $p \in \mathbb{N}^n$, and consider the series of partial derivatives

$$\text{(4)} \qquad \sum_{k=0}^{\infty} \varepsilon_k D^p f_k(x) = \sum_{k<|p|} \varepsilon_k D^p f_k(x) + \sum_{k \geqslant |p|} \varepsilon_k D^p f_k(x)$$

The first term on the right-hand side is just a finite sum. As for the second, we see that $|\varepsilon_k D^p f_k(x)|=0$ if $x \notin B_k$ and is less than 2^{-k} if $x \in B_k$, by the definition of ε_k, so that

$$\forall x \in \mathbb{R}^n, \ \sum_{k \geqslant |p|} \varepsilon_k |D^p f_k(x)| \leqslant \sum_{k \geqslant |p|} 2^{-k} < \infty$$

The series in equation (4) is uniformly convergent for any choice of $p \in \mathbb{N}^n$. It follows that f is C^∞, as announced. ■

Therefore, if we want $f^{-1}(0)$ to be smooth, it is not enough for f itself to be smooth; something more is needed. This is precisely where the inverse function theorem comes in.

PROPOSITION 2

Let $f: X \to Y$ be a C^r map, $r \geq 1$, between Banach spaces. Let $\bar{x}$ be some point in $f^{-1}(0) =: E$. Assume that $f'(\bar{x})$ is surjective and that there exists a continuous projector π† from X onto $\operatorname{Ker} f'(\bar{x})$.

Then there are neighborhoods $\mathcal{V}$ of $\bar{x}$ in X and $\mathcal{U}$ of the origin in X, and a C^r diffeomorphism ρ of $\mathcal{U}$ onto $\mathcal{V}$ such that $\rho(0) = \bar{x}$ and

$$f^{-1}(0) \cap \mathcal{V} = \rho(\operatorname{Ker} f'(\bar{x}) \cap \mathcal{U}) \tag{5}$$ ▲

Proof. We set $N := (I - \pi)X$.
We start by bringing $\bar{x}$ to the origin: $\bar{x} = 0$. Any $x \in X$ can be written $x = x_1 + x_2$, with $x_1 \in N$ and $x_2 \in \operatorname{Ker} f'(\bar{x})$, and this decomposition is unique.

We first apply the implicit function theorem to the map $\psi: N \times \operatorname{Ker} f'(\bar{x}) \to Y$ defined by $\psi(x_1, x_2) = f(x_1 + x_2)$. Here, x_2 is considered a parameter, while x_1 is the true variable. The derivative $\psi'_{x_1}(0, 0)$ is the restriction to N of $f'(0)$, which is an isomorphism between N and Y by the open mapping theorem. It follows that for suitably small x_1 and x_2, there is a unique C^r solution $x_1 = g(x_2)$ of the equation $\psi(x_1, x_2) = 0$ with $g(0) = 0$. In other words, $g: N \to \operatorname{Ker} f'(0)$ satisfies the identity

$$f(g(x_2) + x_2) = 0 \quad \text{near} \quad 0$$

We now apply the inverse function theorem, this time to the map ρ: $x_1 + x_2 \to x_1 + g(x_2) + x_2$, defined on a suitably small neighborhood of the origin in X. Clearly $\rho'(0, 0)$ is invertible, so that ρ^{-1} is well defined and C^r in a neighborhood of the origin in X. Setting $x_1 = 0$, we get $\rho(0, x_2) = g(x_2) + x_2$, so that $f \circ \rho(0, x_2)$ is identically $0 \in Y$. The converse follows from the uniqueness in the inverse function theorem, and the result is proved. ■

It may not seem so at first glance, but proposition 2 gives us a lot of information about the subset E near $\bar{x}$. For instance, the following configurations in Figure 1 are excluded.

The only allowable configuration near $\bar{x}$ is the following:

The map $\rho^{-1}: \mathcal{V} \to \mathcal{U}$ "straightens out" the subset E near $\bar{x}$, turning it into a linear subspace. If this can be done for any choice of $\bar{x}$ in E, we say that E is a submanifold of X. Let us give a formal definition (setting $\rho^{-1} = \psi$).

†A projector is a linear map π such that $\pi^2 = \pi$. In Hilbert spaces, and finite-dimensional spaces, every closed linear subspace is the range of some continuous projector. This is no longer true in general Banach spaces.

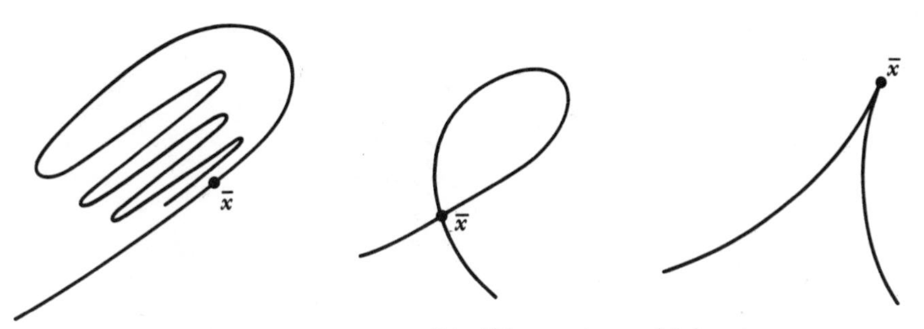

Figure 1. *(a)* Cluster, *(b)* self-intersection, and *(c)* cusp.

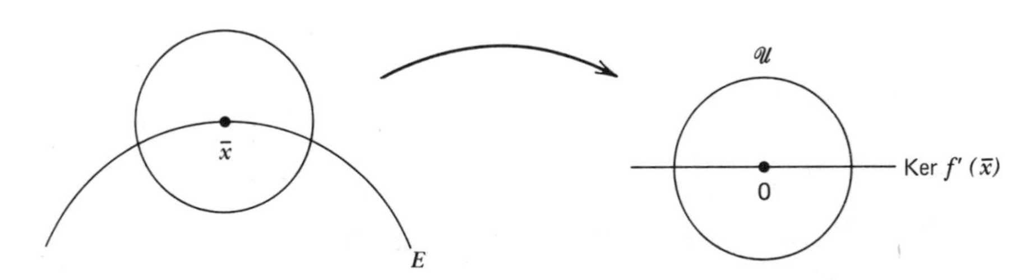

DEFINITION 3

A subset E of X is a C^r submanifold, $r \geqslant 1$ if for any $\bar{x} \in E$, there is a neighborhood $\mathscr{U}$ of $\bar{x}$ in X, a neighborhood $\mathscr{V}$ of 0 in X, a continuous projector π from X onto L, and a C^r diffeomorphism $\psi : \mathscr{U} \to \mathscr{V}$ such that

$$L \cap \mathscr{V} = \psi(E \cap \mathscr{U})$$

If $\dim L = p < \infty$ *for all $\bar{x}$, we say that E has dimension p. If* $\dim (I - \pi)X = q < \infty$ *for all $\bar{x}$, we say that X has codimension q.* ▲

A closed submanifold is the formalization of one's intuitive idea of a smooth surface, without self-intersections and cusps. The requirement that the submanifold be closed helps to eliminate other singularities; for instance, the following spiral (center excluded) is a manifold, but it is not closed (Fig. 2).

The finite dimensional case, $\dim X = n < \infty$, is particularly interesting. In this case, for any $x \in \mathscr{U}$, we can describe $\psi(x) \in \mathscr{U}$ by its n coordinates $(\xi_1, \ldots, \xi_n)$ in some basis of X, which gives us a set of n equations

$$\xi_i = \psi_i(x_1, \ldots, x_n) \qquad 1 \leqslant i \leqslant n$$

If E is p dimensional and modeled on the subspace L, we usually choose the p first basis vectors of X in L, so that $E \cap \mathscr{U}$ is described by the $q = (n-p)$ equations $\psi_{p+1}(x) = \cdots = \psi_n(x) = 0$. We then refer to ψ as a *local chart of* (E, X) *around* $\bar{x}$.

Since ψ is a diffeomorphism, the linear functionals $\psi_i'(\bar{x})$, $1 \leqslant i \leqslant n$, are linearly

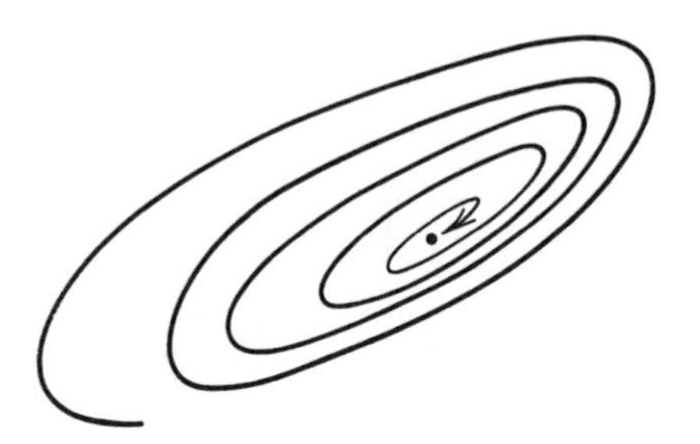

Figure 2. A nonclosed submanifold.

independent on X, so that the q equations $\langle x, \psi_i'(\bar{x})\rangle = 0$ for $p+1 \leqslant i \leqslant n$ define a p-dimensional linear subspace of X. We set, indifferently,

$$T_E(\bar{x}) := T_{\bar{x}}E := \{x \in X \mid < x, \psi_i'(\bar{x}) > = 0, p+1 \leqslant i \leqslant n\}$$

and call it the *tangent space* to E at $\bar{x}$. It is independent of the local chart ψ that has been chosen for (E, X) around $\bar{x}$.

We can also directly define the position of x in $\mathscr{U}$ by the values of the ξ_i, $1 \leqslant i \leqslant n$. We then refer to $(\xi_1, \ldots, \xi_n)$ as a local, curvilinear, coordinate system for X near $\bar{x}$, and $E \cap \mathscr{U}$ is defined by the equations $\xi_1 = \cdots = \xi_p = 0$. They are now linear, but the coordinate system is not!

This finite-dimensional framework enables us to greatly extend proposition 2. We start with a definition.

DEFINITION 4

Let $g: \mathbb{R}^n \to \mathbb{R}^m$ be a C^r map, $r \geqslant 1$, and $M \subset \mathbb{R}^m$ a C^r submanifold of codimension q. Let $\bar{x}$ be a point in $\mathbb{R}^n$. We say that g is transversal to M at $\bar{x}$ if

either $g(\bar{x})$ does not belong to M

or $g(\bar{x})$ belongs to M, and for some local chart ψ of $(M, \mathbb{R}^m)$ around $g(\bar{x})$, the (q, n) matrix $(((\partial/\partial x_i)\psi_j \circ g(\bar{x})))$, $1 \leqslant i \leqslant n$, $m-q+1 \leqslant j \leqslant m$, has rank q.

We say that g is transversal to M if it is transversal to M at every point $\bar{x} \in \mathbb{R}^n$.

▲

It may be easier to understand when g is *not* transversal to M at $\bar{x}$. This means that $g(\bar{x})$ belongs to M and the (q, n) matrix $(((\partial/\partial x_i)\psi_j \circ g(x)))$ has rank $\leqslant q-1$. In other words, the corresponding linear map from $\mathbb{R}^n$ to $\mathbb{R}^q$ is not surjective. Geometrically, this means that there is a vector $\eta \in \mathbb{R}^m$ that is neither in the tangent space $T_{\bar{x}}M$ nor the image by the tangent map $g'(\bar{x})$ of some vector $\xi \in \mathbb{R}^n$ nor a linear combination of both. This property does not depend on the local chart ψ that has been chosen for $(M, \mathbb{R}^m)$ around $\bar{x}$.

THEOREM 5

Let $g: \mathbb{R}^n \to \mathbb{R}^m$ be a C^r map, $r \geqslant 1$, and $M \subset \mathbb{R}^m$ a C^r submanifold of codimension q. Assume g is transversal to M. Then $g^{-1}(M)$ is a C^r submanifold of $\mathbb{R}^n$ with the same codimension q. ▲

Proof. We are going to construct around every point $\bar{x} \in g^{-1}(M)$ a local chart for $(g^{-1}(M), \mathbb{R}^n)$ of a special kind. Set $N = g^{-1}(M)$ and $g(\bar{x}) = \bar{y}$ for the sake of simplicity. Choose a local chart ψ of $(M, \mathbb{R}^m)$ around $\bar{y}$. In some neighborhood $\mathscr{U}$ of $\bar{y}$, we have

$$y \in M \Leftrightarrow \psi_{p+1}(y) = \cdots = \psi_m(y) = 0$$

Choose a global linear coordinate system $(\xi_1, \ldots, \xi_n)$ for $\mathbb{R}^n$, denoting as usual by π_i the ith projection

$$\xi_i = \pi_i(x), \qquad \text{for all } x \in \mathbb{R}^n$$

By assumption, the matrix $(((\partial/\partial x_i)\pi_j \circ g(\bar{x})))$, $1 \leqslant i \leqslant n$, $p+1 \leqslant j \leqslant m$, has rank $q = m - p$. This means that the q rows of that matrix are linearly independent. Introducing the transpose of $g'(\bar{x}) \in \mathscr{L}(\mathbb{R}^n, \mathbb{R}^m)$, we recognize in the jth row, the n components of the linear functional $g'(\bar{x})^* \psi_j'(\bar{y})$.

It follows that we can choose $(n-q)$ of the π_i, say $\pi_1, \ldots, \pi_{n-q}$, in such a way that the n linear functionals

$$(\pi_1, \ldots, \pi_{n-q}, g'(\bar{x})^* \psi_{p+1}'(\bar{y}), \ldots, g'(\bar{x})^* \psi_m'(\bar{y}))$$

are linearly independent. We recognize the n components of the derivative at $\bar{x}$ of the nonlinear map from $\mathbb{R}^n$ to itself

$$(\pi_1, \ldots, \pi_{n-q}, \psi_{p+1} \circ g, \ldots, \psi_m \circ g)$$

Using the inverse function theorem, we see that this map is locally invertible near $\bar{x}$. Moreover, in some appropriate neighborhood of $\bar{x}$, the set $N := g^{-1}(M)$ is described by the q equations $\psi_{p+1} \circ g(x) = \cdots = \psi_m \circ g(x) = 0$. We have thus found a local chart for $(N, \mathbb{R}^n)$ around $\bar{x}$ and thereby proved that N is a submanifold of codimension q in $\mathbb{R}^n$. ■

Of course, if M is a closed submanifold, so is $g^{-1}(M)$, since g is continuous. As a particular case of theorem 5, we obtain the finite-dimensional version of proposition 2.

COROLLARY 6

Let $f: \mathbb{R}^n \to \mathbb{R}^m$ be a C^r map, $r \geq 1$. Assume that for any $x \in f^{-1}(0) =: N$, the derivative $f'(x) \in \mathcal{L}(\mathbb{R}^n, \mathbb{R}^m)$ is onto. Then $N \subset \mathbb{R}^n$ is either empty or a closed C^r submanifold of codimension m. ▲

Proof. This is an easy consequence of proposition 2, because in finite dimension, any closed subspace is the range of a continuous projector. It can also be proved from theorem 5 that the origin $\{0\}$ is a closed submanifold of codimension m in $\mathbb{R}^m$, the tangent space being the linear subspace $\{0\}$. Transversality then means precisely that $f'(x)$ is onto whenever $f(x)=0$. The assumption in theorem 5 is thus satisfied and so is the conclusion. ■

The transversal case being disposed of, we now turn to nontransversal cases. We wish to see what N looks like near a point $\bar{x} \in N$ where $f'(x)$ is *not* surjective. There are, of course, an infinite number of possibilities according to the way f' degenerates at $\bar{x}$. We limit ourselves to investigating the simplest possible degeneracy. From now on, we take $Y = \mathbb{R}$, which means that the set N is described by a single equation.

DEFINITION 7

Let X be a Hilbert space and $f: X \to \mathbb{R}$ a C^r function, $r \geq 2$. Any point $\bar{x}$ where $f'(\bar{x}) = 0$ is called a critical point. *A critical point $\bar{x}$ is* nondegenerate *if its Hessian is nondegenerate as a quadratic form*

$$(f''(\bar{x})y, z) = 0 \text{ for all } z \Leftrightarrow y = 0$$

If all the critical points of f are nondegenerate, we shall say that f is a Morse function. ▲

THEOREM 8 (Morse Lemma)

Let $\bar{x}$ be a nondegenerate critical point of f. Then there exist neighborhoods $\mathcal{V}$ of $\bar{x}$ and $\mathcal{U}$ of the origin and a C^{r-2} diffeomorphism ρ of $\mathcal{U}$ onto $\mathcal{V}$ such that $\rho(0) = \bar{x}$ and

$$\forall y \in \mathcal{U}, \qquad f(\bar{x}) + \tfrac{1}{2}(f''(\bar{x})y, y) = f \circ \rho(y)$$ ▲

We will provide a proof for $r > 2$ only.

PROPOSITION 9 (Hadamard Lemma)

Let X and Y be Banach spaces, $f: X \to Y$ a C^r map, $r \geq 1$, and $\bar{x}$ a point where $f(\bar{x}) = 0$. Then there exists a C^{r-1} map $u: X \to \mathcal{L}(X, Y)$ with $u(\bar{x}) = f'(\bar{x})$ such that

$$f(x) = u(x)(x - \bar{x})$$ ▲

Proof. Let $t \in [0, 1]$ be a real variable. We have

$$\forall x \in X, \qquad f(x)=\int_0^1 \frac{d}{dt} f[\bar{x}+t(x-\bar{x})]dt$$

$$=\int_0^1 <f'[x+t(x-\bar{x})], x-\bar{x}> dt$$

$$=< \int_0^1 f'[\bar{x}+t(x-\bar{x})]dt, x-\bar{x}>$$

Calling the integral $u(x)$ gives the desired result. ■

Proof of Theorem 8. The proof of the Morse lemma now runs as follows. Start with $f: X \to R$ and apply Hadamard's lemma to $f - f(\bar{x})$. We get a C^{r-1} map $u: X \to X$ such that $u(\bar{x})=0$ and

$$f(x)=(u(x), x-\bar{x})+f(\bar{x})$$

Apply Hadamard's lemma once more, this time to u. We get a C^{r-2} map $w: X \to \mathscr{L}(X, X)$ such that

$$\forall x \in X, \; u(x)=w(x)(x-\bar{x})$$

Substituting this into the preceding equation, we obtain

$$\forall x \in X, \qquad f(x)=(w(x)(x-\bar{x}), (x-\bar{x}))+f(\bar{x})$$

Denote by $\bar{w}(x)$ the symmetric part of $w(x)$

$$\bar{w}(x)=\tfrac{1}{2}[w(x)+w(x)^*]$$

Clearly $\bar{w}: X \to \mathscr{L}(X, X)$ is still C^{r-2}, and

$$\forall x \in X, \qquad f(x)=(\bar{w}(x)(x-\bar{x}), (x-\bar{x}))+f(\bar{x})$$

Comparing this with the Taylor expansion of f at $\bar{x}$ and bearing in mind that $f'(x)=0$ and $\bar{w}$ is continuous at $\bar{x}$, we obtain

$$\bar{w}(\bar{x})=2f''(\bar{x})$$

which is invertible, since the Hessian is nondegenerate.

We now introduce the operator $v(x)=(\bar{w}(x)^{-1}\bar{w}(\bar{x}))^{1/2}$ defined by

$$v(x)=I+\sum_{n=1}^{\infty} c_n[\bar{w}(x)^{-1}\bar{w}(\bar{x})-I]^n$$

where the coefficients c_n are defined by $\sqrt{1+t}=1+\sum_{n=1}^{\infty} c_n t^n$ for the real variable t. This series has radius of convergence 1, so $v(x)$ is well defined for $\|\bar{w}(x)^{-1}\bar{w}(\bar{x})-I\|<1$, that is, for x in some neighborhood of $\bar{x}$ and satisfies the relationship

$$v(x)v(x)=\bar{w}(x)^{-1}\bar{w}(\bar{x})$$

It is clear from the series expansion that $v(x)$ will be close to I, and hence invertible, when $\|x\|$ is sufficiently small. Moreover, since $\bar{w}(x)$ is self-adjoint, we have

$$[\bar{w}(x)^{-1}\bar{w}(\bar{x})]^*=\bar{w}(\bar{x})\bar{w}(x)^{-1}$$

so that

$$[\bar{w}(x)^{-1}\bar{w}(\bar{x})]^*\bar{w}(x)=\bar{w}(\bar{x})=\bar{w}(x)[\bar{w}(x)^{-1}\bar{w}(\bar{x})]$$

Summing up the series expansion for $v(x)$ gives

$$v(x)^*\bar{w}(x)=\bar{w}(x)v(x)$$

so that

$$\begin{aligned}v(x)^*\bar{w}(x)v(x)&=\bar{w}(x)v(x)v(x)\\&=\bar{w}(x)\bar{w}(x)^{-1}\bar{w}(\bar{x})\\&=\bar{w}(\bar{x})\end{aligned}$$

In other words, the linear change of variables described by invertible operator $v(x)$ will always bring the variable quadratic form described by $\bar{w}(x)$ to the constant form $\bar{w}(\bar{x})$. We take advantage of this by the change of variables

$$y=v(x)^{-1}(x-\bar{x})$$

The right-hand side has $v(\bar{x})^{-1}$ as its tangent map at $x=\bar{x}$. By the inverse function theorem, it is invertible near $x=\bar{x}$ and $y=0$, and its inverse is precisely the map $\rho(y)$ we are looking for. Indeed, we have

$$\begin{aligned}f(x)&=(\bar{w}(x)(x-\bar{x}),\ (x-\bar{x}))+f(\bar{x})\\&=(\bar{w}(x)v(x)y,\ v(x)y)+f(\bar{x})\\&=(v(x)^*\bar{w}(x)v(x)y,\ y)+f(\bar{x})\\&=(\bar{w}(\bar{x})y,\ y)+f(\bar{x})\\&=\tfrac{1}{2}(f''(\bar{x})y,\ y)+f(\bar{x})\end{aligned}$$

■

This concludes the proof. The Morse lemma is particularly striking in the finite-dimensional case:

COROLLARY 10 (FINITE-DIMENSIONAL MORSE LEMMA)
Assume $f:\mathbb{R}^n\to\mathbb{R}$ is a C^r map, $r\geqslant 2$, and $\bar{x}$ a point where

$$(6)\qquad \begin{cases} \textbf{i.} \quad \dfrac{\partial f}{\partial x_i}(\bar{x})=0, \qquad 1\leqslant i\leqslant n \\ \textbf{ii.} \quad \operatorname{Det}\left(\left(\dfrac{\partial^2 f}{\partial x_i \partial x_j}(\bar{x})\right)\right)\neq 0 \end{cases}$$

Then there exists a local, curvilinear, coordinate system $(\xi_1, \ldots, \xi_n)$ for $\mathbb{R}^n$ near $\bar{x}$ and an integer k, with $0\leqslant k\leqslant n$, such that

$$f(x)=f(\bar{x})-\sum_{i=1}^{k}\xi_i^2+\sum_{k+1}^{n}\xi_i^2$$

for all x in some neighborhood of $\bar{x}$. ▲

DEFINITION 11
The integer k is called the index *of the nondegenerate critical point $\bar{x}$.* ▲

Corollary 10 follows immediately from theorem 8 by finding a suitable (linear) basis for $\mathbb{R}^n$, so that the quadratic form with matrix

$$\left(\left(\frac{1}{2}\frac{\partial^2 f}{\partial x_i \partial x_j}(\bar{x})\right)\right)$$

In the old basis now reads $((\pm\delta_{ij}))$ (recall that the Kronecker symbol δ_{ij} is 0 if $i\neq j$ and 1 if $i=j$). By standard results in linear algebra, this is always possible if the quadratic form is nondegenerate and the number of k times -1 occurs does not depend on the way this diagonalization is performed.

Note that in the coordinate system $(\xi_1, \ldots, \xi_n)$, the function f becomes precisely quadratic: It coincides with its second-order Taylor expansion (there are no first-order forms). This very simple expression for f is achieved at the expense of curving the coordinate system, which is no longer linear.

The index of a nondegenerate critical point is a geometric notion, that is, it is invariant by diffeomorphisms (in particular, change of coordinates) of the base space. The index determines the behavior of f near $\bar{x}$, and, hence, the shape of $N=f^{-1}(0)$ near $\bar{x}$.

For instance, if $k=0$, then f achieves a local minimum at $\bar{x}$. If $k=n$, then f achieves a local maximum at $\bar{x}$. In the one-dimensional case, $n=1$, these are the only possibilities for nondegenerate critical points. When $n=2$, a third possib-

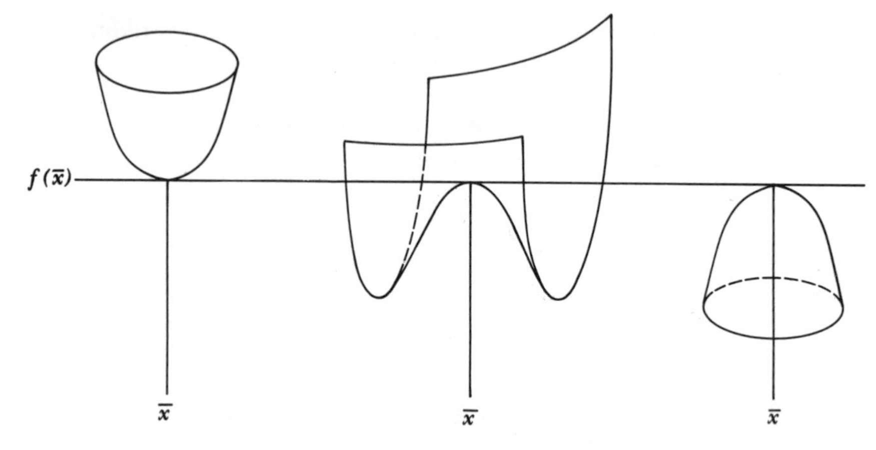

Figure 3. Two-dimensional critical points: *(a)* index 0. *(b)* index 1, and *(c)* index 2.

ility appears, namely, $k=1$. We then have $f(x)=f(\bar{x})+\xi_2^2-\xi_1^2$ in an appropriate coordinate system near $\bar{x}$: We say that $\bar{x}$ is a saddle point (Fig. 3).

Still in the two-dimensional case, we now have a complete picture of $E=f^{-1}(0)$ near a nondegenerate critical point $\bar{x}\in E$ (so that $f'(\bar{x})=0$). By the Morse lemma, there is a neighborhood $\mathscr{U}$ of $\bar{x}$ and a local coordinate system (ξ_1, ξ_2) of $\mathbb{R}^2$, valid in $\mathscr{U}$, such that

($k=0$ case) $\bar{x}$, with coordinates (0, 0), is the only point of E in $\mathscr{U}$.

($k=2$ case) $\bar{x}$, with coordinates (0, 0), is the only point of E in $\mathscr{U}$.

($k=1$ case) all points x in E with coordinates (ξ_1, ξ_2) such that $\xi_1^2-\xi_2^2=0$ belong to $\mathscr{U}$.

The last case is the most interesting. The equation $\xi_1^2-\xi_2^2=0$ splits into two straight lines $\xi_1=\pm\xi_2$, which intersect at the origin. Therefore, N has two smooth branches that intersect at $\bar{x}$. The situation is shown in the following figure.

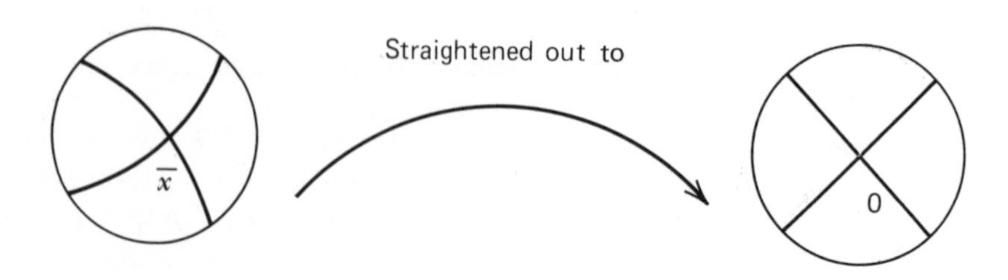

It follows from this analysis that all nondegenerate critical points are isolated. This is a general result, independent of the dimension; indeed, going back to theorem 8, it is clear that $(f\circ\rho)'(y)=\rho'(y)^*f'\circ\rho(y)=f''(\bar{x})y\neq 0$ when $y\neq 0$, so $f'(x)\neq 0$ for x close to $\bar{x}$ but distinct from it.

To conclude with a specific example, let us consider the subset N_λ of $\mathbb{R}^2$ defined by the equation

$$((x_1+1)^2+x_2^2)((x_1-1)^2+x_2^2)=\lambda$$

When λ varies, the N_λ are the level sets of the function

$$f(x_1, x_2)=[(x_1+1)^2+x_2^2][(x_1-1)^2+x_2^2]$$

Let us look for critical points by solving the equations

$$\frac{\partial f}{\partial x_1}(x_1, x_2)=0=\frac{\partial f}{\partial x_2}(x_1, x_2)$$

We obtain

$$2(x_1+1)[(x_1-1)^2+x_2^2]+2(x_1-1)[(x_1+1)^2+x_2^2]=0$$
$$2x_2[((x_1-1)^2+x_2^2)+((x_2+1)^2+x_2^2)]=0$$

The second equation yields $x_2=0$, and substituting it into the first one, we get $x_1=0$, -1, or 1. So we have three critical points, the origin 0 and the points $A(-1, 0)$ and $B(1, 0)$.

The Hessian at the origin turns out to be

$$(f''(0)x, x)=-4x_1^2+4x_2^2$$

The origin, then, is a nondegenerate critical point of index 1. The points A and B turn out to be nondegenerate critical points with index 0. The shape of the level sets N_λ now is easy to understand: They will all be one-dimensional closed

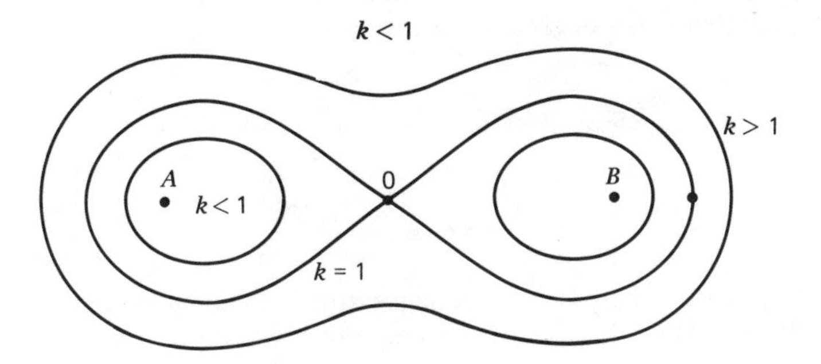

submanifolds, that is, smooth curves, except for those values of λ such that N_λ contains A, B, or 0. This occurs when $\lambda=0$ (N_0 is reduced to the two points A and B) and $\lambda=1$ (N_1 has a double point at the origin, all other points being regular).

Geometrically, N_λ is the set of points M in the plane such that $MA\cdot MB=\sqrt{k}$. The N_1 is known as Bernoulli's lemniscate.

4. BIRTH AND DEATH OF CRITICAL POINTS

Say we have two C^∞ Morse functions, f_0 and f_1, on $\mathbb{R}^n$; then all critical points of f_0 are nondegenerate and so are all critical points of f_1. Now connect f_0 and f_1 by a one-parameter family of smooth functions. By this, we mean a C^∞ map: $f:[0,1]\times\mathbb{R}^n\to\mathbb{R}$ such that

$$(1)\qquad \begin{cases} f(0,x)=f_0(x), & \text{for all } x\in\mathbb{R}^n \\ f(1,x)=f_1(x), & \text{for all } x\in\mathbb{R}^n \end{cases}$$

From now on, we shall write either $f(\lambda, x)$ or $f_\lambda(x)$, so that f_λ is the map $x\to(\lambda, x)$.

Since f_0 and f_1 were chosen at random, they presumably have a different number of critical points. Therefore, the number of critical points of the function f_λ must vary when λ ranges from 0 to 1. This means that for special values of λ, critical points must appear or disappear. We wish to investigate the manner in which this happens.

We begin by two simple results from propositions 1 and 2.

PROPOSITION 1

Assume that for some value $\bar\lambda$ of the parameter, $f_{\bar\lambda}$ is a Morse function. Then for any compact subset K of $\mathbb{R}^n$, there is some $\varepsilon>0$ such that whenever $|\lambda-\bar\lambda|\leqslant\varepsilon$, the function f_λ has no degenerate critical point on K. ▲

Proof. Assume otherwise. Then there is a sequence $\lambda_k\to\bar\lambda$ and a sequence x_k in K such that x_k is a degenerate critical point of f_{λ_k}

$$\frac{\partial f}{\partial x_i}(\lambda_k, x_k)=0 \qquad 1\leqslant i\leqslant n, \qquad \text{for all } k\in\mathbb{N}$$

$$\det\left(\frac{\partial^2 f}{\partial x_i\partial x_j}(\lambda_k, x_k)\right)=0$$

By compactness, there is a subsequence, still denoted by x_k, that converges to some $\bar x$ in K. Taking limits in these equations, we see that $\bar x$ is a degenerate critical point of $f_{\bar\lambda}$ in K, which contradicts the assumption that $f_{\bar\lambda}$ is a Morse function ■

PROPOSITION 2

Assume that for some value $\bar{\lambda}$ of the parameter, f_λ is a Morse function. Let K be any compact subset of $\mathbb{R}^n$. Then there is some $\varepsilon > 0$, some neighborhood $\mathcal{U}$ of K, and a finite number N of C^∞ maps $\xi_j :]\bar{\lambda}-\varepsilon, \bar{\lambda}+\varepsilon[\to \mathcal{U}$, $1 \leqslant i \leqslant N$, such that the $\xi_j(\lambda)$ are precisely the critical points of f in $\mathcal{U}$. ▲

Proof. Since $f_{\bar{\lambda}}$ is a Morse function, all its critical points are isolated, so it can have only a finite number in any compact set. It follows that there is some open neighborhood $\mathcal{U}$ of K such that $\bar{\mathcal{U}}$ is compact and $f_{\bar{\lambda}}$ has no critical point in $\bar{\mathcal{U}} \setminus K$.

Let $\bar{x}$ be a critical point of $f_{\bar{\lambda}}$ in K. It is known to be nondegenerate, so that

$$\det\left(\frac{\partial^2 f}{\partial x_i \partial x_j}(\bar{\lambda}, \bar{x}) \neq 0\right)$$

This means that we can apply the implicit function theorem to the n equations

$$\frac{\partial f}{\partial x_i}(\lambda, x) = 0, \qquad 1 \leqslant i \leqslant n$$

near $(\bar{\lambda}, \bar{x})$: For (λ, x) close enough to $(\bar{\lambda}, \bar{x})$, the only solution to this system is given by $x = \xi(\lambda)$, with ξ a smooth map.

Let $\bar{x}_1, \ldots, \bar{x}_N$ be all the critical points of $f_{\bar{\lambda}}$ in K. For each of them, we proceed as above. We end up with an $\varepsilon > 0$, disjoint open neighborhoods $\mathcal{U}_j$ of the $\bar{x}_j$, and maps ξ_j: $]\bar{\lambda}-\varepsilon, \bar{\lambda}+\varepsilon[\to \mathcal{U}_j$ such that whenever $|\lambda - \bar{\lambda}| < \varepsilon$ and $x \in \cup \mathcal{U}_j$, the equation $f'(\lambda, x) = 0$ implies that $x = \xi_j(\lambda)$ for some j. The proof is now complete, except that we might have $f'(\lambda, x) = 0$ with $x \notin \cup \mathcal{U}_j$. We claim this never happens when $x \in \mathcal{U}$ and $|\lambda - \bar{\lambda}|$ is sufficiently small. Otherwise, there would be sequences x_n in $\mathcal{U}$ and $\lambda_n \to \bar{\lambda}$, with x_n a critical point of f_{λ_n} not belonging to $\cup \mathcal{U}_j$.

Letting $n \to \infty$ and using the compactness of $\bar{\mathcal{U}}$, we obtain in the limit a critical point $\bar{x}$ of $f_{\bar{\lambda}}$ in $\bar{\mathcal{U}}$ not belonging to any $\mathcal{U}_j$ and, hence, different from all the $\bar{x}_j$, $1 \leqslant j \leqslant N$. But this is impossible, since the $\bar{x}_j$ accounted for all the critical points of f in $\bar{\mathcal{U}}$. ■

Propositions 1 and 2 already tell us a lot. If $\bar{\lambda}$ is a value of the parameter for which $f_{\bar{\lambda}}$ is a Morse function, the only way the number of critical points of f_λ can change when λ crosses the value $\bar{\lambda}$ is for critical points to "come in from" or "run away to" infinity. A simple instance of this occurs with the function $f(\lambda, x) = \lambda x^3 - x$ (here, $n = 1$ and $\bar{\lambda} = 0$).

If there is a bounded set K that contains all critical points of f_λ, for $0 \leqslant \lambda \leqslant 1$ and if f_0 and f_1 have different numbers of critical points, then proposition 2 tells us that the f_λ, for $0 < \lambda < 1$, cannot all be Morse functions. The values of λ for which f_λ has degenerate critical points are precisely those for which critical points appear or disappear (at finite distance). By proposition 2, they constitute a closed subset of $[0, 1]$.

Let us investigate this phenomenon more closely. We define the indicatrix of the family f_λ.

DEFINITION 3

Let $f: \mathbb{R} \times \mathbb{R}^n \to \mathbb{R}$ be a C^∞ function. The indicatrix *of f is the subset $I(f)$ of $\mathbb{R} \times \mathbb{R}^n$ defined by*

$$I(f) = \left\{ (\lambda, x) \in \mathbb{R} \times \mathbb{R}^n \,\middle|\, \frac{\partial f}{\partial x_i}(\lambda, x) = 0,\ 1 \leqslant i \leqslant n \right\} \tag{2}$$ ▲

We wish to study $I(f)$ under reasonable assumptions on f. As we have just seen, the assumption that the critical points of f_λ are always nondegenerate would not be reasonable, since it would exclude precisely those families we find most interesting. So we weaken it as follows: Consider the determinants

$$\delta(x, \lambda) = \mathrm{Det}\left(\frac{\partial^2 f}{\partial x_i \partial x_j}(x, \lambda) \right) \tag{3}$$

and

$$\Delta(x, \lambda) = \mathrm{Det}\begin{pmatrix} & & & \frac{\partial^2 f}{\partial \lambda \partial x_1} \\ \frac{\partial^2 f}{\partial x_i \partial x_j} & & & \\ & & & \frac{\partial^2 f}{\partial \lambda \partial x_n} \\ \frac{\partial \delta}{\partial x_1} & \cdots & \frac{\partial \delta}{\partial x_n} & \frac{\partial \delta}{\partial \lambda} \end{pmatrix} \tag{4}$$

Our standing assumption throughout this section is given in assumption A.

ASSUMPTION A

There is no point $(\lambda, x) \in \mathbb{R} \times \mathbb{R}^n$ that satisfies simultaneously the following equations:

$$\begin{cases} \textbf{i.} \quad \dfrac{\partial f}{\partial x_i}(\lambda, x) = 0 \qquad 1 \leqslant i \leqslant n \\ \textbf{ii.} \quad \delta(\lambda, x) = 0 \\ \textbf{iii.} \quad \Delta(\lambda, x) = 0 \end{cases} \tag{5}$$ ▲

Note that this is a system of $(n+2)$ equations with $(n+1)$ unknowns, so that it does seem reasonable to assume that it has no solution. A rigorous argument along this line, showing that assumption A holds for almost all f in C^∞, is possible and relies on transversality theory (see Section 7).

LEMMA 4

Assume A. *Then* $I(f)$ *is either empty or a* C^∞ *one-dimensional closed submanifold of* $\mathbb{R}\times\mathbb{R}^n$. ▲

Proof. All we have to do, according to corollary 6 in the preceding section, is prove that the $n\times(n+1)$ matrix

$$\begin{pmatrix} \dfrac{\partial^2 f}{\partial x_i \partial x_j}(\lambda, x) & \begin{matrix} \dfrac{\partial^2 f}{\partial x_1 \partial \lambda}(\lambda, x) \\ \vdots \\ \dfrac{\partial^2 f}{\partial x_n \partial \lambda}(\lambda, x) \end{matrix} \end{pmatrix} = M(\lambda, x)$$

has rank n for all $(\lambda, x)\in I(f)$. If $\delta(\lambda, x)\neq 0$, the first n columns already have rank n.

If $\delta(\lambda, x)=0$, then $\Delta(\lambda, x)\neq 0$ by assumption A. Computing the determinant $\Delta(\lambda, x)$ from its last line, with $\delta(\lambda, x)=0$, we obtain

$$\Delta(\lambda, x)=\sum_{i=1}^{n}\frac{\partial\delta}{\partial x_i}(\lambda, x)C_i(\lambda, x)\neq 0 \tag{6}$$

So one of the cofactors $C_i(\lambda, x)$ must be nonzero, say $C_1(\lambda, x)\neq 0$. But we recognize in $C_1(\lambda, x)$ the determinant formed by the last columns of the matrix $M(\lambda, x)$. So $M(\lambda, x)$ has rank n again. ■

We shall say that a vector $(\mu, y)\in\mathbb{R}\times\mathbb{R}^n$ is *vertical* if its first coordinate vanishes: $\mu=0$.

LEMMA 5

Assume A *and* $I(f)\neq\varnothing$. *Then the points* $(\bar\lambda, \bar x)\in I(f)$ *where the tangent is vertical are precisely those points where* $\delta(\bar\lambda, \bar x)=0$. *Any such point has a neighborhood* $\mathcal{U}$ *such that* $I(f)\cap\mathcal{U}$ *lies on one side only of the vertical hyperplane through* $(\bar\lambda, \bar x)$. ▲

Proof. Assume first that $\delta(\bar\lambda, \bar x)\neq 0$. We claim that the tangent to $I(f)$ at $(\bar\lambda, \bar x)$ is nonvertical.

Indeed, going back to the proof of lemma 4, this is the case when the first n columns of $M(\lambda, x)$ have rank n. By the implicit function theorem, this means that the n equations

$$\frac{\partial f}{\partial x_i}(\lambda, x)=0, \qquad 1\leqslant i\leqslant n$$

can be solved in terms of λ: There is a neighborhood $\mathcal{U}$ of $(\bar\lambda, \bar x)$ such that $I(f)\cap\mathcal{U}$ is the graph of some smooth map $x=\rho(\lambda)$. So the tangent to $I(f)$ at $(\bar\lambda, \bar x)$ carries the vector $(1, \rho'(\lambda))$, which is certainly nonvertical.

Now assume that $\delta(\bar{\lambda}, \bar{x})=0$. By assumption A, we have $\Delta(\bar{\lambda}, \bar{x})\neq 0$. Suppose we have the case where the last n columns of $M(\lambda, x)$ have rank n. By the implicit function theorem again, this means that the n equations

$$\frac{\partial f}{\partial x_i}(\lambda, x)=0, \qquad 1\leqslant i\leqslant n$$

can be solved in terms of x_1. We shall now express $(x_2, \ldots, x_n, \lambda)$ in terms of a new parameter u, related to x_1 by

$$\frac{dx_1}{du}=C_1=\text{Det}\begin{pmatrix} \dfrac{\partial^2 f}{\partial x_1 \partial x_2} & \cdots & \dfrac{\partial^2 f}{\partial x_1 \partial x_n} & \dfrac{\partial^2 f}{\partial x_1 \partial \lambda} \\ & \cdots & & \\ \dfrac{\partial^2 f}{\partial x_n \partial x_2} & \cdots & \dfrac{\partial^2 f}{\partial x_n \partial x_n} & \dfrac{\partial^2 f}{\partial x_n \partial \lambda} \end{pmatrix}$$

All the quantities involved are to be considered as functions of x_1 through (λ, x), and since $C_i(\bar{\lambda}, \bar{x})\neq 0$, this change of variable is well defined near $x_1=\bar{x}_1$, $u=0$. We now compute $d\lambda/du$ and $d^2\lambda/du^2$.

To do so, we have to differentiate the basic equations, $(\partial f/\partial x_i)(\lambda, x)=0$, $1\leqslant i\leqslant n$, with respect to x_1

$$\sum_j \frac{\partial^2 f}{\partial x_i \partial x_j}\frac{dx_j}{dx_1}+\frac{\partial^2 f}{\partial x_i \partial \lambda}\frac{d\lambda}{dx_1}=0, \qquad 1\leqslant i\leqslant n \tag{7}$$

We rewrite this as

$$\sum_{j>1} \frac{\partial^2 f}{\partial x_i \partial x_j}\frac{dx_j}{dx_1}+\frac{\partial^2 f}{\partial x_i \partial \lambda}\frac{d\lambda}{dx_1}=-\frac{\partial^2 f}{\partial x_i \partial x_1} \tag{8}$$

This is a nonhomogeneous system of n linear equations in n unknowns $dx_2/dx_1, \ldots, dx_n/dx_1, d\lambda/dx_1$. Using Cramer's rules to solve this system gives

$$\frac{dx_j}{dx_1}=\frac{C_j}{C_1} \quad \text{and} \quad \frac{d\lambda}{dx_1}=\frac{\delta}{C_1} \tag{9}$$

and hence,

$$\frac{d\lambda}{du}=\frac{d\lambda}{dx_1}\frac{dx_1}{du}=\delta \tag{10}$$

Turning now to the second derivative, we have

$$\frac{d^2\lambda}{du^2}=\frac{d\delta}{du}=\frac{d\delta}{dx_1}\frac{dx_1}{du}=C_1\frac{d\delta}{dx_1}$$
$$=C_1\left[\frac{\partial\delta}{\partial x_1}+\sum_{j>1}\frac{\partial\delta}{\partial x_j}\frac{dx_j}{dx_1}+\frac{\partial\delta}{\partial\lambda}\frac{d\lambda}{dx_1}\right]$$
$$=C_1\left[\frac{\partial\delta}{\partial x_1}+\sum_{j>1}\frac{C_j}{C_1}\frac{\partial\delta}{\partial x_j}+\frac{\delta}{C_1}\frac{\partial\delta}{\partial\lambda}\right]$$
$$=\sum_{i=1}^{n}C_i\frac{\partial\delta}{\partial x_i}+\delta\frac{\partial\delta}{\partial\lambda}$$
$$=\Delta \text{ by definition of the } C_i$$

Now set $u=0$ to get the point $(\bar{\lambda}, \bar{x})$. We find

$$(11)\qquad \begin{cases} \textbf{i.} & \dfrac{d\lambda}{du}(0)=\delta(\bar{\lambda}, \bar{x})=0 \\ \textbf{ii.} & \dfrac{d^2\lambda}{du^2}(0)=\Delta(\bar{\lambda}, \bar{x})\neq 0 \end{cases}$$

The result follows immediately. ■

We now can visualize the indicatrix in the following figure:

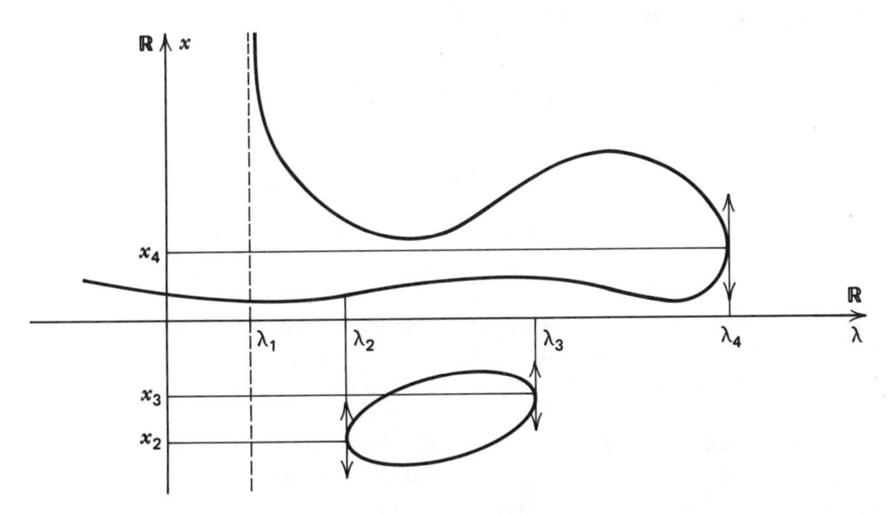

Here, the function f_λ has one critical point for $\lambda<\lambda_1$, two for $\lambda_1<\lambda<\lambda_2$, four for $\lambda_2<\lambda<\lambda_3$, two for $\lambda_3<\lambda<\lambda_4$, and none for $\lambda>\lambda_4$. When λ increases through λ_1, a critical point comes from infinity. When λ increases through λ_2, two critical points are born together at x_2. When λ increases through λ_3 or λ_4, two critical points kill each other at x_3 or x_4. The fact that critical points are born or die in pairs depends only on the concavity $d^2\lambda/du^2$ not vanishing at the points where the tangent is vertical, where $d\lambda/du=0$. We state this fact as a general result in proposition 6.

PROPOSITION 6

Assume A. *Also assume that in the interval* $[\lambda_0, \lambda_1]$, *all critical points of* f_λ *are contained in some bounded subset* K *of* $\mathbb{R}^n$. *Then there is a finite number* $N(\lambda)$ *of them, and there is a finite set* $L \subset [\lambda_0, \lambda_1]$ *of values of* λ *such that* f_λ *is not a Morse function.* $N(\lambda)$ *changes by an even number when* λ *crosses a value in* L *and is constant on all open subintervals determined by* L. *It follows that if* f_{λ_0} *and* f_{λ_1} *are Morse functions, then*

$$N(\lambda_0) \equiv N(\lambda_1) \bmod 2 \tag{12}$$ ▲

The fact that parity is preserved can be very important. For instance, if f has an odd number of critical points, so will f_{λ_1}, which proves that f_{λ_1} has at least one critical point; that is, the equation $f'_{\lambda_1}(x)=0$ has a solution.

If we proceed as we did in the beginning, choosing Morse functions f_0 and f_1 at random, we shall always be able to connect them by a path satisfying A, even if f_0 has an even and f_1 an odd number of critical points. What happens then is that an odd number of critical points are thrown away at infinity or drawn in from infinity. We now turn to another useful tool.

DEFINITION 7

The graphic *of* f *is the subset* $G(f)$ *of* $\mathbb{R}^2$ *defined by*

$$G(f) := \left\{ (\lambda, f(\lambda, x)) \,\middle|\, \frac{\partial f}{\partial x_i}(\lambda, x) = 0 \text{ for } i = 1, \ldots, n \right\} \tag{13}$$ ▲

$G(f)$ is the image of $I(f)$ in the plane by the map

$$\gamma: (\lambda, x) \to (\lambda, f(\lambda, x))$$

LEMMA 8

Let $(\bar{\lambda}, \bar{x})$ *be a point in* $I(f)$ *where* $\delta(\bar{\lambda}, \bar{x}) \neq 0$. *Then there is a neighborhood* $\mathcal{U}$ *of* $(\bar{\lambda}, \bar{x})$ *such that the image of* $I(f) \cap \mathcal{U}$ *by* γ *is a* C^∞ *one-dimensional submanifold of the plane through* $(\bar{\lambda}, f(\bar{\lambda}, \bar{x}))$, *with nonvertical tangent at this point.* ▲

Proof. Since $\delta(\bar{\lambda}, \bar{x}) \neq 0$, lemma 4 tells us there is a neighborhood $\mathcal{U}$ of $(\bar{\lambda}, \bar{x})$ such that $I(f) \cap \mathcal{U}$ is the graph of some C^∞ map $\lambda \mapsto x(\lambda)$ near $\bar{\lambda}$. So the image of $I(f) \cap \mathcal{U}$ by γ is the graph of the C^∞ map $\lambda \mapsto f(\lambda, x(\lambda))$. Hence the result. ■

Setting $\bar{c} = f(\bar{\lambda}, \bar{x})$, we obtain in this way a portion of curve through $(\bar{\lambda}, \bar{c})$ entirely contained within $G(f)$. Now there might well be other points $\bar{x}', \bar{x}'', \ldots$ such that $f(\bar{\lambda}, \bar{x}') = \bar{c} = f(\bar{\lambda}, \bar{x}'') = \ldots$, and, hence, other smooth branches of

$G(f)$ intersecting at $(\bar{\lambda}, \bar{c})$. Because of this, $G(f)$ is not a one-dimensional submanifold of the plane; but there is also another problem, as stated in lemma 9.

LEMMA 9

Assume A. *Let* $(\bar{\lambda}, \bar{x})$ *be a point in* $I(f)$ *where* $\delta(\bar{\lambda}, \bar{x})=0$, *and set* $f(\bar{\lambda}, \bar{x})=\bar{c}$. *Then* $G(f)$ *has a cusp at* $(\bar{\lambda}, \bar{c})$. *The half-tangent at this point is nonvertical and lies inside the cusp.* ▲

Proof. We use the parameter u introduced in lemma 5. By differentiating and remembering that $\partial f/\partial x_i=0$ along $I(f)$, we obtain

$$(14)\qquad \begin{cases} \textbf{i.}\quad \dfrac{df}{du}(u)=\dfrac{\partial f}{\partial \lambda}(\lambda(u), x(u))\dfrac{d\lambda}{du} \\[2ex] \textbf{ii.}\quad \dfrac{df}{du}(0)=\dfrac{\partial f}{\partial \lambda}(\bar{\lambda}, \bar{x})\delta(\bar{\lambda}, \bar{x})=0 \\[2ex] \textbf{iii.}\quad \dfrac{d\lambda}{du}(0)=\delta(\bar{\lambda}, \bar{x})=0 \end{cases}$$

So the curve $u\to(\lambda(u), f(\lambda(u), x(u)))$ does have a singular point for $u=0$. To show that it is a cusp, we have to show that d^2f/du^2 and $d^2\lambda/du^2$ do not vanish simultaneously at $u=0$. Differentiating once more, we obtain

$$\frac{d^2\lambda}{du^2}(0)=\Delta(\bar{\lambda}, \bar{x})\neq 0 \qquad \text{(see lemma 5)}$$

$$\frac{d^2f}{du^2}(0)=\frac{\partial^2 f}{\partial \lambda^2}\left(\frac{d\lambda}{du}\right)^2+\frac{\partial f}{\partial \lambda}\frac{d^2\lambda}{du^2}$$
$$=\frac{\partial f}{\partial \lambda}(\bar{\lambda}, \bar{x})\Delta(\bar{\lambda}, \bar{x})$$

The direction of the half-tangent to the cusp is given by

$$\left(\frac{d^2\lambda}{du^2}(0), \frac{d^2f}{du^2}(0)\right)$$

and its first coordinate is nonzero.

To check the last assertion that the half-tangent lies inside the cusp, we have to prove that the following determinant does not vanish

$$\begin{vmatrix} \dfrac{d^2f}{du^2}(0) & \dfrac{d^3f}{du^3}(0) \\[2ex] \dfrac{d^2\lambda}{du^2}(0) & \dfrac{d^3\lambda}{du^3}(0) \end{vmatrix}$$

The computation finds this determinant to be equal to

$$2\left[\frac{d}{du}\frac{\partial f}{\partial \lambda}(\lambda(u), x(u))\right]_{u=0} \Delta(\bar{\lambda}, \bar{x})^2$$

and the bracketed term is

$$\sum_{i=1}^{n} \frac{\partial^2 f}{\partial \lambda \partial x_i}\frac{dx_i}{du} + \frac{\partial^2 f}{\partial \lambda^2}\frac{d\lambda}{du}$$

If this were zero, multiplying by du/dx_i, we would obtain

$$\sum_{j=1}^{n} \frac{\partial^2 f}{\partial \lambda \partial x_j}\frac{dx_j}{dx_1} + \frac{\partial^2 f}{\partial \lambda^2}\frac{d\lambda}{dx_1} = 0$$

Remembering the n defining equations,

$$\sum_{j=1}^{n} \frac{\partial^2 f}{\partial x_i \partial x_j}\frac{dx_j}{dx_1} + \frac{\partial^2 f}{\partial x_i \partial \lambda}\frac{d\lambda}{dx_1} = 0$$

we find that $dx_1/dx_1 = 1$, $dx_2/dx_1, \ldots, dx_n/dx_1$, $d\lambda/dx_1$ solve a homogeneous system of linear equations, whose matrix is

$$\begin{pmatrix} \dfrac{\partial^2 f}{\partial x_i \partial x_j}(\bar{\lambda}, \bar{x}) & \dfrac{\partial^2 f}{\partial \lambda \partial x_j}(\bar{\lambda}, \bar{x}) \\ \dfrac{\partial^2 f}{\partial x_i \partial \lambda}(\bar{\lambda}, \bar{x}) & \dfrac{\partial^2 f}{\partial \lambda^2}(\bar{\lambda}, \bar{x}) \end{pmatrix}$$

Some algebraic manipulations show that this matrix is invertible if $\delta(\bar{\lambda}, \bar{x}) = 0$ and $\Delta(\bar{\lambda}, \bar{x}) \neq 0$. So all the unknowns should be zero, including dx_1/dx_1, a contradiction. ∎

We now can draw the graphic of f and observe the life of critical points from another point of view. Let us, for instance, observe the same family f_λ as in the preceding figure.

Here $f(\lambda, x(\lambda)) \to \infty$ as $\lambda \to \lambda_1 +$ and $x(\lambda)$ goes to infinity on the vertical branch of $I(\lambda)$, but this might very well not be the case; for example, $f(\lambda, x(\lambda))$ could converge to a finite limit.

The graph tells us that when two critical points are born together, corresponding critical values separate very slowly.

We conclude our investigation by asking what the function $f_{\bar{\lambda}}$ looks like near a point $\bar{x}$ such that $(\bar{\lambda}, \bar{x}) \in I(f)$. If $\delta(\bar{\lambda}, \bar{x}) \neq 0$, then $\bar{x}$ is a nondegenerate critical point of $f_{\bar{\lambda}}$, and the Morse lemma provides the answer: There is a neighborhood

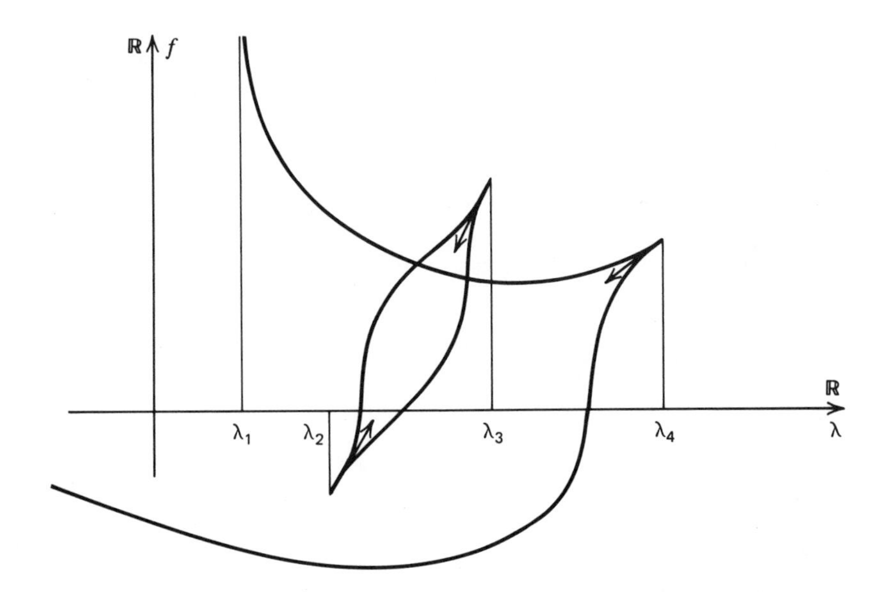

$\mathcal{U}$ of $\bar{x}$, and a local (curvilinear) coordinate system $x=(\xi_1, \ldots \xi_n)$ of $\mathbb{R}^n$, valid in $\mathcal{U}$, such that

$$\bar{x}=(0, \ldots, 0)$$
$$f(\bar{\lambda}, x)=f(\bar{\lambda}, \bar{x})\pm\xi_1^2\pm \cdots \pm\xi_n^2 \qquad \forall x \in \mathcal{U}$$

If $\delta(\bar{\lambda}, \bar{x})=0$, then $\bar{x}$ is a degenerate critical point of $f_{\bar{\lambda}}$, and the Morse lemma fails. However, the assumption $\Delta(\bar{\lambda}, \bar{x})\neq 0$ restricts the possible degeneracies, so that we shall still be able to put $f_{\bar{\lambda}}$ in a canonical form near $\bar{x}$.

LEMMA 10

Assume A. *Let* $(\bar{\lambda}, \bar{x}) \in I(f)$ *be a point where* $\delta(\bar{\lambda}, \bar{x})=0$. *Then there exists a linear coordinate system* $x=(\xi_1, \ldots, \xi_n)$ *of* $\mathbb{R}^n$ *such that*

$$(15) \qquad \begin{cases} \forall x, \ f(\bar{\lambda}, x)=f(\bar{\lambda}, \bar{x})\pm(\xi_1-\bar{\xi}_1)^2\pm \cdots \pm(\xi_{n-1}-\bar{\xi}_{n-1})^2 \\ \qquad + \sum\limits_{i\leqslant j\leqslant k} a_{ijk}(\xi)(\xi_i-\bar{\xi}_i)(\xi_j-\bar{\xi}_j)(\xi_k-\bar{\xi}_k) \end{cases}$$

where $\bar{x}=(\bar{\xi}_1, \ldots, \bar{\xi}_n)$ *and* $a_{nnn}(\bar{\xi})\neq 0$. ▲

Proof. Let $(\xi_1, \ldots, \xi_n)$ be a linear coordinate system that diagonalizes the quadratic form $f''_{\bar{\lambda}}(\bar{x})$. We then have

$$\delta(\bar{\lambda}, \bar{x})=\prod_{i=1}^{n} \frac{\partial^2 f}{\partial \xi_i^2}=0$$

Therefore, one at least of the $\partial^2 f/\partial\xi_i^2$ must be zero; say it is the last one.

By assumption A and lemma 5, we know that $(d\delta/du)(0)=\Delta(\bar{\lambda}, \bar{x})\neq 0$. Computing this derivative with $\partial^2 f/\partial\xi_n^2=0$ gives

$$\frac{d\delta}{du}(0)=\frac{dx_1}{du}(0)\frac{\partial^3 f}{\partial\xi_n^3}\prod_{i<n}\frac{\partial^2 f}{\partial\xi_i^2}$$

No term on the right should vanish. It follows that

$$\frac{\partial^2 f}{\partial\xi_i^2}(\bar{\lambda}, \bar{\xi}_1, \ldots, \bar{\xi}_n)\neq 0 \qquad \text{for} \qquad i\leqslant n-1$$

$$\frac{\partial^2 f}{\partial\xi_n^2}(\bar{\lambda}, \bar{\xi}_1, \ldots, \bar{\xi}_n)=0$$

$$\frac{\partial^3 f}{\partial\xi_n^3}(\bar{\lambda}, \bar{\xi}_1, \ldots, \bar{\xi}_n)\neq 0$$

We can assume the $\partial^2 f/\partial\xi_i^2$ are actually ± 1.

We now use Hadamard's lemma (proposition 9 in the preceding section) three successive times, to prove that

$$(16)\quad f(\bar{\lambda}, x)-f(\bar{\lambda}, \bar{x})-(f''_{\bar{\lambda}}(\bar{x})(x-\bar{x}), x-\bar{x})=\sum a_{ijk}(\xi)(\xi_i-\bar{\xi}_i)(\xi_j-\bar{\xi}_j)(\xi_k-\bar{\xi}_k) \quad \blacksquare$$

LEMMA 11

Assume A. *Let* $(\bar{\lambda}, \bar{x})\in I(f)$ *be a point where* $\delta(\bar{\lambda}, \bar{x})=0$. *Then there exists a neighborhood* $\mathscr{U}$ *of* $\bar{x}$ *and a local (curvilinear) coordinate system* $x=(\eta_1, \ldots, \eta_n)$ *of* $\mathbb{R}^n$ *valid in* $\mathscr{U}$ *such that*

$$\bar{x}=(0, \ldots, 0)$$

$$\eta_i=\xi_i-\bar{\xi}_1 \quad \text{for} \quad 1\leqslant i\leqslant n-1$$

$$f(\bar{\lambda}, x)=f(\bar{\lambda}, \bar{x})\pm\eta_1^2\pm\cdots\pm\eta_{n-1}^2+\eta_n^3+\sum_{\substack{i\leqslant j\leqslant k\\ j\neq n}} b_{ijk}(\eta)\eta_i\eta_j\eta_k \qquad \blacktriangle$$

Proof. Set

$$(17)\qquad \eta_i=\xi_i-\bar{\xi}_i \qquad \text{for} \qquad i=1, \ldots, n-1$$

$$(18)\qquad \eta_n=a_{nnn}(\xi)^{1/3}\left[(\xi_n-\bar{\xi}_n)+\frac{1}{3}\sum_{i\neq n}\frac{a_{inn}(\xi)}{a_{nnn}(\xi)}(\xi_i-\bar{\xi}_i)\right]$$

The determinant of the $\partial\eta_i/\partial\xi_j$ at $\bar{\xi}$ is $a_{nnn}(\bar{\xi})^{1/3}$, which is nonzero. By the inverse function theorem, the map $\xi \mapsto \eta$ is a local diffeomorphism near $\bar{\xi}$, so that $(\eta_1, \ldots, \eta_n)$ is a local coordinate system near $\bar{x}$.

Replacing the ξ_i by the η_i in lemma 10, we obtain

$$f(\bar{\lambda}, x) = f(\bar{\lambda}, \bar{x}) \pm \eta_1^2 \pm \cdots \pm \eta_{n-1}^2 + R(\eta)$$

$$R(\eta) = \sum_{i \leqslant j \leqslant k} a_{ijk}(\xi)(\xi_i - \bar{\xi}_i)(\xi_j - \bar{\xi}_j)(\xi_k - \bar{\xi}_k)$$

$$(19) \qquad = a_{nnn}(\xi)(\xi_n - \bar{\xi}_n)^3 + \sum_{i \neq n} a_{inn}(\xi)(\xi_n - \bar{\xi}_n)^2(\xi_i - \bar{\xi}_i)$$

$$+ \sum_{i,j \neq n} a_{ijk}(\xi)(\xi_i - \bar{\xi}_i)(\xi_j - \bar{\xi}_j)(\xi_k - \bar{\xi}_k)$$

$$= \eta_n^3 - \sum_{i,j \neq n} c_{ijk}(\xi)(\xi_i - \bar{\xi}_i)(\xi_j - \bar{\xi}_j)(\xi_k - \bar{\xi}_k)$$

$$+ \sum_{i,j \neq n} a_{ijk}(\xi)(\xi_i - \bar{\xi}_i)(\xi_j - \bar{\xi}_j)(\xi_k - \bar{\xi}_k)$$

Here the c_{ijk} are coefficients of monomials with degree zero or one in $(\xi_n - \bar{\xi}_n)$ when expanding η_n^3 in equation (18). Replacing the ξ_i by their values in terms of the η_i, we obtain the desired result. ■

PROPOSITION 12

Assume A. *Let* $(\bar{\lambda}, \bar{x})$ *be a point in* $I(f)$ *where* $\delta(\bar{\lambda}, \bar{x}) = 0$. *Then there exists a neighborhood* $\mathscr{U}$ *of* $\bar{x}$ *and a local (curvilinear) coordinate system* $(\zeta_1, \ldots, \zeta_n)$ *of* $\mathbb{R}^n$, *valid in* $\mathscr{U}$, *such that for all* x *in* $\mathscr{U}$

$$f(\bar{\lambda}, x) = f(\bar{\lambda}, \bar{x}) \pm \zeta_1^2 \pm \cdots \pm \zeta_{n-1}^2 + \zeta_n^3 \qquad ▲$$

Proof. Consider the expression for $f(\bar{\lambda}, x)$ obtained from lemma 11. All the monomials in the remainder have joint degree two or three in the combined variables $(\eta_1, \ldots, \eta_{n-1})$. We shall give a procedure for absorbing them into the squares η_i^2.

Let us, for instance, eliminate all remainder terms containing η_1. We first list them.

$$Q(\eta) = b_{111}(\eta)\eta_1^3 + \sum_{k \neq 1} b_{11k}(\eta)\eta_1^2\eta_k$$

$$+ \sum_{1 < j \leqslant k} b_{1jk}(\eta)\eta_1\eta_j\eta_k$$

We then have

$$\begin{aligned}\pm\eta_1^2+Q(\eta)&=\eta_1^2\left[\pm 1+b_{111}\eta_1+\sum_{k\neq 1} b_{11k}\eta_k\right]+\eta_1 \sum_{1<j\leqslant n} b_{1jk}\eta_j\eta_k\\ &=A(\eta)\eta_1^2+B(\eta)\eta_1\\ &=A(\eta)\left[\eta_1+\frac{B(\eta)}{2A(\eta)}\right]^2-\frac{B(\eta)^2}{4A(\eta)^2}\end{aligned}$$

We define a new set of local coordinates $(\zeta_1, \ldots, \zeta_n)$ by

$$\zeta_1=|A(\eta)|^{1/2}\left(\eta_1+\frac{B(\eta)}{2A(\eta)}\right) \qquad \text{and} \qquad \zeta_i=\eta_i \qquad \text{for} \qquad i\geqslant 2.$$

By the inverse function theorem, this is possible since $A(0)=\pm 1$. We have

$$\pm\eta_1^2+Q(\eta)=\pm\zeta_1^2-\frac{B(\eta)^2}{4A(\eta)^2}$$

Here

$$\frac{B(\eta)^2}{4A(\eta)^2}=\frac{1}{4A(\eta)^2}\left[\sum_{1<j\leqslant n} b_{1jk}(\eta)\eta_j\eta_k\right]^2$$

is a polynomial in $(\eta_2, \ldots, \eta_n)$ with variable coefficients. All its monomials have degree of at least two jointly in $(\eta_2, \ldots, \eta_{n-1})$. Turning to the variables $(\zeta_1, \ldots, \zeta_n)$ preserves these properties, so that we obtain

$$f(\bar{\lambda}, x)=f(\bar{\lambda}, \bar{x})\pm\zeta_1^2\pm \cdots \pm\zeta_{n-1}^2+\zeta_n^3+P(\zeta_1, \zeta_2, \ldots, \zeta_n)$$

The new remainder P is a polynomial in $(\zeta_2, \ldots, \zeta_n)$ with variable coefficients. All its monomials have degree of at least two jointly in $(\zeta_2, \ldots, \zeta_{n-1})$. This enables us to iterate the procedure, thus eliminating the variables $\zeta_2, \ldots, \zeta_n$ from the remainder and reaching the desired result. ■

If $n=1$, we have $f(\bar{\lambda}, x)=f(\bar{\lambda}, \bar{x})+(x-\bar{x})^3$ in appropriate coordinates, so that $\bar{x}$ is an inflexion point. In higher dimensions, the points $\bar{x}$ described in proposition 12 will be called *generalized inflexion points.*

5. FURTHER DEGENERACIES: BIFURCATION

In some situations modeled by the equation $f(\lambda, x)=0$, we may observe several solutions in x for parameter values close to $\bar{\lambda}$: a trivial solution, $x_0(\lambda)\equiv 0$, say, and bifurcating solutions $x_1(\lambda), \ldots, x_n(\lambda)$, which coincide with x_0 when $\lambda=\bar{\lambda}$, so that $x_i(\bar{\lambda})=0$ for all i but move away from it when $\lambda \neq \bar{\lambda}$. We wish to gain some mathematical insight into such situations. We begin by a formal definition. We are given a C^k map $f: \Lambda \times X \to Y$ between Banach spaces, with $k \geqslant 1$, such that

$$\forall \lambda \in \Lambda, \qquad f(\lambda, 0)=0$$

DEFINITION 1

We say that $\bar{\lambda}$ is a bifurcation value of the equation $f(\lambda, x)=0$ if every neighborhood of $(\bar{\lambda}, 0)$ in $\Lambda \times X$ contains nonzero solutions. ▲

In other words, there is a sequence $\lambda_n \to \bar{\lambda}$ in Λ and a sequence $x_n \to 0$ in X such that $f(\lambda_n, x_n)=0$ and $x_n \neq 0$ for all n. Of course, bifurcation values are interesting only if they can be reached through nonbifurcation values of the parameter. For the zero map $f(\lambda, x) \equiv 0$, all values of the parameter are trivially bifurcation values.

A few facts are already clear. First, $f'_x(\bar{\lambda}, 0)$ cannot be invertible. Otherwise, by the implicit function theorem, for λ close to $\bar{\lambda}$ and small x, there would be a unique solution of $f(\lambda, x)=0$, namely, $x=0$. Neither could $\text{Im}[f'_x(\bar{\lambda}, 0)]=\text{Im}[f'(\bar{\lambda}, 0)]$ be the whole space Y. Otherwise, assuming that Ker $f'(\bar{\lambda}, 0)$ has a closed complementing subspace, we would apply proposition 3.2. The equation $f(\lambda, x)=0$ would define in some neighborhood $\mathcal{U} \times \mathcal{V}$ of $(\bar{\lambda}, 0)$ a closed submanifold M containing $\mathcal{U} \times \{0\}$ and some additional points (λ_n, x_n). Thus, $M \neq \mathcal{U} \times \{0\}$, so that $\mathcal{U} \times \{0\}$ has empty interior in M, which means precisely that all $\lambda \in \mathcal{U}$ are bifurcation values. Remembering that $\mathcal{U}$ is a neighborhood of $\bar{\lambda}$, we find ourselves in the situation we have described as uninteresting.

So the only nontrivial possibility left is to look for higher degeneracy of f at $(\bar{\lambda}, 0)$; that is, to make $f'_x(\bar{\lambda}, 0)$ nonsurjective. For the problem to be tractable at all, we make the assumption that $f'_x(\bar{\lambda}, 0)$ is a *Fredholm operator*. This means that

(1) $\text{Ker}\, f'_x(\bar{\lambda}, 0)$ has finite dimension d.

(2) $\text{Im}[f'_x(\bar{\lambda}, 0)]$ is closed and has finite codimension r.

In this case, the equation $f(\lambda, x)=0$ in $\Lambda \times Y$ can be reduced to a finite-dimensional system, namely, r equations in $(d+1)$ unknowns, by the so-called *Lyapounov–Schmidt procedure*. To see this, write $X=X_0 \oplus \text{Ker}\, f'_x(\bar{\lambda}, 0)$ and $Y=\text{Im}[f'_x(\bar{\lambda}, 0)] \oplus Y_0$ and denote by π: $Y \to Y_0$ the projection. The equation $f(\lambda, x)=0$ splits into two parts

$$\pi f(\lambda, x)=0 \qquad \text{in } Y_0 \tag{3}$$

$$(I-\pi) f(\lambda, x)=0 \qquad \text{in } \text{Im}[f'_x(\bar{\lambda}, 0)] \tag{4}$$

The restriction of $f'_x(\bar{\lambda}, 0)$ to X_0 is an isomorphism onto $\text{Im}[f'_x(\bar{x}, 0)]$. But it is also the derivative at 0 of the map $x_0 \to (I-\pi)f(\bar{\lambda}, x_0)$, defined on X_0. Applying the inverse function theorem to the map

$$x_0 \to (I-\pi)f(\lambda, x_1 + x_0)$$

depending continuously on the parameters

$$(\lambda, x_1) \in \Lambda \times \operatorname{Ker} f'_x(\bar{\lambda}, 0)$$

we define on some neighborhood $\mathscr{U}$ of $(\bar{\lambda}, 0)$ a C^1 map $(\lambda, x_1) \to x_0(\lambda, x_1)$ such that

$$(I-\pi)f(\lambda, x_1 + x_0(\lambda, x_1)) = 0 \qquad \forall (\lambda, x_1) \in \mathscr{U} \tag{5}$$

The only condition left to satisfy then is

$$\pi f(\lambda, x_1 + x_0(\lambda, x_1)) = 0 \tag{6}$$

Equation (6) is called the bifurcation equation, and it is a system of r equations in the space $\Lambda \times \operatorname{Ker} f'_x(\bar{\lambda}, 0)$. We have thus reduced the problem to a finite-dimensional one: Find the zeros of a function $g: R^{d+1} \to R^r$ near some degenerate zero

$$\begin{cases} \text{solve } g(\lambda, x) = 0 & \text{near} \quad (\bar{\lambda}, 0) \\ \text{with } g(\bar{\lambda}, 0) = 0 & \text{and} \quad g'(\bar{\lambda}, 0) = 0 \end{cases} \tag{7}$$

This is a formidable task! As a matter of fact, it is one purpose of singularity theory to provide partial answers to this question. We have already met this problem in Section 4 and have investigated the simplest case, namely, a single equation ($r=1$) near a nondegenerate critical point, Det $g''(\bar{\lambda}, 0) \neq 0$, when g is C^p with $p \geq 2$. The Morse lemma tells us that there will be some neighborhood $\mathscr{U}$ of $(\bar{\lambda}, 0)$ in $\Lambda \times \operatorname{Ker} f'_x(\bar{\lambda}, 0)$ such that the set of solutions for $g(\lambda, x) = 0$ in $\mathscr{U}$ is C^{p-2} diffeomorphic to some cone C:

$$\sum_{i=1}^{d+1} \pm x_i^2 = 0 \quad \text{in} \quad R^{d+1}$$

the vertex 0 of this cone being carried to $(\bar{\lambda}, 0)$. Composing with the map $(\lambda, x_1) \to (\lambda, x_1 + x_0(\lambda, x))$ and assuming f itself to be C^p, we see that the set of solutions of $f(\lambda, x) = 0$ in $\Lambda \times X$ near $(\bar{\lambda}, 0)$ will also be C^{p-2} diffeomorphic to this cone C near its vertex. By the way, since the set of solutions contains a piece of $\Lambda \times \{0\}$, the cone C must be real, so that $g''(\lambda, 0)$ cannot be positive or negative definite.

The simplest nontrivial case is already very useful;

THEOREM 2

Take $\Lambda = R$ and let $f: R \times X \to Y$ be a C^p function, $p \geq 2$, with $f(\bar{\lambda}, x) = 0$ for all x. Assume that

$$\operatorname{Ker} f'_x(\bar{\lambda}, 0) \quad \text{is one dimensional} \tag{8}$$

$$Y = \operatorname{Im}[f'_x(\bar{\lambda}, 0)] \oplus f''_{\lambda x}(\bar{\lambda}, 0) \operatorname{Ker} f'_x(\bar{\lambda}, 0) \tag{9}$$

Then $\bar{\lambda}$ is an isolated bifurcation value. Precisely, there exist $\varepsilon > 0$ and maps $x: (-\varepsilon, \varepsilon) \to X$ and $\lambda: (-\varepsilon, \varepsilon) \to \mathbb{R}$ such that

$$\begin{aligned} &\textbf{i.} \quad \lambda(0) = \bar{\lambda}, \qquad x(0) = 0, \\ &\textbf{ii.} \quad x'(0) \in \operatorname{Ker} f'_x(\bar{\lambda}, 0), \qquad x'(0) \neq 0 \\ &\textbf{iii.} \quad f(\lambda(s), x(s)) = 0 \qquad \text{for } s \in (-\varepsilon, \varepsilon) \end{aligned} \tag{10}$$

Moreover, there is some neighborhood $\mathcal{U}$ of $(\bar{\lambda}, 0)$ in $\Lambda \times X$ such that whenever $(\lambda, x) \in \mathcal{U}$ solves $f(\lambda, x) = 0$, either $x = 0$ or $\lambda = \lambda(s)$ and $x = x(s)$ for some s. ▲

Proof. We use the Liapounov–Schmidt procedure with $d = 1 = r$. We claim that the function $g(\lambda, \xi) = \pi f(\lambda, \xi + x_0(\lambda, \xi))$ has a nondegenerate critical point at $(\bar{\lambda}, 0)$. Indeed, computing the second derivatives yields the following 2×2 matrix

$$g''(\bar{\lambda}, 0) = \begin{pmatrix} \pi f''_{\lambda\lambda}(\bar{\lambda}, 0) & \pi f''_{\lambda\xi}(\bar{\lambda}, 0) \\ \pi f''_{\lambda\xi}(\bar{\lambda}, 0) & \pi f''_{\xi\xi}(\bar{\lambda}, 0) \end{pmatrix}$$

The upper diagonal term is zero, since $f(\lambda, 0) = 0$ identically. By assumption, there is some $x_1 \in \operatorname{Ker} f'_x(\bar{\lambda}, 0)$ such that $\pi f''_{\lambda\xi}(\bar{\lambda}, 0) x_1 \neq 0$, so that the nondiagonal terms are nonzero. It follows that the determinant is strictly negative.

By the Morse lemma, there is a C^{p-2} diffeomorphism $\lambda = \bar{\lambda}(s, t)$ and $\xi = \bar{\xi}(s, t)$, defined for $|s| < \varepsilon$ and $|t| < \varepsilon$, such that

$$g(\lambda(s, t), \xi(s, t)) = s^2 - t^2$$

It follows that the set $g = 0$ decomposes into $s = t$ and $s = -t$ in (s, t) coordinates. In (λ, ξ) coordinates, this gives two C^{p-2} curves, $s \mapsto (\bar{\lambda}(s, s), \bar{\xi}(s, s))$ and $s \to (\bar{\lambda}(s, -s), \bar{\xi}(s, -s))$. We know that one of these curves must be the straight line $\xi = 0$. So the other one yields the nontrivial solutions of $g(\lambda, \xi) = 0$ near $(\bar{\lambda}, 0)$. Note that both curves intersect transversally at $(\bar{\lambda}, 0)$.

Taking the image of the nontrivial curve by the map $(\lambda, \xi) \to (\lambda, \xi + x_0(\lambda, \xi))$, we obtain a C^{p-2} curve $s \to (\lambda(s), x(s))$, with $(\lambda(0), x(0)) = (\bar{\lambda}, 0)$ and $x'(0) \neq 0$, which describes all nontrivial solutions of $f(\lambda, x)=0$ in some neighborhood of the origin. Everything is now proved, except the fact that $x'(0)$ belongs to $\operatorname{Ker} f'_x(\bar{\lambda}, 0)$. We simply differentiate the identity $f(\lambda(s), x(s))=0$ at $s=0$

$$0=f'(\lambda, 0)\lambda'(0)+f'_x(\lambda, 0)x'(0)=f'_x(\lambda, 0)x'(0)$$ ■

A few examples might be useful. Set $X=R^2=Y$ and consider the map $y=f(\lambda, x)$ given by

$$y_1=\lambda x_1+x_2^3$$
$$y_2=\lambda x_2-x_1^3$$

We can easily check that $x_2y_1-x_1y_2=x_1^4+x_2^4$ for all λ, so $f(x, \lambda)=0$ has $x=0$ as its only solution. So $\lambda=0$ is not a bifurcation value, and yet $f'_x(0, 0)$ is singular. The point is that it vanishes completely, so that $\operatorname{Im}[f'_x(0, 0)]$ has co-dimension 2, and theorem 1 is not applicable. ■

As another example, set $X=R^n=Y$, and let $g: R^n \to R^n$ be a given C^2 function, with $g(0)=0$. Define $f: R\times R^n \to R^n$ by

$$f(\lambda, x)=g(x)-\lambda x$$

We have $f'_x(\lambda, 0)=g'(0)-\lambda I$, which will degenerate whenever λ is an eigen-value of $g'(0)$. Let $\bar{\lambda}$ be such an eigenvalue: Is it a bifurcation value? Theorem 1 gives us sufficient conditions. Taking into account $f''_{\lambda x}=I$, we see that $f'_x(\lambda, 0)$ should have corank 1 and its kernel should not be contained in its range: $(g'(0)-\bar{\lambda} I)^2\bar{x}=0$ if and only if $(g'(0)-\bar{\lambda} I)\bar{x}=0$. In other words, $\bar{\lambda}$ should be a simple eigenvalue. By theorem 1, any simple eigenvalue of $g'(0)$ is a bifurcation value of $f(\lambda, x)=0$. ■

These bifurcation problems take on a special flavor if we consider the differ-ential equation

(11) $$\frac{dx}{dt}=f(\lambda, x)$$

associated with a C^2 map $f: R\times R^n \to R^n$. If $f(\lambda, 0)=0$ for all λ, the origin is an equilibrium point, that is, the equation has the constant solution $x(t)=0$, all $t\in R$. In a physical situation, such equilibria can be observed only if they are asymptotically stable. This means that for any neighborhood $\mathscr{U}$ of the origin, a smaller neighborhood $\mathscr{V}$ can be found such that for all starting points x_0 in $\mathscr{V}$, the solution of the initial-value problem $dx/dt=f(\lambda, x)$, $x(0)=x_0$, stays in $\mathscr{U}$ for all $t>0$ and converges to zero when $t\to\infty$.

It is known that this will certainly be the case if the eigenvalues of $f'_\lambda(\lambda, x)$ have negative real part. On the other hand, it will certainly *not* be the case if one of the eigenvalues has positive real part. The case when some eigenvalues have zero real parts, all others being negative, requires further analysis.

In the case when $\bar{\lambda}$ is a bifurcation value for $f(\lambda, x)=0$, new equilibrium points appear for values of λ close to $\bar{\lambda}$. It is generally the case that the trivial equilibrium $x=0$ loses its stability property when crossing the bifurcation value: If it was stable for $\lambda<\bar{\lambda}$ it becomes unstable for $\lambda>\bar{\lambda}$; and if it was unstable, it becomes stable. It is our purpose now to explain this law of experience, within the limited framework of theorem 1.

In view of differential equation (11), we shall limit ourselves to the case where $X=R^n=Y$. Say $\bar{\lambda}$ is a bifurcation value for $f(\lambda, x)=0$, so that $f'_x(\bar{\lambda}, 0)$ has zero as an eigenvalue. We wish to know what becomes of this eigenvalue for $\lambda\neq\bar{\lambda}$: Assuming the $(n-1)$ other eigenvalues to have negative real parts it is the sign of this remaining one that will determine the stability of the system.

The first question to ask is whether the eigenvalue zero for $f'_x(\bar{\lambda}, 0)$ can be continued into an eigenvalue $\mu(\lambda)$ for $f'_x(\lambda, 0)$.

LEMMA 3

Let A: $R^n\to R^n$ be a linear operator, having μ_0 as a simple eigenvalue and x_0 as an associated eigenvector. Then there is a neighborhood $\mathscr{U}$ of A in $\mathscr{L}(R^n, R^n)$, and C^∞ maps μ: $\mathscr{U}\to R$ and x: $\mathscr{U}\to R^n$ such that

$$(12)\qquad \begin{cases} \textbf{i.}\quad \mu(A)=\mu_0 \quad \textit{and} \quad \mu(B) \textit{ is a simple eigenvalue of } B\in\mathscr{U} \\ \textbf{ii.}\quad x(A)=x_0 \quad \textit{and} \quad x(B) \textit{ is an eigenvector for } \mu(B) \end{cases}$$

The map μ is unique, and so is x if we add the condition that $x(B)-x_0$ belongs to a fixed linear subspace complementing Ker $(A-\mu_0 I)$ *in R^n.* ▲

Proof. Let $R^n=\text{Ker}\,(A-\mu_0 I)\oplus X_1$. We want to solve the equation

$$(B-\mu I)(x_0+x_1)=0$$

with $\mu\in R$ and $x_1\in X_1$. To do this, we apply the implicit function theorem to the map $(\mu, x_1)\to(B-\mu I)(x_0+x_1)$ depending smoothly on the parameter $B\in\mathscr{L}(R^n, R^n)$. The derivative of this map at $(\mu_0, 0)$ for the value A of the parameter

$$(\mu, x_1)\mapsto -\mu x_0+(A-\mu_0 I)x_1$$

This is a linear map from $R\times X_1$ into R^n. Since μ_0 is a simple eigenvalue, R^n splits as the direct sum of Ker $(A-\mu_0 I)$ and Im $(A-\mu_0 I)$. It follows that the preceding linear map is invertible. By the implicit function theorem, there will be neighborhoods $\mathscr{V}$ of A in $\mathscr{L}(R^n, R^n)$ and $\mathscr{W}$ of $(\mu_0, 0)$ in $R\times X^1$, and a C^∞ mapping (μ, x): $\mathscr{V}\to\mathscr{W}$ such that all solutions in $\mathscr{V}\times\mathscr{W}$ of the equation $(B-\mu I)(x_0+x_1)=0$ can be written as $\mu=\mu(B)$, $x_1=x_1(B)$. Thus $\mu(B)$ is an eigenvalue for B and $x_0+x_1(B)$ an associated eigenvector. Note that $\mu(A)=0$ and $x_1(A)=0$.

Since μ_0 is a simple eigenvalue of A, we have $(A-\mu_0 I)x_1 \neq 0$ for all $x_1 \in X_1$, in particular for $\|x_1\|=1$. By compactness, it follows that $(B-\mu(B)I)x_1 \neq 0$ for all $x_1 \in X_1$ with $\|x_1\|=1$ and for all B in some smaller neighborhood $\mathcal{U}$ of A. It follows that $\mu(B)$ must be a simple eigenvalue of B. ■

In the situation in theorem 1, we know that the set of solutions of $f(\lambda, x)=0$ in some neighborhood $\mathcal{U}$ of $(\bar{\lambda}, 0)$ consists of two curves intersecting non-tangentially at $(\bar{\lambda}, 0)$, namely,

$$(-\varepsilon, \varepsilon) \ni s \mapsto (s+\bar{\lambda}, 0)$$

$$(-\varepsilon, \varepsilon) \ni s \mapsto (\lambda(s), x(s))$$

We know that $f'_x(\bar{\lambda}, 0)=0$. Assume that zero is a simple eigenvalue for $f'_x(\bar{\lambda}, 0)$, and apply the preceding lemma. We first choose $x_0=(dx/ds)(0)$ and take $X_1 \subset R^n$ so that:

$$0 \neq x_0 \in \operatorname{Ker} f'_x(\bar{\lambda}, 0) \qquad \text{and} \qquad R^n = \operatorname{Ker} f'_x(\bar{\lambda}, 0) \oplus X_1$$

We then get C^{p-2} maps (μ^0, x_1^0) and (μ^1, x_1^1) from $(-\varepsilon, \varepsilon)$ into $R \times X_1$ such that

$$\text{(13)} \qquad \begin{aligned} &\textbf{i.} \quad (f'_x(s+\bar{\lambda}, 0)-\mu^0(s)I)(x_0+x_1^0(s))=0 \\ &\textbf{ii.} \quad (f'_x(\lambda(s), x(s))-\mu^1(s)I)(x_0+x_1^1(s))=0 \end{aligned}$$

We have $\mu^0(0)=0=\mu^1(0)$, and we wish to know something about $\mu^0(s)$ and $\mu^1(s)$ for $s \neq 0$, at least their sign. Fortunately, we can do so without actually computing the eigenvalues of $f'_x(\lambda, x)$—an enormous task—because the derivatives $(d\mu^\circ/ds)(0)$ and $(d\mu^1/ds)(0)$ are related to each other and to $d\lambda/ds(0)$.

PROPOSITION 4

Assumptions as in theorem 1. *Assume moreover* $X=R^n$ *and the eigenvalue zero for* $f'_x(\bar{\lambda}, 0)$ *is simple. Then*

$$\text{(14)} \qquad \frac{d\mu^0}{ds}(0) \neq 0 \qquad \textit{and} \qquad \frac{d\mu^1}{ds}(0)+\frac{d\lambda}{ds}(0)\frac{d\mu^0}{ds}(0)=0$$

More generally, there is some $\eta>0$ *such that whenever* $|s|<\eta$, *we have* $(d\lambda/ds)(s)=0$ *if and only if* $\mu^1(s)=0$. *Along any sequence* $s_n \to 0$ *such that* $(d\lambda/ds)(s_n)$ *never vanishes, we have*

$$\text{(15)} \qquad \frac{\mu_1(s_n)}{s_n} \Big/ \frac{d\lambda}{ds}(s_n) \to -\frac{d\mu^0}{ds}(0) \qquad \blacktriangle$$

Proof. Let us first work on the branch of trivial solutions. Differentiating equation (13), **i.** with respect to s, we obtain

$$f''_{x\lambda}(\bar{\lambda}, 0)x_0 - f'_x(\bar{\lambda}, 0)\frac{dx_1^0}{ds}(0) = \frac{d\mu^0}{ds}(0)x_0 \tag{16}$$

Now we cannot have zero on the right-hand side, otherwise $f''_{x\lambda}(\bar{\lambda}, 0)x_0$ would fall within the range of $f'_x(\bar{\lambda}, 0)$ in R^n, in contradiction to the assumptions of theorem 1. The first relation is proved.

Consider the branch of nontrivial solutions. We start by differentiating the identity $f(\lambda(s), x(s)) \equiv 0$ and obtain:

$$f'_\lambda(\lambda(s), x(s))\frac{d\lambda}{ds}(s) + f'_x(\lambda(s), x(s))\frac{dx}{ds}(s) \equiv 0 \tag{17}$$

For $s=0$, both terms vanish, as we have seen in the proof of theorem 1. This induces us to look at the higher order terms. Differentiating the preceding identity at $s=0$, and remembering that $(dx/ds)(0) = x_0$, we have

$$2f''_{\lambda x}(\bar{\lambda}, 0)\left(\frac{d\lambda}{ds}(0), x_0\right) + f''_{x^2}(\bar{\lambda}, 0)(x_0, x_0) + f'_x(\bar{\lambda}, 0)\frac{d^2x}{ds^2}(0) = 0 \tag{18}$$

We have another identity to differentiate, namely (13, **ii**), which gives

$$f''_{\lambda x}(\bar{\lambda}, 0)\left(\frac{d\lambda}{ds}(0), x_0\right) + f''_{x^2}(\bar{\lambda}, 0)(x_0, x_0) - \frac{d\mu^1}{ds}(0)x_0 + f'_x(\bar{\lambda}, 0)\frac{dx_1^1}{ds}(0) = 0 \tag{19}$$

Substracting the second equation from the first yields

$$f''_{\lambda x}(\bar{\lambda}, 0)\left(\frac{d\lambda}{ds}(0), x_0\right) + \frac{d\mu^1}{ds}(0)x_0 + f'_x(\bar{\lambda}, 0)\left(\frac{d^2x}{ds^2}(0) - \frac{dx_1^1}{ds}(0)\right) = 0 \tag{20}$$

We now recall equation (16). Multiplying it by $(d\lambda/ds)(0)$, and substracting from the preceding one to eliminate $f''_{\lambda x}(\bar{\lambda}, 0)$, we have

$$f'_x(\bar{\lambda}, 0)\left(\frac{d\lambda}{ds}(0)\frac{dx_1^0}{ds}(0) + \frac{d^2x}{ds^2}(0) - \frac{dx_1^1}{ds}(0)\right) + \left(\frac{d\mu^1}{ds}(0) + \frac{d\lambda}{ds}(0)\frac{d\mu^0}{ds}(0)\right)x_0 = 0 \tag{21}$$

Now, since zero is a simple eigenvalue of $f'_x(\bar{\lambda}, 0)$ and the vector x_0 belongs to its kernel, it cannot belong to its range. It follows that each of the two terms in the foregoing sum must be zero; hence, the second equation.

To treat the case when $(d\mu^1/ds)(0)=0$, we begin by writing for $|s|$ small enough

$$x(s)=sx_0+y_1(s), \quad \text{with} \quad y_1(s)\in X_1 \quad \text{and} \quad \frac{dy_1}{ds}(0)=0 \tag{22}$$

Indeed, we already know that $(dx/ds)(0)=x_0$, which means that we can write

$$x(s)=\phi(s)x_0+y_1(s), \quad \text{with} \quad \frac{d\phi}{ds}(0)=1 \quad \text{and} \quad y_1(s)\in X_1$$

Further identifications give $\phi(0)=0$ and $y_1(0)=0=(dy_1/ds)(0)$. Setting $t=\phi(s)$, and applying the inverse function theorem to $\phi: R\to R$, the preceding identity becomes

$$x(t)=tx_0+y_1\circ\phi^{-1}(t) \quad \text{for} \quad |t| \text{ small enough}$$

Renaming the independent variable, we obtain equation (22).

We start from the identity

$$f'_\lambda(\lambda(s), x(s))\frac{d\lambda}{ds}(s)+f'_x(\lambda(s), x(s))\left(\frac{dy_1}{ds}(s)-x_1^1(s)\right)+\mu^1(s)(x_0+x_1^1(s))=0 \tag{23}$$

We write the Taylor expansions in s

$$\begin{cases} \text{i.} \quad f'_\lambda(\lambda(s), x(s))=sf''_{\lambda x}(\bar\lambda, 0)x_0+s^2r_1(s) \\ \text{ii.} \quad f'_x(\lambda(s), x(s))=f'_x(\bar\lambda, 0)+sr_2(s) \end{cases} \tag{24}$$

Here, r_1 and r_2 are continuous functions of s. Substituting into the preceding identity, we have

$$\left\{\begin{aligned} & s\frac{d\lambda}{ds}(s)\, f''_{\lambda x}(\bar\lambda, 0)x_0+\mu^1(s)x_0+s^2\frac{d\lambda}{ds}(s)r_1(s)x_0+\mu^1(s)x_1^1(s) \\ & +sr_2(s)\left(\frac{dy_1}{ds}(s)-x_1^1(s)\right)=-f'_x(\bar\lambda, 0)\left(\frac{dy_1}{ds}(s)-x_1^1(s)\right) \end{aligned}\right. \tag{25}$$

Now, $(dy_1/ds)(s)-x_1^1(s)$ belongs to X_1 for small $|s|$. Since X_1 complements $\operatorname{Ker} f'_x(\bar\lambda, 0)$ in R^n, the restriction to X_1 of the operator $f'_x(\bar\lambda, 0)$ has a left-inverse. It follows that there are constants k_1 and k_2 such that for $|s|$ small enough

$$\begin{aligned} \left\|\frac{dy_1}{ds}(s)-x_1^1(s)\right\| &\leqslant k_1\left(|\mu_1(s)|+\left|s\frac{d\lambda}{ds}(s)\right|\right) \\ &+sk_2\left(|\mu^1(s)|+\left|s\frac{d\lambda}{ds}(s)\right|+\left\|\frac{dy_1}{ds}(s)-x_1^1(s)\right\|\right) \end{aligned}$$

We subsume everything under a single constant k_3

$$\left\|\frac{dy_1}{ds}(s)-x_1^1(s)\right\|\leqslant k_3\left(|\mu^1(s)|+\left|s\frac{d\lambda}{ds}(s)\right|\right) \tag{26}$$

We now recall, as in the preceding case, that

$$f''_{\lambda x}(\bar{\lambda},0)x_0=f'_x(\bar{\lambda},0)\frac{dx_1^0}{ds}(0)+\frac{d\mu^0}{ds}(0)x_0$$

Since zero is a simple eigenvalue of $f'_x(\bar{\lambda},0)$, its kernel and its range are complementary subspaces of R^n, so that we may take $X_1=\operatorname{Im} f'_x(\bar{\lambda},0)$. With this choice of X_1, relationship (25) projects onto $\operatorname{Ker} f'_x(\bar{\lambda},0)$ as

$$s\frac{d\lambda}{ds}(s)\frac{d\mu^0}{ds}(0)+\mu^1(s)=s\left[s\frac{d\lambda}{ds}(s)r_1(s)x_0+r_2(s)\left(\frac{dy_1}{ds}(s)-x_1^1(s)\right)\right]_0$$

On the right, we have written the first component of the bracketed term, once R^n has been split as $\operatorname{Ker} f'_x(\bar{\lambda},0)+X_1$ and x_0 has been chosen as a base for the one-dimensional subspace $\operatorname{Ker} f'_x(\bar{\lambda},0)$. Using estimate (26), this yields

$$\left|s\frac{d\lambda}{ds}(s)\frac{d\mu^0}{ds}(0)+\mu^1(s)\right|$$
$$\leqslant|s|\left\{\left|s\frac{d\lambda}{ds}(s)\right|\|r_1(s)x_0\|+k_3\|r_2(s)\|\left(|\mu_1(s)|+\left|s\frac{d\lambda}{ds}(s)\right|\right)\right\}$$

Since r_1 and r_2 are continuous at the origin, we can find some constant k_4 such that for $|s|$ small enough

$$\left|s\frac{d\lambda}{ds}(s)\frac{d\mu^0}{ds}(0)+\mu^1(s)\right|\leqslant k_4|s|\left(\left|s\frac{d\lambda}{ds}(s)\right|+|\mu_1(s)|\right) \tag{27}$$

Everything follows from this inequality and the fact that $(d\mu^0/ds)(0)\neq 0$. If $\mu^1(\bar{s})=0$ and $(d\lambda/ds)(\bar{s})\neq 0$, we have

$$\frac{d\mu^0}{ds}(0)\leqslant k_4|\bar{s}|$$

and hence

$$|\bar{s}|\geqslant k_4^{-1}\frac{d\mu^0}{ds}(0)$$

Similarly, we see that whenever $|s|\leqslant k_4^{-1}$, we cannot have $(d\lambda/ds)(s)=0$ and $\mu^1(s)\neq 0$.

Finally, if there were some sequence $s_n \to 0$ such that $s_n(d\lambda/ds)(s_n)(d\mu^0/ds)(0)$ and $\mu^1(s_n)$ have the same sign, we would have

$$\left| s_n \frac{d\lambda}{ds}(s_n) \frac{d\mu^0}{ds}(0) + \mu^1(s_n) \right| = \left| s_n \frac{d\lambda}{ds}(s_n) \right| \left| \frac{d\mu^0}{ds}(0) \right| + |\mu^1(s_n)|$$

$$\geqslant \operatorname{Inf}\left\{1, \left| \frac{d\mu^0}{d\lambda}(\bar{\lambda}) \right| \right\} \left(\left| s_n \frac{d\lambda}{ds}(s_n) \right| + |\mu^1(s_n)| \right)$$

in clear contradiction to inequality (27). This proves that $s_n(d\lambda/ds)(s_n)(d\mu^0/d\lambda)(\bar{\lambda})$ and $\mu^1(s_n)$ have opposite signs, and it is easily seen that their quotient must converge to -1. ■

We now look at the practical implications of proposition 3 for equation (11). The first conclusion, $(d\mu^0/ds)(0) \neq 0$, together with $\mu^0(0)=0$, tells us that the trivial equilibrium $x=0$ cannot remain stable (or remain unstable) when λ crosses its bifurcation value $\bar{\lambda}$ (remember $\lambda = s + \bar{\lambda}$). Assume, for instance, it was asymptotically stable for $\lambda < \bar{\lambda}$. Then $\mu^0(\lambda)$ must have been negative for $\lambda < \bar{\lambda}$ and becomes positive for $\lambda > \bar{\lambda}$, so that $x=0$ becomes an unstable equilibrium.

Let us turn our attention to the branch of nontrivial equilibria $s \to (\lambda(s), x(s))$. If $(d\lambda/ds)(0) > 0$, relationship (14) shows that $(d\mu^1/ds)(0)$ and $(d\mu^0/ds)(0)$ have opposite signs. Recalling that μ^1 and μ^0 coalesce at zero for $\lambda = \bar{\lambda}$, we see that stability is transferred from the trivial to the nontrivial equilibrium when λ crosses its bifurcation value. If, for instance, $x=0$ was asymptotically stable when $\lambda < \bar{\lambda}$, then $x(\lambda)$ was not unstable but will become so for $\lambda > \bar{\lambda}$, whereas $x=0$ becomes unstable. The same conclusions hold if $(d\lambda/ds)(0) < 0$, as the following Figure shows.

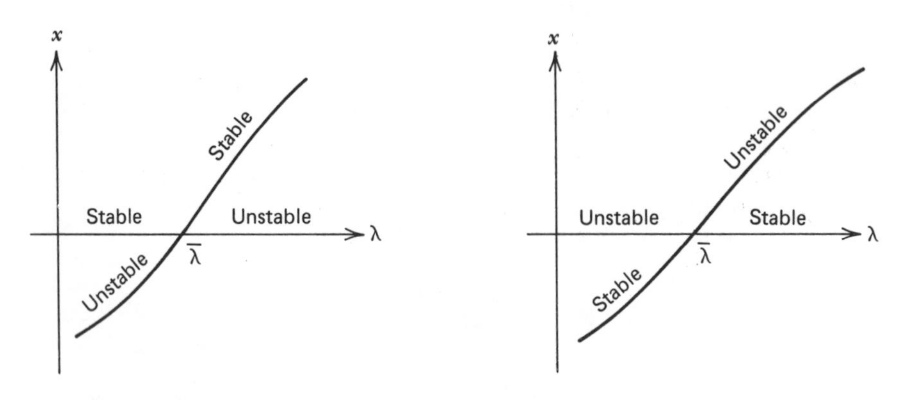

The situation is somewhat more complicated when $(d\lambda/ds)(0)=0$. We then appeal to the more refined equation (15), which implies that $\mu_1(s)$ and $\lambda(s)$ are going to have opposite signs near $s=0$, provided $(d\mu^0/d\lambda)(\bar{\lambda}) > 0$ (they will have the same sign if $(d\mu^0/d\lambda)(\bar{\lambda}) < 0$). The outcome is that whenever there are other equilibria for a certain value of λ beside $x=0$, then all nontrivial equilibria are unstable if $x=0$ is stable, and they are all stable if $x=0$ is unstable. The following figure shows various possibilities.

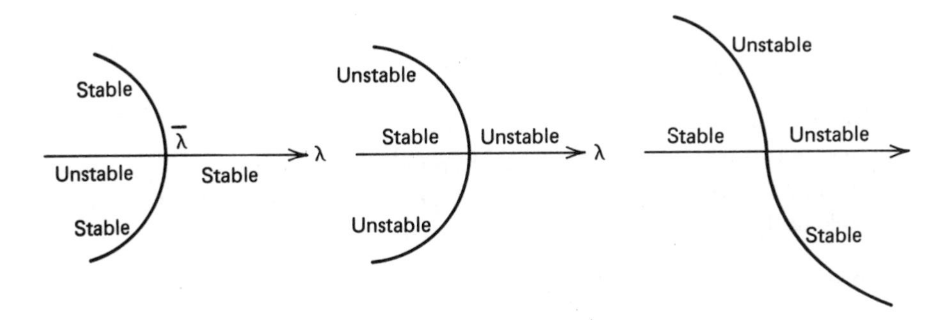

At this point, the reader might think we have just about exhausted stability analysis for the differential equation

$$\frac{dy}{dt}=f(\lambda, y) \in R^n \tag{28}$$

near the trivial equilibrium $x=0$. This is not so. What we have done is to investigate (partly) the case when an eigenvalue of $f'_x(\lambda, 0)$ vanishes. We have proved that in general, that is, except for further degeneracies, when this happens, the trivial equilibrium becomes unstable, and there is a bifurcating branch of nontrivial equilibria that stability is transferred to.

But there is another way for the trivial equilibrium to become unstable: This can happen whenever the real part of an eigenvalue vanishes (and not the eigenvalue itself). We should, therefore, investigate values $\bar{\lambda}$ such that $f'_x(\bar{\lambda}, 0)$ has two purely imaginary eigenvalues. This will not prevent $f'_x(\bar{\lambda}, 0)$ from being invertible, so that the implicit function theorem will apply and no new equilibrium will appear to relieve the trivial one, $x=0$, once it has been destabilized. Thus, any physical system represented by equation (11), which is at rest at $x=0$ for $\lambda<\bar{\lambda}$, cannot stay at rest (even at some other equilibrium) when λ crosses the value $\bar{\lambda}$. If we expect its behavior to be continuous across $\bar{\lambda}$, we must find a stable solution branching off continuously from the equilibrium $x=0$ when $\lambda=\bar{\lambda}$.

The answer is simple and due to Hopf: periodic motions. A function x: $\mathbb{R}\to\mathbb{R}^n$ is periodic if there is some number $T>0$ such that $x(t+T)=x(t)$ for all t; it can then be considered a function on R/TZ. The number $T>0$ is the period.

Take a function x: $R/Z\to R^n$; the function $y(t)=x(t/T)$ then is T periodic, and y is a solution of equation (11) if and only if x satisfies the equation

$$\frac{dx}{dt}=Tf(\lambda, x)$$

Set $X=C^1(R/Z;\ R^n)$ and $Y=C^0(R/Z;\ R^n)$, both Banach spaces, and $\Lambda=R^2$. Define a map $\phi\colon \Lambda\times X\to Y$ by

$$\phi(T,\lambda;x)=\frac{dx}{dt}-Tf(\lambda,x)$$

Equation (11) for $y(t)=x(t/T)$ becomes $\phi(T,\lambda;x)=0$ for x, with T and λ as parameters. Assume the two imaginary eigenvalues $i\omega$ and $-i\omega$ of $f'_x(\bar\lambda,0)$ are simple, and let $\mu(\lambda)$ and $\bar\mu(\lambda)$ be the corresponding (by lemma 2) complex eigenvalues of $f'_x(\lambda,0)$, with $\mu(\bar\lambda)=i\omega$. The Hopf bifurcation theorem states that if suitable nondegeneracy conditions are met, namely,

for no integer $k\neq\pm1$ is $ki\omega$ an eigenvalue of $f'_x(\bar\lambda,0)$

$$\frac{d\mu}{d\lambda}(\bar\lambda)\neq0$$

then $(2\pi\omega^{-1},\bar\lambda)$ is an isolated bifurcation value for ϕ. In other words, bifurcation from the trivial equilibria does not occur within the space of equilibria (i.e., constant solutions), but within the larger space of periodic solutions. We refer to Crandall and Rabinowitz for the proof and detailed analysis, contenting ourselves with depicting the two possible evolutions of a stable trivial equilibrium for $n=2$.

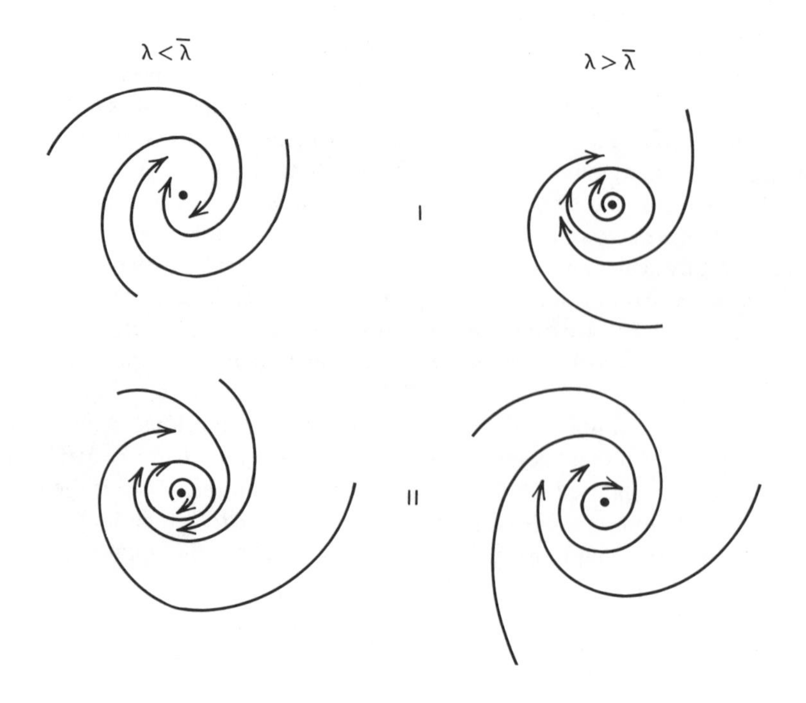

6. TRANSVERSALITY THEORY

The preceding sections have studied the set of solutions of $f(x)=0$, or $f(\lambda, x)=0$, under various sets of assumptions on the derivatives. Even though we have limited ourselves to the very simplest cases, the variety of situations is already considerable. The complexity increases very quickly if we consider higher degeneracy in the derivatives, and we soon reach a stage where classification is impossible.

So the question arises: Which singularities is it really necessary to study? Let us forget about pathological situations and concern ourselves with singularities that are most likely to occur in the class of problems we are investigating.

We owe to René Thom a precise formulation of this idea as well as the mathematical tools to implement it in practical situations. The whole subject is referred to as *transversality theory* and has many ramifications besides the one we are giving.

DEFINITION 1

Let X be a complete metric space and $P(x)$ a statement about points x in X. We say that $P(x)$ is a generic *property if the set of points where it holds true contains a dense G_δ subset of X.* ▲

Recall that a G_δ subset is defined as the intersection of a countable family of open subsets. In other words, $P(x)$ is generic if

$$\{x \in X \mid P(x) \text{ is true}\} = \bigcap_{n=1}^{\infty} \Omega_n$$

with each Ω_n an open and dense subset of X.

Let $P_n(x), n \in \mathbb{N}$, be a countable family of generic properties. Then the property

$$\bigwedge_{n=1}^{\infty} P_n(x) = P_1(x) \quad \text{and} \quad P_2(x) \quad \text{and} \quad \cdots \quad \text{and} \quad P_n(x) \quad \text{and} \quad \cdots$$

is generic also.

This follows immediately from Baire's theorem and would fail miserably if we had defined a generic property to hold on a dense subset of X (the intersection of two dense subsets can be empty).

In an earlier section, we defined transversality of a map $f: \mathbb{R}^n \to \mathbb{R}^n$ to a submanifold $M \subset \mathbb{R}^m$. Let us recall definition 3.4, while extending it to functions defined on Banach manifolds.

DEFINITION 2

Let N be a C^r submanifold of some Banach space and $f: N \to \mathbb{R}^m$ a C^r map. Let

$\bar{x}$ be a point in N. We say that f is transversal to a submanifold M of $\mathbb{R}^m$ at $\bar{x}$ if

$$\text{(1)}\qquad \begin{cases} \text{either } f(\bar{x}) \text{ does not belong to } M \\ \text{or } f(\bar{x}) \text{ belongs to } M \text{ and } \mathbb{R}^m = T_x f\, T_x N + T_{f(x)}M \end{cases}$$ ▲

We now turn our attention to families of smooth maps depending on a parameter u, which belongs to a separable Banach space V. To be precise, let Ω be an open subset of $\mathbb{R}^n$ and

$$f: V \times \Omega \to \mathbb{R}^p$$

be a C^r map, $r \geqslant 1$. Let M be a C^∞ submanifold of $\mathbb{R}^p$ with codimension q. We look for the following property of the parameter value u

$$P(u) = \{f(u, \cdot) \text{ is transversal to } M\}$$

We state Thom's famous transversality theorem, whose proof is deferred to Section 7.

THEOREM 3

Assume $r \geqslant \max(1, n-q+1)$ and $f: V \times \Omega \to \mathbb{R}^p$ is transversal to M. Then $P(u)$ is a generic property on V. ▲

The transversality of $f: V \times \Omega \to \mathbb{R}^p$ to M means that whenever $f(\bar{u}, \bar{x}) = \bar{y} \in M$, any vector $y \in \mathbb{R}^p$ can be written (maybe in several ways) as

$$\text{(2)}\qquad y = f'_u(\bar{u}, \bar{x})u + f'_x(\bar{u}, \bar{x})x + z, \quad \text{with } z \in T_y M$$

In contrast, property $P(\bar{u})$ means that whenever $f(\bar{u}, \bar{x}) = \bar{y} \in M$, any vector $y \in \mathbb{R}^p$ can be expressed as

$$y = f'_x(\bar{u}, \bar{x})x + z, \quad \text{with } z \in T_y M$$

without contribution from variations in the parameter.

The fact that V may be chosen infinite dimensional allows us to pick the function itself as parameter, which can be done as follows. Define $C^r_\infty(\Omega; \mathbb{R}^p)$ to be the space of all C^r functions $f: \Omega \to \mathbb{R}^p$ such that f and all its partial derivatives up to order r are uniformly bounded over Ω. It is a Banach space for the natural norm. Set $V = C^r_\infty(\Omega; \mathbb{R}^p)$ and define a map $\Phi: V \times \mathbb{R}^n \to \mathbb{R}^p$ (called the evaluation map) by

$$\Phi(f, x) = f(x)$$

LEMMA 4

The evaluation map is C^r, and $\Phi'_f(\bar{f}, \bar{x})$ is surjective for all $(\bar{f}, \bar{x})$. ▲

Proof. Note that Φ is continuous and linear with respect to f, and C^r with respect to x. The fact that it is C^r in (f, x) follows immediately. We have, by linearity, $\Phi'_f(\bar{f}, \bar{x})f = f(\bar{x})$, and the map $f \to f(\bar{x})$ from C^r_∞ to $\mathbb{R}^p$ is obviously onto. ■

So Φ is certainly going to be transversal to any submanifold M of $\mathbb{R}^p$. As an easy application, we give corollary 5.

COROLLARY 5

Let M be a C^∞ submanifold of $\mathbb{R}^p$ with codimension q. Then the property

$$P(f) = \{f: \Omega \to \mathbb{R}^p \text{ is transversal to } M\} \tag{3}$$

is generic in $C^r_\infty(\Omega; \mathbb{R}^p)$, for all $r \geqslant \max(1, n+1-q)$. ▲

Just apply Thom's theorem to $\Phi: V \times \Omega \to \mathbb{R}^p$. We might want to have genericity in $C^r(\Omega; \mathbb{R}^p)$ instead of $C^r_\infty(\Omega; \mathbb{R}^p)$; in that case, use lemma 6.

LEMMA 6

Let Ω be an open subset of $\mathbb{R}^n$ and Ω_k, $k \in \mathbb{N}$, a sequence of open subsets such that

$$\begin{cases} \text{i.} \quad \Omega = \bigcup_{k=1}^{\infty} \bar{\Omega}_k \\ \text{ii.} \quad \forall k, \quad \Omega_k \subset \bar{\Omega}_k \subset \Omega_{k+1} \end{cases} \tag{4}$$

Assume a property $P(f)$ is generic in $C^r_\infty(\Omega_k; \mathbb{R}^p)$ for every k. Then it is generic in $C^r(\Omega; \mathbb{R}^p)$. ▲

Proof. Note that $f|_{\Omega_k}$ belongs to $C^r_\infty(\Omega_k; \mathbb{R}^p)$ whenever f belongs to $C^r(\Omega; \mathbb{R}^p)$. More generally, if f belongs to $C^r_\infty(\Omega_j; \mathbb{R}^p)$, then $f|_{\Omega_k}$ belongs to $C^r_\infty(\Omega_k; \mathbb{R}^p)$ for all $k \leqslant j$. Call ϕ_k the map $f \to f|_{\Omega_k}$ from $C^r(\Omega; \mathbb{R}^p)$ to $C^r_\infty(\Omega_k; \mathbb{R}^p)$, and ϕ_{jk} the map $f|_{\Omega_j} \to f|_{\Omega_k}$ from $C^r_\infty(\Omega_j; \mathbb{R}^p)$ to $C^r_\infty(\Omega_k; \mathbb{R}^p)$. For $k \leqslant j$, the diagram

$$\begin{array}{ccc} C^r_\infty(\Omega_j; \mathbb{R}^p) & \xrightarrow{\phi_{jk}} & C^r_\infty(\Omega_k; \mathbb{R}^p) \\ & \nwarrow_{\phi_j} \quad {}^{\phi_k}\nearrow & \\ & C^r(\Omega; \mathbb{R}^p) & \end{array}$$

is commutative

$$\phi_k = \phi_{jk} \circ \phi_j$$

The topology of $C^r(\Omega; \mathbb{R}^p)$ is defined to be the weakest topology that makes all maps ϕ_k continuous. It does not make $C^r(\Omega; \mathbb{R}^p)$ a Banach space, although it does make it a complete metric space, so that Baire's theorem holds.

Now let G_k be the subset of $C^r_\infty(\Omega_k; \mathbb{R}^p)$ where property $P(f)$ holds. By assumption, we have

$$G_k \supset \bigcap_{n=1}^{\infty} U_{k,n}$$

where each $U_{k,n}$ is an open dense subset of $C^r_\infty(\Omega_k; \mathbb{R}^p)$.

Each $\phi_k^{-1}(U_{k,p})$ then is an open subset of $C^r(\Omega; \mathbb{R}^p)$, and

$$G = \bigcap_{k=1}^{\infty} \bigcap_{n=1}^{\infty} \phi_k^{-1}(U_{k,n})$$

is a countable intersection of open subsets of $C^r(\Omega; \mathbb{R}^p)$; that is, a G_δ subset.

We claim that each $\phi_k^{-1}(U_{k,n})$ is dense in $C^r(\Omega; \mathbb{R}^p)$. To see this, pick any open subset U of $C^r(\Omega; \mathbb{R}^p)$. By definition, there is some i and some open subset U_i of $C^r_\infty(\Omega_i; \mathbb{R}^p)$ such that $U = \phi_i^{-1}(U_i)$. If $i = k$,

$$U_i \cap U_{k,n} \neq \varnothing$$

since $U_{k,n}$ is dense in $C^r_\infty(\Omega_i; \mathbb{R}^p)$, and so U meets $\phi_k^{-1}(U_{k,n})$. If $i < k$, we write $U = \phi_k^{-1}(\phi_{ki}^{-1}(U_i))$, and $\phi_{ki}^{-1}(U_i)$, being an open subset of $C^r_\infty(\Omega_k; \mathbb{R}^p)$, has to meet $U_{k,n}$, so that again

$$U \cap \phi_k^{-1}(U_{kn}) \neq \varnothing$$

If $i > k$, pick some $f \in U_i$ and some $\varepsilon > 0$ so small that $\|f - g\| \leqslant \varepsilon$ in $C^r_\infty(\Omega_i; \mathbb{R}^p)$ will imply that $g \in U_i$. Now consider $f|_{\Omega_k} = \phi_{ik}(f)$. Since $U_{k,n}$ is dense in $C^r_\infty(\Omega_k; \mathbb{R}^p)$, there must be some $\tilde{g} \in U_{k,n}$ such that $\|\phi_{ik}(f) - \tilde{g}\| \leqslant \varepsilon/2$ in $C^r_\infty(\Omega_k; \mathbb{R}^p)$. We now extend $\tilde{g}$ to a map $g \in C^r_\infty(\Omega_i; \mathbb{R}^p)$ in such a way that $\|f - g\| \leqslant \varepsilon$.

This will always be possible provided the open sets Ω_k are sufficiently regular —if, for instance, they are all finite unions of balls, which can always be managed. We then have $g \in U_i$ and $\phi_{ik}(g) = \tilde{g} \in U_{k,n}$, so that $U_i \cap \phi_{ik}^{-1}(U_{k,n}) \neq \varnothing$, and so

$$\phi_i^{-1}(U_i) \cap \phi_i^{-1}(\phi_{ik}^{-1}(U_{k,n})) = U \cap \phi_k^{-1}(U_{k,n}) \neq \varnothing$$

So each $\phi_k^{-1}(U_{k,n})$ is open and dense, and by Baire's theorem, G is a dense G_δ subset. ■

Corollary 5 then leads to Corollary 7.

COROLLARY 7

Let M be a C^∞ submanifold of $\mathbb{R}^p$ with codimension q. Then the property

(5) $$P(f)=\{f:\Omega\to\mathbb{R}^p \text{ is transversal to } M\}$$

is generic in $C^r(\Omega;\mathbb{R}^p)$ *for all* $r\geqslant\max(1, n-q+1)$. ▲

It follows, for instance, that if any particular map we are dealing with is not transversal to M, it can be made so by an arbitrarily small C^r perturbation.

For further applications, we use refinements of the evaluation map Φ of lemma 4. The results in propositions 8 and 9 are typical.

PROPOSITION 8

The property

(6) $$P(f)=\{f:\Omega\to\mathbb{R} \text{ is a Morse function}\}$$

is generic in $C^r(\Omega;\mathbb{R})$ *for all* $r\geqslant 2$. ▲

Proof. By lemma 6, it will be enough to prove that $P(f)$ is generic in $C^r_\infty(\Omega;\mathbb{R})$.

Set $V=C^r_\infty(\Omega;\mathbb{R})$ and consider the map $\Psi: V\times\Omega\to\mathbb{R}^n$ defined by

(7) $$\Psi(f,x)=\left(\frac{\partial f}{\partial x_1}(x),\ldots,\frac{\partial f}{\partial x_n}(x)\right)$$

This map is linear in f, and C^{r-1} in x, so that it is jointly C^{r-1}. We have

$$\Psi'_f(\bar{f},\bar{x})f=f'(\bar{x})\in\mathbb{R}^n$$

$$\Psi'_x(\bar{f},\bar{x})=f''(\bar{x})\in\mathscr{L}(\mathbb{R}^n,\mathbb{R}^n).$$

This shows that $\Psi'_f(f,x)\colon V\to\mathbb{R}^n$ is surjective, so Ψ will be transversal to any submanifold M of $\mathbb{R}^n$.

Take $M=\{0\}$, a submanifold of dimension zero. Apply Thom's theorem to Ψ and M. It follows that the property

(8) $$P'(f)=\{\psi(f,\cdot) \text{ is transversal to } M\}$$

is generic in $C^r_\infty(\Omega;\mathbb{R})$. This means that whenever $\Psi(\bar{f},\bar{x})\in M$, the tangent map Ψ'_x is surjective. In other words, whenever $f'(\bar{x})=0$, the matrix $f''(\bar{x})$ is non-degenerate. Therefore, properties $P(f)$ and $P'(f)$ are the same, and the result is proved. ■

Very often, we can guess the correct result by simple counting arguments. For instance, if f is not a Morse function, the system $f'(x)=0$ and $\operatorname{Det} f''(x)=0$ has a solution. But this is a system of $(n+1)$ equations for n unknowns $(x_1,\ldots,x_n)$

and so should have no solution in general. Such heuristics, although incorrect, can give us foresight.

For instance, let us ask whether we can arrange for all critical values to be distinct. The equality of two critical values, corresponding to critical points x_1 and x_2, requires that the system

$$f(x_1)=f(x_2)$$
$$\frac{\partial f}{\partial x_i}(x_1)=0 \qquad 1\leqslant i\leqslant n$$
$$\frac{\partial f}{\partial x_i}(x_2)=0 \qquad 1\leqslant i\leqslant n$$

has a solution. We see that there are $(2n+1)$ equations for $2n$ unknowns, so it would seem to be an exceptional situation. Proposition 9 confirms this observation.

PROPOSITION 9

Let us say that a function $f\colon \Omega\to\mathbb{R}$ is excellent if it is a Morse function on Ω and all its critical values are distinct. The property

$$P(f)=\{f \text{ is excellent}\} \tag{9}$$

is generic in $C^r(\Omega;\mathbb{R})$, all $r\geqslant 2$. ▲

Proof. We shall show that the property

$$Q(f)=\{\text{all critical values of } f \text{ are distinct}\} \tag{10}$$

is generic in $C^r(\Omega;\mathbb{R})$. By proposition 8, the property of being a Morse function is also generic, and by proposition 2, the conjunction of both, namely, $P(f)$, will be generic.

By Lemma 6, it will be enough to prove that $Q(f)$ is generic in $C^r_\infty(\Omega;\mathbb{R})=V$. Introduce the space

$$\tilde{\Omega}=\{(x_1, x_2)\in\Omega\times\Omega \mid x_1\neq x_2\} \tag{11}$$

It is an open subset of $\mathbb{R}^{2n}$. Now define a map $\Psi\colon V\times\tilde{\Omega}\to\mathbb{R}^2\times\mathbb{R}^{2n}$

$$\Psi(f, x_1, x_2)=(f(x_1), f(x_2), f'(x_1), f'(x_2)) \tag{12}$$

It is linear in f, and C^r in (x_1, x_2), so that it is globally C^r, and

$$\Psi'_f(\bar{f}, \bar{x}_1, \bar{x}_2)f=(f(\bar{x}_1), f(\bar{x}_2), f'(\bar{x}_1), f'(\bar{x}_2))$$

So Ψ'_f is surjective, and Ψ will be transversal to any submanifold M of $\mathbb{R}^2 \times \mathbb{R}^{2n}$. Let us choose

$$M = \{(a, a, 0, 0) | a \in \mathbb{R}\}$$

By Thom's theorem, the property

$$Q'(f) = \{\Psi(f, \cdot, \cdot) \text{ is transversal to } M\}$$

is generic.

But $\Psi(f, \cdot, \cdot)$ is a map from $\tilde{\Omega}$, an open subset of $\mathbb{R}^{2n}$, into $\mathbb{R}^{2(n+1)}$. Saying that this map is transversal to M means that whenever $\Psi(\bar{f}, \bar{x}_1, \bar{x}_2) \in M$, any vector in $\mathbb{R}^{2(n+1)}$ can be expressed as the sum of a vector in $(\Psi'_{x_1}, \Psi'_{x_2})(\mathbb{R}^n \times \mathbb{R}^n)$ and a vector in M. But M is one dimensional and $(\Psi'_{x_1}, \Psi'_{x_2})(\mathbb{R}^n \times \mathbb{R}^n)$ is at most $2n$-dimensional, so they cannot make up $\mathbb{R}^{2n+2}$. So $\Psi(f, x_1, x_2)$ cannot belong to M, which means precisely that properties $Q'(f)$ and $Q(f)$ coincide. ■

We now investigate one-parameter families of smooth functions $f(\lambda, x)$ with $\lambda \in \mathbb{R}$ and $x \in \mathbb{R}^n$, as in Section 4. Because of the supplementary variable λ, such families may contain non-Morse functions in a stable way; that is, all neighboring families will contain non-Morse functions, possibly for different values of the parameter λ. What we shall show is that the one-parameter families we investigated in Section 4 were typical; that is, the assumption we made is, in fact, a generic property.

PROPOSITION 10

Recall assumption A *of Section* 4: *There is no point* $(\lambda, x) \in \mathbb{R} \times \Omega$ *such that the* $\partial f/\partial x_i(\lambda, x)$, $1 \leq i \leq n$, $\delta(\lambda, x)$ *and* $\Delta(\lambda, x)$ *vanish simultaneously. The property*

$$P(f) = \{f \text{ satisfies assumption A}\} \tag{13}$$

is generic in $C^r(\mathbb{R} \times \Omega; \mathbb{R})$ *for all* $r \geq 2$. ▲

Proof. Again, it is sufficient to prove that statement (13) is generic in $C^r_\infty(\mathbb{R} \times \Omega; \mathbb{R}) = V$.

Now consider the set M of symmetric $n \times n$ matrices; it is an $n(n+1)/2$-dimensional vector space, and so isomorphic to R^p with $p = n(n+1)/2$. Define a map

$$\Psi: V \times \mathbb{R} \times \Omega \to \mathbb{R}^n \times M$$

$$\Psi(f, \lambda, x) = (f'_x(\lambda, x), f''_{xx}(\lambda, x))$$

It is linear in f, and C^{r-1} in x, so it is jointly C^{r-1} and

$$\Psi'_f(\bar{f}, \bar{\lambda}, \bar{x}) f = (f'_x(\bar{\lambda}, \bar{x}), f''_{xx}(\bar{\lambda}, \bar{x}))$$

so that Ψ'_f is surjective. It follows that Ψ will be transversal to any submanifold of $\mathbb{R}^n \times M$.

Consider in M the subset N of all singular matrices. It is not a submanifold, but it is a finite union of disjoint submanifolds, all of which have codimension $\geqslant 1$. The component of codimension one consists of all symmetric $n \times n$ matrices with rank $(n-1)$; call it N_0.

Now set $F = \{0\} \times N \subset \mathbb{R}^n \times M$. It is a finite union of submanifolds F_{ij}, $i \geqslant (n+1)$, the codimension of F_{ij} being i. We have $\cup_j F_{ij} = \{0\} \times N_0$. Applying Thom's theorem to each of the F_{ij} we see that the property

$$P'(f) = \{\Psi(f, \cdot, \cdot) \text{ is transversal to all the } F_{ij}\}$$

is generic.

That $\Psi(f, \cdot, \cdot)$ is transversal to F_{ij} means that whenever $\Psi(f, \bar{\lambda}, \bar{x}) = z \in F_{ij}$, the image

$$(\Psi'_\lambda(f, \bar{\lambda}, \bar{x}), \Psi'_x(f, \bar{\lambda}, \bar{x}))(\mathbb{R} \times \mathbb{R}^n)$$

and the tangent space $T_z F_{ij}$ span the whole space. Since the former has dimension less than or equal to $\leqslant n+1$, which implies that the codimension of F_{ij} must be $\leqslant n+1$. So $\Psi(f, \cdot, \cdot) = z$ misses the F_{ij} altogether when $i > n+1$. It can only meet $\{0\} \times N_0$.

So $P'(f)$ implies that $f''_{xx}(\lambda, x)$ has rank $\geqslant (n-1)$ for all λ and x; note here that we cannot prove it always has rank n, which would mean $f(\lambda, \cdot)$ is a Morse function throughout, and that would be false in general.

To go farther, we have to use the fact that $\Psi(f, \cdot, \cdot)$ is transversal to $\{0\} \times N_0$. The subset N of M is described by the equation Det $m = 0$, and all points of N_0 are regular for the map Det: $M \to \mathbb{R}$. Near any point $m_0 \in N_0$, there is a local coordinate system $(\xi_1 = \text{Det } m, \xi_2, \ldots, \xi_p)$ for M. So $\{0\} \times N_0$ is defined by the equations $y_1 = \cdots = y_n = 0, \xi_1 = 0$ in $\mathbb{R}^n \times M$, and the transversality of $\Psi(f, \cdot, \cdot) = (f', f'')$ to $\{0\} \times N_0$ means that the tangent map

$$\begin{pmatrix} \dfrac{\partial^2 f}{\partial x_i \partial x_j} & \begin{matrix} \dfrac{\partial^2 f}{\partial x_1 \partial \lambda} \\ \dfrac{\partial^2 f}{\partial x_n \partial \lambda} \end{matrix} \\ \dfrac{\partial \, \text{Det} f''}{\partial x_1} & \dfrac{\partial \, \text{Det} f''}{\partial \lambda} \end{pmatrix}$$

is onto whenever $\Psi(f, \lambda, x) \in N_0$. In other words, $\Delta(\lambda, x) \neq 0$ whenever $f'_x(\lambda, x) = 0$ and $\delta(\lambda, x) = 0$. ∎

We should picture the space $C^r(\Omega; \mathbb{R})$ as being partitioned into open cells by a network of submanifolds of codimension 1, 2, and higher, similar to a beehive.

The open cells consist of Morse functions; if a function is non-Morse, it will lie on the boundary between two or more cells, and a slight perturbation will send it into a cell, that is, make it a Morse function.

A family $f(\lambda, x)$ of functions depending on the real parameter λ is really a map $\lambda \to f(\lambda, \cdot)$ of $\mathbb{R}$ into $C^r(\Omega; \mathbb{R})$, that is, a path in $C^r(\Omega; \mathbb{R})$. If the points $\lambda = 0$ and $\lambda = 1$ lie in two different cells, the path will have to cross some boundary between them at some $\bar{\lambda}$ between zero and one. This situation cannot be altered by slightly changing the path. What can be arranged, though, is for the path to avoid boundaries of codimension 2 or higher, and cross only codimension-1 boundaries. For this reason, unavoidable singularities are called codimension-1 singularities.

Thanks to proposition 10, and to the analysis we carried out in Section 4, we can now state that the singularities of codimension 1 in $C^r(\Omega; \mathbb{R})$ have the form

$$f(x) = f(0) \pm x_1^2 \pm \cdots \pm x_{n-1}^2 + x_n^3$$

in an appropriate local coordinate system.

7. PROOF OF THE TRANSVERSALITY THEOREM AND APPLICATIONS

In the preceding section, we discussed Thom's transversality theorem. We now turn to its proof, following the approach of Stephen Smale, who managed to tie up transversality theory with a classical result in nonlinear analysis, the Sard–Brown theorem. The results of this section, which are mostly due to Smale and range from the very theoretical to the very practical, will show how intimate the connection is.

From the technical viewpoint, in this last section, we abandon (linear) Banach spaces in favor of (nonlinear) Banach manifolds. We shall define a *Banach manifold* to be a submanifold of some Banach space (see definition 3.3). Although this is not the classical definition, it is equivalent to it when the dimension is finite and covers all infinite-dimensional situations arising from functional analysis.

1. The theorem of Brown, Sard and Smale

Henceforward, we shall consider C^r maps, $r \geq 1$, from a Banach manifold X to a Banach manifold Y. We recall that $x \in X$ is a *critical point* of f if the tangent map $T_x f\colon T_x X \to T_{f(x)} Y$ is not surjective. We say that $y \in Y$ is a *critical value* of f if it is the image of some critical point.

Conversely, a point $x \in X$ is *regular* if it is not critical, and $y \in Y$ is a *regular value* for the mapping f if $f^{-1}(y)$ contains no critical point. Note that if $f^{-1}(y)$ is empty, y is considered to be a regular value.

When X and Y have finite dimension, the tangent map can be identified with the Jacobian matrix in some suitable local coordinates. Then x is a critical point for f iff

(1) $$\operatorname{rank} T_x f < \dim Y$$

and y is a regular value for f iff

(2) $$f(x) = y \Rightarrow \operatorname{rank} T_x f = \dim Y$$

THEOREM 1

Assume $f\colon X \to Y$ is a C^r map, $1 \leqslant r \leqslant \infty$, between finite-dimensional separable manifolds, with

(3) $$r > \dim X - \dim Y$$

Then the set of critical values for f is negligible. ▲

By a negligible subset of Y we mean a subset N that is negligible in all local coordinate systems. To be precise, whenever ϕ is a local chart of some open subset U of Y onto an open subset of $\mathbb{R}^p$ when $\phi(U \cap N)$ has Lebesgue measure zero in $\mathbb{R}^p$.

This theorem was first proved by Brown for the case $r = \infty$ and then by Sard for finite r, the proof being significantly more difficult. Condition (3) is automatically satisfied if $r = \infty$ or $\dim Y > \dim X - 1$. Condition (3) requires that the dimension of the target not be too low, namely, $\dim Y > \dim X - r$. Counterexamples are known where this requirement is not heeded.

The Sard–Brown theorem can actually replace Thom's transversality theorem in some simple situations, as shown in corollary 2.

COROLLARY 2

Let H be a linear subspace of $\mathbb{R}^p$ with codimension k, let X an n-dimensional manifold, and $f\colon U \to \mathbb{R}^p$ a C^r map, $r \geqslant 1$. Associate with any $a \in \mathbb{R}^p$ the map

$$f_a\colon x \mapsto f(x) - a$$

Assume $r > n - k$. Then, for almost all $a \in \mathbb{R}^p$, the map f_a is transversal to H. ▲

Proof. Let F be a complementary subspace to H, so that $\mathbb{R}^p = F \oplus H$, and $\pi\colon \mathbb{R}^p \to F$ the associated projection. Saying that f_a is transversal to H means precisely that zero is a regular value of $\pi \circ f_a$, that is, $\pi(a)$ is a regular value of $\pi \circ f$. Let $R \subset F$ be the set of regular values for $\pi \circ f$. By theorem 1, $F \setminus R$ has measure zero, and $\pi^{-1}(R)$ is the set of $a \in \mathbb{R}^p$ such that f_a is transversal to H. ■

The general situation, where f depends on the parameters in a more complicated way, will not yield to this simple-minded approach. This situation motivated Smale to find an infinite-dimensional version of theorem 1 that could be used directly in the function space.

This version requires a few definitions first. Let X and Y be Banach spaces and L: $X \to Y$ a continuous linear map. We say that L is a *Fredholm operator* if both

$$\begin{cases} \textbf{i.} & \text{Ker } L = L^{-1}(0) \text{ is finite dimensional.} \\ \textbf{ii.} & \text{Coker } L = Y/L(X) \text{ is finite dimensional.} \end{cases} \tag{4}$$

The *index* of L is the number

$$\dim(\text{Ker } L) - \text{codim}(L(X)) \in \mathbb{Z} \tag{5}$$

It follows from condition (4) that $L(X)$, the range of L, is closed. If L is a Fredholm operator with index i and K is a compact operator, then $L+K$ is a Fredholm operator of index i. For instance, all operators $A+K$, where A is an isomorphism and K is compact, are Fredholm with index zero.

Now let X and Y be Banach manifolds and f: $X \to Y$ a C^1 map. We shall say that f is a *Fredholm map* if for all $x \in X$, the linear map $T_x f$: $T_x X \to T_{f(x)} Y$ is a Fredholm operator. If X is connected (for instance, if X is a Banach space), the index of $T_x f$ will not depend on the particular choice of the point x in X and is referred to as *the index of* f. Note that if X and Y are finite dimensional, then any C^1 map f: $X \to Y$ is Fredholm with index $\dim X - \dim Y$.

THEOREM 3 (SMALE)

Let f: $X \to Y$ be a C^r Fredholm map between separable Banach manifolds X and Y, with $1 \leqslant r \leqslant \infty$. Assume Y is complete, X connected, and

$$r > \text{index } (f) \tag{6}$$

Then the set of regular values for f contains a dense G_δ subset of Y. ▲

In the finite-dimensional case, theorem 3 reduces to the original Sard–Brown theorem, with one major difference: The property y is a regular value for f is stated to be generic instead of true almost everywhere. These two statements are not equivalent: We can easily find a dense G_δ subset of [0, 1] that is negligible (so its complement has full measure and cannot contain a dense G_δ).‡ However, the genericity statement is the only one to make sense in infinite-dimensional spaces, where there is no analogue to the Lebesgue measure.

‡Let ρ_n, $n \in \mathbb{N}$, be the set of all rationals in [0, 1]. For any $\varepsilon > 0$, define U_ε to be the union of all open intervals $(\rho_n - \varepsilon 2^{-n}, \rho_n + \varepsilon 2^{-n})$, so that meas$(U_\varepsilon) \leqslant 2\varepsilon$. Then $\cap_{\varepsilon > 0} U_\varepsilon$ is a dense G_δ subset with measure zero.

We prove Smale's theorem in two steps.

LEMMA 4

The set of critical points for f is closed in X. ▲

Proof. We shall prove that its complement, the set of regular points, is open. Recall that a point $x \in X$ is regular if $T_x f \colon T_x X \to T_{f(x)} Y$ is surjective. Set $K = \operatorname{Ker} T_x f$, which is finite dimensional, since $T_x f$ is a Fredholm operator, and let $\pi \colon T_x X \to K$ be a continuous projection. By the open mapping theorem, the map $\xi \to (\pi(\xi), f'(x)\xi)$ from $T_x X$ to $K \times T_{f(x)} Y$ is an isomorphism; call it i. Set $y = f(x)$ to simplify notations.

Let $p \colon K \times T_y Y \to T_y Y$ be the projection. With any map $u \colon T_x X \to K \times T_y Y$ we associate $p \circ u$. In this way we define a continuous linear map L from $\mathscr{L}(T_x X, K \times T_y Y)$ into $\mathscr{L}(T_x X, T_y Y)$. It is clearly surjective, so it is open, using the open mapping theorem again. We have just seen that $T_x f = p \circ i$, with $i \in \mathscr{U}$, the set of isomorphisms in $\mathscr{L}(T_x X, K \times T_y Y)$. So $T_x f \in L(\mathscr{U})$. Now $L(\mathscr{U})$ is open since $\mathscr{U}$ is open, and consists only of surjective maps (they can all be written as $p \circ u$, with u an isomorphism). Hence, the result. ■

LEMMA 5

The restriction of f to a suitably small neighborhood of any point maps closed subsets onto closed subsets. ▲

Proof. Take any point $x \in X$. Set $K = \operatorname{Ker} T_x f$ and $R = T_x f\, X$. Let $\pi \colon T_x X \to K$ and $p \colon T_y Y \to R$ be continuous projections. By the open mapping theorem, the map $\xi \to (\pi(\xi), f'(x)\xi)$ is an isomorphism of $T_x X$ onto $K \times R$.

This is the tangent map to the nonlinear mapping $\xi \to (\pi(\xi), p \circ f(\xi))$ at $\xi = x$. By the inverse function theorem, we can use it as a local coordinate system for X near $\xi = x$. In these new coordinates, the map $p \circ f$ now reads

$$p \circ f \colon (\xi_1, \ldots, \xi_n, \eta) \to \eta$$

and the map f itself

$$f \colon (\xi_1, \ldots, \xi_n, \eta) \to (\phi_1(\xi, \eta), \ldots, \phi_p(\xi, \eta), \eta)$$

Here $n = \dim K$ and $p = \operatorname{codim} R$, so that $n - p = \operatorname{index}(f)$.

Now let U be a closed bounded neighborhood of x where this local chart is valid and x^k a sequence in U such that $f(x^k)$ converges to some $\bar{y} \in Y$.

Reading off the coordinates $x^k = (\xi^k, \eta^k)$, we see that $p \circ f(\xi^k, \eta^k) = \eta^k$ converges to $p(\bar{y})$. As for the ξ^k, they stay within $\pi(U)$, a bounded set in the finite-dimensional space K, so a suitable subsequence will converge. The corresponding subsequence of x^k converges also. ■

The proof of Smale's theorem is as follows. Choose local charts as in lemma 5 and call ϕ the map with components $(\phi_1, \ldots, \phi_p)$. Then $x=(\xi, \eta) \in U$ will be a critical point of f if and only if ξ is a critical point of $\phi(\cdot, \eta)$, a map from $\pi(U) \subset \mathbb{R}^n$ into $\mathbb{R}^p$. And $y \in f(U)$ will be a critical value if and only if $y-p(y)$ is a critical value of $\phi(\cdot, \eta)$, where $\eta = p(y)$.

By Theorem 1 for every $\eta \in R$, the set of critical values of $\phi(\cdot, \eta)$ has measure zero. It follows that it has empty interior. In other words, for each $\eta \in R$, the set of $y \in p^{-1}(\eta)$ that are critical values of f over U has empty interior. Then the set C_U of critical values of f over U has empty interior.

By lemmas 4 and 5, the set C_U is also closed. We now find a countable family of U covering the whole space X. The set of critical values for f over X is the union of the C_U, and its complement is a dense G_δ by Baire's theorem, since Y is a complete metric space.

2. Proof of the Transversality Theorem

We shall restate the transversality theorem in a nonlinear framework. Now Ω is a connected separable n-dimensional manifold (instead of an open subset of $\mathbb{R}^n$), Y a separable p-dimensional manifold (instead of $\mathbb{R}^p$), and V a separable Banach space.

THEOREM 6

Let $f: V \times \Omega \to Y$ be a C^r map, $1 \leq r \leq \infty$, and M a submanifold of Y. Assume that f is transversal to M and that

$$r > \dim \Omega - \operatorname{codim} M \tag{7}$$

We denote by f_u the map $\omega \to f(u, \omega)$. Then the property

$$P(u) = \{f_u : \Omega \to Y \text{ is transversal to } M\} \tag{8}$$

is generic in V. ▲

Let us introduce the set $E = f^{-1}(M)$ and the map $\pi: E \to V$, the restriction to E of the first projection $(u, \omega) \to u$. Since f is transversal to M, the set E is a closed submanifold of $V \times \Omega$, and the map π is C^r. The geometric situation is shown in the following figure.

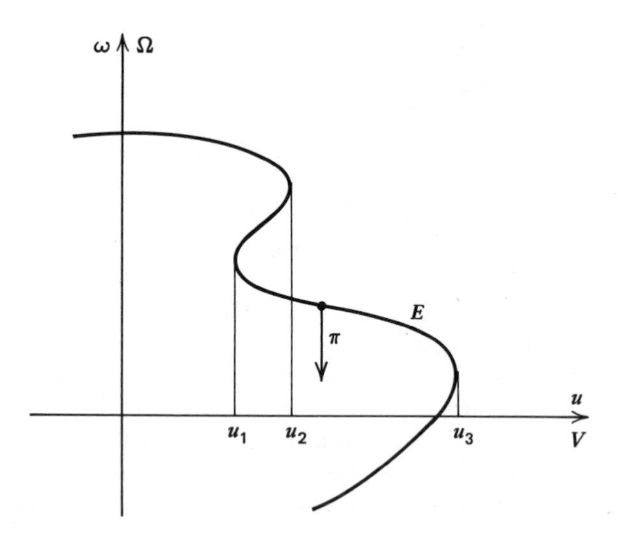

$\{u_1, u_2, u_3\}$ is the apparent contour of E. It is the set of critical values for π.

We shall denote by $x=(u, \omega)$ the points of E.

LEMMA 7

u is a regular value for π if and only if f_u is transversal to M on Ω. ▲

Proof. If $f_u(\Omega)$ does not meet M, then f_u is transversal to M, and there is no point x in $E:=f^{-1}(M)$ with $\pi(x)=u$, so u is a regular value.

Now assume $f_u(\Omega)\cap M\neq\varnothing$, and let $\bar{x}=(\bar{u}, \bar{\omega})\in E$. We claim that $\bar{x}$ is a regular point for π if and only f_u is transversal to M at $\bar{x}$. The result will then follow. Set $\bar{u}=0$ for the sake of convenience.

Assume first that f_0 is transversal to M at x. Proceeding as in theorem 3.5, we can find local coordinates $(\psi_1, \ldots, \psi_k)$ in Y around $\bar{y}=f(0, \bar{x})$, such that M is locally defined by k equations $\psi_1(y)=\cdots=\psi_k(y)=0$, and corresponding local coordinates $(\phi_1, \ldots, \phi_{n-k}, \psi_1\circ f_0, \ldots, \psi_k\circ f_0)$ in Ω around ω.

By the implicit function theorem, the equations $\psi_1\circ f(u, \omega)=\cdots=\psi_k\circ f(u, \omega)=0$ can be solved near $(0, \bar{\omega})$ in terms of ϕ_i and u. In other words, the ϕ_i, $1\leqslant i\leqslant n-k$, and $\pi\colon E\to V$ constitute a local coordinate system for E near $\bar{x}$, which implies $\bar{x}$ is a regular point for π.

Conversely, assume that $x=(0, \omega)$ is a regular point for π in E. This means that the map $T_x\pi\colon T_xE\to T_xV$ is onto. Now T_xE is closed linear subspace of $T_0V\times T_\omega\Omega$, and $T_x\pi$ is the restriction to T_xE of the first projection. It follows that

$$T_x(V\times\Omega)=T_xE+T_\omega\Omega \tag{9}$$

Here we have identified $T_x(V\times\Omega)$ with $T_0V\times T_\omega\Omega$, and $T_\omega\Omega$ with $T_x(\{0\}\times\Omega)$; the sum on the right is not direct, that is, we are not claiming that $T_xE\cap T_\omega\Omega=\{0\}$.

Applying $T_x f$ to both sides

$$T_x f\, T_x(V \times \Omega) = T_x f T_x E + T_x f T_\omega \Omega$$

Since f is transversal to M, we also have

$$T_x f T_x(V \times \Omega) + T_y M = T_y Y, \quad \text{with } y = f(0, \omega)$$

Replacing the first term by its value from the preceding equation,

$$T_x f T_x E + T_x f T_\omega \Omega + T_y M = T_y Y$$

The first term on the left is contained in the third, $T_x f T_x E \subset T_y M$, since $f(E) = M$. The second term is easily identified with $T_\omega f_0 T_\omega \Omega$. Finally, we have

$$T_x f_0 T_\omega \Omega + T_y M = T_y Y$$

This means precisely that f_0 is transversal to M at x, as desired. ■

LEMMA 8
π is a Fredholm map, with constant index $n-k$. ▲

Proof. As we just pointed out, $T_x E$ is a closed linear subspace of $T_0 V \times T_\omega \Omega$, with codimension k, and $T_x \pi$ is the restriction to $T_x E$ of the first projection. We introduce spaces N_1, L, K, F, and N_2 such that

$$N_1 = \text{Ker } T_x \pi = T_x E \cap T_\omega \Omega$$

$$T_x E = N_1 \oplus L$$

$$T_0 V \times T_\omega \Omega = T_x E \oplus K \qquad (\dim K = k)$$

$$N_2 = K \cap T_\omega \Omega$$

$$K = N_2 \oplus F$$

Clearly, $N_1 \oplus N_2 = T_\omega \Omega$. We have

$$T_0 V \times T_\omega \Omega = L \oplus N_1 \oplus N_2 \oplus F = L \oplus T_\omega \Omega \oplus F$$

So

$$T_x \pi(T_0 V \times T_\omega \Omega) = T_x \pi(L) \oplus T_x \pi(F)$$

It follows that $T_x \pi(F)$, which is isomorphic to F itself, is a supplementary subspace to $T_x \pi(L) = T_x \pi(T_x E)$.

$$\text{codim } T_x\pi(T_xE) = \dim F = k - \dim N_2$$
$$\dim (T_x\pi)^{-1}(0) = \dim N_1 = n - \dim N_2$$

So the index of $T_x\pi$ is $n-k$. ■

The proof of the transversality theorem now follows from theorem 3 applied to $\pi: E \to V$, noting that

$$\dim \Omega - \dim Y = n - k$$

3. Newton's method revisited

We conclude this chapter by applying the latest results we obtained to the first problem we started with, namely, the problem of solving the equation

$$f(x) = 0 \tag{10}$$

in some domain B of $\mathbb{R}^n$. For the sake of convenience, we shall take B to be the unit ball, with boundary S. The function f maps a bounded open subset $\Omega \subset \mathbb{R}^n$ containing B into $\mathbb{R}^n$.

At this point, we are looking for a priori conditions that will ensure that equation (10) has a solution and for a practical procedure to solve it numerically. Smale popularized a method that achieves both at one stroke.

Smale's method

Choose some point $x_0 \in S$. If $f(x_0) \neq 0$, construct the straight line $D = \mathbb{R}f(x_0)$, and the set

$$C = B \cap f^{-1}(D)$$ ▲

Clearly, C is not empty, since it contains x_0, and C contains any zero of f in B, since $0 \in D$. The merit of considering C lies in the following two properties, (11) and (13), which we shall require of f

$$f \text{ is transversal to } D \text{ and } f^{-1}(D) \text{ to } S \tag{11}$$

This tells us that $f^{-1}(D)$ is a closed submanifold of Ω with codimension $(n-1)$, that is, a smooth curve. For all $x \in f^{-1}(D)$, define $\lambda(x) \in \mathbb{R}$ by

$$f(x) = \lambda(x) f(x_0) \tag{12}$$

By the implicit function theorem, we can use λ as a local coordinate for $f^{-1}(D)$ in the neighborhood of any point $\bar{x}$ where $f'(\bar{x})$ is invertible, that is, $\operatorname{Det} f'(\bar{x}) \neq 0$. On the other hand, when $\operatorname{Det} f'(\bar{x}) = 0$, one of the projections in

$\mathbb{R}^n$, $x \to x_i$, say, will serve as a local coordinate near $\bar{x}$ for $f^{-1}(D)$, since the latter is a one-dimensional submanifold. Our second property (13) follows (all derivatives taken at $x_i = \bar{x}_i$ with the preceding definition of i).

$$f(x) \in D, \quad \text{Det } f'(\bar{x}) = 0 \Rightarrow \frac{d^2\lambda}{dx_i^2} \neq 0, \quad \frac{d}{dx_i} \text{Det } f'(x) \neq 0 \tag{13}$$

Differentiating the equation $f(x) = \lambda f(x_0)$ at $\bar{x}$, we have

$$f'(\bar{x}) \frac{dx}{dx_i} = f(x_0) \frac{d\lambda}{dx_i} \tag{14}$$

Since $\text{Det} f'(\bar{x}) = 0$, the map $f'(x)$ is not surjective. We know from assumption (11) that $f(x_0)$ and the image of $f'(\bar{x})$ span the whole space $\mathbb{R}^n$, so $f(x_0)$ cannot belong to the image of $f'(\bar{x})$. Equation (14) then implies that

$$\frac{d\lambda}{dx_i} = 0 \qquad \text{at } \bar{x}$$

Assumption (13) now tells us that $d\lambda/dx_i$ and $\text{Det} f'(x)$ change signs simultaneously: The critical point of f cuts $f^{-1}(D)$ into arcs on which $\lambda(x)$ is monotone, alternatively increasing and decreasing on consecutive arcs. We can now prove existence results constructively.

PROPOSITION 9

Assume (11) *and* (13). *Assume* $f(x) = x$ *for all* $x \in S$. *Then* f *has a zero in the interior of* B. ▲

Proof. The set $f^{-1}(D)$ is a closed C^1 submanifold of Ω. By assumption (13), the set of critical points of f that belong to D is discrete: If $\text{Det} f'(\bar{x}) = 0$ and $\bar{x} \in f^{-1}(D)$, neighboring points of D have all $\text{Det} f'(x) \neq 0$. By compactness, it follows that there can be only finitely many critical points of f on $f^{-1}(D) \cap B$.

Now investigate the connected components of $f^{-1}(D) \cap \bar{B}$ containing x_0. It is an arc contained in B, starting at x_0 transversally to S; call it $\mathscr{C}$.

Let $x_1, \ldots, x_N$ be the (finitely many) critical points of $f(x)$ on $\mathscr{C}$. Delete the portion between x_0 and x_N from $\mathscr{C}$ and consider the remaining arc, originating at x_N. On this arc, there are no more critical points of $f(x)$, so λ is monotone, and we can use it as a global coordinate. Using the compactness of $\bar{B}$, we see that λ is bounded, and since it is monotone, it converges to a limit $\bar{\lambda}$, and the corresponding point x of $\mathscr{C}$ converges to a limit $\bar{x}$. Since $\mathscr{C}$ is closed, it must contain $\bar{x}$. So $\mathscr{C}$ ends at $\bar{x}$. If $\bar{x}$ were an interior point of B, this would contradict the fact that $f^{-1}(D)$ is a submanifold. So $\bar{x}$ belongs to S.

We cannot have $\bar{x} = x_0$, or x_0 would be a self-intersection point of $f^{-1}(D)$, contradicting again the fact that it is a submanifold. So $\bar{x} \neq x_0$, and yet $f(\bar{x}) =$

$\lambda f(x_0)$; since $\bar{x}$ belongs to S, this reduces to $\bar{x}=\lambda x_0$. The only possibility is $\bar{x}=-x_0$, so $\lambda=-1$. Since we started with $\lambda=1$ at x_0, we must have $\lambda=0$ somewhere on $\mathscr{C}$. ■

This proposition can easily be extended in two different directions.

COROLLARY 10
Assume (11) *and* (13). *Assume f does not vanish on S, and consider the map $\bar{f}: S \to S$ defined by*

$$\bar{f}(x)=f(x)\|f(x)\|^{-1} \tag{15}$$

If there is an odd number d of points x such that $\bar{f}(x)=\bar{f}(x_0)$, then f has a zero in B. ▲

Proof. This contains the preceding result as a particular case: $\bar{f}(x)=f(x)=x$ in S, and x_0 is the only antecedent of $f(x_0)$, so $d=1$.

The proof is essentially the same. Take x_0 and let x_1 be the endpoint of the corresponding arc, so $x_1 \neq x_0$ and $f(x_1)=\lambda f(x_0)$. If $\lambda<0$, we are done, since λ must change sign between x_0 and x_1. If $\lambda>0$, we have $\bar{f}(x_1)=\bar{f}(x_0)$. Since d is odd, and hence $\neq 2$, we can take another starting point x_2, with $\bar{f}(x_2)=(\bar{f}(x_0)$, and get a corresponding endpoint x_3, with $f(x_3)=\lambda f(x_2)$. If $\lambda<0$, we are done; if $\lambda>0$, we can continue since $\lambda \neq 4$. Since d is odd, it will be impossible to appariate all the antecedents of $\bar{f}(x_0)$, so we shall eventually find $\bar{f}(x_n)\neq \bar{f}(x_{n+1})$ for some suitable starting point x_n. ■

COROLLARY 11
Assume (11) *and* (13). *Assume the scalar product $(x, f(x))$ does not vanish on S. Then f has a zero in B.* ▲

Proof. Say $(x, f(x))$ is always positive on S. This enables us to extend f to the ball $\Delta=2B$, with boundary $\Sigma=2S$, in such a way that the extension g: $\Delta \to \mathbb{R}^n$ is the identity on Σ and never vanishes on $\Delta \setminus B$. This is done by finding a smooth function $\phi: [1, 2] \to [0, 1]$ such that $\phi(1)=0$ and $\phi(2)=1$ and

$$\begin{cases} g(x)=\phi(\|x\|)x+(1-\phi(\|x\|))f\left(\dfrac{x}{\|x\|}\right) & \text{if } 1 \leqslant \|x\| \leqslant 2 \\ g(x)=f(x) & \text{if } \|x\| \leqslant 1 \end{cases}$$

Clearly, $g(x)=x$ if $\|x\|=2$, and $(g(x), x)>0$ for $1 \leqslant \|x\| \leqslant 2$, so $g(x) \neq 0$ in that region. We then apply Smale's method to g on Δ, starting from a suitable point on Σ. The corresponding curve may break when crossing Σ (it remains continuous, but may lose differentiability), but the argument in proposition 9 otherwise works to give a point $\bar{x} \in \Delta$ where $g(\bar{x})=0$. Since g does not vanish outside B, we have $\bar{x} \in B$ and $f(\bar{x})=g(\bar{x})=0$. ■

At this point, the reader may wonder about the role of properties (11) and (13). As a matter of fact, they are generic in $C^r_\infty(\Omega)$, all $r \geqslant 2$. For (11), it follows immediately from Thom's transversality theorem. For (13) (once $x_0 \in S$ is prescribed), it is also true but more intricate.

So corollaries (10) and (11) extend by density to all of $C^r_\infty(\Omega)$: Assumptions (11) and (13) can be dropped and so can the differentiability assumption. Corollary (11) extends to any continuous f with $(f(x), x) \neq 0$ on the boundary, while corollary (10) requires an appropriate definition of d, which is now called the *degree* of the map $\bar{f}: S \to S$.

But at this point, we are more interested in the computational problem, and this is where (11) and (13) are handy. We are supposed to find the curve $f(x) = \lambda f(x_0)$. We do this by starting at x_0 and integrating numerically the corresponding differential equation

$$f'(x)\frac{dx}{d\lambda} = f(x_0)$$

or

$$\frac{dx}{d\lambda} = f'(x)^{-1} f(x_0) \tag{16}$$

which will break down when the curve approaches a critical point $\bar{x}$, where $f'(\bar{x})$ is singular. We should then take another coordinate along the curve x_1, say. Equation (16) is then replaced by

$$f'(x)\frac{dx}{dx_1} = f(x_0)\frac{d\lambda}{dx_1} \tag{17}$$

which can be solved for dx/dx_1 and $d\lambda/dx_1$. Once the critical value is crossed, we can resume using equation (16).

All this is done automatically if we replace equation (16) by the system

$$\begin{cases} \dfrac{dx}{dt} = \text{Cof } f'(x) \cdot f(x_0) \\ \dfrac{d\lambda}{dt} = \text{Det } f'(x) \end{cases} \tag{18}$$

Here, Cof $f'(x)$ is the matrix of cofactors of $f'(x)$

$$f'(x) \cdot \text{Cof } f'(x) = \text{Det } f'(x) \cdot I$$

Cof $f'(x) \cdot f(x_0)$ never vanishes along the curve $f^{-1}(D)$, otherwise f would not be transversal to D. So the trajectory of the first equation is the whole of $f^{-1}(D)$.

The second equation reminds us that the points where Det $f'(x)=0$ separate $f^{-1}(D)$ into arcs where λ is monotone, alternatively increasing and decreasing.

Finally, note that with $f(x)=\lambda f(x_0)$, the original equation (16) can be written

$$\lambda \frac{dx}{d\lambda}=f'(x)^{-1}f(x)$$

This is a continuous version of the discrete algorithm (18) of Chapter 2 for solving $f(x)=0$ by Newton's method

$$x_{n+1}-x_n=f'(x_n)^{-1}f(x_n)$$

which is why Smale also refers to it as Newton's method.

We shall later give another set of assumptions ensuring both the existence of a zero for f and the convergence of Newton's method in a more general setting.

CHAPTER 3

Set-Valued Maps

When X and Y are Hausdorff topological spaces, we face the problem of defining continuity of set-valued maps. In the case of single-valued maps f from X to Y, continuous functions are characterized by two equivalent properties

(a) For any neighborhood $\mathcal{N}(f(x_0))$ of $f(x_0)$, there exists a neighborhood $\mathcal{N}(x_0)$ of x_0 such that $f(\mathcal{N}(x_0)) \subset \mathcal{N}(f(x_0))$.

(b) For any generalized sequence of elements x_μ converging to x_0, the sequence $f(x_\mu)$ converges to $f(x_0)$.

These two properties can be adapted to the case of strict set-valued maps from X to Y; they become

(A) For any neighborhood $\mathcal{N}(F(x_0))$ of $F(x_0)$, there exists a neighborhood $\mathcal{N}(x_0)$ of x_0 such that $F(\mathcal{N}(x_0)) \subset \mathcal{N}(F(x_0))$.

(B) For any generalized sequence of elements x_μ converging to x_0 and for any $y_0 \in F(x_\mu)$, there exists a sequence of elements $y_\mu \in F(x_\mu)$ that converges to y_0.

In the case of set-valued maps, these two properties are no longer equivalent. We call *upper semicontinuous* maps those that satisfy property (A), *lower semicontinuous* maps those that satisfy property B, and *continuous* maps the ones that satisfy both properties (A) and (B).

In the first section of Chapter 3 we review elementary properties of semicontinuous maps and present a list of examples.

1. If W is a function from $X \times Y$ to R, we set $W_x: y \to W_x(y) := W(x, y)$ and study the upper semicontinuity of the map

$$x \to \text{epigraph}\,(W_x)$$

2. If f maps $X \times U$ to Y, we give sufficient conditions for the set-valued map F defined by

$$F(x) := \{f(x, u) | u \in U\}$$

to be upper or lower semicontinuous.

3. If f is a map from $K \times Y$ to Z, U and T are set-valued maps from K to Y and Z, respectively, and we prove that the set-valued map C defined by

$$C(x) := \{ y \in U(x) | f(x, y) \in T(x) \}$$

is lower semicontinuous under a convenient set of assumptions.

4. We consider a map T sending elements $x \in X$ to closed convex cones $T(x)$ of R^n. We prove that T is lower semicontinuous if and only if the graph of the map $x \to T(x)^-$ is closed.

5. Finally, we investigate the continuity properties of the *marginal function* V of a family of maximization problems

$$V(y) := \sup_{x \in G(y)} W(x, y)$$

depending on a parameter y, as well as the upper semicontinuity of the *marginal map* associating with the parameter y the set of maximizers

$$M(y) := \{ x \in G(y) | V(y) = W(x, y) \}$$

In Section 2, we single out an important class of set-valued maps, namely, maps with *closed convex values*. Such maps F from X to a Banach space Y can be characterized by their support functions

$$\sigma(F(x), p) := \sup_{y \in F(x)} \langle p, y \rangle$$

since the Hahn–Banach separation theorem tells us that

$$F(x) = \{ y \in Y | \forall p \in Y^\star, \langle p, y \rangle \leqslant \sigma(F(x), p) \}$$

These support functions are very easy to manipulate. For instance, we shall observe that the upper semicontinuity of F implies the upper semicontinuity of the functions $x \to \sigma(F(x), p)$ when p ranges over $Y^\star$. So, we shall select maps enjoying the latter property, which we call *upper hemicontinuous maps*.

A theorem due to Castaing states that any upper hemicontinuous map with *compact* convex values is, conversely, upper semicontinuous.

Upper hemicontinuous maps with closed convex values enjoy many fixed point and surjectivity properties, which are presented in Chapter 6, Section 4.

We devote the third section to studying maps with convex graphs as well as maps whose graphs are cones (called *processes*). Convex processes whose graphs are convex cones are the set-valued analogues of linear operators and share some of their properties.

Convex processes will be used for defining derivatives of set-valued maps (see Chapter 4, Section 2 and Chapter 7, Section 7). They also enjoy spectral properties (eigenvalues and eigenvectors), which will be studied in Section 4.

When A is a (continuous) linear operator and $P \subset X$ and $Q \subset Y$ are (closed) convex sets, the set-valued map F defined by

$$F(x) := \begin{cases} Ax - Q & \text{when } x \in P \\ \varnothing & \text{when } x \notin P \end{cases}$$

provides an example of a (closed) convex map. The inverse of this map is defined by

$$\forall y \in Y, \qquad F^{-1}(y) = \{x \in P \mid Ax \in Q + y\}$$

These subsets $F^{-1}(y)$ provide the main class of subsets on which we minimize or maximize functions in optimization theory. When P and Q are (closed) convex cones, the map F is a (closed) convex process.

We can adapt to closed convex maps the Banach open mapping and closed graph theorems, and we shall prove that such maps are lower semicontinuous on the interior of their domain.

We have an even stronger result: If $x_0 \in \operatorname{Int} \operatorname{Dom} F$ and $y_0 \in F(x_0)$ are chosen, there exists $\gamma > 0$ such that

$$\begin{cases} \forall x \in x_0 + \gamma B, \qquad \forall y \in \operatorname{Im} F \\ d(y, F(x)) \leqslant \dfrac{1}{\gamma}\, d(x, F^{-1}(y))(1 + \|y - y_0\|) \end{cases}$$

This theorem implies that *any closed convex process from a Banach space X to a Banach space Y whose domain is the whole space X is Lipschitz*: There exists $\gamma > 0$ such that

$$\forall x_1, x_2 \in X, \qquad F(x_2) \subset F(x_1) + \frac{1}{\gamma}\|x_1 - x_2\|$$

We also deduce that if F is closed, convex, and locally bounded (for every $x \in \operatorname{Int} \operatorname{Dom} F$, the image of some neighborhood is bounded), then F is locally Lipschitz on the interior of its domain. (When F is the inverse of a surjective continuous linear operator A, this is the Banach open mapping principle.)

We shall use this theorem in many crucial instances; for example, for proving the nontrivial formulas of convex analysis.

We then define the transpose $F^\star$ of a closed convex process F, which generalizes the usual transpose of continuous linear operators. If A is a continuous linear operator, we prove the expected formulas

$$(FA)^\star = A^\star F^\star, \qquad (AF)^\star = F^\star A^\star$$

and

$$(F_1 + F_2)^\star = F_1^\star + F_2^\star$$

Contrary to the case of a continuous linear operator, these formulas are not always valid (nor obvious). Actually, we shall even define the transpose of any closed convex map. Indeed, properties of the transpose are used for proving "closedness" theorems, such as the one stating that the sum of two closed convex maps is still closed.

We prove in the last section of this chapter that several spectral properties of positive matrices can be extended to positive set-valued maps with closed convex graph. Let us review the theorems we plan to generalize.

PERRON–FROBENIUS THEOREM

Let G be a positive matrix, $g_i^j > 0$ for all i and j.

i. *It has a positive eigenvalue δ that is larger than or equal to the absolute value of any other eigenvalue of G.*

ii. *It is the only eigenvalue of G to which there corresponds a nonzero nonnegative eigenvector.*

iii. *$\mu - G$ is invertible, and $(\mu - G)^{-1}$ is positive if and only if $\mu > \delta$.* ▲

***M*-MATRICES**

Let $H := (h_i^j)$ be a matrix satisfying

$$(\star) \qquad \forall i \neq j, \qquad h_i^j \leqslant 0$$

Then the following conditions are equivalent:

i. $\forall i = 1, \ldots, n,\ \exists q^j > 0$ *such that* $\sum_{j=1}^{n} h_j^i q^j > 0$

ii. *H is invertible and H^{-1} is positive.* ▲

Matrices H satisfying condition $(\star)$, and either one of the equivalent conditions **i** or **ii**, are called M matrices.

VON NEUMANN–KEMENY THEOREM

Consider two matrices $F := (f_i^j)$ and $G := (g_i^j)$ from R^n to R^m satisfying

$$\begin{cases} \textbf{i.}\ \ \forall i, j, \quad g_i^j \geqslant 0 & (G \textit{ is nonnegative.}) \\ \textbf{ii.}\ \ \forall i = 1, \ldots, m, & \sum_{j=1}^{n} g_i^j > 0 \\ \textbf{iii.}\ \ \forall j = 1, \ldots, n, & \sum_{i=1}^{n} f_i^j > 0 \end{cases}$$

Then there exist $\delta>0$, $\bar{x}\in R^n_+$, $\bar{x}\neq 0$ *and* $\bar{p}\in R^m_+$, $\bar{p}\neq 0$ *satisfying*

$$(\star\star)\qquad \begin{cases} \text{i.} & \delta F\bar{x}\leqslant G\bar{x} \\ \text{ii.} & \delta F^\star\bar{p}\geqslant G^\star\bar{p} \\ \text{iii.} & \delta\langle\bar{p}, F\bar{x}\rangle=\langle\bar{p}, G\bar{x}\rangle \end{cases}$$

Furthermore, for all $\mu>\delta$ *and for all* $y\in \operatorname{Int}(R^m_+)$, *there exists* $x\in R^n_+$ *such that*

$$\mu Fx-Gx\leqslant y$$

▲

Von Neumann devised the following economic interpretation. The economy is assumed to have n production processes that produce and consume m goods. The entry f_i^j denotes the quantity of good i consumed by process j when operating at unit intensity, and entry g_i^j denotes the corresponding quantity produced. We assume constant returns to scale, so that the pair of matrices (F, G) completely describes the production possibilities of the economy. The operation of the economy is further specified by a vector $x\in R^n$, representing the intensities at which the n production processes are operated, and by $p\in R^{m\star}$, representing the price systems on the commodity space R^m. A triple $(\bar{x}, \bar{p}, \delta)$ satisfying condition $(\star\star)$ is called an *equilibrium.*

A δ such that $\delta Fx\leqslant Gx$ can be regarded as a growth rate, whereas a number δ such that $\delta F^\star p\geqslant G^\star p$ can be regarded as an interest rate. This theorem states the existence of intensities and prices for which both rates coincide.

In Section 4, we not only prove these theorems, but actually deduce them from analogous statements for set-valued maps with closed convex graphs.

We prove the existence of solutions $\bar{x}\in R^n_+$ such that

$$\delta F(\bar{x})\in G(\bar{x})-R^n_+$$

when F is a convex operator and G a positive set-valued map with closed convex graph from which we deduce equilibrium theorems of the von Neumann type. When $R^n=R^m$, we add more specific requirements that imply a generalization of the Perron–Frobenius theorem. We finally define M convex processes and prove that in some sense, they map the positive cone R^n_+ onto itself.

1. UPPER AND LOWER SEMICONTINUITY OF SET-VALUED MAPS

Let F be a set-valued map from a Hausdorff topological space X to another Y.

DEFINITION 1

We say that F is upper semicontinuous (*in short, u.s.c.*) *at $x_0 \in X$ if for any neighborhood $\mathcal{N}(F(x_0))$ of $F(x_0)$, there exists a neighborhood $\mathcal{N}(x_0)$ of x_0 such that*

$$\forall x \in \mathcal{N}(x_0), \qquad F(x) \subset \mathcal{N}(F(x_0)) \tag{1}$$

We say that F is upper semicontinuous *if F is u.s.c. at every point $x \in X$.* ▲

Remark

We recall that when a subset K is a compact subset of a metric space X, the subsets

$$B(K, \eta) := \{y \in X | d(K, y) \leq \eta\} \tag{2}$$

form a fundamental basis of neighborhoods of K, in the sense that any neighborhood of K contains $B(K, \eta)$ for some $\eta > 0$.

Therefore, if F is a *compact-valued* map F from a metric space X to a metric space Y, F is upper semicontinuous at $x_0 \in X$ if and only if for all $\varepsilon > 0$, there exists $\eta > 0$ such that

$$\forall x \in B(x_0, \eta), \qquad F(x) \subset B(F(x_0), \varepsilon) \tag{3}$$

Note that if $F(x_0)$ is not compact, this property may be true even if F is not upper semicontinuous at x_0: Consider, for instance, the set-valued map from R to R^2 defined by $F(\xi) := \{(x, y) | x = \xi\}$. Property (3) obviously holds when $\varepsilon = \eta$. Moreover, F is not upper semicontinuous in the usual sense: Indeed, the subset $\mathcal{N} := \{(x, y) | \, |y| < 1/|x|\}$ is a neighborhood of $F(0)$, but for every $x \neq 0$, $F(x)$ is not contained in $\mathcal{N}$. ■

In the case of single-valued maps, as we have remarked, continuity can also be characterized in terms of generalized sequences. This second point of view leads to definition 2.

DEFINITION 2

We say that F is lower semicontinuous (*in short l.s.c.*) *at $x^0 \in X$ if for any $y^0 \in F(x^0)$ and any neighborhood $\mathcal{N}(y^0)$ of y^0, there exists a neighborhood $\mathcal{N}(x^0)$ of x^0 such that*

$$\forall x \in \mathcal{N}(x^0), \qquad F(x) \cap \mathcal{N}(y^0) \neq \emptyset \tag{4}$$

We say that F is lower semicontinuous *if it is lower semicontinuous at every $x^0 \in X$.* ▲

Definition 2 could be phrased as follows: Given any generalized sequence x_μ converging to x^0 and any $y^0 \in F(x^0)$, there exists a generalized sequence

$y_\mu \in F(x_\mu)$ that converges to y^0. When X and Y are metric, this last characterization holds true with usual (i.e., countable) sequences.

Examples

The mapping F_+, from R into its subsets, defined by $F_+(0)=[-1, +1]$ and $F_+(x)=\{0\}$, $x \neq 0$, is u.s.c., while it is not l.s.c. The mapping F_-, defined by $F_-(0)=\{0\}$, $F_-(x)=[-1, +1]$, $x \neq 0$, is l.s.c., while it is not u.s.c.

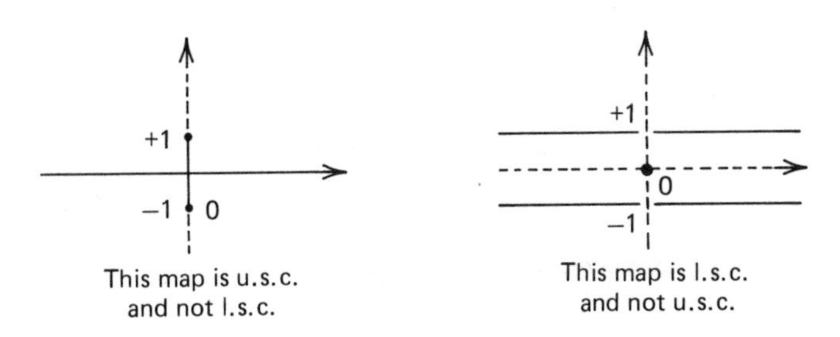

This map is u.s.c. and not l.s.c.

This map is l.s.c. and not u.s.c.

DEFINITION 3

A set-valued map F from X to Y is said to be continuous at $x_0 \in X$ if it is both u.s.c. and l.s.c. at x_0. It is said to be continuous if it is continuous at every point $x \in X$. ▲

An important class of continuous set-valued maps is provided by Lipschitz maps.

DEFINITION 4

Let F be a strict set-valued map from a metric space X to a metric space Y. We say that F is Lipschitz around $x_0 \in X$ if there exist a neighborhood $\mathcal{N}(x_0)$ and a constant $c>0$ (the Lipschitz constant) such that

$$\forall x, y \in \mathcal{N}(x_0), \qquad F(x) \subset B(F(y), cd(x, y)) \tag{5}$$

It is locally Lipschitz if it is Lipschitz around every $x_0 \in X$. It is Lipschitz if there exists $c>0$ such that

$$\forall x, y \in X, \qquad F(x) \subset B(F(y), cd(x, y)) \tag{6}$$ ▲

Note that definition 4 is symmetric in x and y.

We shall use a weaker nonsymmetric statement in definition 5.

DEFINITION 5

We shall say that a strict set-valued map from a metric space X to a metric space Y is upper locally Lipschitz at $x_0 \in X$ if there exist a neighborhood $\mathcal{N}(x_0)$ of x_0

and a constant $c>0$ such that

$$\forall x \in \mathcal{N}(x_0), \qquad F(x) \subset B(F(x_0), cd(x_0, x)) \tag{7}$$ ▲

We begin by stating that the composition of two upper semicontinuous maps is upper semicontinuous.

PROPOSITION 6

Let F and G be two set-valued maps from X to Y and from Y to Z, respectively. Define GF by

$$GF(x) := \bigcup_{y \in F(x)} G(y) \tag{8}$$

If F and G are upper semicontinuous, so is GF. ▲

PROPOSITION 7

Let F be an upper semicontinuous set-valued map from X to Y with closed values. Then F is closed (*i.e., its graph is closed*). ▲

Proof. Let us consider a sequence of elements (x_μ, y_μ) of the graph of F that converges to some $(x, y) \in X \times Y$. Since F is upper semicontinuous, we can associate to any closed neighborhood $\mathcal{N}(F(x))$ an index μ_0 such that $\forall \mu > \mu_0$, $y_\mu \in \mathcal{N}(F(x))$. Hence, y belongs to every neighborhood of $F(x)$. ■

There is a partial converse to this result.

THEOREM 8

Let F and G be two set-valued maps from X to Y such that $\forall x \in X$, $F(x) \cap G(x) \neq \emptyset$. We suppose that

$$\begin{cases} \textbf{i.} & F \text{ is upper semicontinuous at } x_0. \\ \textbf{ii.} & F(x_0) \text{ is compact.} \\ \textbf{iii.} & G \text{ is closed.} \end{cases} \tag{9}$$

Then the set-valued map $F \cap G: x \rightarrow F(x) \cap G(x)$ is upper semicontinuous at x_0. ▲

Proof. Let $\mathcal{N} := \mathcal{N}(F(x_0) \cap G(x_0))$ be an open neighborhood of $F(x_0) \cap G(x_0)$. We have to find a neighborhood $\mathcal{N}(x_0)$ such that

$$\forall x \in \mathcal{N}(x_0), \qquad F(x) \cap G(x) \subset \mathcal{N}$$

If $\mathcal{N} \supset F(x_0)$, this follows from the upper semicontinuity of F. If not, then we introduce the subset

$$K := F(x_0) \setminus \mathcal{N}$$

which is compact (since $F(x_0)$ is compact). Let $P:=\text{graph}\,(G)$, which is closed. For any $y \in K$, we have $y \notin G(x_0)$ and thus $(x_0, y) \notin P$. Since P is closed, there exist open neighborhoods $\mathcal{N}_y(x_0)$ and $\mathcal{N}(y)$ such that $P \cap (\mathcal{N}_y(x_0) \times \mathcal{N}(y)) = \varnothing$. Therefore,

$$\text{(10)} \qquad \forall x \in \mathcal{N}_y(x_0), \qquad G(x) \cap \mathcal{N}(y) = \varnothing$$

Since K is compact, it can be covered by n neighborhoods $\mathcal{N}(y_i)$. The union $\mathcal{M} := \cup_{i=1}^{n} \mathcal{N}(y_i)$ is a neighborhood of K and $\mathcal{M} \cup \mathcal{N}$ is a neighborhood of $F(x_0)$. Since F is upper semicontinuous at x_0, there exists a neighborhood $\mathcal{N}_0(x_0)$ of x_0 such that

$$\text{(11)} \qquad \forall x \in \mathcal{N}_0(x_0), \qquad F(x) \subset \mathcal{M} \cup \mathcal{N}$$

We set $\mathcal{N}(x_0) := \mathcal{N}_0(x_0) \cap \bigcap_{i=1}^{n} \mathcal{N}_{y_i}(x_0)$. Hence, when $x \in \mathcal{N}(x_0)$, properties (10) and (11) imply that

$$\text{(12)} \qquad \begin{cases} \textbf{i.} \quad F(x) \subset \mathcal{M} \cup \mathcal{N} \\ \textbf{ii.} \quad G(x) \cap \mathcal{M} = \varnothing \end{cases}$$

Therefore, $F(x) \cap G(x) \subset \mathcal{N}$ when $x \in \mathcal{N}(x_0)$. ■

COROLLARY 9

Let G be a closed set-valued map from X to a compact space Y. Then G is upper semicontinuous. ▲

Proof. We take F to be defined by $F(x) := Y$ for all $x \in X$, and we apply theorem 8. ■

Corollary 9 provides a very useful tool for proving that a given set-valued map is upper semicontinuous. Note, though, that the assumption of the compactness of Y is essential. Consider the map F from R into the subsets of R^2, defined by $F(\xi) = \{(x, y) : y = \xi x\}$. Then F has closed graph: Assume that

$$\xi_n \to \xi^\star, \qquad (x_n, y_n) \to (x^\star, y^\star), \qquad (x_n, y_n) \in F(\xi_n)$$

Then $y_n = \xi_n x_n$, and passing to the limit, $y^\star = \xi^\star x^\star$, that is, $(x^*, y^*) \in F(\xi^\star)$. However, consider, for instance, $\xi^0 = 0$. For no $\varepsilon > 0$ and for no $\xi \neq 0$ is $F(\xi) \subset F(0) + \varepsilon B$. ■

When X is a complete metric space and Y a compact metric space, a map with closed graph is not only upper semicontinuous but also almost continuous.

THEOREM 10
Let X be a complete metric space, Y a compact metric space, and F a strict set-valued map with closed graph. The subset of points at which F is continuous is residual. ▲

Proof. Since Y is a compact metric space (and thus separable), there exists a countable family of open subsets U_n such that for every open set $\mathcal{U}$ of E and for every $x \in U$, there exists U_n such that $x \in U_n \subset U$.

We set

$$K_n := \{x \in X | F(x) \cap \bar{U}_n \neq \emptyset\}$$

These sets are *closed*, because the graph of F is closed and the subsets $\bar{U}_n$ are compact. We claim that the set of points x at which F is not lower semicontinuous is contained in the union $M := \bigcup_{n=1}^{\infty} \partial K_n$ of the boundaries ∂K_n of the closed subsets K_n. Indeed, to say that F is lower semicontinuous at x amounts to saying that whenever K_n contains x, then x belongs to the interior of K_n. Hence, if F is not lower semicontinuous at x, there exists K_n containing x such that x does not belong to the interior of K_n, that is, such that x belongs to the boundary ∂K_n of K_n. Since K_n is closed, the interior of ∂K_n is empty. Hence, Baire's theorem implies that the interior of $\bigcup_{n=1}^{\infty} \partial K_n$ is also empty. Therefore, the interior of the set of points of discontinuity of F is empty. ■

Remark
We shall see in Section 3 that when X and Y are Banach spaces and the graph of F is *closed and convex*, then F is lower semicontinuous on the interior of its domain.

Finally, we have the following result in proposition 11.

PROPOSITION 11
Let F be a strict upper semicontinuous map with compact values from a compact space X to Y. Then $F(X)$ is compact. ▲

We proceed now with several examples of upper or lower semicontinuous maps.

Example 12
Let X and Y be Hilbert spaces and $A \in \mathcal{L}(X, Y)$. To say that the set-valued map $F := A^{-1}$ is lower semicontinuous amounts to saying that there exists a constant $c > 0$ such that

$$\text{(13)} \qquad \forall y \in Y, \quad d(0, A^{-1}(y)) \leq c\|y\|$$

The fundamental Banach isomorphism theorem (also called the open mapping principle) states that when A is a continuous surjective linear operator from a Banach space X to a Banach space Y, the set-valued map A^{-1} is lower semicontinuous. There is no wonder that extensions of this theorem play an important role: Banach's theorem will be extended to maps with closed convex graph. ■

Example 13

Let X be a Hilbert space, V a proper function from X to $R \cup \{+\infty\}$ and $\mathbf{V}_+$ be the set-valued map defined by

$$\mathbf{V}_+(x) := V(x) + R_+ \quad \text{when} \quad V(x) < +\infty \quad \text{and} \quad \mathbf{V}_+(x) := \varnothing \quad \text{when} \quad V(x) = +\infty$$

It is a well-known fact that V is *lower semicontinuous if and only if* $\mathbf{V}_+$ *has a closed* graph. ■

It will be useful to use the upper semicontinuity of the map that associates to x the epigraph of a function depending on this parameter x.

PROPOSITION 14

Let X and Y be Hausdorff topological spaces and $W: X \times Y \to R$ a lower semicontinuous function. We denote by W_x the function $y \to W_x(y) := W(x, y)$ and by Ep W_x *its epigraph. If for all $x \in X$, $y \to V(x, y)$ is continuous, and if Y is compact, then*

$$x \to \text{Ep } W_x \text{ is upper semicontinuous} \tag{14}$$

▲

Proof. Let $\mathcal{U}$ be an open neighborhood of Ep W_{x_0}. Take $y \in Y$, and fix $\varepsilon_y > 0$ such that the pair $(y, W(x_0, y) - 2\varepsilon_y)$ belongs to $\mathcal{U}$. Then the set

$$\mathcal{U}(y) = \{(z, \lambda) | W(x_0, z) - \lambda < \varepsilon_y\}$$

is open and nonempty. Let $\mathcal{M}(y)$ be the projection of $\mathcal{U}(y)$ onto Y, which is open. Since W is lower semicontinuous, there exist neighborhoods $\mathcal{N}(y) \subset \mathcal{M}(y)$ and $\mathcal{N}_y(x_0)$ such that

$$\forall x \in \mathcal{N}_y(x_0), \qquad \forall z \in \mathcal{N}(y), \qquad W(x_0, y) \leqslant W(x, z) + 2\varepsilon_y \tag{15}$$

Since Y is compact, it can be covered by n neighborhoods $\mathcal{N}(y_i)$. We set $\mathcal{N}_0(x_0) := \bigcap_{i=1}^n \mathcal{N}_{y_i}(x_0)$. Therefore, for all $x \in \mathcal{N}_0(x_0)$ and any $y \in Y$, there exists y_i such that $y \in \mathcal{N}(y_i)$; so by (15), $W(x_0, y_i) \leqslant W(x, y) + 2\varepsilon_{y_i}$. In other words, for all $y \in Y$ and $\lambda \geqslant W(x, y)$, there exists $y_i \in Y$ such that the pair (y_i, λ) satisfies $W(x_0, y_i) - \varepsilon y_i < \lambda$, that is, belongs to the set $\mathcal{U}(y_i)$, which is contained in $\mathcal{U}$. Hence, Ep $W_x \subset \mathcal{U}$ for all $x \in \mathcal{N}_0(x_0)$. ■

Remark

When V is a convex function from an open subset $\mathscr{U}$ of a Hilbert space X to R, we shall prove that the set-valued map $x \to \partial V(x)$ from $\mathscr{U}$ to X^* associating with x its subdifferential $\partial V(x)$ is upper semicontinuous when X^* is supplied with the weak * topology $\sigma(X^*, X)$. This result remains true when V is locally Lipschitz and $\partial V(x)$ is the generalized gradient. ■

Control theory provides set-valued maps of the following type. We consider three sets X, Y, and U and a map f from $X \times U$ to Y. We associate with it the set-valued map F from X to Y defined by

$$F(x) := \{f(x, u) | u \in U\} \tag{16}$$

We say that F is "parametrized."

PROPOSITION 15

Assume that X and Y are Hausdorff topological spaces.

a. *If we suppose that*

$$\forall u \in U, \quad x \to f(x, u) \text{ is continuous} \tag{17}$$

then F is lower semicontinuous.

b. *If we suppose that*

$$\begin{cases} \textbf{i.} & U \text{ is a compact topological space} \\ \textbf{ii.} & f \text{ is continuous from } X \times U \text{ to } Y \end{cases} \tag{18}$$

then F is upper semicontinuous. ▲

Proof.

a. The first statement is obvious.

b. Let $\mathscr{N}$ be an open neighborhood of $F(x_0)$. For any $u \in U$, $\mathscr{N}$ is a neighborhood of $f(x_0, u)$. The continuity of f implies that there exist open neighborhoods $\mathscr{M}_u(x_0)$ and $\mathscr{N}(u)$ such that $f(x, v)$ belongs to $\mathscr{N}$ whenever x belongs to $\mathscr{M}_u(x_0)$ and v to $\mathscr{N}(u)$. We can cover the compact set U by n such open sets $\mathscr{N}(u_i)$. Hence

$$\mathscr{M}(x_0) := \bigcap_{i=1}^{n} \mathscr{M}u_i(x_0)$$

is still a neighborhood, and if v is chosen in U, hence, in some set $\mathscr{N}(u_i)$, then $f(x, v)$ belongs to $\mathscr{N}$. Therefore, $F(x)$ is contained in $\mathscr{N}$ whenever x ranges over $\mathscr{M}(x_0)$. ■

The following theorem allows the construction of lower semicontinuous maps.

THEOREM 16
Let K be a Hausdorff topological space, Y and Z Banach spaces, $f: K\times Y\to Z$ a continuous map that is affine with respect to the second variable. Let U and T be lower semicontinuous maps from K to Y and Z, respectively, with convex values. We suppose that U is locally bounded (U is bounded on some neighborhood of each point). We assume that there exists $\gamma>0$ such that

$$\text{(19)}\qquad \begin{cases}\forall x\in K,\quad \forall z\in Z,\quad \textit{satisfying } \|z\|\leqslant\gamma,\quad \exists y\in U(x)\quad \textit{such that}\\ \qquad f(x,y)+z\in T(x)\end{cases}$$

Then the set-valued map C defined by

$$\text{(20)}\qquad C(x):=\{y\in U(x)\,|\,f(x,y)\in T(x)\}$$

is lower semicontinuous (with convex values). ▲

Proof. Let $y\in C(x_0)$ and $\varepsilon>0$. We have to prove that there exists a neighborhood $\mathcal{N}(x_0)$ of $x_0\in K$ such that

$$\text{(21)}\qquad \forall x\in N(x_0),\quad \exists\bar{y}_x\in C(x)\cap(y_0+\varepsilon B)$$

Let $\mathcal{N}_0(x_0)$ be the neighborhood on which U is bounded, δ equal to

$$\sup\{\text{diam}\,(U(x))|x\in\mathcal{N}_0(x_0)\}$$

which is finite, and $\varepsilon<2\delta$. We set $\alpha:=\gamma\varepsilon/(2\delta-\varepsilon)$. Since f is continuous and U and T are lower semicontinuous, we can find a neighborhood

$$\mathcal{N}(x_0)\subset\mathcal{N}_0(x_0)\qquad\text{and}\qquad \eta<\varepsilon/2$$

such that $\forall x\in\mathcal{N}(x_0)$,

$$\text{(22)}\qquad \begin{cases}\textbf{i.}\quad \textit{If}\ \|y_0-y\|<\eta,\quad \text{then}\ \|f(x,y)-f(x_0,y_0)\|<\dfrac{\alpha}{2}\\ \textbf{ii.}\quad y\in U(x)+\eta B\\ \textbf{iii.}\quad f(x_0,y)\in T(x)+\dfrac{\alpha}{2}B\end{cases}$$

Hence,

$$\text{(23)}\qquad \forall x\in\mathcal{N}(x_0),\qquad \exists y_x\in U(x)\quad\text{such that}\quad f(x,y_x)\in T(x)+\alpha B$$

Assumption (19) can be written

$$\forall x \in K, \qquad \gamma B \subset T(x) - f(x, U(x)) \tag{24}$$

Let us set $\theta := \gamma/(\alpha+\gamma) < 1$ and multiply inclusion (23) by θ. By noticing that $\theta\alpha = (1-\theta)\gamma$, we obtain

$$\theta f(x, y_x) \in \theta T(x) + (1-\theta)\gamma B \tag{25}$$

We multiply inclusion (24) by $(1-\theta)$ and obtain

$$(1-\theta)\gamma B \subset (1-\theta)T(x) - (1-\theta)f(x, U(x))$$

by using this result in (25), we obtain

$$\theta f(x, y_x) \in \theta T(x) + (1-\theta)T(x) - (1-\theta)f(x, U(x)) \tag{26}$$

Since $T(x)$ is convex and f is affine with respect to y, we have proved the existence of $y_1 \in U(x)$ such that

$$\theta f(y, \theta y_x + (1-\theta)y_1) \in T(x)$$

that is, such that

$$\bar{y}_x := \theta y_x + (1-\theta)y_1 \in C(x)$$

It remains to see that

$$\|y_0 - \bar{y}_x\| \leqslant \|y_0 - y_x\| + (1-\theta)\|y_x - y_1\| \leqslant \frac{\varepsilon}{2} + \frac{\alpha}{\alpha+\gamma}\|y_x - y_1\| \leqslant \frac{\varepsilon}{2} + \frac{\alpha}{\alpha+\gamma}\delta = \varepsilon$$

So, C is lower semicontinuous. ■

It is much easier to get upper semicontinuous maps in this way.

PROPOSITION 17

Let K be a Hausdorff topological space, Y and Z Banach spaces, f: $K \times Y \to Z$ a continuous single-valued map. Let U and T be set-valued maps from K to Y and X, respectively, both with closed graph. Then take set-valued map C: $K \to Y$ defined by

$$C(x) := \{y \in U(x) \mid f(x, y) \in T(x)\}$$

has a closed graph. ▲

Proof. The proof is an easy exercise. ∎

We recall that given a cone T, we denote by T^- its negative polar cone

(27) $$T^- := \{p \in X^\star \mid \forall v \in T, \langle p, v\rangle \leq 0\}$$

PROPOSITION 18

Let X be a finite dimensional space and $T(\cdot)$ be a set-valued map associating with any $x \in K$ a closed convex cone $T(x)$. Let $N(\cdot)$ be the set-valued map $x \to N(x) := T(x)^-$. The following conditions are equivalent:

(28) *The set-valued map $N(\cdot)$ has a closed graph.*

(29) *The set-valued map $T(\cdot)$ is lower semicontinuous.* ▲

Proof. **a.** Let us assume that $T(\cdot)$ is lower semicontinuous. Let $(x_n, p_n) \in$ graph $(N(\cdot))$ be a sequence converging to (x, p). To prove that $p \in N(x)$, let us choose any $v \in T(x)$ and check that $\langle p, v\rangle \leq 0$. Since $T(\cdot)$ is lower semicontinuous, $v = \lim_{n\to\infty} v_n$ where $v_n \in T(x_n)$. Since $\langle p_n, v_n\rangle \leq 0$ (for $p_n \in N(x_n)$), we deduce that $\langle p, v\rangle = \lim_{n\to\infty} \langle p_n, v_n\rangle \leq 0$. Hence, (x, p) belongs to graph $(N(\cdot))$, and thus N has a closed graph.

b. Let us assume that the graph of $N(\cdot)$ is closed; let x_n converge to x and $v \in T(x)$. We shall prove that the projection

$$v_n := \pi_{T(x_n)}(v) \in T(x_n)$$

converges to v. Indeed, we know that

$$p_n := \pi_{N(x_n)}(v) \in N(x_n)$$

and that

$$\|p_n\| \leq \|v\|$$

Since X is a finite dimensional space, some subsequence (again denoted by) p_n converges to $p \in X$. Since the graph of $N(\cdot)$ is closed, we deduce that $p \in N(x)$. Hence, the subsequence $v_n = v - p_n$ converges to $v - p$. Since $\langle p_n, v_n\rangle = 0$, we deduce that $\langle p, v-p\rangle = 0$, that is, $\|p\|^2 - \langle p, v\rangle \leq 0$. Thus $p = 0$ and $v = \lim_{n\to\infty} v_n$. ∎

Let X and Y be two sets, G a set-valued map from Y to X, and W a real-valued function defined on $X \times Y$. We consider the family of maximization problems

(30) $$V(y) := \sup_{x \in G(y)} W(x, y)$$

that depend on the parameter y. The function V is called the *marginal function* and the set-valued map M associating to the parameter $y \in Y$ the set

$$M(y) := \{x \in G(y) \mid V(y) = W(x, y)\} \tag{31}$$

of solutions to the maximization problem $V(y)$ is called the *marginal map*.

By the *stability* of this family of optimization problems, we usually mean the study of various continuity properties of the marginal function V and the marginal map M.

We begin with proposition 19.

PROPOSITION 19

Suppose that

$$\begin{cases} \text{i.} & W \text{ is lower semicontinuous on } X \times Y. \\ \text{ii.} & G \text{ is lower semicontinuous at } y_0. \end{cases} \tag{32}$$

Then the marginal function V is lower semicontinuous at y_0. ▲

Proof. Fix $\varepsilon > 0$, and choose $x_0 \in G(y_0)$ such that $V(y_0) - \varepsilon/2 \leqslant W(x_0, y_0)$. Since W is lower semicontinuous, there exist neighborhoods $\mathcal{N}(x_0)$ and $\mathcal{M}_1(y_0)$ such that

$$\forall x \in \mathcal{N}(x_0), \quad \forall y \in \mathcal{M}_1(y_0), \quad W(x_0, y_0) - \frac{\varepsilon}{2} \leqslant W(x, y) \tag{33}$$

Since G is lower semicontinuous, there exists a neighborhood $\mathcal{M}_2(y_0)$ such that

$$\forall y \in \mathcal{M}_2(y_0), \quad G(y) \cap \mathcal{N}(x_0) \neq \varnothing \tag{34}$$

So we take

$$\mathcal{M}(y_0) := \mathcal{M}_1(y_0) \cap \mathcal{M}_2(y_0)$$

For any $y \in \mathcal{M}$, we know that there exists $x \in G(y)$ such that

$$W(x, y) \geqslant W(x_0, y_0) - \frac{\varepsilon}{2} \geqslant V(y_0) - 2\frac{\varepsilon}{2}$$

Hence V is lower semicontinuous at y_0. ■

We single out a useful consequence in corollary 20.

COROLLARY 20

Let Y be a metric space. If $G: X \to Y$ is lower semicontinuous, then the function

$$(x, y) \in X \times Y \to d(G(x), y) \tag{35}$$

is upper semicontinuous. ▲

We consider now the case when W and G are upper semicontinuous.

PROPOSITION 21

Suppose that

$$\begin{cases} \textbf{i.} & W \text{ is upper semicontinuous on } X \times Y. \\ \textbf{ii.} & G(y_0) \text{ is compact, and } G \text{ is upper semicontinuous at } y_0. \end{cases} \tag{36}$$

Then the marginal function V is upper semicontinuous at y_0. ▲

Proof. We have to prove that for any $\varepsilon > 0$, there exists a neighborhood $\mathcal{N}(y_0)$ of y_0 such that

$$\forall y \in \mathcal{N}(y_0), \qquad V(y) \leq V(y_0) + \varepsilon \tag{37}$$

Since W is upper semicontinuous, we can associate with any $x \in X$ open neighborhoods $\mathcal{N}(x)$ of x and $\mathcal{N}_x(y_0)$ such that

$$\forall x' \in \mathcal{N}(x), \qquad \forall y \in \mathcal{N}_x(y_0), \qquad W(x', y) \leq W(x, y_0) + \varepsilon \tag{38}$$

Since $G(y_0)$ is compact, it can be covered by n neighborhoods $\mathcal{N}(x_i)$. Therefore,

$$\mathcal{N} := \bigcup_{i=1}^{n} \mathcal{N}(x_i) \text{ is a neighborhood of } G(y_0) \tag{39}$$

The set-valued G being upper semicontinuous, there exists a neighborhood $\mathcal{N}_0(y_0)$ such that

$$\forall y \in \mathcal{N}_0(y_0), \qquad G(y) \subset \mathcal{N} \tag{40}$$

We consider the following neighborhood of y_0

$$\mathcal{N}(y_0) := \mathcal{N}_0(y_0) \cap \bigcap_{i=1}^{n} \mathcal{N}_{x_i}(y_0)$$

When y belongs to $\mathcal{N}(y_0)$ and x belongs to $G(y)$, then x belongs to $\mathcal{N}$ and thus to some $\mathcal{N}(x_i)$. Hence, since y belongs to $\mathcal{N}_{x_i}(y_0)$, we deduce from (38) that

$$W(x, y) \leq W(x_i, y_0) + \varepsilon \leq V(y_0) + \varepsilon$$

Hence, by taking the supremum when x ranges over $G(y)$, we obtain inequality (37). ■

We point out the following often used consequence in corollary 22.

COROLLARY 22
Let X be a Hausdorff topological space, Y a compact space, and W an upper semicontinuous function from $X \times Y$ to R. Then the marginal function V defined on Y by

$$V(y) := \sup_{y \in Y} W(x, y) \tag{41}$$

is upper semicontinuous. ▲

When the function W and the set-valued map G are continuous, so is the marginal function V, thanks to the preceding theorems. We now investigate the behavior of the sets of maximizers

$$M(y) := \{x \in G(y) \mid V(y) = W(x, y)\}$$

PROPOSITION 23
Suppose that

$$\begin{cases} \textbf{i.} & W \textit{ is continuous on } X \times Y. \\ \textbf{ii.} & G \textit{ is continuous with compact values.} \end{cases} \tag{42}$$

Then the marginal function V is continuous, and the marginal set-valued map M is upper semicontinuous. ▲

Proof. We note that $M(y) = G(y) \cap K(y)$ where

$$K(y) := \{x \in X \mid V(y) = W(x, y)\}$$

Since V and W are continuous functions, the graph of K is closed. The subsets $M(y)$ are obviously nonempty. Since G is upper semicontinuous, theorem 8 implies that M is also upper semicontinuous. ■

Finally, we consider the case when W and G are both Lipschitz.

PROPOSITION 24
Suppose that

$$(43) \qquad \begin{cases} \textbf{i.} & W \textit{ is Lipschitz on } X \times Y \textit{ with Lipschitz constant } \ell. \\ \textbf{ii.} & G \textit{ is Lipschitz with Lipschitz constant } c. \end{cases}$$

Then the marginal function V is Lipschitz with Lipschitz constant $\ell(c+1)$. ▲

Proof. We take $y_i \in Y$ and we choose $x_1 \in G(y_1)$ satisfying $V(y_1) \leqslant W(x_1, y_1) + \varepsilon$. By assumption (43), there exists $x_2 \in G(y_2)$ such that

$$\begin{aligned} \|x_1 - x_2\| &\leqslant d(x_1, G(y_2)) + \varepsilon \\ &\leqslant c\|y_1 - y_2\| + \varepsilon \end{aligned}$$

Since $V(y_2) \geqslant W(x_2, y_2)$, we deduce that

$$\begin{aligned} V(y_1) - V(y_2) &\leqslant W(x_1, y_1) - W(x_2, y_2) + \varepsilon \\ &\leqslant \ell(\|x_1 - x_2\| + \|y_1 - y_2\|) + \varepsilon \\ &\leqslant \ell(c+1)\|y_1 - y_2\| + (\ell + 1)\varepsilon \end{aligned}$$

Since ε is arbitrary, it follows that

$$V(y_1) - V(y_2) \leqslant \ell(c+1)\|y_1 - y_2\|$$ ■

2. MAPS WITH CLOSED CONVEX VALUES

Let X be a Hausdorff topological space and Y a Hausdorff locally convex vector space. We associate with a set-valued map F from X to Y its *support function* defined by

$$(1) \qquad \forall x \in X, \qquad \forall p \in Y^\star, \qquad \sigma(F(x), p) := \sup_{y \in F(x)} \langle p, y \rangle$$

DEFINITION 1
We say that F is upper hemicontinuous at $x_0 \in \operatorname{Dom} F$ if for any $p \in Y^\star$, the function $x \to \sigma(F(x), p)$ is upper hemicontinuous at x_0. The map F is said to be upper hemicontinuous if it is upper hemicontinuous at any point of its domain. ▲

Upper semicontinuous maps are upper hemicontinuous.

PROPOSITION 2

We supply Y with the weak topology $\sigma(Y, Y^\star)$. Any set-valued map from X to Y upper semicontinuous at x_0 is upper hemicontinuous at x_0. ▲

Proof. Let $p \neq 0$ belong to $Y^\star$ and $\varepsilon > 0$ be fixed. The subset

$$B_p(\varepsilon) := \{ y \in Y \mid \langle p, y \rangle \leqslant \varepsilon \}$$

is a neighborhood of 0 for the weak topology. Since F is upper semicontinuous at x_0, there exists a neighborhood $\mathcal{N}(x_0)$ of x_0 such that

$$\forall x \in \mathcal{N}(x_0), \qquad F(x) \subset F(x_0) + B_p(\varepsilon)$$

By taking the support functions, we obtain the inequality

$$\forall x \in \mathcal{N}(x_0), \qquad \sigma(F(x), p) \leqslant \sigma(F(x_0), p) + \varepsilon$$

which expresses the upper semicontinuity of the function $x \to \sigma(F(x), p)$ at x_0. ■

Remark

The converse is not true. For example, consider the map F from R to R^2 defined by

$$F(x) := \{ (y_1, y_2) \mid y_2 \geqslant (1 + x) y_1 \}$$

Then F is upper hemicontinuous at $x = 0$. Indeed, fix $p = (p_1, p_2)$. Either $p_2 > 0$, and there is nothing to prove, because $\sigma(F(0), p) = +\infty$, or $p_2 < 0$. In the latter case, an easy calculation shows that

$$\sigma(F(x), p) = \frac{3}{4} p_1^2 [p_2(1 + x)]^{-1}$$

is continuous at $x = 0$. However, it is easy to see that F is not upper semicontinuous at $x = 0$. ■

Actually, the converse becomes true if we make the additional assumption that the images $F(x)$ of F are *convex and weakly compact*. This theorem, due to Castaing, is proved in the last part of Section 2.

We now mention some useful elementary facts.

PROPOSITION 3

A finite sum and a finite product of upper hemicontinuous set-valued maps is upper hemicontinuous. ▲

PROPOSITION 4
If X is compact and F is upper hemicontinuous with bounded values, then $F(X)$ is bounded in Y. ▲

Proof. Since the images of F are bounded, the functions $x \to \sigma(F(x), p)$ are finite. Since X is compact and F is upper hemicontinuous, then the function ϕ defined on $Y^\star$ by

$$(2) \qquad \forall p \in Y^\star, \qquad \phi(p) := \sup_{x \in X} \sigma(F(x), p) < +\infty$$

is a *finite* positively homogeneous lower semicontinuous convex function: It is the support function of a *bounded* closed convex subset K of Y. ■

PROPOSITION 5
The graph of any upper hemicontinuous map with closed convex values is closed in $X \times Y$ (when Y is supplied with the weak topology). ▲

Proof. Let us consider a generalized sequence of elements (x_μ, y_μ) of graph (F) converging to (x, y) in $X \times Y$. Since for all $p \in Y^\star$, $\langle p, y_\mu \rangle \leqslant \sigma(F(x_\mu), p)$ and $x \to \sigma(F(x), p)$ is upper semicontinuous, we deduce by taking the limit that

$$\langle p, y \rangle = \lim_\mu \langle p, y_\mu \rangle \leqslant \limsup_\mu \sigma(F(x_\mu), p) \leqslant \sigma(F(x), p)$$

Hence,

$$y \in \overline{\text{co}}(F(x) = F(x)$$ ■

The following theorem plays an important role in the theory of differential inclusions as well as in other domains.

THEOREM 6 (CONVERGENCE THEOREM)
Let F be an upper hemicontinuous set-valued map with closed convex values from a Banach space X to a Banach space Y. Let Ω be a measured space. We consider two sequences of functions $x_n(\cdot)$ belonging to $L^p(\Omega, X)$ and $y_n(\cdot) \in L^q(\Omega, Y)(p, q \geqslant 1)$ satisfying

$$(3) \qquad \begin{cases} \text{for almost all } \omega \in \Omega, \text{ for every } \varepsilon > 0, \text{ there exists an integer} \\ N := N(\omega, \varepsilon) \text{ such that} \\ \forall n \geqslant N, \qquad d((x_n(\omega), y_n(\omega)), \text{graph }(F)) \leqslant \varepsilon \end{cases}$$

If we assume that

$$(4) \quad \begin{cases} \textbf{i.} & x_k(\cdot) \text{ converges strongly to } x(\cdot) \text{ in } L^p(\Omega, X). \\ \textbf{ii.} & y_k(\cdot) \text{ converges weakly to } y(\cdot) \text{ in } L^q(\Omega, Y). \end{cases}$$

then we conclude that

$$(5) \quad \text{for almost all } \omega \in \Omega,\ y(\omega) \in F(x(\omega)) \qquad ▲$$

Proof. **a.** We recall that for convex subsets of a Banach space (here, $L^q(\Omega, Y)$), the strong closure coincides with the weak closure (for the weakened topology $\sigma(L^q(\Omega, Y), L^{q^*}(\Omega, Y^\star))$ with $1/q+1/q^\star=1$). We use this fact. For any integer n, the weak limit $y(\cdot)$ belongs to the weak closure of the convex hull $\text{co}\{y_m(\cdot)\}_{m\geq n}$ of the subsets $\{y_m(\cdot)\}_{m>n}$. Hence, we can choose functions

$$(6) \quad z_n(\cdot) = \sum_{m=n}^{\infty} a_n^m y_m(\cdot)$$

(where the coefficients a_n^m are nonnegative, equal to zero except for a finite number, and $\sum_{m=n}^{\infty} a_n^m = 1$) belonging to this convex hull such that

$$\|y(\cdot) - z_n(\cdot)\|_{L^q(\Omega, Y)} \leq \frac{1}{n}$$

In other words, the sequence $z_n(\cdot)$ converges to $y(\cdot)$ *strongly* in $L^q(\Omega, Y)$.
b. Since $x_n(\cdot)$ converges strongly to $x(\cdot)$ in $L^p(\Omega, X)$ and $z_n(\cdot)$ converges strongly to $y(\cdot)$ in $L^q(\Omega, Y)$, there exist subsequences (again denoted by) $x_n(\cdot)$ and $z_n(\cdot)$ that converge to $x(\cdot)$ and $y(\cdot)$ for almost all $\omega \in \Omega$.

Let us choose $\omega \in \Omega$ such that $x_n(\omega)$ converges to $x(\omega)$, $z_n(\omega)$ converges to $y(\omega)$, and assumption (3) holds true. Let us fix $\varepsilon > 0$ and $p \in Y^\star$. Since F is upper hemicontinuous, there exists $\eta \in]0, \varepsilon/2\|p\|_\star[$ such that

$$(7) \quad \|u - x(\omega)\| \leq 2\eta \Rightarrow \sigma(F(u), p) \leq \sigma(F(x(\omega), p) + \frac{\varepsilon}{2}$$

Assumption (3) implies that we can associate to η an integer N such that

$$(8) \quad \begin{cases} \forall m \geq N, \quad \exists (u_m, v_m) \in \text{graph}(F) \text{ such that} \\ \|x_n(\omega) - u_n\| \leq \eta \quad \text{and} \quad \|z_m(\omega) - v_m\| \leq \eta \end{cases}$$

Since $x_m(\omega)$ converges to $x(\omega)$, there exists $N_1 \geq N$ such that

$$\|x_m(\omega) - x(\omega)\| \leq \eta \quad \text{when} \quad m \geq N_1$$

Hence, (8) implies that

$$\forall m \geqslant N_1, \quad \langle p, y_n(\omega)\rangle \leqslant \langle p, w_m\rangle + \eta\|p\|_\star$$
$$\leqslant \sigma(F(u_m, p) + \eta\|p\|_\star$$

and (7) implies that

$$\sigma(F(u_m), p) \leqslant \sigma(F(x(\omega)), p) + \frac{\varepsilon}{2}$$

because $\|\dot{u}_n - x(t)\| \leqslant 2\eta$. Therefore, since $\eta \leqslant \varepsilon/2\|p\|_\star$ we obtain

$$(9) \qquad \forall m \geqslant N_1, \qquad \langle p, y_m(\omega)\rangle \leqslant \sigma(F(x(\omega), p) + \varepsilon$$

Let n be larger that N_1. By multiplying inequalities (9) by $a_n^m \geqslant 0$ for all $m \geqslant n$ and adding them together, we deduce that

$$(10) \qquad \forall n \geqslant N_1, \qquad \langle p, z_n(\omega)\rangle \leqslant \sigma(F(x(\omega), p) + \varepsilon$$

By taking the limit when $n \to \infty$ and $\varepsilon \to 0$, we obtain

$$\forall p \in Y^\star, \qquad \langle p, y(\omega)\rangle \leqslant \sigma(F(x(\omega), p)$$

that is, $y(\omega) \in \overline{\text{co}}\, F(x(\omega)) = F(\omega)$. ■

As a first consequence, we deduce the following corollary.

COROLLARY 7

Let F be an upper hemicontinuous map from a Banach space X to a Banach space Y and Ω a measured space. We supply $L^p(\Omega, X)$ with the strong topology and $L^q(\Omega, Y)$ with the weak topology. Then the graph of the map $\mathscr{F}$ from $L^p(\Omega, X)$ to $L^q(\Omega, Y)$ defined by

$$(11) \qquad \mathscr{F}(x(\cdot)) := \{y \in L^q(\Omega, Y) \mid \textit{for almost all } \omega \in \Omega, \ y(\omega) \in F(x(\omega))\}$$

is closed in $L^p(\Omega, X) \times L^q(\Omega, Y)$. ▲

Proof. We take a sequence of elements $(x_n(\cdot), y_n(\cdot))$ belonging to graph $(\mathscr{F})$ satisfying (4). Since property (3) holds true, we deduce that the limit $(x(\cdot), y(\cdot))$ belongs to graph $(\mathscr{F})$. ■

COROLLARY 8

We consider two Banach spaces X and Y, a compact convex subset K of Y and a *lower semicontinuous function* $W: X \times K \to R$ *satisfying*

(12) $\quad \forall x \in X, \quad y \to W(x, y)$ *is convex and continuous*

Let us consider sequences of functions $x_n \in L^p(\Omega, X)$, $y_n \in L^q(\Omega, K)$ *and* $w_n \in L^q(\Omega)$ *such that*

(13)
$$\begin{cases} \text{i.} & x_n \text{ converges strongly to } x \text{ in } L^p(\Omega, X). \\ \text{ii.} & y_n \text{ converges weakly to } y \text{ in } L^q(\Omega, Y). \\ \text{iii.} & w_n \text{ converges weakly to } v \text{ in } L^q(\Omega). \end{cases}$$

We assume that

(14)
$$\begin{cases} \text{for almost all } \omega \in \Omega, \text{ for every } \varepsilon > 0, \text{ there exists an integer} \\ N \text{ such that } W(x_n(\omega), y_n(\omega)) \leqslant w_n(\omega) + \varepsilon \text{ for all } n \geqslant N \end{cases}$$

Then

(15) $\quad$ *for almost all* $\omega \in \Omega, \quad W(x(\omega), y(\omega)) \leqslant v(\omega)$ ▲

Proof. We apply theorem 6 to the set-valued map F from X to $Y \times R$ defined by

$$F(x) := \mathrm{Ep}\,(W_x)$$ ■

A slight variant of the proof of corollary 8 by means of the convergence theorem and proposition 14 yields the following useful proposition.

PROPOSITION 9

We consider two Banach spaces X and Y, a compact subset K of Y and a nonnegative lower semicontinuous function $W: X \times K \to R_+$ *satisfying*

(16) $\quad \forall x \in X, \quad y \to W(x, y)$ *is convex and continuous*

We define the following functional $\underset{\sim}{W}$ *on* $L^p(\Omega, X) \times L^q(\Omega, Y)$

(17)
$$\underset{\sim}{W}(x(\cdot), y(\cdot)) = \int_\Omega W(x(\omega), y(\omega))\,d\mu(\omega) \in [0, +\infty]$$

This functional is lower semicontinuous when $L^p(\Omega, X)$ *is supplied with the strong topology and* $L^q(\Omega, Y)$ *is supplied with the weak topology.* ▲

Proof. We take a sequence of functions $x_n(\cdot)$ converging strongly to $x(\cdot)$ in $L^p(\Omega, X)$ and a sequence of functions $y_n(\cdot)$ converging weakly to $y(\cdot)$ in $L^q(\Omega, Y)$, then consider $v:=\liminf_{n\to\infty} \underset{\sim}{W}(x_n(\cdot), y_n(\cdot))$. If $v=+\infty$, the theorem is true. If not, there exist subsequences (again denoted by) $x_n(\cdot)$ and $y_n(\cdot)$ such that

$$\forall n, \qquad \underset{\sim}{W}(x_n(\cdot), y_n(\cdot)) \leqslant v + \frac{1}{n} \tag{18}$$

Introduce the sequence of functions $z_n(\cdot)$ defined by (6) that converges *strongly* to $y(\cdot)$ in $L^q(\Omega, Y)$. Therefore, there exist subsequences (again denoted by) $x_n(\cdot)$ and $z_n(\cdot)$ such that for almost all $\omega \in \Omega$, $x_n(\omega)$ converges to $x(\omega)$ and $z_n(\omega)$ converges to $y(\omega)$.

Let W_x denote the function $y \to W_x(y) := W(x, y)$. Proposition 14 implies that the map $x \to \text{Ep}\,(W_x)$ is upper semicontinuous. Therefore, for every $\eta > 0$, there exists $N := N(\eta, x(\omega))$ such that

$$\forall n \geqslant N, \qquad (y_n(\omega), W(x_n(\omega), y_n(\omega))) \in \text{Ep}\,(W_x) + \eta(B \times B)$$

where $B \times B$ denotes the unit ball in $Y \times R$.

Since the epigraph of W_x is convex by assumption (16), we obtain by setting

$$v_n(\omega) := \sum_{m=n}^{\infty} a_n^m W(x_m(\omega), y_m(\omega)) \tag{19}$$

that

$$\forall n \geqslant N, \qquad (z_n(\omega), v_n(\omega)) \in \text{Ep}\,(W_x) + \eta(B \times B) \tag{20}$$

We introduce

$$v(\omega) := \liminf_{n \to \infty} v_n(\omega) \tag{21}$$

We can associate to ε an $\eta \leqslant \varepsilon$ such that

$$W(x(\omega), y(\omega)) \leqslant W(x(\omega), z) + \varepsilon \quad \text{when} \quad \|z - y(\omega)\| \leqslant 2\eta$$

Let $N_1 \geqslant N$ be chosen such that $\|z_n(\omega) - y(\omega)\| \leqslant \eta$ when $n \geqslant N_1$. By (20), there exist $\bar{z}_n \in z_n(\omega) + \eta B$ such that

$$W(x(\omega), \bar{z}_n) \leqslant v_n(\omega) + \eta \leqslant v_n(\omega) + \varepsilon$$

Finally, by (21), there exists $n \geqslant N_1$ such that $v_n(\omega) \leqslant v(\omega) + \varepsilon$. Therefore,

$$W(x(\omega), y(\omega)) \leqslant v(\omega) + 2\varepsilon$$

By letting ε converge to zero, we have proved that

$$\text{(22)} \qquad \text{for almost all } \omega \in \Omega, \qquad W(x(\omega), y(\omega)) \leqslant \liminf_{n \to \infty} v_n(\omega)$$

We integrate this inequality and apply Fatou's lemma, which is possible because the function W is nonnegative.

We obtain

$$\underset{\sim}{W}(x(\cdot), y(\cdot)) \leqslant \int_\Omega \liminf_{n \to \infty} v_n(\omega) d\mu(\omega)$$

$$\leqslant \liminf_{n \to \infty} \int_\Omega v_n(\omega) d\mu(\omega)$$

But, we observe that by (18) and (19),

$$\int_\Omega v_n(\omega) d\mu(\omega) = \sum_{m=n}^{\infty} a_n^m \underset{\sim}{W}(x_m(\cdot), y_m(\cdot)) \leqslant v + \frac{1}{n}$$

so that

$$\underset{\sim}{W}(x(\cdot), y(\cdot)) \leqslant v := \liminf_{n \to \infty} \underset{\sim}{W}(x_n(\cdot), y_n(\cdot)) \qquad \blacksquare$$

We now prove the announced partial converse of proposition 2.

THEOREM 10

Let F be a strict upper hemicontinuous map from X to Y. If $F(x_0)$ is convex and weakly compact, F is upper semicontinuous at x_0. ▲

We now need lemma 11.

LEMMA 11

We posit the assumptions of theorem 10. Let $\mathscr{U}$ be a weakly open set containing $F(x^0)$. Then there is a finite set of pairs (p_i, ε_i), such that

$$\mathscr{W} = \{ y \mid \langle p_i, y \rangle \leqslant \sigma(F(x^0), p_i) + \varepsilon_i : i = 1, \ldots, m \}$$

is contained in $\mathscr{U}$. ▲

Proof of Theorem 10. Let $\mathscr{U}$ be a neighborhood of $F(x_0)$. Since $F(x_0)$ is weakly compact, there exists a finite set of pairs (p_i, ε_i) such that $\mathscr{W}$ is contained in $\mathscr{U}$ by lemma 11. So, it remains to note that since F is upper hemicontinuous at x_0, there exists a neighborhood $\mathcal{N}(x_0)$ of x_0 such that for all $x \in \mathcal{N}(x_0)$, $F(x) \subset \mathcal{W}$, and, consequently, $F(x) \subset \mathcal{U}$. ■

Proof of Lemma 11. We proceed in two steps.

a. *Case when* $Y = R^n$. We can replace $\mathcal{U}$ by $F(x_0) + B$, since $F(x_0)$ is compact. Set K to be the complement of $\mathcal{U}$ in the closure of $F(x^0) + 2B$: K is a compact set.

Fix any $p \in Y^\star$, $\|p\| = 1$ and any $\varepsilon \in]0, 1[$. The set

$$K(p, \varepsilon) := \{y | \langle p, y\rangle \leqslant \sigma(F(x^0), p) + \varepsilon\} \cap K$$

is (compact and) nonempty: Consider a point $y_m \in F(x^0)$ such that

$$\langle p, y_m\rangle = \inf\{\langle p, y\rangle | y \in F(x^0)\}$$

and add a vector $(-2z)$, where z is such that $\langle p, z\rangle = \|z\| = 1$. Then the distance of $y_m - 2z$ from $F(x^0)$ is exactly 2; hence, it belongs to K. Assume that lemma 11 is false. Then the family $\{K(p, \varepsilon)\}$ has the finite intersection property, that is, no matter which finite set of pairs (p_i, ε_i) we choose, the set

$$\bigcap_i K(p_i, \varepsilon_i) \neq \varnothing$$

Since K is compact, the intersection over all the $K(p, \varepsilon)$ must be nonempty. Let ξ be in this intersection. By a basic separation argument, there are $\bar{p}$, $\bar{\varepsilon}$ such that $\langle \bar{p}, \xi\rangle > \sigma(F(x^0), \bar{p}) + \bar{\varepsilon}$. Hence, $\xi \notin K(\bar{p}, \bar{\varepsilon})$, a contradiction. This proves the lemma.

b. *General case.* For every $y \in F(x_0)$, let $(P, E)_y$ be a finite set of pairs (p_i, ε_i) such that $y + N(P, E)_y \subset \mathcal{U}$, where

$$N(P, E)_y := \{z | \, |\langle p_i, z\rangle| < \varepsilon_i \text{ for all } i\}$$

Since $F(x_0)$ is compact, let $\{y_j + N(P, E)_j\}$ be a finite subcover of $F(x_0)$. [We set $N(P, E)_j := N(P, E)_{y_j}$.] Consider the finite set of all functionals $\{p_{ij}\}$ for all points y_i, and write $Y = M + N$, where M is a finite dimensional space and N the intersection of all the kernels of these functionals. Since M is a finite dimensional space, the result is true on M. Set Π to be the projection onto M. For each j, $\Pi(y_j + N(P, E)_j)$ is open, and the union over j covers $\Pi(F(x^0))$.

On M, consider the bounded closed convex set $\Pi F(x^0)$, its open neighborhood $\cup_j \Pi(y_j + N(P, E)_j)$, and let $\sigma^M(\Pi F(x^0), p)$ be its support function. Then there is a finite set of pairs (P^M, E) with $p_i^M \in M^\star$, such that the set

(23) $\{y \in M | \langle p_i^M, y\rangle < \sigma^M(\Pi F(x^0), p_i^M) + \varepsilon_i\}$

is contained in $\bigcup_j (y_j + N(P, E)_j)$.

Extend each p_i^M to a $p_i \in Y^\star$ by

$$\langle p_i, y\rangle = \langle p_i, m+n\rangle := \langle p_i^M, m\rangle$$

Then it is easy to verify that

$$\sigma(F(x^0), p_i) = \sigma^M(\Pi F(x^0), p_i^M)$$

We claim that the set

$$\mathscr{W} := \{y \mid \langle p_i, y\rangle < \sigma(F(x^0), p_i) + \varepsilon_i : i = 1, \ldots, m\}$$

is contained in $\mathscr{U}$. Let $\bar{y}$ belong to it, and consider $\Pi\bar{y}$. For every i,

$$\langle p_i^M, \Pi\bar{y}\rangle = \langle p_i, \bar{y}\rangle < \sigma(F(x^0), p_i) + \varepsilon_i = \sigma^M(\Pi F(x^0), p_i^M) + \varepsilon_i$$

By (23), there exists some $j^\star$ such that

$$\Pi\bar{y} \in \Pi(y_{j^*} + N(P, E)_{j^*}), \quad \text{that is,} \quad \Pi\bar{y} = y_{j^*} + \Pi v$$

with $v \in N(P, E)_{j^*}$. Fix $(p_{ij^*}, \varepsilon_i)$ in $(P, E)_{j^*}$. Note that since N is contained in the null space of p_{ij^*} for every vector ξ, $\langle p_{ij^*}, \xi\rangle = \langle p_{ij^*}, \Pi\xi\rangle$. Hence,

$$\langle p_{ij^*}, \bar{y}\rangle = \langle p_{ij^*}, \Pi\bar{y}\rangle = \langle p_{ij^*}, \Pi y_{j^*} + \Pi v\rangle < \langle p_{ij^*}, y_{j^*}\rangle + \varepsilon_j$$

This means that

$$\bar{y} \in y_{j^*} + N(P, E)_{j^*} \subset \mathscr{U} \qquad \blacksquare$$

3. MAPS WITH CLOSED CONVEX GRAPHS

We observe that the graph of a set-valued map F from a Banach space X to a Banach space Y is *convex* if and only if for any convex combination $\sum_{i=1}^n \lambda_i x_i$ of elements $x_i \in X$, we have

$$\sum_{i=1}^n \lambda_i F(x_i) \subset F\left(\sum_{i=1}^n \lambda_i x_i\right)$$

Set-valued maps satisfying the opposite inclusion

$$\sum_{i=1}^n \lambda_i F(x_i) \supset F\left(\sum_{i=1}^n \lambda_i x_i\right)$$

for all convex combinations were introduced by Ioffe, who called them *fans*.

Let F be a map with convex graph; then its images $F(x)$, its domain, and its range are convex subsets. The inverse of a map with convex graph has a convex graph. We observe that a convex map is closed if and only if it is weakly closed, because in Banach spaces, closed convex subsets and weakly closed convex subsets do coincide.

Examples of convex maps are provided by convex functions V from X to $R \cup \{+\infty\}$: The function V is convex if and only if the set-valued map $\mathbf{V}_+$ defined by

$$\mathbf{V}_+(x) := \begin{cases} V(x) + R_+ & \text{when } V(x) < +\infty \\ \varnothing & \text{when } V(x) = +\infty \end{cases}$$

is convex (because the graph of $\mathbf{V}_+$ is the epigraph of V).

More generally, if P is a closed convex cone of a Banach space Y, we associate with a single-valued map from its domain $D(A)$ to Y the set-valued map $\mathbf{A}_+$ defined by

$$\mathbf{A}_+(x) := \begin{cases} A(x) + P & \text{when } x \in D(A) \\ \varnothing & \text{when } x \notin D(A) \end{cases}$$

and we observe that the following conditions are equivalent:

i. The map $\mathbf{A}_+$ is convex.

ii. For all convex combination, $\sum_{i=1}^{n} \lambda_i A(x_i) \in A\left(\sum_{i=1}^{n} \lambda_i x_i\right) + P$

We say that such maps A are *P-convex operators.*

We shall pay special attention to the *convex processes*, which are set-valued maps whose graphs are *convex cones*, that is, the set-valued maps satisfying the properties

i. $\forall x_1, x_2 \in X, \quad F(x_1) + F(x_2) \subset F(x_1 + x_2)$

ii. $\forall \lambda > 0, \quad \forall x \in X, \quad F(\lambda x) = \lambda F(x)$

A *closed convex process* is a set-valued map whose graph is a closed convex cone.

Convex processes are set-valued analogues of linear operators, since a strict single-valued map F is a convex process if and only if F is linear.

Closed convex set-valued maps from Banach spaces to Banach spaces are lower semicontinuous on the interior of their domains. This important result is an extension of the closed graph theorem for continuous linear operators. Since it is customary to prove the closed graph theorem from the Banach open mapping principle, we state the Robinson–Ursescu theorem in the following way in theorem 1.

THEOREM 1 (CLOSED GRAPH)
Let X and Y be Banach spaces, F a proper closed convex map from X to Y, whose range has a nonempty interior.

Let $y_0 \in \text{Int}(\text{Im } F)$ *and* $x_0 \in F^{-1}(y_0)$ *be chosen. There exists* $\gamma > 0$ *such that*

$$(1) \quad \forall x \in \text{Dom } F, \forall y \in y_0 + \gamma B, \quad d(x, F^{-1}(y)) \leq \frac{1}{\gamma} d(y, F(x))(1 + \|x - x_0\|) \quad \blacktriangle$$

In particular, F^{-1} is lower semicontinuous on Int $F(K)$. By taking $x = x_0$, we have

$$(2) \qquad \forall y \in y_0 + \gamma B, \qquad d(x_0, F^{-1}(y)) \leq \frac{1}{\gamma} d(y, F(x_0))$$

COROLLARY 2
Let X and Y be Banach spaces, F a proper closed convex map from X to Y, whose range has a nonempty interior. Assume furthermore that F^{-1} *is locally bounded. Then* F^{-1} *is locally Lipschitz on* Int Im *F*. ▲

When F is a closed convex process, theorem 1 yields corollary 3.

COROLLARY 3
Let F be a closed convex process from a Banach space X to a Banach space Y. If

$$(3) \qquad \text{Im } F = Y \quad (\textit{i.e., } F \textit{ is surjective})$$

then F^{-1} *is a Lipschitz: There exists* $\gamma > 0$ *such that*

$$(4) \qquad \forall y_1, y_2 \in Y, \quad F^{-1}(y_2) \subset F^{-1}(y_1) + \frac{1}{\gamma} \|y_1 - y_2\| B$$

▲

Proof. Since the image of a convex process is a convex cone, then 0 belongs to the interior of this image if and only if it coincides with the whole space Y. We take $x_0 = x = 0$ and $y_0 = 0$ in inequality (2). We obtain

$$\forall y \in \gamma B, \quad d(0, F^{-1}(y)) \leq \frac{1}{\gamma} d(y, F(0)) \leq \frac{1}{\gamma} \|y\|$$

Since F^{-1} is positively homogeneous, this inequality applied to $\gamma(y/\|y\|)$ implies that

$$(5) \qquad \forall y \in Y, \qquad d(0, F^{-1}(y)) \leq \frac{1}{\gamma} \|y\|$$

Consider now y_1 and y_2 in Y and $x_2 \in F^{-1}(y_2)$. For all $\varepsilon>0$, we can choose $x_\varepsilon \in F^{-1}(y_1-y_2)$ such that

$$\|x_\varepsilon\| \leqslant d(0, F^{-1}(y_1-y_2))+\varepsilon \leqslant \frac{1}{\gamma}\|y_1-y_2\|+\varepsilon$$

Hence, $x_{1\varepsilon} := x_2 + x_\varepsilon$ belongs to

$$F^{-1}(y_2+y_1-y_2)=F^{-1}(y_1)$$

(because F is a convex process) and consequently, $x_2 = x_{1\varepsilon} - x_\varepsilon$ belongs to

$$F^{-1}(y_1)+\left(\frac{1}{\gamma}\|y_1-y_2\|+\varepsilon\right)B \qquad \text{for all } \varepsilon>0$$

By letting ε converge to 0, we deduce that inclusion (4) holds true. ■

Let us apply theorem 1 to the map F, defined by

$$F(x):=\begin{cases} Ax-M & \text{when } x\in L \\ \varnothing & \text{when } x\notin L \end{cases} \tag{6}$$

COROLLARY 4
Let $L\subset X$ and $M\subset Y$ be nonempty closed convex subsets of Banach spaces X and Y, and let A belong to $\mathscr{L}(X, Y)$. We set

$$\forall y\in Y, \qquad F^{-1}(y):=\{x\in L \mid Ax\in M+y\} \tag{7}$$

Let us assume that

$$\operatorname{Int}(A(L)-M)\neq\varnothing \tag{8}$$

Then for all $y_0\in \operatorname{Int}(A(L)-M)$ and $x_0\in F^{-1}(y_0)$, there exists $\gamma>0$ such that

$$\forall x\in L, \qquad \forall y\in y_0+\gamma B, \qquad d(x, F^{-1}(y))\leqslant \frac{1}{\gamma}\, d_M(Ax-y)(1+\|x-x_0\|) \tag{9}$$ ▲

Proof. The graph of this map is obviously closed and convex, its image is equal to $A(L)-M$, its inverse is defined by (7), and $d(y, F(x))=d_M(Ax-y)$. These remarks made, corollary 4 follows from theorem 1. ■

When L and M are cones, corollary 3 implies the following statement in corollary 5.

COROLLARY 5

Let $P \subset X$ and $Q \subset Y$ be closed convex cones of Banach spaces X and Y, and let A be a continuous linear operator from X to Y. We assume that

$$Y = A(P) + Q \tag{10}$$

Let F^{-1} be the map defined from Y to X by

$$\forall y \in Y, \quad F^{-1}(y) := \{x \in P \mid Ax \in Q + y\} \tag{11}$$

Then F^{-1} is Lipschitz: There exists $\gamma > 0$ such that

$$\forall y_1, y_2 \in Y, \quad F^{-1}(y_2) \subset F^{-1}(y_1) + \frac{1}{\gamma}\|y_1 - y_2\| B \tag{12}$$ ▲

By taking $P := X$ and $Q := \{0\}$, we obtain the Banach open mapping principle stated in corollary 6.

COROLLARY 6

Let A be a surjective continuous linear operator from a Banach space X to a Banach space Y. Then the inverse A^{-1} is a Lipschitz set-valued map from Y to X. ▲

Another consequence of theorem 1 is the following property of lower semicontinuous convex functions stated in corollary 7.

COROLLARY 7

Any proper lower semicontinuous, convex function V from a Banach space X to $R \cup \{+\infty\}$ is locally Lipschitz on the interior of its domain. ▲

Proof. We recall that Baire's theorem implies that V is continuous on the interior of its domain. We apply theorem 1 to the inverse $F^{-1} := \mathbf{V}_+^{-1}$ of the map $\mathbf{V}_+ : X \to R$ defined by

$$\mathbf{V}_+(x) := V(x) + R_+ \text{ when } x \in \text{Dom } V \qquad \text{and} \qquad \mathbf{V}_+(x) = \varnothing \quad \text{otherwise}$$

Let

$$x_0 \in \text{Int Dom } V = \text{Int Im } F \qquad \text{and} \qquad y_0 = V(x_0) \in F^{-1}(x_0)$$

be given. Since V is continuous at x_0, there exists $\gamma_0 \leq \gamma$ such that $|V(x) - V(x_0)| \leq 1$ when $x \in x_0 + \gamma_0 B$. Let x_1 and x_2 belong to $x_0 + \gamma_0 B$. Assume that $V(x_1) > V(x_2)$; then, we apply theorem 1, which states that since x_1 belongs to $x_0 + \gamma_0 B \subset x_0 + \gamma B$

and $V(x_2)$ belongs to Dom F, we have

$$|V(x_1)-V(x_2)|=V(x_1)-V(x_2)=\inf_{\lambda \in F^{-1}(x_1)} |\lambda - V(x_2)|$$

$$=d(F^{-1}(x_1), V(x_2))\leqslant\frac{1}{\gamma} d(x_1, F(V(x_2)))(1+|V(x_2)-V(x_0)|)$$

$$\leqslant\frac{2}{\gamma} \|x_1-x_2\| \qquad \blacksquare$$

The proof of theorem 1 is rather involved. We begin by deducing it from a more concise statement, given in proposition 8.

PROPOSITION 8

We posit the assumptions of theorem 1. *Let* $y_0 \in \operatorname{Int} \operatorname{Im}(F)$ *and* $x_0 \in F^{-1}(y_0)$ *be given and B denote the unit ball of X. Then after setting* $K:=\operatorname{Dom} F$, *we have*

$$y_0 \in \operatorname{Int} F(K\cap(x_0+B)) \tag{13}$$ ▲

We shall decompose the proof of proposition 8 into three lemmas. But first, we deduce theorem 1 from proposition 8.

Proof of Theorem 1 *from Proposition* 8. Let $x \in L$ be fixed. By proposition 8, there exists $\gamma>0$ such that

$$y_0+2\gamma B\subset F(K\cap(x_0+B)) \tag{14}$$

Let $y \in y_0+\gamma B$ be given. If x belongs to $F^{-1}(y)$, then $d(x, F^{-1}(y))=0$ and the conclusion is satisfied. If not, for all $\varepsilon>0$, there exists $z \in F(x)$ such that

$$\|y-z\|\leqslant d(y, F(x))(1+\varepsilon)$$

Since $y+\gamma B\subset F(K\cap(x_0+B))$, we obtain

$$\frac{\gamma(y-z)}{\|y-z\|} \in F(K\cap(x_0+B))-y \tag{15}$$

Let us set $\lambda:=\|y-z\|/(\gamma+\|y-z\|)$, which belongs to $]0, 1[$. We can write inclusion (15) in the form

$$(1-\lambda)(y-z) \in \lambda F(K\cap(x_0+B))-\lambda y \tag{16}$$

Since $(1-\lambda)z \in (1-\lambda)F(x)$ and the graph of F is convex, we obtain

$$(17) \qquad y \in \lambda F(K \cap (x_0+B)) + (1-\lambda)F(x) \subset F(\lambda(K \cap (x_0+B)) + (1-\lambda)x)$$

Then there exists $x_1 \in K \cap (x_0+B)$ such that, by setting $x_y := \lambda x_1 + (1-\lambda)x$, we have $x_y \in F^{-1}(y)$. Furthermore, we observe that

$$\begin{aligned}\|x_y - x\| &= \|\lambda x_1 + (1-\lambda)x - x\| \\ &= \lambda\|x_1 - x\| \leqslant \lambda(\|x - x_0\| + \|x_1 - x_0\|) \leqslant \lambda(\|x - x_0\| + 1)\end{aligned}$$

because $x_1 \in x_0 + B$.

Now

$$\lambda := \frac{\|y-z\|}{\gamma + \|y-z\|} \leqslant \frac{d(y, F(x))(1+\varepsilon)}{\gamma}$$

and thus

$$(18) \qquad d(x, F^{-1}(y)) \leqslant \|x - x_y\| \leqslant \frac{d(y, F(x))(1+\varepsilon)}{\gamma}(\|x - x_0\| + 1)$$

By letting ε converge to zero, we obtain

$$(19) \qquad d(x, F^{-1}(y)) \leqslant \frac{1}{\gamma}\, d(y, F(x))(\|x - x_0\| + 1)$$

■

The proof of proposition 8 is analogous to the proof of the open mapping theorem. We use lemma 9.

LEMMA 9

Let T be a subset of a Banach space Y satisfying

$$(20) \qquad \frac{1}{2}\sum_{k=0}^{\infty} 2^{-k}T \subset T$$

If zero belongs to the interior of the closure of T, it actually belongs to the interior of T. ▲

Proof. By assumption, there exists $\gamma > 0$ such that $2\gamma B \subset \bar{T}$. Hence, for every $k > 1$, we have $2 \cdot 2^{-k}\gamma B \subset 2^{-k}\bar{T}$.

Let $y \in \gamma B$. Then there exists $v_0 \in T$ such that

$$2y - v_0 \in 2 \cdot 2^{-1}\gamma B \subset 2^{-1}\bar{T}$$

since $2y \in \bar{T}$.

Let us assume that we have constructed a sequence of elements $v_k \in T$ $(0 \leqslant k \leqslant n-1)$ such that

$$2y - \sum_{k=0}^{n-1} 2^{-k} v_k \subset 2 \cdot 2^{-n} \gamma B \tag{21}$$

Since $2 \cdot 2^{-n} \gamma B \subset 2^{-n} \bar{T}$, we can find $v_n \in T$ such that

$$2y - \sum_{k=0}^{n-1} 2^{-k} v_k - 2^{-n} v_n \in 2^{-(n+1)} \gamma B \subset 2^{-(n+1)} \bar{T}$$

In this way, we have constructed a sequence of elements $v_k \in T$ such that

$$y = \frac{1}{2} \sum_{k=0}^{\infty} 2^{-k} v_k \in \frac{1}{2} \sum_{k=0}^{\infty} 2^{-k} T$$

By assumption, y belongs to T. So, we have proved $\gamma B \subset T$. ■

We apply this lemma to the subset

$$T := F(K \cap (x_0 + B)) - y_0 \tag{22}$$

Lemma 10 proves that zero belongs to the interior of the closure of T and lemma 11 states that this subset T satisfies property (20). Hence, we shall conclude that zero belongs to the interior T, that is, the conclusion of proposition 8.

LEMMA 10

We posit the assumptions of proposition 8. *Then zero belongs to the interior of the closure of the subset* $T := F(K \cap (x_0 + B)) - y_0$. ▲

Proof. We set $K_n := K \cap (x_0 + nB)$. Hence, $T = F(K_1) - y_0$.

We remark that $K = \bigcup_{n=1}^{\infty} K_n$ and thus $F(K) = \bigcup_{n=1}^{\infty} F(K_n)$. We note also that

$$\left(1 - \frac{1}{n}\right) x_0 + \frac{1}{n} K_n \subset K_1$$

The graph of F being convex, we deduce that

$$\left(1 - \frac{1}{n}\right) F(x_0) + \frac{1}{n} F(K_n) \subset F(K_1) \tag{23}$$

Since $0 \in \text{Int}\,(F(K) - y_0)$, there exists $\gamma > 0$ such that

$$\gamma B \subset F(K) - y_0 = \bigcup_{n=1}^{\infty} (F(K_n) - y_0)$$

But $y_0 \in F(x_0)$, and (23) implies that

$$\left(1-\frac{1}{n}\right)y_0+\frac{1}{n}F(K_n)\subset F(K_1)$$

that is,

$$F(K_n)-y_0\subset n(F(K_1)-y_0)=nT$$

Therefore, $\gamma B\subset\bigcup_{n=1}^{\infty} n\bar{T}$. Baire's theorem implies that the interior of some subset $n\bar{T}$ is nonempty. Then there exists $x_0 \in \bar{T}$ and $\delta>0$ such that $x_0+\delta B\subset\bar{T}$. Since $-\gamma x_0/\|x_0\|$ belongs to γB, and thus to the union of the $n\bar{T}$'s, there exists n such that $-\gamma x_0/n\|x_0\|$ belongs to $\bar{T}$. Let $\lambda:=\gamma/(\gamma+n\|x_0\|)\in\,]0, 1[$. Hence,

$$\lambda\delta B=\lambda x_0-(1-\lambda)\frac{\gamma x_0}{n\|x_0\|}+\lambda\delta B\subset\lambda\bar{T}+(1-\lambda)\bar{T}\subset\bar{T}$$

because $\bar{T}$ is convex. We have shown that zero belongs to the interior of $\bar{T}$.

It remains to check that T satisfies property (20). ■

LEMMA 11

We posit the assumptions of proposition 8. Then the subset $T:=F(K\cap(x_0+B))-y_0$ *satisfies the property*

$$\frac{1}{2}\sum_{k=0}^{\infty} 2^{-k}T\subset T$$ ▲

Proof. We take $y_0=0$ for the sake of simplicity. Let $y_\star$ belong to $\frac{1}{2}\sum_{k=0}^{\infty}2^{-k}T$. We set $\alpha_n:=1/\sum_{k=0}^{n} 2^{-k}$ so that $\alpha_n\to\frac{1}{2}$ and $y_n:=\alpha_n\sum_{k=0}^{n} 2^{-k}v_k$ where $v_k\in T$ so that y_n converges to $y_\star$. By definition of T, we can find $u_k\in K\cap(x_0+B)$ such that $v_k\in F(u_k)$. Since the graph of F is convex, we deduce that

$$\text{(24)}\qquad y_n\in\alpha_n\sum_{k=0}^{n} 2^{-k}F(u_k)\subset F\left(\alpha_n\sum_{k=0}^{n} 2^{-k}u_k\right)$$

Let us consider the sequence of elements $x_n:=\alpha_n\sum_{k=0}^{n} 2^{-k}u_k$. Since $u_k\in x_0+B$, we deduce that it is a Cauchy sequence, which converges to some $x_\star$ for X is complete. Since $K\cap(x_0+B)$ is convex, x_n belongs to $K\cap(x_0+B)$. This set also being closed, $x_\star$ belongs to $K\cap(x_0+B)$. Inclusion (24) says that the sequence of elements (x_n, y_n) belongs to the graph of F. Since it is closed, it follows that $(x_\star, y_\star)\in\operatorname{graph} F$, that is, $y_\star$ belongs to $F(x_\star)\subset F(K\cap(x_0+B))=:T$. ■

We can adapt the concept of transpose to set-valued maps.

DEFINITION 12
Let F be a set-valued map from a Banach space X to a Banach space Y. We associate with F the convex process $F^\star$ from $Y^\star$ to $X^\star$ defined in the following way

$$\text{(25)}\qquad \begin{cases} p \in F^\star(q) \textit{ if and only if } (p, -q) \textit{ belongs to the barrier cone} \\ \mathrm{b}(\text{graph}\,(F)) \textit{ of } \text{graph}\,(F) \end{cases}$$

We say that $F^\star$ is the transpose of F. ▲

In other words,

$$\text{(26)}\qquad p \in F^\star(q) \quad \text{if and only if} \quad \sup_{x \in X} \sup_{y \in F(x)} (\langle p, x\rangle - \langle q, y\rangle) < +\infty$$

When F is a process, then its transpose is the *closed convex* process defined by

$$\text{(27)}\qquad \begin{cases} p \in F^\star(q) \text{ if and only if } (p, -q) \text{ belongs to the negative} \\ \text{polar cone } (\text{graph}\,(F))^- \text{ of graph } F \end{cases}$$

or, equivalently,

$$\text{(28)}\qquad p \in F^\star(q) \quad \text{if and only if} \quad \forall x \in X, y \in F(x), \quad \langle p, x\rangle \leqslant \langle q, y\rangle$$

since the barrier cone of a cone is its negative polar cone.

When G is a set-valued map from $Y^\star$ to $X^\star$, we define the convex process $G^\star$ from X to Y by

$$\text{(29)}\qquad y \in G^\star(x) \quad \text{if and only if} \quad (x, -y) \in \mathrm{b}(\text{graph}\,(G))$$

[instead of requiring that $(-x, y)$ belongs to the barrier cone of graph (G)].

Remark
When F is a continuous linear operator from X to Y, its transpose as a continuous linear operator coincides with its transpose as a closed convex process. This is why the minus sign appears in the definition of $F^\star$. It could have appeared in front of x instead of y; it is a matter of convenience. ■

We now list formulas allowing the characterization of transposes.

The transpose of F^{-1}, the inverse of F, is given by

$$\text{(30)}\qquad (F^{-1})^\star(p) = -(F^\star)^{-1}(-p)$$

PROPOSITION 13
Let X, Y, and Z be Banach spaces, F a set-valued map from X to Y, and $B \in \mathscr{L}(Y, Z)$.

Then

$$(BF)^\star = F^\star B^\star \tag{31}$$ ▲

Proof. Indeed, the graph of BF is equal to $(1 \times B)$ graph (F). By formula (25) of Section 5, Chapter 1

$$\mathrm{b}((1 \times B) \text{ graph } (F)) = (1 \times B)^{\star -1} \mathrm{b}(\text{graph } (F))$$

so that $(p, -q)$ belongs to b(graph (BF)) if and only if $(p, -B^\star q)$ belongs to b(graph (F)); that is, if p belongs to $F^\star(B^\star q)$. ■

When $A \in \mathscr{L}(X_0, X)$, the set-valued map is proper if and only if the intersection of Im A and Dom F is nonempty, that is, if and only if zero belongs to Im $A -$ Dom F. We shall make a stronger assumption in proposition 14.

PROPOSITION 14

Let X_0, X, Y be Banach spaces, F a closed convex map from X to Y, and $A \in \mathscr{L}(X_0, X)$. If

$$0 \in \text{Int } (\text{Im } A - \text{Dom } F) \tag{32}$$

then

$$(FA)^\star = A^\star F^\star \tag{33}$$ ▲

Proof. Indeed, graph $(FA) = (A \times 1)^{-1}$ graph (F) and, consequently, graph $(FA)^\star = \mathrm{b}((A \times 1)^{-1}$ graph $(F))$. We apply formula (30) of Section 5, Chapter 1. For that purpose, we have to check that condition

$$(0, 0) \in \text{Int } (\text{Im } (A \times 1) + \text{graph } (F)) \tag{34}$$

holds true. It follows from assumption (32). Indeed, let $\gamma > 0$ be such that $\gamma B \subset \text{Im} A - \text{Dom } F$. Let (x, y) belong to the ball of radius $\gamma > 0$ in $X \times Y$.

Since x can be written $x = Ax_0 - x_1$, where x_0 belongs to X_0 and x_1 belongs to the domain of F, then we can write $(x, y) = (Ax_0, y_0) - (x_1, y_1)$ where $y_1 \in F(x_1)$ and $y_0 = y + y_1$. Hence,

$$(x, y) \in \text{Im } (A \times 1) - \text{graph } (F)$$

Therefore, formula (18) of Section 5, Chapter 1 implies equality

$$\text{graph } ((FA)^\star) = \mathrm{b}(\text{graph } (FA)) = (A^\star \times 1)\mathrm{b}(\text{graph } (F)) \tag{35}$$

and thus,

(36) $\quad r \in (FA)^{\star}(q)$ if and only if there exists $p \in F^{\star}(q)$ such that $\gamma = A^{\star}p$ ■

COROLLARY 15

Let V, X, Y, and Z be Banach spaces, F a closed convex map from X to Y, let A belong to $\mathscr{L}(V, X)$, and B to $\mathscr{L}(Y, Z)$. If

$$0 \in \operatorname{Int}(\operatorname{Im} A - \operatorname{Dom} F) \tag{37}$$

then

$$(BFA)^{\star} = A^{\star}F^{\star}B^{\star} \tag{38}$$

▲

COROLLARY 16

Let X, Y, and Z be Banach spaces, $A \in \mathscr{L}(X, Y)$, F a closed convex map from X to Z, and G a closed convex process from Y to Z. If

$$0 \in \operatorname{Int}(A \operatorname{Dom} F - \operatorname{Dom} G) \tag{39}$$

then

$$(F + GA)^{\star} = F^{\star} + A^{\star}G^{\star} \tag{40}$$

▲

Proof. Let us set

$$H(x) := F(x) + G(Ax), \qquad B(y, z) := y + z$$

and

$$(F \times G)(x, y) := F(x) \times G(y) \quad \text{and} \quad (1 \times A)(x) = (x, Ax)$$

Then

$$H := B(F \times G)(1 \times A) \tag{41}$$

Since

$$\operatorname{Dom}(F \times G) = \operatorname{Dom} F \times \operatorname{Dom} G$$

assumption (39) implies that

$$0 \in \operatorname{Int}(\operatorname{Im}(1 \times A) - \operatorname{Dom}(F \times G)) \tag{42}$$

Indeed, if $(u, v) \in X \times Y$, then there exist $\varepsilon > 0$, $x \in \mathrm{Dom}\, F$ and $y \in \mathrm{Dom}\, G$ such that

$$\varepsilon(Au - v) = -Ax + y$$

By setting $z = x + \varepsilon u$, we see that

$$\varepsilon(u, v) = (z - x, Az - y) \in (1 \times A)z - \mathrm{Dom}\,(F \times G)$$

Hence, assumption (42) holds true. Therefore, the preceding corollaries imply that $H^\star = (1 \times A)^\star(F \times G)^\star B^\star$. Since $(F \times G)^\star(p) = F^\star(p) \times G^\star(p)$ and $(1 \times A)^\star = 1 + A^\star$, we deduce that

$$H^\star(p) = F^\star(p) + A^\star G^\star(p)$$ ■

COROLLARY 17

Let F be a closed convex map from X to Y and $K \subset X$ be a closed convex subset. We assume that

$$0 \in \mathrm{Int}(K - \mathrm{Dom}\, F) \tag{43}$$

Then the transpose of the restriction $F|_K$ of F to K is defined by

$$(F|_K)^\star(q) = F^\star(q) + b(K) \tag{44}$$ ▲

Proof. We apply corollary 16 with A = identity and G defined by $G(x) = 0$ when $x \in K$; $G(x) = \varnothing$ when $x \notin K$, whose domain is K, whose graph is $K \times \{0\}$, and whose transpose $G^\star$ is the constant map defined by $G^\star(q) = b(K)$. ■

PROPOSITION 18

Let F_1 and F_2 be two closed convex maps from X to Y. If

$$0 \in \mathrm{Int}(\mathrm{graph}(F_1) - \mathrm{graph}(F_2)) \tag{45}$$

then

$$(F_1 \cap F_2)^\star(q) = F_1^\star(q_1) + F_2^\star(q_2) \qquad \textit{where } q = q_1 + q_2 \tag{46}$$ ▲

Proof. Since the graph of $F_1 \cap F_2$ is the intersection of the graphs of F_1 and F_2, assumption (45) and formula (29) in Section 5, Chapter 1 imply that

$$b(\mathrm{graph}(F_1) \cap \mathrm{graph}(F_2)) = b(\mathrm{graph}(F_1)) + b(\mathrm{graph}(F_2))$$

Therefore, the graph of $(F_1 \cap F_2)^\star$ is equal to the sum of the graphs of $F_1^\star$ and $F_2^\star$. ■

The formula $(\text{Im } A)^{\perp} = \text{Ker } A^{\star}(= A^{\star -1}(0))$ satisfied by continuous linear operators can be adapted to a set-valued map and yields the same surjectivity conditions.

PROPOSITION 19

Let F be a set-valued map from X to Y. Then

$$b(\text{Im } F) = -F^{\star -1}(0) \tag{47}$$ ▲

Proof. It is obvious that an element q belongs to the barrier cone of Im F if and only if the pair $(0, q)$ belongs to the barrier cone of the graph of F; that is, if $(0, -q)$ belongs to the graph of $F^{\star}$. ■

We deduce the following interesting result in proposition 20.

PROPOSITION 20

a. *Let F be a proper closed convex map from a Banach space X to a Banach space Y. Then the image of F is dense if and only if*

$$F^{\star -1}(0) = \{0\} \tag{48}$$

b. *Let F be a closed convex process from X to Y. Then F is surjective if and only if the image of F is closed and* $F^{\star -1}(0) = \{0\}$. ▲

Proof. **a.** Since the image of F is convex, then $\overline{\text{Im } F}$ is equal to Y if and only if its barrier cone is equal to $\{0\}$, that is, if and only if $F^{\star -1}(0) = \{0\}$ by proposition 19.
b. Since the image of F is a closed convex cone, then Im $F = (\text{Im } F)^{--} = (b(\text{Im } F))^{-} = Y$. ■

COROLLARY 21

Let F be a set-valued map from X to Y and let $K \subset X$ *be a closed convex subset such that*

$$0 \in \text{Int}(\text{Dom } F - K) \tag{49}$$

Then

$$b(F(K)) = -F^{\star -1}(-b(K)) \tag{50}$$ ▲

Proof. We apply proposition 9 to the map $F|_K$, since its image is equal to $F(K)$. Then q belongs to the barrier cone of $F(K)$ if and only if zero belongs to $(F|_K)^{\star}(-q)$, which is equal to $F^{\star}(-q) + b(K)$ by corollary 17. This amounts to saying that $-q$ belongs to $F^{\star -1}(-b(K))$. ■

As a consequence, we obtain the following extension of Farkas's lemma to closed convex processes.

COROLLARY 22
Let F be a closed convex process from X to Y and let K be a closed convex cone of X such that

$$\text{Dom } F - K = X \tag{51}$$

Then

$$F(K)^- = -F^{\star -1}(-K^-) \tag{52}$$

If $F(K)$ is closed, then

$$F(K) = -(F^{\star -1}(-K^-))^- \tag{53}$$

▲

We shall use proposition 20 to prove an extension of the Lax–Milgram theorem t_0 closed convex processes.

DEFINITION 23
We shall say that a set-valued map F from X to $X^\star$ is X elliptic if

$$\begin{cases} \exists c > 0 \text{ such that for any two } (x^i, y^i) \in \text{graph}(F) \\ \langle y^1 - y^2, x^1 - x^2 \rangle \geqslant c\|x^1 - x^2\|^2 \end{cases} \tag{54}$$

▲

LEMMA 24
The image of an X-elliptic map F with closed graph is closed, and its inverse is single valued and Lipschitz with constant c^{-1}. ▲

Proof. The fact that F^{-1} is single valued follows from (54) by taking $y = y^1 = y^2$ an x_1, x_2 in $F^{-1}(y)$.

Inequality (54) also implies that

$$c\|F^{-1}(y_1) - F^{-1}(y_2)\|^2 \leqslant c\|y_1 - y_2\| \, \|F^{-1}(y_1) - F^{-1}(y_2)\|$$

To prove that $\text{Im}(F)$ is closed, let us consider a Cauchy sequence of elements $p_n \in \text{Im } F$. Let us take x_n in $F^{-1}(p_n)$. Since F is X elliptic, we deduce that

$$c\|x_n - x_m\|^2 \leqslant \langle p_n - p_m, x_n - x_m \rangle \leqslant \|x_n - x_m\| \, \|p_n - p_m\|$$

and, therefore, that the sequence of elements x_n is a Cauchy sequence. Then the sequence of elements $(x_n, p_n) \in \text{graph}(F)$ converges to some (x, p), which belongs to the graph of F, since the latter is closed. Hence, p belongs to $\text{Im}(F)$. We have proved that $\text{Im}(F)$ is complete and, thus, closed. ■

We deduce a surjectivity criterion analogous to the Lax–Milgram theorem on X-elliptic continuous linear operators.

PROPOSITION 25

Let F be an X-elliptic closed convex process from X to $X^\star$. If $(\mathrm{Dom}\ F)^- \subset \mathrm{Im}\ F$ and if the domain of F is closed, then F is surjective, and its inverse is a single-valued Lipschitz map from $X^\star$ to X. ▲

Proof. By assumption, $-F^{\star-1}(0) = (\mathrm{Im}\ F)^-$ (by proposition 19) is contained in $(\mathrm{Dom}\ F)^{--} = \mathrm{Dom}\ F$, since the domain of F is closed. Let us pick $x_0 \in F^{\star-1}(0)$, and choose $y_0 \in F(-x_0)$. Since $(0, -x_0)$ belongs to graph $(F)^-$, we deduce that $\langle 0, x_0\rangle - \langle x_0, y_0\rangle \leqslant 0$. Since F is a X-elliptic process, we deduce that

$$c\|x_0\|^2 = c\|-x_0 - 0\|^2 \leqslant \langle -x_0 - 0, y_0 - 0\rangle = -\langle x_0, y_0\rangle \leqslant 0$$

Hence, $x_0 = 0$. Therefore, $F^{\star-1}(0)$ is equal to $\{0\}$. Since $\mathrm{Im}\ F$ is closed by lemma 24, proposition 20 implies that F is surjective. ■

COROLLARY 26

Any X-elliptic, closed convex process F whose domain is X is surjective and F^{-1} is a single-valued Lipschitz map from X to $X^\star$. ▲

COROLLARY 27 (LAX–MILGRAM COROLLARY)

Any X-elliptic continuous linear operator from X to $X^\star$ is an isomorphism. ▲

It is clear that if F is a closed convex set-valued map from X to Y and if $A \in \mathscr{L}(X_0, X)$, then FA is still a closed convex set-valued map from X_0 to X. If $B \in \mathscr{L}(X, Z)$, the set-valued map BF is convex, but not necessarily closed. We denote by $\overline{BF}$ the set-valued map whose graph is the closure of the graph of BF.

PROPOSITION 28

Let X, Y, and Z be reflexive Banach spaces, F a closed convex map from X to Y, and $B \in \mathscr{L}(Y, Z)$ a continuous linear operator. If

$$0 \in \mathrm{Int}(\mathrm{Im}\ B^\star - \mathrm{Dom}\ F^\star) \tag{55}$$

then the graph of BF is closed and convex. ▲

Proof. We have to prove that the graph of BF is closed. Since graph $(BF) = (1 \times B)$ graph (F), we can apply theorem 1.5.5. We easily check that assumption (55) implies that

$$(0, 0) \in \mathrm{Int}(1 \times B)^\star + \mathrm{b}(\mathrm{graph}\ (F)) \tag{56}$$

Since graph (F) is convex and closed, it is weakly closed. Hence, theorem 1.5.5 implies that $(1 \times B)$ graph $(F)=\text{graph}(BF)$ is closed. ■

COROLLARY 29

Let V, X, Y, and Z be reflexive Banach spaces, F a closed convex map from X to Y, let A belong to $\mathscr{L}(V, X)$, and B to $\mathscr{L}(Y, Z)$. If

$$(57) \qquad 0 \in \operatorname{Int}(\operatorname{Im} B^\star - \operatorname{Dom} F^\star)$$

then BFA is a closed convex map from X to V. ▲

COROLLARY 30

Let F_1 and F_2 be two closed convex set-valued maps from a reflexive Banach space X to a reflexive Banach space Y. If

$$(58) \qquad 0 \in \operatorname{Int}(\operatorname{Dom} F_1^\star - \operatorname{Dom} F_2^\star)$$

Then the set-valued map F_1+F_2 has a closed convex graph. ▲

Proof. Let $A \in \mathscr{L}(X, X \times X)$ be the map defined by $Ax:=(x, x)$ and $B \in \mathscr{L}(Y \times Y, Y)$ the map defined by $B(y_1, y_2):=y_1+y_2$. Let $\vec{F}$ be the set-valued map defined by $\vec{F}(x_1, x_2)=F_1(x_1)\times F_2(x_2)$. Then, $F_1+F_2=B\vec{F}A$. Since $\vec{F}^\star(p_1, p_2)=F_1^\star(p_1)\times F_2^\star(p_2)$ and thus, $\operatorname{Dom} \vec{F}^\star=\operatorname{Dom} F_1^\star \times \operatorname{Dom} F_2^\star$, we see that assumption (58) implies that $0 \in \operatorname{Int}(\operatorname{Im} B^\star - \operatorname{Dom} \vec{F}^\star)$. Hence, $F_1+F_2=B\vec{F}A$ is closed and convex by corollary 29. ■

4. EIGENVALUES OF POSITIVE MAPS WITH CLOSED CONVEX GRAPHS

We set

$$(1) \qquad \Sigma^n:=\left\{x \in R^n_+ \,\middle|\, \sum_{i=1}^{n} x_i=1\right\}$$

We introduce the following items:

$$(2) \qquad \begin{cases} \text{a single-valued map } f \text{ from } \Sigma^n \text{ to } R^m \text{ whose components} \\ f_i \text{ are lower semicontinuous and convex} \end{cases}$$

and

$$(3) \qquad \begin{cases} \text{a strict upper semicontinuous convex map } G \\ \text{with compact images from } \Sigma^n \text{ to } R^m_+ \end{cases}$$

We posit the following positivity conditions:

(4) $$\forall p \in \Sigma^m, \qquad \exists x \in \Sigma^n \text{ such that } \sigma(G(x), p) > 0$$

and

(5) $$\exists \tilde{p} \in \Sigma^m \text{ such that } \forall x \in \Sigma^n, \qquad \langle \tilde{p}, f(x) \rangle > 0$$

Then we can define the *positive* number δ by

(6) $$\frac{1}{\delta} := \sup_{p \in \Sigma^m} \inf_{x \in \Sigma^n} \frac{\langle p, f(x) \rangle}{\sigma(G(x), p)}$$

PROPOSITION 1

We posit assumptions (2), (3), (4), *and* (5). *Then there exists a solution* $\bar{x} \in \Sigma^n$ *to the inclusion*

(7) $$\delta f(\bar{x}) \in G(\bar{x}) - R^m_+$$

and there exists $\bar{p} \in \Sigma^m$ *such that the minimax property holds*

(8) $$\frac{1}{\delta} = \frac{\langle \bar{p}, f(\bar{x}) \rangle}{\sigma(G(\bar{x}), \bar{p})} = \inf_{x \in \Sigma^n} \frac{\langle \bar{p}, f(x) \rangle}{\sigma(G(x), \bar{p})} = \sup_{p \in \Sigma^m} \frac{\langle p, f(\bar{x}) \rangle}{\sigma(G(\bar{x}), p)}$$

Furthermore, when $x \in \Sigma^n$ *and* $\mu > 0$ *satisfy the inclusion*

(9) $$\mu f(x) \in G(x) - R^n_+$$

then μ *is not larger than* δ. ▲

Remark

We can say that when a pair (μ, x) satisfies (9), μ *is an eigenvalue* and x *is an eigenvector* of $G(\cdot) - R^n_+$ with respect to f. So proposition 1 states the existence of a positive eigenvalue that is the largest of the nonnegative eigenvalues. ■

Proof. **a.** We set $F_+(x) := f(x) + R^m_+$. Since the functions f_i are convex, the set-valued map F_+ is convex, and, consequently, the set-valued map $G - \delta F_+$ is also convex. Hence

$$(G - \delta F_+)(\Sigma^n) = \operatorname{Im}(G - \delta F_+)$$

is a convex subset. The continuity properties (2) and (3) imply that

(10) $$(G - \delta F_+)(\Sigma^n) \text{ is a closed convex subset.}$$

Indeed, let $z_k \in G(x_k) - \delta F_+(x_k)$ belong to $(G - \delta F_+)(\Sigma^n)$ that converge to some z in R^n. For all $p \in R^m_+$, we deduce that

$$\langle p, z_k \rangle \leqslant \sigma(G(x_k), p) - \delta \sum_{i=1}^{m} p_i f_i(x_k)$$

Since Σ^n is compact, a subsequence of elements $x_{k'}$ converges to some $x \in \Sigma^n$. The upper semicontinuity of

$$x \to \sigma(G(x), p) - \delta \sum_{i=1}^{n} p_i f_i(x)$$

implies that

$$\begin{aligned}\langle p, z \rangle &= \lim_{k' \to \infty} \langle p, z_{k'} \rangle \\ &\leqslant \limsup_{k' \to \infty} \left(\sigma(G(x_{k'}), p) - \delta \sum_{i=1}^{m} p_i f_i(x_{k'}) \right) \\ &\leqslant \sigma(G(x), p) - \delta \sum_{i=1}^{m} p_i f_i(x) = \sigma(G(x) - \delta f(x), p)\end{aligned}$$

Since this inequality holds true for all $p \in R^m_+$, we deduce that

$$z \in G(x) - \delta f(x) - R^m_+ = G(x) - \delta F_+(x)$$

b. Next, we prove that

$$0 \in (G - \delta F_+)(\Sigma^n)$$

If not, we can apply the separation theorem.

There exists $p_0 \in R^m$ and $\varepsilon > 0$ such that for all $x \in \Sigma^n$,

$$\sigma(G(x), p_0) - \delta \langle p_0, f(x) \rangle + \sigma(-R^m_+, p_0) \leqslant -\varepsilon$$

This implies that $p_0 \in R^m_+$ and that $\sigma(-R^m_+, p_0) = 0$. Since $p_0 \neq 0$, we have $\bar{p}_0 := p_0 / \sum_{i=1}^{m} p_0^i \in \Sigma^m$ and we deduce that for all $x \in \Sigma^n$

$$\frac{1}{\delta} \leqslant \frac{\langle \bar{p}_0, f(x) \rangle}{\sigma(G(x), \bar{p}_0)} - \frac{\varepsilon}{\delta \sigma(G(x), \bar{p}_0)} \leqslant \frac{\langle \bar{p}_0, f(x) \rangle}{\sigma(G(x), \bar{p}_0)} - \frac{\varepsilon}{\delta c}$$

where $c := \sup_{x \in \Sigma^n} \sigma(G(x), p_0)$. Therefore, we obtain the contradiction

$$\frac{1}{\delta} \leqslant \inf_{x \in \Sigma^n} \frac{\langle \bar{p}_0, f(x) \rangle}{\sigma(G(x), \bar{p}_0)} - \frac{\varepsilon}{\delta c} \leqslant \frac{1}{\delta} - \frac{\varepsilon}{\delta c}$$

c. Hence, there exists $\bar{x} \in \Sigma^n$ such that

$$0 \in G(\bar{x}) - \delta f(\bar{x}) - R^m_+$$

Consequently, for all $p \in \Sigma^m$, we have

$$0 \leqslant \sigma(G(\bar{x}), p) - \delta \langle p, f(\bar{x}) \rangle$$

and thus,

$$\sup_{p \in \Sigma^m} \frac{\langle p, f(\bar{x}) \rangle}{\sigma(G(\bar{x}), p)} \leqslant \frac{1}{\delta}$$

This implies the minimax equality

$$\frac{1}{\delta} = \sup_{p \in \Sigma^m} \frac{\langle p, f(\bar{x}) \rangle}{\sigma(G(\bar{x}), p)} = \inf_{x \in \Sigma^n} \sup_{p \in \Sigma^m} \frac{\langle p, f(x) \rangle}{\sigma(G(x), p)}$$

Since $p \to \langle p, f(\bar{x}) \rangle / \sigma(G(\bar{x}), p)$ is upper semicontinuous and Σ^m is compact, there exists $\bar{p} \in \Sigma^m$ such that

$$\frac{1}{\delta} = \frac{\langle \bar{p}, f(\bar{x}) \rangle}{\sigma(G(\bar{x}), \bar{p})} = \inf_{x \in \Sigma^n} \frac{\langle \bar{p}, f(x) \rangle}{\sigma(G(x), \bar{p})}$$

c. Let us assume that there exist $\hat{x} \in \Sigma^n$ and $\mu > 0$ satisfying $\mu f(\hat{x}) \in G(\hat{x}) - R^m_+$. Thus, for all $p \in \Sigma^m$, we would have

$$\mu \langle p, f(\hat{x}) \rangle \leqslant \sigma(G(\hat{x}), p)$$

$$\frac{1}{\delta} = \inf_{x \in \Sigma^n} \sup_{p \in \Sigma^m} \frac{\langle p, f(x) \rangle}{\sigma(G(x), p)} \leqslant \sup_{p \in \Sigma^m} \frac{\langle p, f(\hat{x}) \rangle}{\sigma(G(\hat{x}), p)} \leqslant \frac{1}{\mu}$$

that is, $\mu \leqslant \delta$. ■

Remark

We can replace Σ^n by any convex compact subset and replace assumption (3) on G by the weaker assumptions

(11)
- **i.** The graph of G is convex.
- **ii.** The images of G are closed and bounded above.
- **iii.** $\forall p \in \Sigma^m, \quad x \to \sigma(G(x), p)$ is upper semicontinuous.

Proposition 1 implies the existence of a solution $\bar{x}$ to the inclusion

$$0 \in G(\bar{x}) - \delta f(\bar{x}) - R^m_+$$

■

Now, we shall prove that when $\mu > \delta$, for every $y \in \text{Int } R^m_+$, there exist a negative number β and a solution $\hat{x} \in \Sigma^n$ to the inclusion

$$\beta y \in G(\hat{x}) - \mu f(\hat{x}) - R^m_+ \tag{12}$$

PROPOSITION 2

We posit assumptions (2), (3), (4), *and* (5) *of proposition* 1. *We associate with any* $\mu > \delta$ *and* $y \in \text{Int } (R^m_+)$ *the negative number*

$$\beta := \inf_{p \in \Sigma^m} \sup_{x \in \Sigma^n} \frac{\sigma(G(x) - \mu f(x), p)}{\langle p, y \rangle} \tag{13}$$

Then there exists $\hat{x} \in \Sigma^n$, *a solution to inclusion* (12). ▲

Proof. **a.** We begin by checking that $\beta < 0$. Indeed, since $\mu > \delta$, then

$$\frac{1}{\mu} < \frac{1}{\delta} = \inf_{x \in \Sigma^n} \frac{\langle \bar{p}, f(x) \rangle}{\sigma(G(x), \bar{p})} \quad \text{for some } \bar{p} \in \Sigma^m$$

Hence

$$\beta \leqslant \sup_{x \in \Sigma^n} \frac{\sigma(G(x) - \mu f(x), \bar{p})}{\langle \bar{p}, y \rangle} < 0$$

b. Next, we prove that

$$\beta y \in (G - \mu F_+)(\Sigma^n) \tag{14}$$

If not, we can apply the separation theorem. There exist $p_0 \in R^m$ and $\varepsilon > 0$ such that, for all $x \in \Sigma^n$,

$$\sigma(G(x), p_0) - \mu \langle p_0, f(x) \rangle + \sigma(-R^m_+, p_0) \leqslant \beta \langle p_0, y \rangle - \varepsilon$$

We deduce that $p_0 \in R^m_+$ and since $p_0 \neq 0$, that

$$\hat{p}_0 := p_0 \Big/ \sum_{i=1}^{n} p_0^i \in \Sigma^m$$

Then,

$$\beta \leqslant \sup_{x \in \Sigma^n} \frac{\sigma(G(x) - \mu f(x), \hat{p}_0)}{\langle \hat{p}_0, y \rangle} \leqslant \beta - \frac{\varepsilon}{\langle \hat{p}_0, y \rangle}$$

which is a contradiction.

c. It follows that there exists $\hat{x} \in \Sigma^n$ such that

$$\beta y \in G(\hat{x}) - \mu f(\hat{x}) - R^n_+$$

and that for all $p \in \Sigma^m$,

$$\beta \leqslant \frac{\sigma(G(\hat{x}) - \mu f(\hat{x}), p)}{\langle p, y \rangle}$$

so that

$$\beta \leqslant \sup_{x \in \Sigma^n} \inf_{p \in \Sigma^m} \frac{\sigma(G(x) - \mu f(x), p)}{\langle p, y \rangle} \leqslant \beta \qquad \blacksquare$$

Remark

We have proved that the minimax property

$$(15) \qquad \beta = \sup_{x \in \Sigma^n} \inf_{p \in \Sigma^m} \frac{\sigma(G(x) - \mu f(x), p)}{\langle p, y \rangle}$$

holds true. Furthermore, since Σ^n is compact, we deduce that there exists $\hat{p} \in \Sigma^m$ such that

$$(16) \qquad \beta = \sup_{x \in \Sigma^n} \frac{\sigma(G(x) - \mu f(x), \hat{p})}{\langle \hat{p}, y \rangle} = \frac{\sigma(G(\hat{x}) - \mu f(\hat{x}), \hat{p})}{\langle \hat{p}, y \rangle}$$

that is, such that

$$(17) \qquad \begin{cases} \textbf{i.} & \forall x \in \Sigma^n, \quad \sigma(G(x) - \mu f(x) - \beta y, \hat{p}) \leqslant 0 \\ \textbf{ii.} & \sigma(G(\bar{x}) - \mu f(\bar{x}) - \beta y, \hat{p}) = 0 \end{cases} \qquad \blacksquare$$

We consider the case when the convex map G is defined by $G(x) := g(x) - R^n_+$, where g is a single-valued concave operator.

PROPOSITION 3 (KY FAN)

Let f and g be single-valued maps from Σ^n to R^m satisfying

$$(18) \qquad \begin{cases} \textbf{i.} & \textit{The components } f_i \textit{ are convex and lower semicontinuous.} \\ \textbf{ii.} & \textit{The components } g_i \textit{ are non negative, concave, and} \\ & \textit{upper semicontinuous.} \\ \textbf{iii.} & \exists \tilde{p} \in \Sigma^m \quad \textit{such that} \quad \forall x \in \Sigma^n, \quad \langle \tilde{p}, f(x) \rangle > 0 \\ \textbf{iv.} & \exists \tilde{x} \in \Sigma^n \quad \textit{such that} \quad g_i(\tilde{x}) > 0 \quad \textit{for} \quad i = 1, \ldots, m \end{cases}$$

Then there exist $\delta>0$, $\bar{x}\in\Sigma^n$ and $\bar{p}\in\Sigma^m$ such that

(19)
$$\begin{cases} \text{i.} & \delta f(\bar{x})\leqslant g(\bar{x}), \\ \text{ii.} & \forall x\in\Sigma^n, \quad \langle g(x)-\delta f(x), \bar{p}\rangle\leqslant 0 \\ \text{iii.} & \delta\langle\bar{p}, f(\bar{x})\rangle=\langle\bar{p}, g(\bar{x})\rangle \end{cases}$$

Furthermore, for all $\mu>\delta$ and for all $y\in Int(R^m_+)$, there exist $\beta_0<0$ and $\hat{x}\in\Sigma^n$ such that

(20)
$$\beta_0 y\leqslant g(\hat{x})-\mu f(\hat{x})$$ ▲

When g and f are linear operators from R^n to R^m, we obtain the following solution to a problem posed by von Neumann.

PROPOSITION 4

Let $f:=(f^j_i)$ and $g:=(g^j_i)$ be two matrices satisfying

(21)
$$\begin{cases} \text{i.} & g^j_i\geqslant 0 \quad \textit{for all } i, j \quad (g \textit{ is nonnegative}) \\ \text{ii.} & \forall i=1,\dots,m, \quad \sum_{j=1}^{n} g^j_i>0 \\ \text{iii.} & \forall j=1,\dots,n, \quad \sum_{i=1}^{m} f^j_i>0 \end{cases}$$

Then there exist $\bar{x}\in\Sigma^n$, $\bar{p}\in\Sigma^m$ and $\delta>0$ such that

(22)
$$\begin{cases} \text{i.} & \delta f\bar{x}\leqslant g\bar{x} \\ \text{ii.} & \delta f^\star\bar{p}\geqslant g^\star\bar{p} \\ \text{iii.} & \delta\langle\bar{p}, f\bar{x}\rangle=\langle\bar{p}, g\bar{x}\rangle \end{cases}$$

Furthermore, for all $\mu>\delta$ and for all $y\in$ Int (R^m_+), there exists $\tilde{x}\in R^n_+$ such that

(23)
$$\mu f\tilde{x}-g\tilde{x}\leqslant y$$ ▲

When the dimensions n and m are equal, a boundary condition on f implies that the solutions $\bar{x}\in\Sigma^n$ and $\hat{x}\in\Sigma^n$ to the inclusions

(24)
$$\delta f(\bar{x})\in g(\bar{x})-R^n_+ \quad\text{and}\quad \beta y\in g(\hat{x})-\mu f(\hat{x})-R^n_+$$

are actually solutions to the equations

(25)
$$\delta f(\bar{x})=g(\bar{x}) \quad\text{and}\quad \beta y=g(\hat{x})-\mu f(\hat{x})$$

THEOREM 5
Let f be a single-valued map from Σ^n to R^n satisfying the properties

$$(26)\quad \begin{cases} \textbf{i.} & \text{The components } f_i \text{ of } f \text{ are convex and lower semicontinuous.} \\ \textbf{ii.} & \exists \tilde{p} \in \operatorname{Int}(R^n_+) \quad \text{such that} \quad \forall x \in \Sigma^n, \ \langle \tilde{p}, f(x) \rangle > 0 \\ \textbf{ii.} & \text{when } x_i = 0, \quad \text{then } f_i(x) \leqslant 0 \quad \text{(boundary condition)} \end{cases}$$

and let g be a singlevalued map from Σ^n to R^n satisfying

$$(27)\quad \begin{cases} \textbf{i.} & \text{The components } g_i \text{ of } g \text{ are concave and upper semicontinuous.} \\ \textbf{ii.} & \forall x \in \Sigma^n, \quad \forall i = 1, \ldots, n, \quad g_i(x) > 0 \end{cases}$$

a. *Let $\delta > 0$ be defined by*

$$(28)\qquad \frac{1}{\delta} := \sup_{p \in \Sigma^n} \inf_{x \in \Sigma^n} \frac{\langle p, f(x) \rangle}{\langle p, g(x) \rangle}$$

Then there exist $\bar{x} \in \operatorname{Int} R^n_+$ and $\bar{p} \in \operatorname{Int} R^n_+$ such that

$$(29)\qquad \begin{cases} \textbf{i.} & \delta f(\bar{x}) = g(\bar{x}) \\ \textbf{ii.} & \forall x \in \Sigma^n, \quad \langle g(x) - \delta f(x), \bar{p} \rangle \leqslant 0 \end{cases}$$

b. *Let $\mu > \delta$ and $y \in \operatorname{Int}(R^n_+)$ be given and $\beta < 0$ defined by*

$$(30)\qquad \beta := \inf_{p \in \Sigma^n} \sup_{x \in \Sigma^n} \frac{\langle p, g(x) - \mu f(x) \rangle}{\langle p, y \rangle}$$

Then there exists a solution $\hat{x} \in \operatorname{Int} R^n_+$ to the equation

$$(31)\qquad \beta y = g(\hat{x}) - \mu f(\hat{x}) \qquad \blacktriangle$$

Proof. **a.** We denote by e^j the jth element of the canonical basis: $e^j_j = 1$ and $e^j_k = 0$ when $k \neq j$.

The boundary condition (26, iii) on f implies that

$$(32)\qquad \forall k \neq j, \qquad f_k(e^j) \leqslant 0$$

and joined with the positivity condition (26, ii) on f, it implies that

$$(33)\qquad \forall j = 1, \ldots, n, \qquad f_j(e^j) > 0$$

because there exists $\tilde{p} \in \Sigma^n \cap \operatorname{Int}(R^n_+)$ such that

$$\tilde{p}^j f_j(e^j) + \sum_{k \neq j} \tilde{p}^k f_k(e^j) = \langle \tilde{p}, f(e^j) \rangle > 0$$

The conclusions of theorem 5 actually follow from these two properties (32) and (33).

b. By proposition 3, there exist $\bar{x} \in \Sigma^n$ and $\bar{p} \in \Sigma^n$ satisfying

$$(34) \quad \begin{cases} \textbf{i.} & \delta f(\bar{x}) \leqslant g(\bar{x}) \\ \textbf{ii.} & \forall x \in \Sigma^n, \quad \langle \bar{p}, g(x) - \delta f(x) \rangle \leqslant 0 \\ \textbf{iii.} & \langle \bar{p}, g(\bar{x}) - \delta f(\bar{x}) \rangle = 0 \end{cases}$$

We check first that $\bar{p}$ belongs to Int (R^n_+). By taking $x := e^j$ in inequality (34, ii) and using positivity assumption (27, ii), we deduce that

$$\bar{p}^j \delta f_j(e^j) + \sum_{k \neq j} \bar{p}^k \delta f_k(e^j) \geqslant \langle \bar{p}, g(e^j) \rangle > 0$$

Then inequalities (32) and (33) imply that $\bar{p}^j$ is strictly positive.

Let $\bar{z} = g(\bar{x}) - \delta f(\bar{x})$ belong to R^n_+ by (34, **i**). Then property (34, **iii**) implies that $\langle \bar{p}, \bar{z} \rangle = 0$. Since $\bar{p}$ belongs to Int (R^n_+), then $\bar{z}$ is equal to zero. Therefore, for all $i = 1, \ldots, n$, $\delta f_i(\bar{x}) = g_i(\bar{x})$, and since $g_i(\bar{x})$ is positive, $f_i(\bar{x})$ is also positive for all $i = 1, \ldots, n$. We thus deduce from the boundary condition (26, iii) that $\bar{x}_i > 0$ for all $i = 1, \ldots, n$.

c. The proof of the second statement of theorem 5, which is analogous to the proof of the first one, is left as an exercise. ■

The identity mapping obviously satisfies assumption (26) required on the map f. In this case, we state the following corollary on eigenvalues of positive concave operators.

COROLLARY 6

Let g be a single-valued map from Σ^n to Int R^n_+ *whose components are concave and upper semicontinuous.*

a. *Let $\delta > 0$ be defined by*

$$(35) \quad \frac{1}{\delta} := \sup_{p \in \Sigma^n} \inf_{x \in \Sigma^n} \frac{\langle p, x \rangle}{\langle p, g(x) \rangle}$$

There exist $\bar{x} \in$ Int R^n_+ *and $\bar{p} \in$* Int R^n_+ *such that*

$$(36) \quad \begin{cases} \textbf{i.} & \delta \bar{x} = g(\bar{x}) \\ \textbf{ii.} & \forall x \in \Sigma^n, \quad \langle g(x) - \delta x, \bar{p} \rangle \leqslant 0 \end{cases}$$

b. *Let $\mu > \delta$ and $y \in$* Int R^n_+. *Then there exist $\beta < 0$ and $\hat{x} \in$* Int R^n_+ *satisfying*

$$(37) \quad \beta y = g(\hat{x}) - \mu \hat{x}$$ ▲

When G is a positive matrix, we obtain the Perron–Frobenius theorem.

THEOREM 7

Let G be a positive matrix.

a. *Then G has a positive eigenvalue δ and a corresponding eigenvector $\bar{x}$ with positive components.*

b. *δ is the only eigenvalue of G for which there corresponds an eigenvector $\bar{x} \in \Sigma^n$.*

c. *δ is larger than or equal to the absolute value of any other eigenvalue of G.*

d. *The matrix $\mu - G$ is invertible and $(\mu - G)^{-1}$ is positive if and only if $\mu > \delta$.* ▲

Proof. **a.** By corollary 6, we know that there exist $\delta > 0$, $\bar{x} \in \text{Int } R^n_+$ and $\bar{p} \in \text{Int } R^n_+$ such that $\delta\bar{x} = G(\bar{x})$ and $G^\star\bar{p} - \delta\bar{p} \leqslant 0$. Actually, $G^\star\bar{p} - \delta\bar{p} = 0$ because $\langle \bar{x}, G^\star\bar{p} - \delta\bar{p} \rangle = \langle \bar{p}, G^\star\bar{x} - \delta\bar{x} \rangle = 0$ and because the components of $\bar{x}$ are strictly positive.

b. Let $x \in \Sigma^n$ and μ satisfy $\mu x = G(x)$. We deduce that

$$\mu\langle \bar{p}, x \rangle = \langle \bar{p}, G(x) \rangle = \langle G^\star\bar{p}, x \rangle = \delta\langle \bar{p}, x \rangle$$

Since $\langle \bar{p}, x \rangle$ is strictly positive because x belongs to Σ^n and $\bar{p}$ belongs to Int R^n_+, we can divide by $\langle \bar{p}, x \rangle$ and observe that $\mu = \delta$. Hence, the second statement is proved.

c. Let λ be an eigenvalue of G and let $z \in R^n$ be an associated eigenvalue. Equalities

$$\lambda z_i = \sum_{j=1}^{n} g_i^j z_j \qquad (i = 1, \ldots, n) \tag{38}$$

imply that

$$|\lambda|\,|z_i| \leqslant \sum_{j=1}^{n} g_i^j |z_j| \tag{39}$$

Let $|z|$ denote the vector of components $|z_j|$. Then $|\lambda|\,|z|$ belongs to $G|z| - R^n_+$ and thus $|\lambda| \leqslant \delta$.

d. We know that when $\mu > \delta$, the matrix $\mu - G$ is invertible because δ is the largest eigenvalue of G. We know that for all $y \in \text{Int } R^n_+$, the solution $(\mu - G)^{-1}y$ belongs to Int R^n_+ by the second statement of corollary 6. This implies that $(\mu - G)^{-1}$ is positive. Conversely, assume that $\mu - G$ is invertible and $(\mu - G)^{-1}$ is positive. We cannot have the inequality $\mu \leqslant \delta$, because we would deduce that $\mu\bar{x} \leqslant \delta\bar{x} = G\bar{x}$ and thus that

$$-\bar{x} = (\mu - G)^{-1}(G\bar{x} - \mu\bar{x}) \in R^n_+$$

since $\mu - G$ is invertible, $(\mu - G)^{-1}$ is positive, and $G\bar{x} - \mu\bar{x}$ is a positive vector. Hence, $\mu > \delta$. ■

We now recall the definition of a M-matrix $M=(m_i^j)$: It is a matrix that satisfies

(40) $$\forall i \neq j, \qquad m_i^j \leq 0$$

and

(41) $$\forall x \in \Sigma^n, \qquad \exists q \in \Sigma^n \ \text{such that} \ \langle q, Mx \rangle > 0$$

Let $b > \max_{i=1,\dots,n} m_i^i$. Then it is clear that condition (40) is equivalent to

(42) $$\forall x \in \Sigma^n, \qquad bx \in Mx + R_+^n$$

We shall extend the concept of M matrix to the concept of M-convex operator in the following way.

DEFINITION 8

Let H be a convex process from R_+^n to R^n. We say that H is a M-convex process if it satisfies both properties (43) *and* (44):

(43) $$\exists b \in R \ \text{such that} \ \forall x \in R_+^n, \qquad bx \in H(x) + R_+^n$$

and

(44) $$\forall x \in \Sigma^n, \qquad \exists q \in \text{Int } R_+^n \ \text{such that} \ \inf_{y \in H(x)} \langle q, y \rangle > 0 \qquad \blacktriangle$$

PROPOSITION 9

Let H be a strict upper semicontinuous, convex process with compact images from R_+^n to R^n. If H is a M-convex process, then

(45) $$\forall y \in \text{Int } R_+^n, \qquad \exists x \in R_+^n \ \text{such that} \ y \in H(x) + R_+^n \qquad \blacktriangle$$

Proof. Let μ be larger than the number b involved in assumption (43). We introduce the map G from Σ^n to R^n defined by

(46) $$\forall x \in \Sigma^n, \qquad G(x) := \mu x - H(x)$$

and we take F to be the identity mapping.

Therefore, we can use proposition 1: There exist $\delta > 0$ and $\bar{x} \in \Sigma^n$ satisfying

(47) $$\delta \bar{x} \in G(\bar{x}) - R_+^n = \mu \bar{x} - H(\bar{x}) - R_+^n$$

We make use of assumption (44): Let $\bar{q} \in \text{Int } R_+^n$ such that $\inf_{y \in H(\bar{x})} \langle \bar{q}, y \rangle > 0$.

Since $(\mu-\delta)\bar{x}$ belongs to $H(\bar{x})+R^n_+$ by (44), we deduce that

$$(\mu-\delta)\langle\bar{q}, \bar{x}\rangle \geqslant \inf_{y \in H(\bar{x})} \langle\bar{q}, y\rangle > 0$$

then $\mu-\delta>0$ because $\langle\bar{q}, \bar{x}\rangle$ is strictly positive.

Now, we apply proposition 2. We can associate to any $y \in \text{Int } R^n_+$ elements $\hat{x} \in \Sigma^n$ and $\beta<0$ such that

$$\beta y \in G(\hat{x})-\mu\hat{x}-R^n_+ = -H(\hat{x})-R^n_+$$

We divide this inclusion by $-\beta$, and we set $x := \hat{x}/(-\beta)$. Then $y \in H(x)+R^n_+$. ■

We consider now the case when

$$H(x)=h(x)+R^n_+$$

where h is a positively homogeneous concave operator.

DEFINITION 10

Let h be a single-valued map from R^n_+ to R^n whose components satisfy

(48) $\forall i=1, \ldots, n$, h_i *is convex, lower semicontinuous, and positively homogeneous*

We say that it is a M-operator if

(49) $$\exists b \in R \quad \text{such that} \quad \forall x \in R^n_+, \quad bx_i \geqslant h_i(x)$$

and

(50) $$\forall x \in \Sigma^n, \qquad \exists q \in \Sigma^n \quad \text{such that} \quad \langle q, h(x)\rangle > 0$$ ▲

THEOREM 11

Let h be a single-valued map from R^n_+ to R^n satisfying properties (48), (49), *and* (50). *Then h maps* Int R^n_+ *onto itself*

(51) $$\forall y \in \text{Int } R^n_+, \qquad \exists x \in \text{Int } R^n_+ \quad \text{such that} \quad hx=y$$ ▲

When h is a matrix, we obtain the characterization of M matrices.

THEOREM 12

Let h be a matrix satisfying (40). *Then the following statements are equivalent*:

a. *h is a M matrix.*

b. *h is invertible, and h^{-1} is positive.*

c. *$h^\star$ is invertible, and h^{-1} is positive.* ▲

Proof. The implication **a**$\Rightarrow$**b** follows from the Perron–Frobenius theorem. The implication **b**$\Rightarrow$**c** is obvious. It remains to check that **c** implies **a**. Let $p \in \operatorname{Int} R^n_+$ be the solution to $h^\star p = \mathbb{1}$, where $\mathbb{1}$ is the vector of components 1. Then, for all $x \in \Sigma^n$,

$$\langle p, hx \rangle = \langle h^\star p, x \rangle = \sum_{i=1}^{n} x_i = 1$$

Therefore, property (41) is satisfied and h is a M matrix. ∎

CHAPTER 4

Convex Analysis and Optimization

The main objective of this chapter is the study of convex minimization problems

$$(*) \qquad W(y) := \inf_{x \in X} V(x, y)$$

depending on a parameter y when the function V is convex.

Besides sufficient conditions implying the existence of solutions x_y to these minimization problems, we are looking for equations or inclusions that characterize those solutions (variational principles), and we are studying differentiable properties of the "marginal function" W defined by (*) and the set of minimizers x_y of $x \to V(x, y)$ with respect to the parameter y.

Characterization of solutions to minimization problems as solutions to equations (or inclusions) is a very old problem, since Fermat discovered the famous rule

$$\text{if } \bar{x} \text{ minimizes } U, \quad \text{then } U'(\bar{x}) = 0$$

for algebraic functions in 1637, and later in 1684, Leibniz extended it to differentiable functions.

This "Fermat rule" is still the object of recent works, when the function U is no longer differentiable. Indeed, usual differentiability is not stable for the pointwise supremum, so that in the framework of optimization and game theory, we very naturally meet nondifferentiable functions. Before considering the most general case in Chapter 7, we shall restrict our attention to the pointwise suprema of affine continuous functions, which are convex, lower semicontinuous functions (and, as we shall see, which make up the whole class of convex, lower semicontinuous functions).

Consider the simplest example: We minimize the convex function $x \to |x|$. It achieves its minimum at $x = 0$, and we are unable to write the Fermat rule, because this function is not differentiable at this point. But $x \to |x|$ is the supremum of the affine functions $x \to ax + b$ with $a \in [-1, +1]$ and $b \leqslant 0$. When x is negative, there is only one such affine function passing through $(x, |x|)$, that is, its tangent $x \to -x$, whose slope is -1: It is the gradient of the function at this point. When x is positive, there is still a smaller affine function passing through $(x, |x|)$; that is, its tangent $x \to x$, whose slope is 1. When $x=0$, there is no tangent but "subtangents"; that is, the smaller affine functions $x \to ax$, $a \in [-1, +1]$, passing through (0, 0). The revolutionary idea was to suggest taking the set $[-1, +1]$ of all the gradients of these affine functions—called subgradients—as a candidate for replacing the missing concept of gradient; this set is called the subdifferential of $x \to |x|$ at zero. The Fermat rule still holds true in this case, because zero belongs to the subdifferential $[-1, +1]$. This idea works in the general case; the price to pay to hold the Fermat rule true for convex minimization problems was to accept dealing with "set-valued gradients"—so to speak—of convex functions.

The adaptation of the Fermat rule and a decent "subdifferentiable calculus" made convex analysis more and more indispensable not only for studying convex programs (convex minimization problems in finite dimensional spaces) but also for problems in calculus of variations and optimal control. The ideas of Euler, Lagrange, and Hamilton can be adapted to nonsmooth problems as long as they are convex.

Convex analysis also deals with set-valued maps with closed convex graph and convex sets. It is customary to begin the presentation of convex analysis with the study of convex function and then deduce the properties of normal cones to subsets, and so on. We shall follow the inverse route: In the first section, we begin with the study of tangent and normal cones to convex subsets, then give some examples and devise a set of formulas allowing the characterization of tangent and normal cones. In the second section, we adapt to the case of set-valued maps the ancient geometrical concept of the derivative of a real-valued function, whose graph is the tangent to the graph of the function.

When F is a set-valued map from X to Y with a closed convex graph and (x_0, y_0) belongs to the graph of F, we define the derivative of F at (x_0, y_0) as the closed convex process $DF(x_0, y_0)$ from X to Y, whose graph is the tangent cone to the graph of F at (x_0, y_0). It possesses most of the virtues we expect from a derivative. Its transpose $DF(x_0, y_0)^*$, a closed convex process from Y^* to X^*, called the codifferential of F at (x_0, y_0), naturally plays an important role.

When V is a single-valued map from X to $R \cup \{+\infty\}$ and we are interested in only minimization properties where the order relation of R plays a crucial role, we associate with V the set-valued map $\mathbf{V}_+$ from X to R defined by

$$\mathbf{V}_+(x) := V(x) + R_+ \quad \text{if } V(x) < +\infty, \qquad \mathbf{V}_+(x) := \varnothing \quad \text{if } V(x) = +\infty$$

whose graph is the epigraph of V.

Hence, $\mathbf{V}_+$ has a closed convex graph if and only if V is lower semicontinuous and convex. In section 3, we observe that the graph of the derivative $D\mathbf{V}_+(x, V(x))$ of the set-valued map $\mathbf{V}_+$ at $(x, V(x))$ is the epigraph of a function we shall denote by $D_+V(x)$, called the epiderivative, defined by

$$D_+V(x)(v) = \liminf_{\substack{h \to 0_+ \\ u \to v}} \frac{V(x+hu) - V(x)}{h}$$

We also observe that the transpose $D\mathbf{V}_+(x, V(x))^*$, a closed convex process from R to X^*, satisfies

$$D\mathbf{V}_+(x, V(x))^*(\lambda)=\begin{cases}\lambda\partial V(x) & \text{when } \lambda\geqslant 0\\ \varnothing & \text{when } \lambda<0\end{cases}$$

where

$$\partial V(x):=D\mathbf{V}_+(x, V(x))^*(1)$$

is the subdifferential of the function V at x. In summary, we present a unified treatment of convex analysis; starting with the definition and properties of tangent cones to convex sets, deducing the definition and the properties of derivatives to set-valued maps with convex graphs, which are convex processes, and then the definition and properties of the epiderivative of a convex function.

But there is more to that. In the fourth section, where we consider the specific class of perturbations of a convex minimization problem

$$\forall p\in X^*, \quad W(p):=\inf_{x\in X}\,[V(x)-\langle p, x\rangle]$$

we observe that the function V^* defined by $V^*(p):=-W(p)$ is also a convex, lower semicontinuous function on the dual of X, which is called the *conjugate function*:

$$V^*(p):=\sup_{x\in X}\,[\langle p, x\rangle - V(x)]$$

We define the biconjugate function V^{**} by

$$V^{**}(x):=\sup_{p\in X^*}\,[\langle p, x\rangle - V^*(p)]$$

which is also convex and lower semicontinuous.

The condition $V=V^{**}$, which is necessary for V to be convex and lower semicontinuous, happens to be also *sufficient*. This shows that the correspondence $V\to V^*$ is a one to one correspondence between convex, lower semicontinuous functions on X and its dual. This duality result will play an important role. We also observe that this result implies that any lower semicontinuous, convex function V can be obtained as the supremum of the continuous affine functions $x\to\langle p, x\rangle - V^*(p)$ as p ranges over X^*.

It is easy to check that the subdifferential

$$\partial V(x):=\{p\in X^* | \langle p, x\rangle = V(x)+V^*(p)\}$$

is the set of gradients p of the smaller affine functions $x\to\langle p, x\rangle - V^*(p)$ passing through $(x, V(x))$. The symmetry of this formula implies at once that

$$p \in \partial V(x) \Leftrightarrow x \in \partial V^*(p)$$

that is, the set-valued map $p \to \partial V^*(p)$ is the inverse of the map $x \to \partial V(x)$

$$\partial V^* = (\partial V)^{-1}$$

This formula shows that the conjugate function plays the same role in the convex framework as the *Legendre transform* in the framework of smooth analysis; actually, they coincide when V is both convex and smooth. Now, if we return to the framework of the minimization problem

$$v := \inf_{x \in X} V(x)$$

when V is a convex, lower semicontinuous function, we observe that

(a) $\inf_{x \in X} V(x) = -V^*(0)$

(b) $\bar{x}$ minimizes V on $X \Leftrightarrow 0 \in \partial V(\bar{x})$

(c) The set of minimizers of V is $\partial V^*(0)$.

Statement (b) is the Fermat rule, and statement (c) allows us to characterize the set of minimizers as the subdifferential of V^* at zero. Therefore, conditions on the conjugate function V^* (implying that V^* is subdifferentiable at zero) are sufficient conditions for the existence of a minimizer of V. This is the essence of the duality theory for the convex minimization problems.

In Section 5, we consider a family of optimization problems

$$W(y) := \inf_{x \in X} V(x, y)$$

where V is a proper lower semicontinuous, convex function from $X \times Y$ to $R \cup \{+\infty\}$. We associate with these families of problems the function h^* defined on $X^* \times Y$ by

$$h^*(p, y) := \sup_{x \in X} [\langle p, y \rangle - V(x, y)]$$

which is convex with respect to p and concave with respect to y. Assume that $\bar{x}$ is a solution to the minimization problem $W(\bar{y})$. We shall prove that the following statements are equivalent.

(a) $\bar{q} \in \partial W(\bar{y})$

(b) $(0, \bar{q}) \in \partial V(\bar{x}, \bar{y})$

(c) $\bar{x} \in \partial_p h^*(0, \bar{y})$ and $\bar{q} \in \partial_y(-h^*)(0, \bar{y})$

The elements $\bar{q}$ given by one of these equivalent statements are called *Lagrange multipliers* of the minimization problems. Formula **(c)** shows how the pairs $(\bar{x}, \bar{q})$ of solutions and Lagrange multipliers evolve with respect to the parameter y: They depend on the regularity of the set-valued map

$$y \to \partial_p h^*(0, y) \times \partial_y(-h^*)(0, y)$$

We shall consider the particular case when the minimization problems are of the form

$$W(y) := \inf_{x \in F^{-1}(y)} V(x)$$

where $U: X \to R \cup \{+\infty\}$ is a proper lower semicontinuous function and F is a set-valued map from X to Y with a closed convex graph. Let $\bar{x} \in F^{-1}(\bar{y})$ be a solution to the problem $W(\bar{y})$. Then we shall prove that the condition

$$0 \in \text{Int}\,(\text{Dom}\, F - \text{Dom}\, U)$$

implies the formula

$$\partial W(\bar{y}) = D(F^{-1})(\bar{y}, \bar{x})^* \partial U(\bar{x})$$

between the subdifferentials ∂W and ∂U and the codifferential of the set-valued map F defining the constraints.

We consider in the sixth section more specific problems of the form

$$W(y) := \inf_{x \in X} \left[U(x) - \langle p, x\rangle + V(Ax + y)\right]$$

where $U: X \to R \cup \{+\infty\}$ and $V: Y \to R \cup \{+\infty\}$ are proper lower semicontinuous, convex functions and A is a continuous linear operator from X to Y. We shall associate to it its "dual problem"

$$W^{\circledast}(p) := \inf_{q \in Y^*} \left[U^*(-A^*q + p) + V^*(q) - \langle q, y\rangle\right]$$

We shall prove that the condition

$$p \in \text{Int}\,(A^*\, \text{Dom}\, V^* + \text{Dom}\, U^*)$$

on the dual problem implies that there exist solutions $\bar{x}$ to the problem $W(y)$ and that the condition

$$y \in \text{Int}\,(\text{Dom}\, V - A\, \text{Dom}\, U)$$

on the initial problem implies the existence of solutions $\bar{q}$ to the dual problem (called Lagrange multipliers of the initial problem). When both assumptions are satisfied, we prove that the set of solutions to $W(y)$ is the subdifferential $\partial W^{\circledast}(p)$ and that the set of solutions to the dual problem $W^{\circledast}(p)$ is the subdifferential $\partial W(y)$: A very interesting phenomenon for economists of the marginal school. The solutions $\bar{x}$ and $\bar{q}$ to the minimization problems $W(y)$ and $W^{\circledast}(p)$ are solutions to the system of inclusions

$$\left.\begin{aligned} p &\in \partial U(\bar{x}) + A^*\bar{q} \\ y &\in -A\bar{x} + \partial V^*(\bar{q}) \end{aligned}\right\} \text{(abstract Hamiltonian system)}$$

By eliminating $\bar{q}$ in these inclusions, we observe that the solutions $\bar{x}$ to $W(y)$ are solutions to the inclusions

$$p \in \partial U(\bar{x}) + A^*\partial V(A\bar{x}+y) \qquad \text{(abstract Euler–Lagrange equation)}$$

and by eliminating $\bar{x}$, we note that the solutions $\bar{q}$ to $W^*(p)$ are solutions to the inclusion

$$y \in \partial V^*(\bar{q}) - A\partial U^*(p - A^*\bar{q}) \qquad \text{(dual Euler–Lagrange equation)}$$

We shall define in Chapter 7 (on nonsmooth analysis) the concepts of generalized second derivatives, derivatives of set-valued maps and devise inverse function theorems for set-valued maps. Applied to our minimization problems, they imply that if the matrix of closed convex processes

$$\begin{pmatrix} \partial^2 U & A^* \\ -A & \partial^2 V^* \end{pmatrix}$$

from $X \times Y^*$ to $X^* \times Y$ is surjective, then the solutions $(\bar{x}, \bar{q})$ of the problems $W(y)$ and $W^{\circledast}(p)$ depend in a Lipschitz manner on the parameters y and p, and the marginal variations δx, δq of the solutions depend on the marginal variations of the parameters through the formula

$$\begin{pmatrix} \delta x \\ \delta q \end{pmatrix} \in \begin{pmatrix} \partial^2 U & A^* \\ -A & \partial^2 V^* \end{pmatrix}^{-1} \begin{pmatrix} \delta p \\ \delta y \end{pmatrix}$$

In the last section, we consider minimization problems of the form

$$v := \inf_{x \in X} L(x, Ax)$$

where L is a proper lower semicontinuous function from $X \times Y$ to $R \cup \{+\infty\}$ and A belongs to $\mathscr{L}(X, Y)$. We associate to it its *dual problem*

$$v^* := \inf_{q \in Y^*} L^*(-A^*q, q)$$

and the nonnegative function $\mathscr{A}$ defined on $X \times Y^*$ by

$$\mathscr{A}(x, q) := L(x, Ax) + L^*(-A^*q, q)$$

We observe that $\bar{x} \in X$ is a solution to v and $\bar{q} \in Y^*$ is a solution to v^* and $v + v^* = 0$ if and only if $\mathscr{A}(\bar{x}, \bar{q}) = 0$.

We associate to the function L, which plays the role of a Lagrangian in the calculus of variations, the function H defined on $X \times Y^*$ by

$$H(x, q) := \sup_{y \in Y} [\langle q, y\rangle - L(x, y)]$$

which plays the role of an Hamiltonian, which is concave with respect to x and convex with respect to q. We shall observe that the solutions $(\bar{x}, \bar{q})$ of the minimization problems v and v^* are the solutions to the Hamiltonian inclusion

$$-A^*\bar{q} \in \partial_x(-H)(\bar{x}, \bar{q}) \qquad \text{and} \qquad A\bar{x} \in \partial_q H(\bar{x}, \bar{q})$$

and that the solutions $\bar{x}$ to the minimization problem v are solutions to the Euler–Lagrange inclusion

$$0 \in (1 \oplus A^*)\partial L(\bar{x}, A\bar{x})$$

where $1 \oplus A^*$ is the continuous linear operator from $X^* \times Y$ to X^* defined by

$$(1 \oplus A^*)(p, q) := p + A^*q$$

If we assume that the Hamiltonian is convex and, thus, the Lagrangian is concave with respect to x and convex with respect to y, we can still write the Hamiltonian inclusion in the form

$$(A^*\bar{q}, A\bar{x}) \in \partial H(\bar{x}, \bar{q})$$

and the Euler–Lagrange inclusion in the form

$$0 \in -\partial_x(-L)(\bar{x}, A\bar{x}) + A^*\partial_y L(\bar{x}, A\bar{x})$$

Since the Lagrangian is no longer convex, there is no hope of solving the minimization problem v. But we shall prove that solutions $(\bar{x}, \bar{q})$ to the Hamiltonian inclusion are the solutions to the equation

$$\mathscr{B}(\bar{x}, \bar{q}) = 0$$

where $\mathscr{B}$ is the nonnegative function defined on $X \times Y^*$ by

$$\mathscr{B}(x, q) := (H(x, q) - \langle q, Ax\rangle) + (H^*(A^*q, Ax) - \langle A^*q, x\rangle)$$

We observe that the solutions $(\bar{x}, \bar{q})$ of the minimization problems

$$w := \inf_{(x,q) \in X \times Y^*} (H(x, q) - \langle q, Ax \rangle)$$

and

$$w^* := \inf_{(x,q) \in X \times Y^*} (H^*(A^*q, Ax) - \langle A^*q, x \rangle)$$

are also solutions to the Hamiltonian inclusion. The problem w plays the role of the *least action principle*, whereas problem w^* can be said to be the *dual least action principle.*

We shall apply these ideas to solving problems in calculus of variations in the last chapter.

1. TANGENT AND NORMAL CONES TO CONVEX SUBSETS

Let X be a normed space.

DEFINITION 1

Let $K \subset X$ be a convex subset and $x \in K$. We denote by

$$S_K(x) := \bigcup_{h>0} \frac{1}{h}(K - x) \tag{1}$$

the cone spanned by $K - x$ and by

$$T_K(x) := \mathrm{cl}\left(\bigcup_{h>0} \frac{1}{h}(K - x)\right) \tag{2}$$

its closure. $T_K(x)$ is called the tangent cone to K at x. ▲

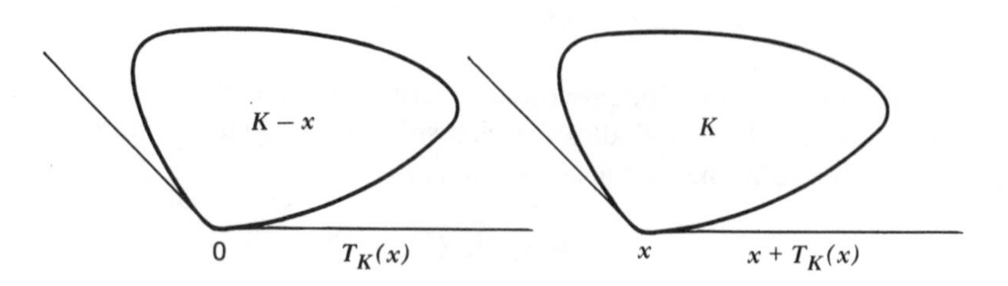

We observe that

$$\forall v \in S_K(x), \quad \exists h > 0 \quad \text{such that} \quad \forall t \in [0, h], \quad x + tv \in K \tag{3}$$

Indeed

$$x+tv=\left(1-\frac{t}{h}\right)x+\frac{t}{h}(x+hv)$$

is a convex combination of elements belonging to the convex set K. We can interpret this remark by saying that when v belongs to $S_K(x)$, the beginning of the "half-curve" starting from x defined by

$$\forall t \in [0, h], \qquad \phi(t)=x+tv$$

is contained in K. Hence, the cone $S_K(x)$ is the set of directions v such that the half-lines $x+R_+v$ intersect K. This subsumes the idea lying behind the concept of tangent. We point out that v belongs to the tangent cone $T_K(x)$ if and only if

$$\text{(4)} \qquad \forall \varepsilon>0, \quad \exists u \in v+\varepsilon B, \quad \exists k>0 \quad \text{such that} \quad x+ku \in K$$

or, equivalently,

(5) there exist sequences of vectors $u_n \in X$ converging to U and of numbers $h_n>0$ such that $x+h_nu_n \in K$ for all n.

Condition (3) implies that if condition (4) is satisfied for some $k>0$, it is also satisfied for all $h<k$. It follows that

$$\text{(6)} \qquad T_K(x)=\bigcap_{\varepsilon>0}\bigcap_{\alpha>0}\bigcup_{h\in]0,\alpha]}\left(\frac{1}{h}(K-x)+\varepsilon B\right)$$

We shall see in Chapter 7 on nonsmooth analysis that the right-hand side of this formula defines the contingent cone to K at x.

We observe that

$$\text{(7)} \qquad \forall x \in K, \qquad T_K(x)=T_{\bar{K}}(x)$$

and that

$$\text{(8)} \qquad \text{if } x \in \text{Int}(K), \quad \text{then } S_K(x)=T_K(x)=X$$

since $K-x$ contains a neighborhood of the origin. It is also clear that $T_K(x)$ is contained in the closed vector subspace $M(K)=\text{cl}[\{\alpha K-\beta K\}_{\alpha,\beta \in R}]$ *spanned by* K. We note that

$$\text{(9)} \qquad K \subset x+S_K(x) \subset x+T_K(x)$$

PROPOSITION 2

The cones $S_K(x)$ and $T_K(x)$ are convex. ▲

Proof. Indeed, if v_1 and v_2 belong to $S_K(v)$, then $x+h_i v_i \in K$ for $i=1, 2$; let $h=\min(h_1, h_2)$. Then by (3), $x+hv_i \in K$ for $i=1, 2$. Hence,

$$x+h(\alpha v_1+(1-\alpha)v_2) \in K \quad \text{when} \quad \alpha \in [0, 1]$$

Since the closure of a convex cone is still a convex cone, we have proved the proposition. ■

A normal to a smooth submanifold at a point x is any vector orthogonal to the tangent space. In the case of tangent cones to convex subsets, the concept of an orthogonal subspace to the tangent space has to be replaced by the negative polar cone to the tangent cone, which will be called the normal cone. The introduction of these two concepts, tangent and normal cones, allows the use of duality relations.

DEFINITION 3

Let K be a nonempty convex subset of X. The normal cone $N_K(x)$ to K at x is the negative polar cone to the tangent cone. ▲

PROPOSITION 4

The normal cone $N_K(x)$ is equal to

(10) $$N_K(x) := \{p \in X^\star \text{ such that } \langle p, x\rangle = \max\{\langle p, y\rangle \mid y \in K\} := \sigma_K(p)\}.$$ ▲

Note that $T_K(x)=N_K(x)^-$, since $T_K(x)$ is a closed convex cone.

Proof. If $p \in T_K(x)^-$, then $\langle p, y-x\rangle \leqslant 0$ for all $y \in K$, since $v=y-x \in T_K(x)$ when $y \in K$.
Conversely, let p satisfy $\langle p, x\rangle=\sigma_K(p)$ and $v=\lim_{n\to\infty} \lambda_n(y_n-x) \in T_K(x)$, where $\lambda_n \geqslant 0$ and $y_n \in K$. Hence, $\langle p, v\rangle \leqslant 0$, since $\langle p, \lambda_n(y_n-x)\rangle=\lambda_n\langle p, y_n-x\rangle \leqslant 0$ for all $n \geqslant 0$. ■

PROPOSITION 5

Assume that X is a Hilbert space. Let π_K be the projection of best approximation onto a closed convex subset K. Then $\pi_K^{-1}(x)=x+N_K(x)$ and v belong to $T_K(x)$ if and only if

$$\langle y-x, v\rangle \leqslant 0 \qquad \forall y \in \pi_K^{-1}(x)$$ ▲

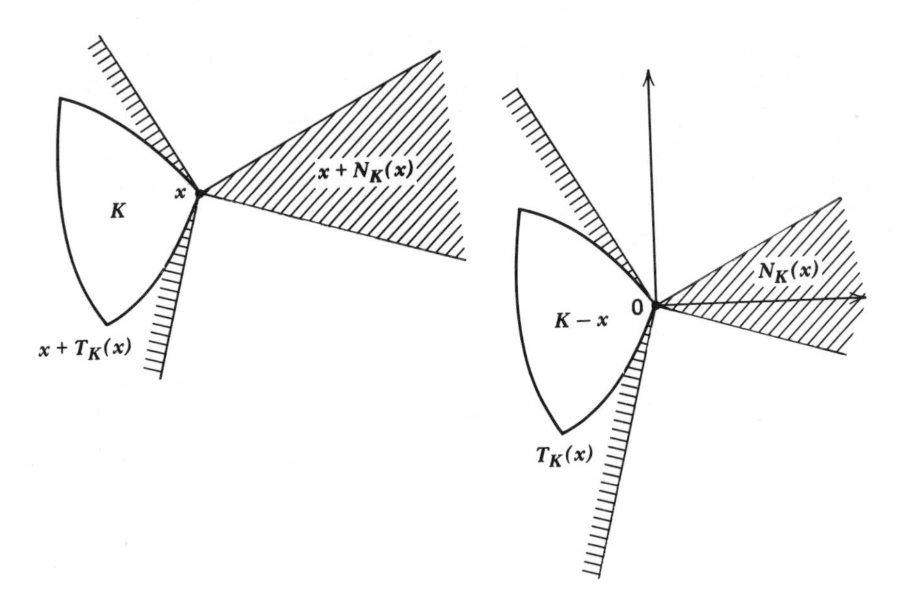

PROPOSITION 6

Let K be a closed subset of a normed space x. Then

$$x \to N_K(x) \text{ has a closed graph.} \tag{11}$$

If X is finite dimensional, then

$$x \to T_K(x) \text{ is lower semicontinuous.} \tag{12}$$

▲

Proof. **a.** Let (x_n, p_n) be a sequence of elements of the graph of $N(\cdot)$ converging to (x, p). For all $y \in K$, we have $\langle p_n, y\rangle \leq \langle p_n, x_n\rangle$. Hence, letting $n \to \infty$, we deduce that $\langle p, y\rangle \leq \langle p, x\rangle$. Thus, (x, p) belongs to the graph of $N(\cdot)$.
b. The second statement is equivalent to the first by proposition 3.1.18. ■

Remark

We can prove that $T_K(\cdot)$ is lower semicontinuous when X is a Hilbert space.

PROPOSITION 7

If K has a nonempty interior, then for all $x \in K$, the interior of the tangent cone is nonempty and spanned by Int $K - x$

$$\text{Int } T_K(x) = \bigcup_{h>0} \frac{1}{h}(\text{Int } K - x) \tag{13}$$

Furthermore, the set-valued map $x \to \text{Int } T_K(x)$ has an open graph. ▲

Proof. **a.** The cone $\bigcup_{h>0}(1/h)(\text{Int } K-x)$ is open, being a union of open subsets. Hence, it is contained in Int $T_K(x)$. Since Int $T_K(x)=$ Int $S_K(x)$, it suffices to prove that any $v \in$ Int $S_K(x)$ belongs to some $(1/h)(\text{Int } K-x)$. Let $\eta>0$ be such that $v+\eta B\subset S_K(x)$. If $x+v$ belongs to Int K, the proof is finished. If not, let x_0 belong to Int K and let us set $v_0:=x_0-x$. Hence, $v-(\eta/\|v_0\|)v_0$ belongs to $S_K(x)$, and, consequently, there exists $h>0$ such that $x+h(v-(\eta/\|v_0\|)v_0)$ belongs to K. Let us set $\alpha:=h\eta/(h\eta+\|v_0\|)$. We observe that

$$x+(1-\alpha)hv=\alpha x_0+(1-\alpha)\left(x+h\left(v-\frac{\eta}{\|v_0\|}v_0\right)\right)$$

Since x_0 belongs to Int K and $x+h(v-(\eta/\|v_0\|)v_0)$ belongs to K and since α belongs to $]0, 1[$, then $x+(1-\alpha)hv$ also belongs to the interior of K.

This proves that v belongs to

$$\frac{1}{(1-\alpha)h}(\text{Int } K-x)$$

b. Let $v_0 \in$ Int $T_K(x_0)$. Therefore, $v_0 \in (1/h_0)(\text{Int } K-x_0)$ for some $h_0>0$; hence, there exists $\varepsilon>0$ such that

$$x_0+h_0v_0+\varepsilon B=x_0+h_0\left(v_0+\frac{\varepsilon}{h_0}B\right)\subset \text{Int } K$$

Take $x \in x_0+\varepsilon/2B$ and $v \in v_0+\varepsilon/2h_0B$. Then

$$x+h_0v \in x_0+h_0v_0+\varepsilon B\subset \text{Int } K$$

and, therefore, $v \in$ Int $T_{K'}(x)$. Hence, the graph of $x\to$ Int $T_K(x)$ is open. ■

We now list a few examples.

PROPOSITION 8

Let B be the unit ball of a Hilbert space and x belong to B. Then

$$(14)\quad \begin{cases} \textbf{i.} & T_B(x)=X \quad \text{if} \quad x\in \text{Int } B, \quad \text{and} \quad T_B(x)=\{x\}^- \quad \text{if} \quad \|x\|=1 \\ \textbf{ii.} & N_B(x)=\{0\} \quad \text{if} \quad x\in \text{Int } B, \quad \text{and} \quad N_B(x)=R_+x \quad \text{if} \quad \|x\|=1 \end{cases} \blacktriangle$$

Proof. We take $\|x\|=1$. Then $p \in N_K(x)$ if and only if $\|p\|_*=\sup_{y\in B}\langle p, y\rangle=\langle p, x\rangle$. By the Cauchy–Schwarz inequality, this is equivalent to $p=\lambda x$ with $\lambda>0$. By polarity, we deduce the formula for the tangent cone. ■

PROPOSITION 9

Let $K \subset X$ be a closed convex cone. Then $N_K(x) = K^- \cap \{x\}^\perp$, and thus

(15) $\quad v \in T_K(x)$ *if and only if* $\langle p, v\rangle \leq 0$ *for all* $p \in K^-$ *satisfying* $\langle p, x\rangle = 0$

If K is a closed subspace, then $T_K(x) = K$ and $N_K(x) = K^\perp$. ▲

Proof. It is clear that $K^- \cap \{x\}^\perp$ is contained in $N_K(x)$. *Conversely*, if $p \in N_K(x)$, then $\langle p, x\rangle = \max_{y \in K} \langle p, y\rangle$. Since K is a cone, we deduce that $\langle p, x\rangle = 0$ and $p \in K^-$. ■

PROPOSITION 10

Let $A \in \mathscr{L}(X, Y)$ and $K = A^{-1}(y)$ be an affine subspace. Then if $Ax = y$,

$$T_{A^{-1}(y)}(x) = \operatorname{Ker} A \tag{16}$$ ▲

Proof. **a.** If $v \in \operatorname{Ker} A$, then $v + x \in A^{-1}(y) = K$, and thus $v = v + x - x$ belongs to $T_K(x)$.

b. Conversely, if $v = \lim_{n\to\infty} v_n \in T_K(x)$, where $v_n = \lambda_n(x_n - x)$ with $x_n \in K$ and $\lambda_n > 0$, then $v_n \in \operatorname{Ker} A$ and thus $v \in \operatorname{Ker} A$.

PROPOSITION 11

Let

$$\Sigma^n := \left\{ x \in R^n_+ \,\middle|\, \sum_{i=1}^n x_i = 1 \right\}$$

and

$$I(x) := \{ i = 1, \ldots, n \mid x_i = 0 \}$$

Then

(17) $\quad v \in T_{R^n_+}(x) \quad$ *if and only if* $\quad v_i \geq 0 \quad$ *for all* $\quad i \in I(x)$

and

(18) $\quad v \in T_{\Sigma^n}(x) \quad$ *if and only if* $\quad v_i \geq 0 \quad$ *for all* $\quad i \in I(x) \quad$ *and* $\quad \sum_{i=1}^n v_i = 0$

▲

Proof. **a.** If $K = R^n_+$, the first statement follows from proposition 9. Indeed, if $p \in R^n$ satisfies

$$\sum_{i=1}^n p_i x_i = \sum_{i \notin I(x)} p_i x_i = 0$$

then $p_i=0$ whenever $i \notin I(x)$; hence, $v \in T_{R^n_+}(x)$ if $\sum_{i \in I(x)} p_i v_i \leqslant 0$ for all $p \in R^n_+$, that is, if and only if (17) holds.

b. Let v satisfy $v_i \geqslant 0$ if $i \in I(x)$ and $\sum_{i=1}^n v_i=0$. If $v_i=0$ for all $i \notin I(x)$, then $v=0$. If not, let

$$\lambda = \min_{\substack{i \notin I(x) \\ v_i \neq 0}} \frac{x_i}{|v_i|} > 0$$

Therefore, $x+\lambda v \in \Sigma^n$ since

$$x_i+\lambda v_i=\lambda_i v_i \geqslant 0 \quad \text{if} \quad i \in I(x)$$

$$x_i+\lambda v_i \geqslant x_i-\lambda|v_i| \geqslant x_i-x_i=0 \quad \text{if} \quad i \notin I(x)$$

and

$$\sum_{i=1}^n (x_i+\lambda v_i)=1+0=1$$

Hence,

$$v \in \frac{1}{\lambda}(\Sigma^n-x) \in T_{\Sigma^n}(x)$$

c. If $v=\lambda(y-x)$, where $y \in \Sigma^n$ and $\lambda>0$, then we deduce that

$$v_i=\lambda(y_i-x_i)=\lambda y_i \geqslant 0 \quad \text{when} \quad i \in I(x)$$

and $\sum_{i=1}^n v_i=0$. Therefore

$$\bigcup_{\lambda>0} \lambda(\Sigma^n-x) \subset \left\{v \in T_{R^n_+}(x) \,\middle|\, \sum_{i=1}^n v_i=0\right\}$$

Since the latter subset is closed, we deduce that

$$T_{\Sigma^n}(x)=\left\{v \in T_{R^n_+}(x) \,\middle|\, \sum_{i=1}^n v_i=0\right\} \qquad \blacksquare$$

The concept of tangent and normal cones is useful only if enough formulas allow us to characterize (or compute) tangent cones. For instance, we would like to know tangent cones to products, intersections, direct or inverse images by linear operators of sets whose tangent cones are known. We begin by stating obvious properties. All the subsets involved are naturally assumed to be convex.

PROPOSITION 12

a. *If* $x \in K \subset L$, *then*

$$T_K(x) \subset T_L(x) \quad \text{and} \quad N_L(x) \subset N_K(x) \tag{19}$$

b. *Let* $K := \bigcap_{i \in J} K_i$ *and let* $J(x) = \{i | x_i \notin \text{Int } K_i\}$
Then

$$T_K(x) \subset \bigcap_{i \in J(x)} T_{K_i}(x) \tag{20}$$ ▲

PROPOSITION 13

Let $K := \prod_{i=1}^{n} K_i$ *and* $\vec{x} = (x_1, \ldots, x_n) \in \vec{K}$. *Then*

$$T_{\vec{K}}(\vec{x}) = \prod_{i=1}^{n} T_{K_i}(x_i) \quad \text{and} \quad N_{\vec{K}}(\vec{x}) = \prod_{i=1}^{n} N_{K_i}(x_i) \tag{21}$$ ▲

Proof. It is obvious that $T_{\vec{K}}(\vec{x}) \subset \prod_{i=1}^{n} T_{K_i}(x_i)$. *Conversely*, let $v_i \in T_{K_i}(x_i)$ for $i = 1, \ldots, n$. Then there exist sequences of elements v_i^k converging to v_i and of $h_i^k > 0$ such that $x_i + h_i^k v_i^k \in K_i$ for all $i = 1, \ldots, n$. We set $h^k := \min_{i=1,\ldots,n} h_i^k > 0$. Since the subsets K_i are convex, $x^i + h_i^k v_i^k \in K_i$ for all i, that is, $\vec{x} + h\vec{v}^k \in \vec{K}$. Hence, $\vec{v} \in T_{\vec{K}}(\vec{x})$. We deduce the formula on normal cones by polarity. ■

PROPOSITION 14

Let $A \in \mathscr{L}(X, Y)$ *and* $K \subset X$. *Then*

$$\forall x \in K, \qquad T_{A(K)}(Ax) = \text{cl}\,(A T_K(x)) \tag{22}$$

and

$$N_{A(K)}(x) = A^{*-1}(N_K(x)) \tag{23}$$ ▲

Proof. Since

$$\langle p, Ax \rangle = \max_{y \in K} \langle p, Ay \rangle = \max_{y \in K} \langle A^* p, y \rangle = \langle A^* p, x \rangle$$

we obtain the formula for the normal cones and deduce it by polarity for tangent cones. ■

COROLLARY 15

Let K *and* L *be two closed convex subsets,* $x \in K$ *and* $y \in L$. *Then*

$$T_{K+L}(x+y) = \text{cl}\,(T_K(x) + T_L(y)) \quad \text{and} \quad N_{K+L}(x+y) = N_K(x) \cap N_L(y) \tag{24}$$ ▲

THEOREM 16

Let $A \in \mathscr{L}(X, Y)$ be a continuous linear operator and let $L \subset X$ and $M \subset Y$ be closed convex sets. We set

$$K := \{x \in L \mid Ax \in M\} = L \cap A^{-1}(M) \tag{25}$$

Assume that $K \neq \varnothing$, that is, $0 \in A(L) - M$, and choose x in K. The inclusions

$$T_K(x) \subset T_L(x) \cap A^{-1}(T_M(Ax)) \quad \text{and} \quad N_L(x) + A^* N_M(Ax) \subset N_K(x)$$

are always true. If we assume that

$$0 \in \operatorname{Int}(A(L) - M) \tag{26}$$

then the equalities

$$T_K(x) = T_L(x) \cap A^{-1}(T_M(Ax)) \quad \text{and} \quad N_K(x) = N_L(x) + A^* N_M(Ax) \tag{27}$$

hold true. ▲

Proof. **a.** The first inclusion is obvious. The equality follows from the closed graph theorem (see theorem 3.3.1.). Let $x_0 \in K$ and $v_0 \in T_L(x_0) \cap A^{-1} T_M(Ax_0)$. There exist sequences of elements $v_n \in X$ and $u_n \in Y$ converging to v_0 and Av_0, respectively such that for all n, $x_0 + h_n^1 v_n \in L$ and $Ax_0 + h_n^2 u_n \in M$. We set $h_n := \min(h_n^1, h_n^2, 1) > 0$. Since L and M are *convex*, we deduce that

$$\text{for all } n, \quad x_n := x_0 + h_n v_n \in L \quad \text{and } y_n := Ax_0 + h_n u_n \in M \tag{28}$$

The theorem is proved if $u_n = Av_n$ for an infinite subset of indices. If not, we apply corollary 3.3.4. to the set-valued map F defined from L to Y by

$$F(x) := Ax - M$$

We take $y_0 = 0$ and $x_0 \in F^{-1}(0) = K$. By assumption (26), $y_0 \in \operatorname{Int} F(L) = \operatorname{Int}(A(L) - K)$. Hence, there exists $\gamma > 0$ such that $\forall y \in y_0 + \gamma B$

$$\forall x \in L, \quad d(x, F^{-1}(y)) \leqslant \frac{1}{\gamma} d(y, F(x))(\|x_0 - x\| + 1)$$

We take $y = 0$ and $x = x_0 + h_n v_0$. So $\|x_0 - x\| = h_n \|v_0\|$, $d(0, F(x_0 + h_n v_0)) = d(Ax_0 + h_n Av_0, M) \leqslant d(Ax_0 + h_n u_n, M) + h_n \|Av_0 - u_n\| = h_n \|Av_0 - u_n\|$. Therefore,

$$\begin{aligned} \frac{d(x_0 + h_n v_0, K)}{h_n} &\leqslant \frac{1}{\gamma h_n} d(0, F(x_0 + h_n v_0))(\|x - x_0\| + 1) \\ &\leqslant \frac{1}{\gamma} \|Av_0 - u_n\|(h_n \|v_0\| + 1) \end{aligned}$$

Hence, since u_n converges to Av_0, we deduce that

$$\inf_{h_n>0} \frac{d(x_0+h_n v_0, K)}{h_n}=0$$

This means that $v_0 \in T_K(x_0)$.

b. By polarity, we deduce that

$$N_K(x)=\mathrm{cl}\,(N_L(x)+A^* N_M(Ax)) \tag{29}$$

Assumption (26) also implies that $N_L(x)+A^*N_M(Ax)$ *is closed.* For that purpose, let $r_n=p_n+A^*q_n$ be a sequence converging to r, where $p_n \in N_L(x)$ and $q_n \in N_M(Ax)$. We prove that

$$\forall v \in Y, \sup_n \langle q_n, v\rangle < +\infty$$

Indeed, there exist $\lambda>0$, $y \in L$, and $z \in M$ such that $v=\lambda(z-Ay)$ by assumption (26). Hence,

$$\begin{aligned}
\langle q_n, v\rangle &=\lambda\langle q_n, z-Ay\rangle=\lambda(\langle q_n, z\rangle-\langle A^*q_n, y\rangle)\\
&=\lambda(\langle p_n, y\rangle+\langle q_n, z\rangle-\langle r_n, y\rangle)\\
&\leqslant(\langle p_n, x\rangle+\langle q_n, Ax\rangle-\langle r_n, y\rangle) \quad (\text{since } p_n \in N_L(x) \text{ and } q_n \in N_M(Ax))\\
&=\lambda\langle r_n, x-y\rangle \leqslant \lambda\|r_n\|\,\|x-y\|<+\infty
\end{aligned}$$

since the converging sequence r_n is bounded. Therefore, the sequence of elements q_n is bounded by the uniform boundedness theorem and thus relatively compact in the weak-$\star$ topology. Some subsequence (again denoted by) q_n converges to $q \in N_M(Ax)$. Hence, $p_n=r_n-A^*q_n$ converges to $p=r-A^*q \in N_L(x)$, and thus $r=p+A^*q$ belongs to $N_L(x)+A^*N_M(x)$. ■

COROLLARY 17

If L is a closed subset of X and if $A \in \mathscr{L}(X, Y)$, then for any $y \in \mathrm{Int}\,A(L)$ and $x \in L\cap A^{-1}(y)$, we have

$$\begin{cases} T_{L\cap A^{-1}(y)}(x)=T_L(x)\cap \mathrm{Ker}\,A\\ N_{L\cap A^{-1}(y)}(x)=N_L(x)+\mathrm{Im}\,A^* \end{cases} \tag{30}$$

▲

COROLLARY 18

Let $P\subset X$ be a closed convex cone and $p_0 \in P^-$ such that the subset $K=\{x \in P/\langle p_0, x\rangle=-1\}$ is not empty. Then

(31) $v \in T_K(x)$ *if and only if* $\langle p_0, v\rangle=0$ *and* $\langle p, v\rangle\leqslant 0$
for all $p \in P^-$ *satisfying* $\langle p, x\rangle=0$

and

(32) $$N_K(x) = \text{Ker}\, p_0 + (P^- \cap \{x\}^\perp)$$ ▲

By taking $X = Y$ and A to be the identity, we obtain corollary 19.

COROLLARY 19

Let K and L be two closed convex subsets of X. If

(33) $$0 \in \text{Int}\,(K - L)$$

then

(34) $$\forall x \in K \cap L, \qquad T_{K \cap L}(x) = T_K(x) \cap T_L(x)$$ ▲

The equality between the tangent cone of a finite intersection and the intersection of tangent cones holds true under the following assumptions.

PROPOSITION 20

Let $K := \bigcap_{i=1}^n K_i$ *be the intersection of n closed convex subsets; we assume that*

(35) $$\exists \gamma > 0 \quad \textit{such that} \quad \forall v_i \in \gamma B \ (i = 1, \ldots, n), \ \bigcap_{i=1}^n (K_i - v_i) \neq \varnothing$$

then

(36) $$\forall x \in K, \qquad T_K(x) = \bigcap_{i=1}^n T_{K_i}(x)$$ ▲

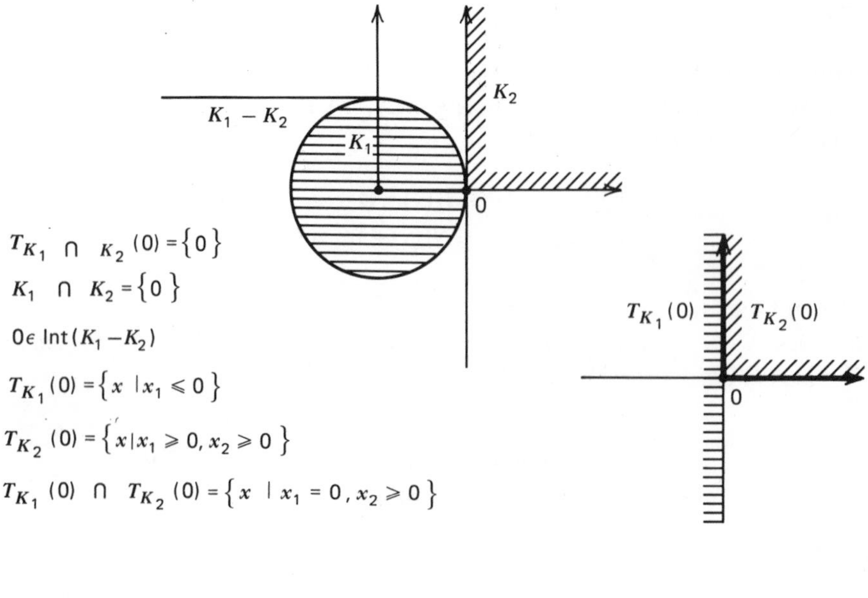
K₂
K₁ − K₂
K₁
0
T_{K₁ ∩ K₂}(0) = {0}
K₁ ∩ K₂ = {0}
0 ∈ Int(K₁ − K₂)
T_{K₁}(0) = {x | x₁ ≤ 0}
T_{K₂}(0) = {x | x₁ ≥ 0, x₂ ≥ 0}
T_{K₁}(0) ∩ T_{K₂}(0) = {x | x₁ = 0, x₂ ≥ 0}
T_{K₁}(0)
T_{K₂}(0)
0

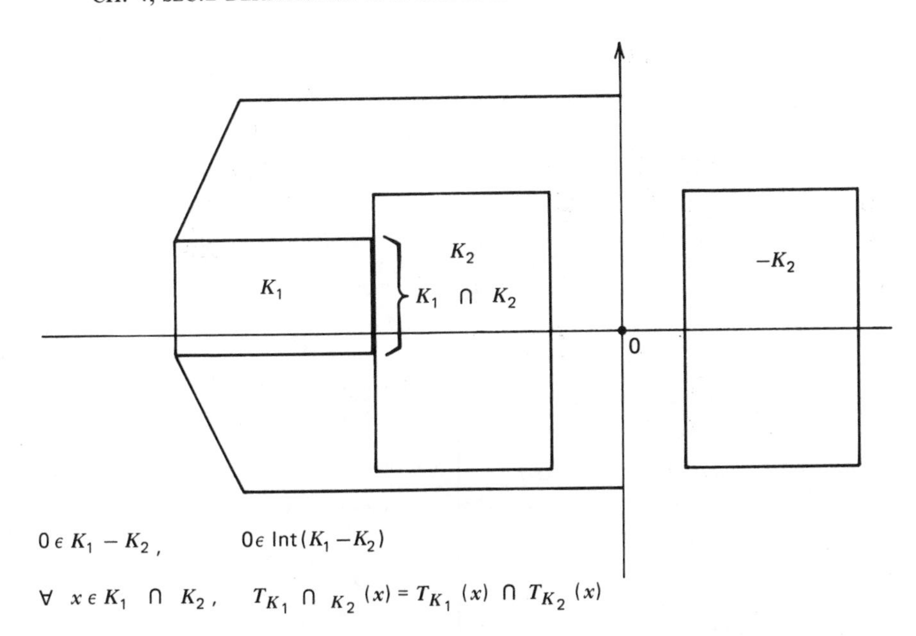

Proof. Let $\vec{D}$ be the closed vector subset of X^n of constant sequences $\vec{x}=(x, x, \ldots, x)$. Then K is identified with $\vec{D}\cap\prod_{i=1}^n K_i$. Assumption (35) implies that

$$0 \in \operatorname{Int}\left(\prod_{i=1}^{n} K_i-\vec{D}\right) \tag{37}$$

Since $T_D(x)=\vec{D}$, proposition 13 and corollary 19 imply that

$$T_{\vec{D}\cap\prod_{i=1}^n K_i}(x)=\vec{D}\cap T_{\prod_{i=1}^n K_i}(\vec{x})=\vec{D}\cap\prod_{i=1}^{n} T_{K_i}(x)$$

This proves that $T_K(x)$ is equal to the intersection of the tangent cones $T_{K_i}(x)$. ∎

2. DERIVATIVES AND CODIFFERENTIALS OF SET-VALUED MAPS WITH CONVEX GRAPHS

We now propose to develop a differential calculus for convex set-valued maps. We proceed as in elementary calculus, when the derivatives of real-valued functions are defined from the tangents to the graph. Let F be a convex set-valued map from a Banach space X to a Banach space Y. We fix a point (x_0, y_0) in the graph of F.

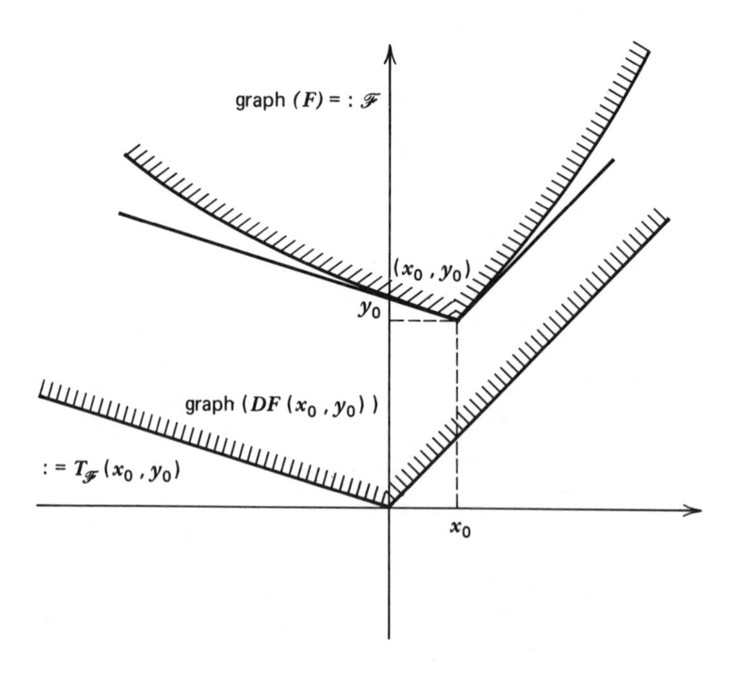

The tangent cone $T_{\text{graph}(F)}(x_0, y_0)$ to the graph of F at (x_0, y_0) is a closed convex cone of $X \times Y$. We regard this cone as the graph of a *closed convex process*, denoted by $DF(x_0, y_0)$ and call it the derivative of F at $x_0 \in X$ and $y_0 \in F(x_0)$.

DEFINITION 1

The derivative $DF(x_0, y_0)$ *of a set-valued map with convex graph at* $(x_0, y_0) \in$ graph(F) *is the closed convex process whose graph is the tangent cone to the graph of* F *at* (x_0, y_0). *We shall say that the transpose* $DF(x_0, y_0)^*$ *of* $DF(x_0, y_0)$ *is the* codifferential *of the convex set-valued map* F *at* (x_0, y_0). *In other words,*

$$(1) \quad \begin{cases} \text{i.} & v \in DF(x_0, y_0)(u) \Leftrightarrow (u, v) \in T_{\text{graph}(F)}(x_0, y_0) \\ \text{ii.} & p \in DF(x_0, y_0)^*(q) \Leftrightarrow (p, -q) \in N_{\text{graph}(F)}(x_0, y_0) \end{cases}$$ ▲

We begin by pointing out the following obvious (but useful) formulas.

$$(2) \quad \begin{cases} \text{i.} & D(F^{-1})(y_0, x_0) = DF(x_0, y_0)^{-1} \\ \text{ii.} & D(F^{-1})(y_0, x_0)^*(p) = -(DF(x_0, y_0)^*)^{-1}(-p). \end{cases}$$

Also, since the normal cone is the negative polar cone of the tangent cone, we have the following equivalent definitions.

$$(3) \quad \begin{cases} \text{i.} & p \in DF(x_0, y_0)^*(q) \\ \text{ii.} & \forall x \in X, \quad \forall y \in F(x), \quad \langle q, y_0 - y \rangle \leq \langle p, x_0 - x \rangle \\ \text{iii.} & \forall u \in X, \forall v \in DF(x_0, y_0)(u), \quad \langle p, u \rangle \leq \langle q, v \rangle \end{cases}$$

We point out the following *monotonicity property*:

$$\text{(4)} \qquad \begin{cases} \text{If } p_i \in DF(x_i, y_i)^*(q_i) \quad (i=1, 2), \quad \text{then} \\ \langle q_1 - q_2, y_1 - y_2 \rangle \leqslant \langle p_1 - p_2, x_1 - x_2 \rangle \end{cases}$$

which follows obviously from (3, **ii**).

We observe that

$$\text{(5)} \qquad \begin{cases} \textbf{i.} \quad \operatorname{Dom} DF(x_0, y_0) \subset T_{\operatorname{Dom}(F)}(x_0) \\ \textbf{ii.} \quad \operatorname{Im} DF(x_0, y_0) \subset T_{\operatorname{Im}(F)}(y_0) \end{cases}$$

Example

Let ϕ_K be the set-valued map from X to Y defined by

$$\text{(6)} \qquad \phi_K(x) := 0 \quad \text{when} \quad x \in K \quad \text{and} \quad \phi_K(x) = \varnothing \quad \text{when} \quad x \notin K$$

Then

$$\text{(7)} \qquad \forall x \in K, \qquad D\phi_K(x, 0) = \phi_{T_K(x)} \qquad \blacktriangle$$

Indeed, graph $D\phi_K(x, 0) = T_{\operatorname{graph}\phi_K}(x, 0) = T_{K \times \{0\}}(x, 0) = T_K(x) \times \{0\} = \operatorname{graph}(\phi_{T_K(x)})$. ∎

These remarks having been made, we now prove an analytical characterization of the derivative to a convex set-valued map that captures the idea of a derivative as a suitable limit of "differential quotients."

PROPOSITION 2

Let (x_0, y_0) belong to the graph of a set-valued map F with convex graph from X to Y. Then $v_0 \in DF(x_0, y_0)(u_0)$ if and only if

$$\text{(8)} \qquad \liminf_{u \to u_0} \inf_{h > 0} d\left(v_0, \frac{F(x_0 + hu) - y_0}{h}\right) = 0 \qquad \blacktriangle$$

Proof. Indeed, $(u_0, v_0) \in T_{\operatorname{graph}(F)}(x_0, y_0)$ if and only if for all $\varepsilon_1 > 0$, $\varepsilon_2 > 0$, there exist $u_{\varepsilon_1} \in X$ and $v_{\varepsilon_2} \in Y$ satisfying $\|u_{\varepsilon_1}\| \leqslant \varepsilon_1$ and $\|v_{\varepsilon_2}\| \leqslant \varepsilon_2$ and $h_0 > 0$ such that

$$(x_0 + h(u_0 + u_{\varepsilon_1}), \quad y_0 + h(v_0 + v_{\varepsilon_2})) \in \operatorname{graph}(F)$$

for all $h \in]0, h_0[$. Hence

$$v_0 \in \frac{F(x_0 + h(v_0 + u_{\varepsilon_1})) - y_0}{h} - v_{\varepsilon_2}$$

and, consequently,

$$\inf_{0<h<h_0} d\left(v_0, \frac{F(x_0+h(u_0+u_{\varepsilon_1}))-y_0}{h}\right) \leqslant \varepsilon_2$$

The convexity of the graph of F implies that for all $y \in F(x)$, when $h_1 \leqslant h_2$, then

$$\frac{F(x+h_2u)-y}{h_2} \subset \frac{F(x+h_1u)-y}{h_1}$$

Indeed, for all $y \in F(x)$, we can write

$$\frac{h_1}{h_2} F(x+h_2v)+\left(1-\frac{h_1}{h_2}\right) y \subset F\left(\frac{h_1}{h_2}(x+h_2v)+\left(1-\frac{h_1}{h_2}\right)x\right)=F(x+h_1v)$$

This amounts to saying that the function $\theta \to d(v,(F(x+\theta u)-y)/\theta)$ is increasing. So, we can write

$$\text{(9)} \qquad \lim_{h\to 0^+} d\left(v_0, \frac{F(x_0+hu)-y_0}{h}\right) = \inf_{h>0} d\left(v_0, \frac{F(x_0+hu)-y_0}{h}\right)$$

Therefore,

$$\text{(10)} \qquad \begin{aligned} &\inf_{0<h<h_0} d\left(v_0, \frac{F(x_0+h(u_0+u_{\varepsilon_1}))-y_0}{h}\right) \\ &= \inf_{h>0} d\left(v_0, \frac{F(x_0+h(u_0+u_{\varepsilon_1}))-y_0}{h}\right) \\ &= \lim_{h\to 0^+} d\left(v_0, \frac{F(x_0+h(u_0+u_{\varepsilon_1}))-y_0}{h}\right) \end{aligned}$$

We have proved that

$$\text{(11)} \qquad \inf_{\|u-u_0\|\leqslant \varepsilon_1} \inf_{h>0} d\left(v_0, \frac{F(x_0+hu)-y_0}{h}\right) \leqslant \varepsilon_2$$

By letting ε_1 and ε_2 converge to zero, we obtain formula (8). ■

PROPOSITION 3

Let x_0 and x belong to the domain of a set-valued map F with convex graph. For any $y_0 \in F(x_0)$, we have

$$\text{(12)} \qquad F(x)-y_0 \subset DF(x_0, y_0)(x-x_0)$$ ▲

Proof. Indeed, for any $h \in]0, 1[$ and any $y \in F(x)$, we have

$$(1-h)y_0 + hy \subset F(x_0 + h(x - x_0))$$

Hence,

$$y - y_0 \in \frac{F(x_0 + h(x - x_0)) - y_0}{h}$$

This implies that $y - y_0 \in DF(x_0, y_0)(x - x_0)$. ■

Let $P \subset Y$ be a closed convex cone of Y defining a preorder. We say that $x_0 \in K$ achieves the minimum of set-valued map $F: K \to Y$ at $y_0 \in F(x_0)$ if

$$\forall x \in K, \quad F(x) \subset y_0 + P \tag{13}$$

Proposition 3 allows us to characterize the minimum of a set-valued map (variational principle).

PROPOSITION 4

Let F be a set-valued map from K to Y with convex graph. Then $x_0 \in K$ achieves the minimum of F on K at $y_0 \in F(x_0)$ if and only if one of the equivalent conditions

$$\begin{cases} \textbf{i.} & \forall u_0 \in X, \quad DF(x_0, y_0)(u_0) \subset P \\ \textbf{ii.} & \forall q \in P^+, \quad 0 \in DF(x_0, y_0)^*(q) \end{cases} \tag{14}$$

hold true. ▲

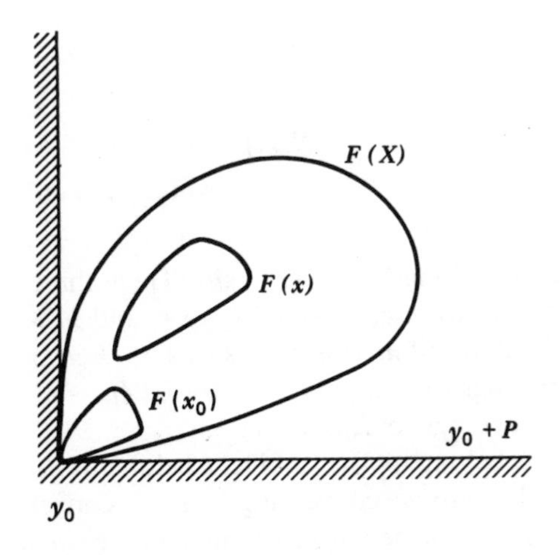

Proof. **a.** Let $x \in K$. By proposition 3, property (12) implies that

$$F(x) \subset y_0 + DF(x_0, y_0)(x - x_0) \subset y_0 + P$$

Hence, x_0 achieves the minimum of F at $y_0 \in F(x_0)$.
b. Let x_0 achieve the minimum of F at $y_0 \in F(x_0)$ and let $v_0 \in DF(x_0, y_0)(u_0)$. For all $\varepsilon > 0$, there exist $u \in u_0 + \varepsilon B$ and $h > 0$ such that

$$v_0 \in \frac{F(x_0 + hu) - y_0}{h} + \varepsilon B \subset P + \varepsilon B \tag{15}$$

by (14, **ii**). Since P is closed, we deduce that v_0 belongs to P by letting ε converge to zero.
c. We prove that (14, **i**) implies (14, **ii**). Indeed, for all $u \in \operatorname{Dom} DF(x_0, y_0)$, $v \in DF(x_0, y_0)(u) \subset P$, $q \in P^+$, we have $\langle 0, u\rangle - \langle q, v\rangle \leq 0$. Hence, $0 \in DF(x_0, y_0)^*(q)$.
d. Conversely, assume that $0 \in DF(x_0, y_0)^*(q)$ for all $q \in P^+$. Let v_0 belong to $DF(x_0, y_0)(u_0)$. Then we have $\langle 0, u_0\rangle - \langle q, v_0\rangle \leq 0$ for all $q \in P^+$, that is, $v_0 \in P$. ∎

PROPOSITION 5

Let X and Y be Banach spaces, $L \subset X$ and $M \subset Y$ closed convex subsets, and $A \in \mathscr{L}(X, Y)$ a continuous linear operator. We define the set-valued map F on X by

$$F(x) := \begin{cases} Ax - M & \text{when } x \in L \\ \varnothing & \text{when } x \notin L \end{cases} \tag{16}$$

For any $x \in L$ and $y \in Ax - M$, we have

$$DF(x, y)(u) = \begin{cases} Au - T_M(Ax - y) & \text{when } u \in T_L(x) \\ \varnothing & \text{when } u \notin T_L(x) \end{cases} \tag{17}$$

and

$$DF(x, y)^*(q) = \begin{cases} A^*q + N_L(x) & \text{when } q \in N_M(Ax - y) \\ \varnothing & \text{when } q \notin N_M(Ax - y) \end{cases} \tag{18}$$
▲

Proof. **a.** Let v belong to $DF(x, y)(u)$. Then there exist sequences of elements u_n converging to u, v_n converging to v, and $h_n > 0$ such that $y + h_n v_n$ belongs to $F(x + h_n u_n)$ for all n. This means that $x + h_n u_n$ belongs to L for all n (and, thus, that u belongs to $T_L(x)$) and $Ax - y + h_n(Au_n - v_n)$ belongs to M for all n. Therefore, $Au - v$ belongs to $T_M(Ax - y)$.
b. Conversely, let u belong to $T_L(x)$ and let v belong to $Au - T_M(Ax - y)$. There exist sequences of elements u_n converging to u, w_n converging to $Au - v$, and $h_n^1, h_n^2 > 0$ such that $x + h_n^1 u_n$ belongs to L and $Ax - y + h_n^2 w_n$ belongs to M for all $n > 0$. We set $h_n := \min(h_n^1, h_n^2)$ and $v_n := Au_n - w_n$, which converges to v. We observe that $y + h_n v_n$ belongs to $F(x + h_n u_n)$ and, consequently, v belongs to $DF(x, y)(u)$.

c. By definition, p belongs to $DF(x, y)^*(q)$ if and only if

$$\sup_{u \in T_L(x)} \sup_{w \in T_M(Ax-y)} [\langle p, u\rangle + \langle -q, Au - w\rangle]$$

$$= \sup_{u \in T_L(x)} \langle p - A^*q, u\rangle + \sup_{w \in T_M(Ax-y)} \langle q, w\rangle \leqslant 0$$

This amounts to saying that $p - A^*q$ belongs to $N_L(x)$ and that q belongs to $N_M(Ax-y)$. ■

We now study calculus for derivatives and codifferentials of set-valued maps with convex graph. Let us start with the chain rule.

THEOREM 6

Let F be a set-valued map from X to Y with closed convex graph and let A belong to $\mathscr{L}(Z, X)$. We assume that

$$0 \in \operatorname{Int}(\operatorname{Im} A - \operatorname{Dom} F) \tag{19}$$

Then the following chain rule formulas hold:

$$\begin{cases} \textbf{i.} & D(FA)(z_0, y_0) = DF(Az_0, y_0)A \\ \textbf{ii.} & D(FA)(z_0, y_0)^* = A^*DF(Az_0, y_0)^* \end{cases} \tag{20}$$

▲

Proof. We know that the graph $\mathscr{G}$ of $G = FA$, which is closed and convex, is equal to $(A \times 1)^{-1}\mathscr{F}$, where $\mathscr{F}$ is the graph of F. The assumption $0 \in \operatorname{Int}(\operatorname{Im} A - \operatorname{Dom} F)$ obviously implies that in $X \times Y$, zero belongs to $\operatorname{Int}(\operatorname{Im}(A \times 1) - \mathscr{F})$. So by theorem 1.16, we know that

$$T_{\mathscr{G}}(z_0, y_0) = (A \times 1)^{-1} T_{\mathscr{F}}(Az_0, y_0) \quad \text{and that}$$
$$N_{\mathscr{G}}(z_0, y_0) = (A^* \times 1) N_{\mathscr{F}}(Az_0, y_0).$$

This implies formulas (20, **i** and **ii**). ■

Note that *assumption* (19) *is always satisfied when A is surjective* or *when* $\operatorname{Im} A \cap \operatorname{Int} \operatorname{Dom} F \neq \varnothing$.

Let us consider a set-valued map F with convex graph from X to Y and $B \in \mathscr{L}(Y, Z)$. The graph of the set-valued map BF from X to Z is still convex.

PROPOSITION 7

Let F be a set-valued map from X to Y with convex graph and let B belong to $\mathscr{L}(Y, Z)$. Then

$$\begin{cases} \textbf{i.} & D(BF)(x_0, By_0) = \overline{BDF(x_0, y_0)} \\ \textbf{ii.} & D(BF)(x_0, By_0)^* = DF(x_0, y_0)^*B^* \end{cases} \tag{21}$$

▲

Proof. We note that the graph $\mathscr{G}$ of $G:=BF$ is $(1\times B)\mathscr{F}$, where $\mathscr{F}$ is the graph of F. We deduce from proposition 1.14 that $T_{\mathscr{G}}(x_0, By_0)=\text{cl}((1\times B)T_{\mathscr{F}}(x_0, y_0))$ and $N_{\mathscr{G}}(x_0, By_0)=((1\times B)^{*-1}N_{\mathscr{F}}(x_0, y_0)$. So formulas (21, i and ii) ensue. ■

The two preceding results are summarized in theorem 8.

THEOREM 8

Let U, X, Y, and Z be Banach spaces, F a map with closed convex graph from X to Y. Let A belong to $\mathscr{L}(U, Y)$ and B to $\mathscr{L}(Y, Z)$; let $u_0 \in U$ and $y_0 \in F(Au_0)$ be fixed. If

$$0 \in \text{Int}(\text{Im } A - \text{Dom } F), \tag{22}$$

then

$$\begin{cases} \textbf{i.} & D(BFA)(u_0, By_0)=\overline{BDF(Au_0, y_0)A} \\ \textbf{ii.} & D(BFA)(u_0, By_0)^*=A^*DF(Au_0, y_0)^*B^* \end{cases} \tag{23}$$ ▲

As a consequence, we obtain theorem 9.

THEOREM 9

Let X, Y, and Z be Banach spaces, F a set-valued map from X to Z, G a set-valued map from Y to Z, and let A belong to $\mathscr{L}(X, Y)$. We assume that the graphs of F and G are closed and convex and that

$$0 \in \text{Int}\,(A \text{ Dom } F - \text{Dom } G) \tag{24}$$

Let $x \in \text{Dom } F \cap A^{-1}\text{ Dom } G$, $y \in F(x)$, and $z \in G(Ax)$ be given. Then,

$$\begin{cases} \textbf{i.} & D(F+GA)(x, y+z)=\overline{DF(x, y)+DG(Ax, z)A} \\ \textbf{ii.} & D(F+GA)^*(x, y+z)=DF(x, y)^*+A^*DG(Ax, z)^* \end{cases} \tag{25}$$ ▲

Proof. We observe that $F+GA$ can be written as $\vec{B}\vec{G}\vec{A}$ where

a. $\vec{A}: x \in X \to (x, Ax) \in X \times Y$
b. $\vec{F}: (x, y) \in X \times Y \to F(x) \times G(y) \subset Z \times Z$
c. $\vec{B}: (z_1, z_2) \in Z \times Z \to z_1 + z_2$

We also observe that assumption (24) implies that $0 \in \text{Int}(\text{Im } \vec{A} - \text{Dom } \vec{F})$. It then suffices to check that $D\vec{F}(x, y, z_1, z_2)=DF(x, z_1)\times DG(y, z_2)$. ■

COROLLARY 10

Let F and G be two set-valued maps with closed convex graphs from a Banach space X to a Banach space Y. Assume that

$$0 \in \text{Int}(\text{Dom } F - \text{Dom } G) \tag{26}$$

and take $x \in \text{Dom } F \cap \text{Dom } G$, $y \in F(x)$, *and* $z \in G(x)$. *Then,*

$$(27) \quad \begin{cases} \textbf{i.} & D(F+G)(x, y+z) = \overline{DF(x, y) + DG(x, z)} \\ \textbf{ii.} & D(F+G)(x, y+z)^* = DF(x, y)^* + DG(x, z)^* \end{cases}$$ ▲

We can compute the derivative of the restriction of a set-valued map with a closed convex graph. We note that $F|_K = F + \phi_K$, where ϕ_K is defined by (6).

COROLLARY 11

Let F be a set-valued map from X to Y with closed convex graph and let $F + \phi_K$ *be its restriction to the closed convex subset* $K \subset X$. *If*

$$(28) \quad 0 \in \text{Int } (\text{Dom } F - K)$$

then for all $x \in K$ *and* $y \in F(x)$,

$$(29) \quad \begin{cases} \textbf{i.} & D(F|_K)(x, y) \text{ is the restriction of } DF(x, y) \text{ to } T_K(x). \\ \textbf{ii.} & D(F|_K)(x, y)^*(q) = DF(x, y)^*(q) + N_K(x) \end{cases}$$ ▲

Proof. We apply corollary 10 with $G := \phi_K$, since $F|_K = F + G$. ■

COROLLARY 12

Let F be a set-valued map from X to Y with closed convex graph and let $P \subset Y$ *be a closed convex cone. We assume that the subsets* $F_+(x) := F(x) + P$ *are closed for all x. Then for all* $x \in \text{Dom } F$ *and for all* $y \in F(x)$

$$(30) \quad \begin{aligned} &\textbf{i.} \quad DF_+(x, y)(u) = DF(x, y)(u) + P \\ &\textbf{ii.} \quad DF_+(x, y)^*(q) = \begin{cases} DF(x, y)^*(q) & \text{if } q \in P^+ \\ \varnothing & \text{if } q \notin P^+ \end{cases} \end{aligned}$$

[$DF_+(x, y)^*$ *is the restriction of* $DF(x, y)^*$ *to* P^+.] ▲

Proof. We apply corollary 10 with G defined by $G(x) := P$ for all $x \in X$. ■

We now compute the derivative of an intersection.

PROPOSITION 13

Let us consider n set-valued maps F_i *from X to Y with closed convex graph. We assume that*

$$(31) \quad x_0 \in \bigcap_{i=1}^{n} \text{Int Dom } F_i$$

and that

$$(32) \quad \exists \gamma > 0 \quad \textit{such that} \quad \forall (u_i, v_i) \in \gamma(B \times B), \ \bigcap_{i=1}^{n} (F_i(x+u_i) - v_i) \neq \varnothing$$

Then for any $x \in \bigcap_{i=1}^n \operatorname{Dom} F_i$ *and any* $y \in \bigcap_{i=1}^n F_i(x)$, *we have*

$$D\left(\bigcap_{i=1}^{n} F_i\right)(x, y) = \bigcap_{i=1}^{n} DF_i(x, y) \tag{33}$$

▲

Proof. The graph of $\bigcap_{i=1}^n F_i$ being the intersection of the graphs of the set-valued maps F_i, we use proposition 1.20. We observe that assumptions (31) and (32) imply that there exists $\delta > 0$ such that

$$\forall (u_i, v_i) \in \delta(B \times B), \bigcap_{i=1}^{n} (\operatorname{graph}(F_i) - (u_i, v_i)) \neq \varnothing \tag{34}$$

Indeed, we take x_0 in the intersection of the domains F_i and $\delta \leqslant \gamma$ such that $x_0 + \delta B \subset \bigcap_{i=1}^n \operatorname{Dom} F_i$. Then we choose y_0 in $\bigcap_{i=1}^n (F_i(x_0 + u_i) - v_i)$, which exists by assumption (32). Hence, (x_0, y_0) belongs to the intersection of the $(\operatorname{graph} F_i - (u_i, v_i))$. Therefore, proposition 13 follows from proposition 1.20.

3. EPIDERIVATIVES AND SUBDIFFERENTIALS OF CONVEX FUNCTIONS

Let V be a real-valued function defined on a convex subset K of X. We extend it to X by setting

$$V(x) := \infty \quad \text{when } x \notin K \quad (\text{we say that } K = \operatorname{Dom} V) \tag{1}$$

We associate to V the set-valued map from X to R defined by

$$\mathbf{V}_+(x) := V(x) + R_+ \quad \text{if} \quad x \in K, \mathbf{V}_+(x) = \varnothing \quad \text{if} \quad x \notin K \tag{2}$$

whose graph is the epigraph of V. The latter is convex if and only if the function V is convex. Therefore, we can define the derivative $D\mathbf{V}_+(x, V(x))$ of $\mathbf{V}_+$ at $(x, V(x))$, which is a closed convex process from X to R. Since $D\mathbf{V}_+(x, V(x))(u)$ is either $\bar{R}$, or a half-line $[a, \infty[$ or $\{+\infty\}$, this set is characterized by its lower bound.

DEFINITION 1
The epiderivative *of* V *at* x *is*

$$D_+V(x)(u) := \inf\{v \mid v \in D\mathbf{V}_+(x, V(x))(u)\} \in \bar{R} \tag{3}$$

▲

We can check that

$$D_+V(x_0)(u_0) = \lim_{u \to u_0} \inf_{h \to 0^+} \frac{V(x_0 + hu) - V(x_0)}{h} \tag{4}$$

We note that the convexity of V implies that

$$\lim_{h\to 0^+} \frac{V(x_0+hu)-V(x_0)}{h} = \inf_{h>0} \frac{V(x_0+hu)-V(x_0)}{h} \tag{5}$$

The transpose of the derivative $D\mathbf{V}_+(x, V(x))$ is a convex process from R to X^*, whose domain is contained in R_+. Then for any $q \geqslant 0$, we have

$$D\mathbf{V}_+(x, V(x))^*(q) = qD\mathbf{V}_+(x, V(x))^*(1) \tag{6}$$

This brings us to definition 2.

DEFINITION 2

Let V be a convex function form X to $R \cup \{+\infty\}$ and x belong to Dom V. *The subset $\partial V(x) := D\mathbf{V}_+(x, V(x))^*(1)$ of the dual X^* of X is called the* subdifferential *of V at x.* ▲

We immediately give an analytical characterization of the subdifferential in proposition 3.

PROPOSITION 3

Let V: $X \to R \cup \{+\infty\}$ be a convex function. The following statements are equivalent:

(a) $p_0 \in \partial V(x_0)$

(b) $\forall x \in X,\ V(x_0) - V(x) \leqslant \langle p_0, x_0 - x\rangle$

(c) $\forall u \in X,\ \langle p_0, u\rangle \leqslant D_+V(x_0)(u)$ ▲

When V is a proper convex function, we obtain the following characterization of a minimizer of V.

PROPOSITION 4

Let V be a convex function from X to $]-\infty, +\infty]$ whose domain is nonempty.

a. *For any x_0, $x \in K$, we have*

$$V(x) - V(x_0) \geqslant D_+V(x_0)(x - x_0) \tag{7}$$

b. *x_0 minimizes V on K if and only if*

$$\forall u \in \text{Dom}\ D_+V(x_0), \qquad 0 \leqslant D_+V(x_0)(u) \tag{8}$$

or, equivalently,

$$0 \in \partial V(x_0) \tag{9}$$ ▲

Theorem 2.6 implies the following property of the epiderivatives and subdifferentials of lower semicontinuous convex functions, which plays an important role in optimization theory.

THEOREM 5

Let X and Y be Banach spaces, $V: X \to R \cup \{+\infty\}$ and $W: Y \to R \cup \{+\infty\}$ proper lower semicontinuous, convex functions, and let A belong to $\mathscr{L}(X, Y)$. We assume that

$$0 \in \operatorname{Int}(A \operatorname{Dom} V - \operatorname{Dom} W) \tag{10}$$

Then,

$$\begin{cases} \textbf{i.} & D_+(V+WA)(x) = D_+V(x) + D_+W(Ax)A \\ \textbf{ii.} & \partial(V+WA)(x) = \partial V(x) + A^*\partial W(Ax) \end{cases} \tag{11}$$

Consequently, $\bar{x}$ is a minimizer of $x \to V(x) + W(Ax)$ if and only if $\bar{x}$ is a solution to the inclusion

$$0 \in \partial V(\bar{x}) + A^*\partial W(A\bar{x}) \tag{12}$$

that is, if and only if there exists $\bar{q} \in Y^*$ such that

$$-A^*\bar{q} \in \partial V(\bar{x}), \qquad \bar{q} \in \partial W(A\bar{x}) \tag{13}$$ ▲

It is useful to list several consequences of this theorem.

COROLLARY 6

a. *Let $A \in \mathscr{L}(X, Y)$ be a continuous linear operator from a Banach space X to a Banach space Y and let $W: Y \to R \cup \{+\infty\}$ be a proper lower semicontinuous convex function. If*

$$0 \in \operatorname{Int}(\operatorname{Im} A - \operatorname{Dom} W) \tag{14}$$

then

$$\begin{cases} \textbf{i.} & D_+(WA)(x) = D_+W(Ax)A \\ \textbf{ii.} & \partial(WA)(x) = A^*\partial W(Ax) \end{cases} \tag{15}$$

b. *Let V and W be proper lower semicontinuous, convex functions from a Banach space X to $R \cup \{+\infty\}$. If*

$$0 \in \operatorname{Int}(\operatorname{Dom} V - \operatorname{Dom} W) \tag{16}$$

then

$$\text{(17)} \quad \begin{cases} \textbf{i.} & D_+(V+W)(x)=D_+V(x)+D_+W(x) \\ \textbf{ii.} & \partial(V+W)(x)=\partial V(x)+\partial W(x) \end{cases}$$

c. *Let V be a proper lower semicontinuous, convex function from a Banach space X to $R\cup\{+\infty\}$ and let $K\subset X$ be a closed convex subset. If*

$$\text{(18)} \quad 0\in \operatorname{Int}(K-\operatorname{Dom} V)$$

then

$$\text{(19)} \quad \begin{cases} \textbf{i.} & D_+(V|_K)(x)=D_+V(x)|_{T_K(x)} \\ \textbf{ii.} & \partial(V|_K)(x)=\partial V(x)+N_K(x) \end{cases}$$

Consequently, $\bar{x}\in K$ minimizes V on K if and only if $0\in\partial V(\bar{x})+N_K(\bar{x})$

d. *Let V be a proper lower semicontinuous, convex function from a Banach space X to $R\cup\{+\infty\}$, let A belong to $\mathscr{L}(X, Y)$, and M be a closed convex subset of Y. If*

$$\text{(20)} \quad 0\in \operatorname{Int}(M-A\operatorname{Dom} V)$$

then

$$\text{(21)} \quad \begin{cases} \textbf{i.} & D_+(V|_{A^{-1}(M)})(x)=D_+V(x)|_{A^{-1}T_M(Ax)} \\ \textbf{ii.} & \partial(V|_{A^{-1}(M)})=\partial V(x)+A^*N_M(Ax) \end{cases}$$

Consequently, $\bar{x}\in A^{-1}(M)$ minimizes V on $A^{-1}(M)$ if and only if $0\in\partial V(\bar{x})+A^*N_M(A\bar{x})$. ▲

More generally, we introduce

(22) a proper lower semicontinuous, convex function $L\colon X\times Y\to R\cup\{+\infty\}$

to which we associate the lower semicontinuous, convex function V defined by

$$\text{(23)} \quad \forall x\in X,\ V(x):=L(x, Ax) \quad \text{where} \quad A\in\mathscr{L}(X, Y)$$

We denote by $A\oplus -1\in\mathscr{L}(X\times Y, Y)$ and $1\oplus A^*\in\mathscr{L}(X^*\times Y^*, X^*)$ the continuous linear operators defined by

$$\text{(24)} \quad (A\oplus -1)(x, y)=Ax-y, \quad (1\oplus A^*)(p, q)=p+A^*q$$

We observe that

(25) $$\text{Dom } V \neq \varnothing \quad \text{if and only if} \quad 0 \in (A \oplus -1) \text{ Dom } L$$

because $x \in \text{Dom } V$ if and only if there exists $(u, v) \in \text{Dom } L$ such that $x = u$ and $Ax = v$, that is, if and only if $0 = Au - v \in (A \oplus -1) \text{ Dom } L$.

THEOREM 7

Let L satisfy (22) *and V be defined by* (23). *If we posit the assumption*

(26) $$0 \in \text{Int } ((A \oplus -1) \text{ Dom } L)$$

then

(27) $$\partial V(x) = (1 \oplus A^*) \partial L^*(x, Ax)$$

Consequently, $\bar{x} \in X$ *is a minimizer of V if and only if*

(28) $$0 \in (1 \oplus A^*) \partial L(\bar{x}, A\bar{x})$$

that is, if and only if there exists $\bar{q} \in Y^*$ *such that*

(29) $$(-A^*\bar{q}, \bar{q}) \in \partial L(\bar{x}, A\bar{x})$$ ▲

Proof. Let $1 \otimes A$ be the map: $x \to (x, Ax)$ from X to $X \times Y$ whose transpose is $1 \oplus A^*$. The function V defined by (23) is $L(1 \otimes A)$. We point out that assumption (26) implies that $(0, 0) \in \text{Int } (\text{Im}(1 \otimes A) - \text{Dom } L)$. Indeed, let $(u, v) \in X$. By assumption (26), there exist $\lambda > 0$ and $(x, y) \in \text{Dom } L$ such that $\lambda(v - Au) = Ax - y$. Set $z := \lambda u + x$. Then $Az = \lambda v + y$ and

$$\lambda(u, v) = (z, Az) - (x, y) \in \text{Im } (1 \otimes A) - \text{Dom } L$$

Then formula (27) follows from formula (15, **ii**), and the second equivalent statements follow from **b** of proposition 7. ■

PROPOSITION 8

Let us consider n convex lower semicontinuous functions V_i *from X to* $]-\infty, +\infty]$. *Let V be defined by*

(30) $$V(x) := \max_{i=1,\ldots,n} V_i(x)$$

and

(31) $$J(x) := \{i = 1, \ldots, n \mid V_i(x) = V(x)\}$$

Let us assume that the n functions V_i are continuous at x. Then

$$
(32)\quad \begin{aligned} &\textbf{i.}\quad D_+V(x)(v)=\max_{i\in J(x)} D_+V_i(x)(v)\\ &\textbf{ii.}\quad \partial V(x)=\overline{\text{co}}\left(\bigcup_{i\in J(x)}\partial V_i(x)\right)\end{aligned} \qquad \blacktriangle
$$

Proof. We observe that $\mathbf{V}(x)=\bigcap_{i=1}^n \mathbf{V}_{i+}(x)$, hence, our result follows from proposition 2.13. We know that x belongs to the intersection of the interiors of the domains of the maps $\mathbf{V}_{i+}$. Assumption (2.32) is obviously satisfied. Then by proposition 2.13,

$$
D\mathbf{V}_+(x, V(x))=\bigcap_{i=1}^n D\mathbf{V}_{i+}(x, V(x))
$$

When i belongs to $J(x)$, then

$$
D\mathbf{V}_{i+}(x, V(x))(u)=D\mathbf{V}_{i+}(x, V_i(x))(u)=[D_+V_i(x)(u), \infty[
$$

When i does not belong to $J(x)$, then $(x, V(x))$ belongs to the interior of the graph of $\mathbf{V}_{i+}$, so that $D\mathbf{V}_{i+}(x, V(x))(u)=R$ for all $u\in X$. Then

$$
D\mathbf{V}_+(x, V(x))(u)=\bigcap_{i\in J(x)} D\mathbf{V}_{i+}(x, V(x))(u)
$$

so that formulas (32) ensue. ■

Let us point out the monotonicity property of the subdifferential of a convex function.

PROPOSITION 9
Let V be a proper convex function from X to $R\cup\{+\infty\}$. Then the set-valued map $x\in X\to\partial V(x)\in X^*$ is monotone

$$
(33)\qquad \forall(x, p),\quad \forall(y, q)\in graph\,(\partial V),\quad \langle p-q, x-y\rangle\geq 0 \qquad \blacktriangle
$$

Proof. Indeed, since $p\in\partial V(x)$ implies that $V(x)-V(y)\leq\langle p, x-y\rangle$ and since $q\in\partial V(y)$ implies that $V(y)-V(x)\leq\langle q, y-x\rangle$, we obtain formula (33) by adding these two inequalities. ■

We proceed by computing the subdifferential of convex functions of the norm in a Banach space.

PROPOSITION 10
Let X be a Banach space. Then the norm is subdifferentiable and

$$
(34)\qquad \forall x\in X,\ \partial\|\cdot\|(x):=\{p\in X^*|\langle p, x\rangle=\|x\| \quad\text{and}\quad \|p\|=1\}
$$

If $\alpha>1$, *then*

$$(35)\quad \begin{cases} \forall x\in X, \partial\left(\frac{1}{\alpha}\|\cdot\|^{\alpha}\right)(x)=\|\cdot\|^{\alpha-1}\partial(\|\cdot\|)(x) \\ \qquad =\{p\in X^*\,|\,\|p\|_*=\|x\|^{\alpha-1} \text{ and } \langle p,x\rangle=\|x\|^{\alpha}\} \end{cases} \blacktriangle$$

Proof. **a.** The Hahn–Banach extension theorem implies that we can associate with any $x_0\in X$ a continuous linear functional $p_0\in X^*$ such that

$$\langle p_0, x_0\rangle=\|x_0\| \quad \text{and} \quad \|p_0\|_*=1$$

Indeed, we extend the continuous linear functional defined on Rx_0 by $p(\alpha x_0):=\alpha\|x_0\|$, whose norm is equal to one, by a continuous linear form p_0 on X whose norm $\|p_0\|_*$ is still equal to one.

Then

$$\langle p_0, x-x_0\rangle\leqslant\|p_0\|_*\|x\|-\langle p_0, x_0\rangle=\|x\|-\|x_0\|$$

This implies that $p_0\in\partial(\|\cdot\|)(x_0)$. Conversely, if $p_0\in\partial(\|\cdot\|)(x_0)$, then $\|x_0\|-\|x\|\leqslant\langle p_0, x_0-x\rangle$. By taking $x=\lambda x_0$, this implies that

$$(1-\lambda)(\langle p_0, x_0\rangle-\|x_0\|)\geqslant 0$$

By successively taking $\lambda>1$ and $\lambda<1$, we deduce that $\langle p_0, x_0\rangle=\|x_0\|$. Therefore, $\langle p_0, x\rangle\leqslant\|x\|$ for all $x\in X$ and $\langle p_0, x_0\rangle=x_0$. This implies that $\|p_0\|_*=1$.

b. The second statement follows obviously from the first. ■

Actually, any lower semicontinuous, convex function on a Hilbert space is subdifferentiable on a dense subset of its domain. This is also true in Banach spaces, as we shall see in theorem 4.4.3 in the next chapter.

THEOREM 11

Let X *be a Hilbert space and* V *a proper lower semicontinuous, convex function from* X *to* $R\cup\{+\infty\}$.

a. *The domain* Dom $(\partial V(\cdot))$ *of the set-valued map* $\partial V(\cdot)$ *is dense in* Dom V.

b. *For any* $x\in X$ *and* $\lambda>0$, *there exists a unique solution* x_λ *to the inclusion*

$$(36)\qquad x\in x_\lambda+\lambda\partial V(x_\lambda) \qquad \blacktriangle$$

Proof. We associate to each positive λ the minimization problem

$$(37)\qquad V_\lambda(x):=\inf_{y\in X}\left[V(y)+\frac{1}{2\lambda}\|y-x\|^2\right]$$

As we will see later on (theorem 4.2), a proper lower semicontinuous, convex function is bounded below by an affine function: There exist $p \in X^*$ and $a \in R$ such that $\forall y \in X$, $V(y) \geqslant \langle p, y \rangle + a$. Therefore,

$$V(y) + \frac{1}{2\lambda} \|y - x\|^2 \geqslant \frac{1}{2\lambda} (\|y - x + \lambda p\|^2 - \lambda^2 \|p\|^2) - a + \langle p, x \rangle$$

so that $V_\lambda(x) \geqslant -(\lambda/2)\|p\|^2 + a + \langle p, x \rangle$ is a finite number.

a. We begin by proving that there exists x_λ such that

$$V_\lambda(x) = V(x_\lambda) + \frac{1}{2\lambda} \|x_\lambda - x\|^2 \tag{38}$$

To this end, we consider a minimizing sequence of elements y^n, which satisfy $V(y^n) + (1/2\lambda)\|y^n - x\|^2 \leqslant V_\lambda(x) + (1/n)$. It is a Cauchy sequence because

$$\begin{aligned}
\|y^n - y^m\|^2 &= 2\|y^n - x\|^2 + 2\|y^m - x\|^2 - 4 \left\| \frac{y^n + y^m}{2} - x \right\|^2 \\
&\leqslant 4\lambda \left[\frac{1}{n} + \frac{1}{m} + 2V_\lambda(x) - V(y^n) - V(y^m) \right] + 8\lambda \left[V\left(\frac{y^n + y^m}{2} \right) - V_\lambda(x) \right] \\
&\leqslant 4\lambda \left[\frac{1}{n} + \frac{1}{m} + 2V\left(\frac{y^n + y^m}{2} \right) - V(y^n) - V(y^m) \right] \\
&\leqslant 4\lambda \left(\frac{1}{n} + \frac{1}{m} \right) \qquad \text{(for } V \text{ is convex)}
\end{aligned}$$

Hence, y^n converges to an element x_λ. Since V is lower semicontinuous, it follows that

$$V(x_\lambda) + \frac{1}{2\lambda} \|x_\lambda - x\|^2 \leqslant \liminf_{n \to \infty} V(y^n) + \lim_{n \to \infty} \frac{1}{2\lambda} \|y^n - x\|^2 \leqslant V_\lambda(x)$$

b. Let x_λ be a solution to the minimization problem (37). Then by taking $y = x_\lambda + \theta(x_\lambda - z)$, we deduce that

$$\begin{aligned}
V(x_\lambda) + \frac{1}{2\lambda} \|x_\lambda - x\|^2 &\leqslant (1 - \theta)V(x_\lambda) + \theta V(z) \\
&\quad + \frac{1}{2\lambda} (\|x_\lambda - x\|^2 + 2\theta \langle x_\lambda - x, x_\lambda - z \rangle + \theta^2 \|x_\lambda - z\|^2)
\end{aligned}$$

After simplification and division by $\theta > 0$, we obtain

$$V(x_\lambda) - V(z) \leqslant \frac{1}{\lambda} \langle x_\lambda - x, x_\lambda - z \rangle + \theta \|x_\lambda - z\|^2$$

By letting $\theta \to 0$, we have proved that

$$\frac{1}{\lambda}(x - x_\lambda) \in \partial V(x_\lambda) \tag{39}$$

that is, V is subdifferentiable at x_λ, a solution to inclusion (36).

c. We now prove that the domain of the map $\partial V(\cdot)$ is dense in the domain of V. Let x belong to Dom V and p belong to the domain of its conjugate function V^*, which is not empty (see theorem 4.2 below). Since

$$\frac{1}{2\lambda}\|x_\lambda - x\|^2 + V(x_\lambda) = V_\lambda(x) \leq V(x) \tag{40}$$

and since

$$-V(x_\lambda) \leq -a - \langle p, x_\lambda \rangle$$

we deduce that

$$\begin{aligned} \frac{1}{2\lambda}\|x_\lambda - x\|^2 &\leq V(x) - a - \langle p, x \rangle + \langle p, x - x_\lambda \rangle \\ &\leq \frac{1}{4\lambda}\|x - x_\lambda\|^2 + V(x) - a - \langle p, x \rangle + \lambda \|p\|^2 \end{aligned}$$

(for $ab \leq a^2/4\lambda + b^2\lambda$). Hence, when λ converges to zero,

$$\|x - x_\lambda\|^2 \leq 4\lambda(V(x) - a - \langle p, x \rangle + \lambda \|p\|^2) \to 0$$

Since x_λ belongs to the domain of $\partial V(\cdot)$, we have proved the required statement. ■

Remark

By taking $y = x$ in the definition of V_λ, we see that $V(x_\lambda) \leq V(x)$ for all $\lambda > 0$. Since x_λ converges to $x \in \text{Dom } V$ and since V is lower semicontinuous, we deduce that

$$\forall x \in \text{Dom } V, \qquad V(x) = \lim_{\lambda \to 0^+} V_\lambda(x) \tag{41}$$ ■

Remark

Proposition 9 states that $\partial V(\cdot)$ is a monotone map, and theorem 11 states that $1 + \lambda \partial V$ is surjective. By Minty's theorem, we know that the subdifferential map $\partial V(\cdot)$ is a *maximal monotone* set-valued map; properties of maximal monotone maps are studied in Chapter 6, Section 7. We know in particular that

(42)
$(1+\lambda\partial V)^{-1}$ is a single-valued nonexpansive map defined on the whole space X

which is called the resolvent $J_\lambda := (1+\lambda\partial V)^{-1}$ of ∂V. The map A_λ defined by

$$A_\lambda(x) := \frac{1}{\lambda}(x - J_\lambda x) \tag{43}$$

is Lipschitz with constant $1/\lambda$: It is called the *Yosida approximation of* $\partial V(\cdot)$. In our case, $A_\lambda(x)$, the Yosida approximation of ∂V, coincides with the Fréchet derivative of the convex function V_λ. ∎

PROPOSITION 12

The Yosida approximation V_λ of a proper lower semicontinuous, convex function $V\colon X \to R \cup \{+\infty\}$ is a differentiable function defined on X, which converges pointwise to V. ▲

Proof. We already observed that V_λ does converge pointwise to V; the convexity of V_λ is obvious. We remark that $J_\lambda x$ is the solution x_λ to inclusion (36), and we set $y_\lambda := J_\lambda y$. We shall prove that

$$A_\lambda x := \frac{1}{\lambda}(x - x_\lambda) = \nabla V_\lambda(x)$$

On one hand,

$$\begin{aligned}
V_\lambda(x) - V_\lambda(y) &= V(x_\lambda) - V(y_\lambda) + \frac{1}{2\lambda}\|x_\lambda - x\|^2 - \frac{1}{2\lambda}\|y_\lambda - y\|^2 \\
&\leqslant \left\langle \frac{1}{\lambda}(x - x_\lambda), x_\lambda - y_\lambda \right\rangle + \frac{1}{2\lambda}\|x_\lambda - x\|^2 - \frac{1}{2\lambda}\|y_\lambda - y\|^2 \\
&\leqslant \left\langle \frac{1}{\lambda}(x - x_\lambda), x - y \right\rangle + \frac{1}{\lambda}\langle (x - x_\lambda), x_\lambda - x - (y_\lambda - y)\rangle \\
&\quad + \frac{1}{2\lambda}\|x_\lambda - x\|^2 - \frac{1}{2\lambda}\|y_\lambda - y\|^2 \\
&\leqslant \left\langle \frac{1}{\lambda}(x - x_\lambda), x - y \right\rangle - \frac{1}{2\lambda}\|x_\lambda - x\|^2 - \frac{1}{2\lambda}\|y_\lambda - y\|^2 \\
&\quad + \frac{1}{\lambda}\|x_\lambda - x\|\,\|y_\lambda - y\| \\
&\leqslant \left\langle \frac{1}{\lambda}(x - x_\lambda), x - \lambda \right\rangle
\end{aligned}$$

Hence, $(1/\lambda)(x-x_\lambda) \in \partial V_\lambda(x)$.

On the other hand, since $(1/\lambda)(y-y_\lambda) \in \partial V_\lambda(y)$, we obtain

$$\begin{aligned}
V_\lambda(x)-V_\lambda(y) &\geq \left\langle \frac{1}{\lambda}(y-y_\lambda), x-y \right\rangle \\
&\geq \left\langle \frac{1}{\lambda}(x-x_\lambda), x-y \right\rangle - \left\langle \frac{1}{\lambda}(x-x_\lambda) - \frac{1}{\lambda}(y-y_\lambda), x-y \right\rangle \\
&\geq \left\langle \frac{1}{\lambda}(x-x_\lambda), x-y \right\rangle - \left\| \frac{1}{\lambda}(x-x_\lambda) - \frac{1}{\lambda}(y-y_\lambda) \right\| \|x-y\| \\
&\geq \left\langle \frac{1}{\lambda}(x-x_\lambda), x-y \right\rangle - \frac{1}{\lambda}\|x-y\|^2
\end{aligned}$$

because $A_\lambda x = (1/\lambda)(x-x_\lambda)$ is Lipschitz with constant $1/\lambda$.

So we have proved that

$$-\frac{1}{\lambda}\|x-y\| \leq \frac{V_\lambda(x)-V_\lambda(y)-\langle (1/\lambda)(x-x_\lambda), x-y\rangle}{\|x-y\|} \leq 0$$

This proves that $A_\lambda x = \nabla V_\lambda(x)$. ■

Before proving that a continuous convex function is subdifferentiable, we show that convex continuous functions are locally Lipschitz.

PROPOSITION 13

Let $V\colon X \to R \cup \{+\infty\}$ be a proper convex function. The following conditions are equivalent:

$$(44) \quad \begin{cases} \textbf{i.} & V \textit{ is bounded above on an open subset (contained in } \mathrm{Dom}\, V). \\ \textbf{ii.} & V \textit{ is locally Lipschitz on } \mathrm{Int\,Dom}\, V. \end{cases}$$

▲

Proof. **a.** It is clear that (44, **ii**) implies (44, **i**).

b. Let us assume that V is bounded by a on a ball $x_0+\eta B \subset \mathrm{Dom}\, V$. We associate with any $x \in X$ the element

$$y := \frac{x_0-(1-\theta)x}{\theta}, \qquad \text{where} \qquad \theta = \frac{\|x-x_0\|}{\|x-x_0\|+\eta} < 1$$

Hence, $\|y-x_0\| = \eta$ and, consequently, $V(y) \leq a$. The convexity of V implies that

$$V(x_0) = V(\theta y + (1-\theta)x) \leq \theta a + (1-\theta)V(x)$$

Thus,

$$V(x_0) \leq V(x) + \frac{\theta}{1-\theta}(a - V(x_0)) = V(x) + \frac{a-V(x_0)}{\eta}\|x-x_0\|$$

Now, take $x \in x_0 + \eta B$ and $y = (x-(1-\theta)x_0)/\theta$, where $\theta = \|x_0 - x\|/\eta \leqslant 1$. Thus, $\|y - x_0\| = \eta$, and, consequently, $V(y) \leqslant a$. The convexity of V implies that

$$V(x) = V(\theta y + (1-\theta)x_0) \leqslant \theta a + (1-\theta)V(x_0) = \theta(a - V(x_0)) + V(x)$$
$$= \frac{a - V(x_0)}{\eta} \|x - x_0\| + V(x)$$

Therefore, when $x \in x_0 + \eta B$, then

$$(45) \qquad |V(x) - V(x_0)| \leqslant \frac{a - V(x_0)}{\eta} \|x - x_0\|$$

c. Let us prove that V is Lipschitz on the ball $x_0 + \beta B$, where $\beta \in]0, \eta[$. Let x_1, x_2 belong to $x_0 + \beta B$. We choose an integer $n \geqslant \|x_1 - x_2\|/(\eta - \beta)$. For $j = 0, \ldots, n$, we introduce the elements $y_j = x_1 + (j/n)(x_2 - x_1)$. Then $y_0 = x_1$, $y_n = x_2$, $\|y_{j+1} - y_j\| = \|x_1 - x_2\|/n \leqslant \eta - \beta$, $\|x_1 - x_2\| = \sum_{j=0}^{n-1} \|y_{j+1} - y_j\|$, and for $j = 0, \ldots, n$, y_j belongs to $x_0 + \beta B$.

Hence by (45), $a - V(y_j) \leqslant 2(a - V(x_0))$ and since V is bounded above by a on $y_j + (\eta - \beta)B$, we deduce from (45) (where x_0 is replaced by y_j) that

$$|V(y_{j+1}) - V(y_j)| \leqslant \frac{2(a - V(x_0))}{\eta - \beta} \|y_{j+1} - y_j\|$$

Consequently,

$$|V(x_1) - V(x_2)| \leqslant \sum_{j=0}^{n-1} |V(y_{j+1}) - V(y_j)|$$
$$\leqslant \frac{2(a - V(x_0))}{\eta - \beta} \sum_{j=0}^{n-1} \|y_{j+1} - y_j\| = \frac{2(a - V(x_0))}{\eta - \beta} \|x_1 - x_2\|$$

d. Finally, we shall prove that for any $x_1 \in \text{Int}(\text{Dom } V)$, V is bounded above on a neighborhood of x_1 and thus by the preceding result V is Lipschitz on a neighborhood of x_1. Since there exists γ such that $x_1 + \gamma B \subset \text{Dom } V$, then

$$x_2 := x_0 + \frac{1}{1-\lambda}(x_1 - x_0) = \frac{x_1 - \lambda x_0}{1 - \lambda} \in \text{Dom } V$$

since

$$\|x_2 - x_1\| = \frac{\lambda}{1-\lambda} \|x_1 - x_0\| = \gamma \qquad \text{when} \qquad \lambda := \frac{\gamma}{\gamma + \|x_1 - x_0\|} < 1$$

Let $y \in x_1 + \lambda\eta$. Then the element

$$z := \frac{1}{\lambda}(y + \lambda x_0 - x_1) = \frac{1}{\lambda}(y - (1-\lambda)x_2)$$

satisfies $\|z - x_0\| = (1/\lambda)\|y - x_1\| \leqslant \eta$. Consequently, $V(z) \leqslant a$, and the convexity of V implies that

$$V(y) = V(\lambda z + (1-\lambda)x_2) \leqslant \lambda V(z) + (1-\lambda)V(x_2) \leqslant \lambda a + (1-\lambda)V(x_2) =: b$$

Hence, V is bounded above on a neighborhood of x_1. ■

We deduce the following important consequences.

COROLLARY 14

If the interior of the domain of a convex function $V: R^n \to R \cup \{+\infty\}$ is nonempty, then V is locally Lipschitz on Int(Dom V). ▲

Proof. Let $B(x_0, \eta)$ be a ball with center x_0 and radius η contained in Dom V. We can then find n points $x_i \in B(x_0, \eta)$ such that the vectors $x_i - x_0$ are linearly independant. The subset S of convex combinations $\sum_{i=0}^n \lambda_i x_i$, where $\lambda_i > 0$ for all i, is open and contained in Dom V. Consequently, since $V(\sum_{i=0}^n \lambda_i x_i) \leqslant \sum_{i=0}^n \lambda_i V(x_i)$, the convex function V is bounded above on S. The statement follows from proposition 13. ■

COROLLARY 15

If the interior of the domain of a lower semicontinuous, convex function V from a Banach space to $R \cup \{+\infty\}$ is nonempty, then V is locally Lipschitz on Int Dom V. ▲

Proof. By a consequence of Baire's theorem, V, a lower semicontinuous, real-valued function defined on the open set Int(Dom V), is bounded above on a nonempty open subset. The statement follows from proposition 13. ■

Continuous convex functions are epidifferentiable and subdifferentiable on the interior of their domain.

THEOREM 16

Let x_0 belong to the interior of the domain of a lower semicontinuous, convex function $V: X \to R \cup \{+\infty\}$. Then

$$(46) \qquad \forall u \in X, \quad D_+V(x_0)(u) = \lim_{h \to 0^+} \frac{V(x_0 + hu) - V(x_0)}{h} \quad \textit{is finite}$$

There exists $\eta > 0$ *such that*

$$(47) \qquad \forall u \in \eta B, \quad V(x_0) - V(x_0 - u) \leqslant D_+ V(x_0)(u) \leqslant V(x_0 + u) - V(x_0)$$

In addition, for some constant $c > 0$ *we have*

$$(48) \quad \begin{cases} \textbf{i.} & D_+ V(x_0)(u) \leqslant c\|u\| \\ \textbf{ii.} & (x, u) \in \operatorname{Int} \operatorname{Dom} V \times X \to D_+ V(x)(u) \text{ is upper semicontinuous.} \end{cases}$$

▲

Proof. Let $\eta > 0$ such that the ball $x_0 + \eta B$ is contained in Dom V. Since

$$x_0 = \frac{1}{1+h}(x_0 + hu) + \frac{h}{1+h}(x_0 - u)$$

we deduce from the convexity of V that

$$-\infty < V(x_0) - V(x_0 - u) \leqslant \frac{V(x_0 + hu) - V(x_0)}{h}$$

Since $h \to [V(x_0 + hu) - V(x_0)]/h$ is increasing, it follows that

$$V(x_0) - V(x_0 - u) \leqslant \inf_{h > 0} \frac{V(x_0 + hu) - V(x_0)}{h} = \lim_{h \to 0^+} \frac{V(x_0 + hu) - V(x_0)}{h}$$

We also know that there exists $\eta > 0$ such that V is Lipschitz on the ball $x_0 + \eta B$. If $c > 0$ denotes the Lipschitz constant, then

$$D_+ V(x_0)(u) \leqslant \frac{V(x_0 + hu) - V(x_0)}{h} \leqslant c\|u\|$$

Finally, $D_+ V(x)(u)$ being the infimum of the continuous functions $(x, u) \to [V(x + hu) - V(x)]/h$ on Int Dom $V \times X$, it is upper semicontinuous. ■

We now translate these results into terms of the subdifferential.

THEOREM 17

Let x_0 *belong to the interior of the domain of a lower semicontinuous, convex function* $V: X \to]-\infty, +\infty]$. *Then*

(49) $\quad \partial V(x_0)$ *is a nonempty, bounded, closed, convex subset of* X^* *(and thus weakly compact)*

and

(50) $\quad x \in \text{Int Dom } V \to \partial V(x)$ *is upper hemicontinuous on* Int Dom V.

Furthermore,

$$D_+ V(x_0)(u) = \sigma(\partial V(x_0), u) \tag{51}$$

that is, $D_+ V(x)(\cdot)$ *is the support function of* $\partial V(x)$. *If* $\partial V(x_0)$ *contains exactly one element, this element is the gradient* $\nabla V(x_0)$ *of* V *at* x_0. ▲

Proof. Since

$$\partial V(x_0) = \{p \in X^* | \forall u \in X, \langle p, u \rangle \leqslant D_+ V(x_0)(u)\}$$

and $D_+ V(x)(\cdot)$ is convex, lower semicontinuous, and positively homogeneous, it is the support function of a nonempty, closed convex, which is $\partial V(x_0)$. Since

$$\sigma(V(x_0), u) = D_+ V(x_0)(u) \leqslant c\|u\| = \sigma(cB_*, u)$$

we deduce that $\partial V(x_0) \subset cB_*$, that is, it is bounded. By theorem 16, $\partial V(\cdot)$ is upper semicontinuous from Int Dom V to X^*.

If $\partial V(x_0) = \{p_0\}$ contains exactly one point, then

$$u \to D_+ V(x_0)(u) = \sigma(\{p_0\}, u) = \langle p_0, u \rangle$$

is a continuous linear functional. Hence, $p_0 = \nabla V(x_0)$ is the gradient of V at x_0. ■

4. CONJUGATE FUNCTIONS

We can introduce the conjugate function V^* of a proper function $V: X \to R \cup \{+\infty\}$ in many ways. We choose the following point of view. Instead of studying simply the minimization problem: $\inf_{x \in X} V(x)$, we study the whole family of perturbed problems

$$-V^*(p) := \inf_{x \in X} (V(x) - \langle p, x \rangle)$$

when the function V is perturbed by the continuous linear functionals $x \to \langle p, x \rangle$. (They constitute the class of *simplest* perturbations that can be considered.) This defines a function $p \to V^*(p)$ on the dual X^* of X, which is obviously *convex and lower semicontinuous*, since V^* is the pointwise supremum of the continuous affine functions $p \to \langle p, x \rangle - V(x)$. Observe that if $V: X \to R \cup \{+\infty\}$ has a nonempty domain, then V^* maps X^* into $R \cup \{+\infty\}$, because if x_0 belongs to Dom V, then for all $p \in X^*$, we have

$$-\infty < \langle p, x_0 \rangle - V(x_0) \leqslant \sup_x (\langle p, x \rangle - V(x)) =: V^*(p)$$

DEFINITION 1
Let V be a proper function from X to $R\cup\{+\infty\}$. The function V^* from X^* to $R\cup\{+\infty\}$ defined by

$$\forall p \in X^*, \quad V^*(p) := \sup_{x \in X} (\langle p, x\rangle - V(x)) \tag{1}$$

is called the conjugate function *of V.*

Similarly, if W is a proper function from X^* to $R\cup\{+\infty\}$, we define W^* from X to $R\cup\{+\infty\}$ by

$$W^*(x) := \sup_{p \in X^*} (\langle p, x\rangle - W(x)) \tag{2}$$

and consequently, the biconjugate V^{**} *of V is defined by $V^{**} := (V^*)^*$.* ▲

We observe that both V^* and V^{**} are *lower semicontinuous, convex functions* and that

$$\forall x, \quad V^{**}(x) \leqslant \sup_{p \in X^*} (\langle p, x\rangle - (\langle p, x\rangle - V(x))) \leqslant V(x) \tag{3}$$

The crucial point is that the conjugacy operation $V \to V^*$ is a one to one correspondence between proper lower semicontinuous, convex functions defined on X and X^*, respectively, which allows us to use the convenient interchange properties of duality: We shall have each time the choice of working with either a function V or a function V^* according to the properties needed to solve the problem at hand.

THEOREM 2
A proper function V: $X \to R\cup\{+\infty\}$ is convex and lower semicontinuous if and only if $V = V^{**}$. In this case, the domain of V^* is nonempty. ▲

Proof. Since V is convex and lower semicontinuous,

$$Ep(V) := \{(x, \lambda) \in X \times R / V(x) - \lambda \leqslant 0\}$$

is a closed convex subset of $X \times R$.

a. We assume $a < V(x)$. Since the pair (x, a) does not belong to $Ep(V)$, there exists a continuous linear functional $(p, -\alpha) \in X^* \times R$ that strictly separates (x, a) from $Ep(v)$. Then there exists $\varepsilon > 0$ such that

$$\forall y \in \text{Dom } V, \quad \forall \lambda \geqslant 0, \quad \langle p, y\rangle - \alpha V(y) - \alpha\lambda \leqslant \langle p, x\rangle - \alpha a - \varepsilon \tag{4}$$

By taking the supremum when $\lambda \geqslant 0$, we deduce that $\alpha \geqslant 0$ and

$$\forall y \in \text{Dom } V, \quad \langle p, y\rangle - \alpha V(y) \leqslant \langle p, x\rangle - \alpha a - \varepsilon \tag{5}$$

b. When $\alpha>0$, we can divide by α and set $\bar{p}=p/\alpha$. Inequality (5) becomes

$$\forall y \in \operatorname{Dom} V, \quad \langle \bar{p}, y\rangle - V(y) \leqslant \langle \bar{p}, x\rangle - a - \frac{\varepsilon}{\alpha} \tag{6}$$

and thus by taking the supremum when y ranges over Dom V,

$$V^*(\bar{p}) \leqslant \langle \bar{p}, x\rangle - a - \frac{\varepsilon}{\alpha} \tag{7}$$

The first consequence is that Dom $V^* \neq \varnothing$, that is, V^* is proper. The second consequence is that when $a < V(x)$, then $a < \langle \bar{p}, x\rangle - V^*(\bar{p}) \leqslant V^{**}(x)$. So by letting a converge to $V(x)$, we deduce that $V(x)=V^{**}(x)$.
c. We observe that $\alpha>0$ whenever $x \in \operatorname{Dom} V$; indeed, we take $y=x$ in inequality (5) and find that $\alpha(a-V(x)) \leqslant -\varepsilon$, $\alpha>0$. When $x \notin \operatorname{Dom} V$, both situations $\alpha>0$ and $\alpha=0$ can occur. It remains to study the latter case.
d. When $\alpha=0$ and $x \notin \operatorname{Dom} V$, inequality (5) implies that

$$\forall y \in \operatorname{Dom} V, \quad \langle p, y-x\rangle + \varepsilon \leqslant 0 \tag{8}$$

Let $\bar{p} \in \operatorname{Dom} V^*$, which exists by (b) and (c). By multiplying (8) by $n>0$ and adding to it inequality $\langle \bar{p}, y\rangle - V(y) - V^*(\bar{p})$, we obtain

$$\forall y \in \operatorname{Dom} V, \quad \langle \bar{p}+np, y\rangle - n\langle p, x\rangle + n\varepsilon - V^*(\bar{p}) - V(y) \leqslant 0 \tag{9}$$

which implies that by taking the supremum with respect to $y \in \operatorname{Dom} V$,

$$V^*(\bar{p}+np) - n\langle p, x\rangle + n\varepsilon - V^*(\bar{p}) \leqslant 0$$

By adding and substracting $\langle \bar{p}, x\rangle$, we deduce that

$$n\varepsilon + \langle \bar{p}, x\rangle - V^*(\bar{p}) \leqslant \langle \bar{p}+np, x\rangle - V^*(\bar{p}+np) \leqslant V^{**}(x)$$

By letting $n\to\infty$, we conclude that $V^{**}(x)=\infty$. ■

The very definition of the conjugate function V^* implies that for all $x \in X$, $p \in X^*$, inequalities

$$\langle p, x\rangle \leqslant V(x) + V^*(p) \tag{10}$$

known as Fenchel inequalities, hold true. It is quite natural to distinguish the pairs (x, p) such that inequality (10) is actually an equality.

PROPOSITION 3
Let $V\colon X \to R \cup \{+\infty\}$ *be a proper convex function and* $x \in \operatorname{Dom} V$ *be fixed.*

The following statements are equivalent:

(a) $p \in \partial V(x)$

(b) $\langle p, x\rangle = V(x) + V^*(p)$ ▲

In other words, $\partial V(x)$ is the set of gradients p of the affine functions $x \to \langle p, x\rangle - V^*(p)$ that pass through $(x, V(x))$ and are below V.

Proof. Indeed, $p \in \partial V(x)$ if and only if by proposition 3.3,

$$\forall y \in X, \qquad \langle p, y\rangle - V(y) \leqslant \langle p, x\rangle - V(x)$$

that is, if and only if

$$V^*(p) := \sup_{y \in X} \left[\langle p, y\rangle - V(y)\right) = \langle p, x\rangle - V(x)$$ ■

When V is lower semicontinuous, we deduce the following reciprocity formula in theorem 4.

THEOREM 4

Let V be a proper lower semicontinuous, convex function from X to $R \cup \{+\infty\}$. Then the inverse of the set-valued map $\partial V(\cdot)\colon X \to X^*$ is ∂V^*.

(11) $$p \in \partial V(x) \quad \textit{if and only if} \quad x \in \partial V^*(p)$$ ▲

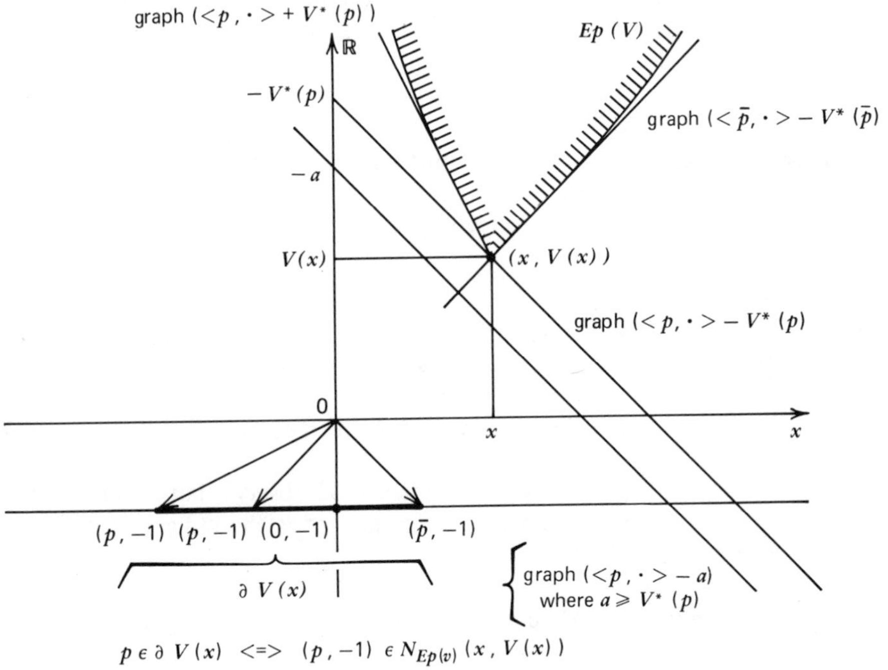

$p \in \partial V(x) \iff (p, -1) \in N_{Ep(v)}(x, V(x))$

Proof. It follows obviously from proposition 3 applied to V^* and equality $V=V^{**}$. ■

As a consequence, we obtain proposition 5.

PROPOSITION 5

Let V be a proper lower semicontinuous, convex function. Then $\partial V^*(p)$ is the set of minimizers of the perturbed function $y \to V(y) - \langle p, y \rangle$. ▲

We now proceed with some examples of conjugate functions.

PROPOSITION 6

The support function of a subset K is the conjugate of its indicator. Conversely, if K is closed and convex, $\psi_K = \sigma_K^*$. ▲

Proof. Indeed

$$\sigma_K(p) := \sup_{x \in K} \langle p, x \rangle = \sup (\langle p, x \rangle - \psi_K(x)) =: \psi_K^*(p)$$

When K is closed and convex, ψ_K is convex and lower semicontinuous, and thus $\psi_K = \psi_K^{**} = \sigma_K^*$. ■

In particular, we mention this important example: The conjugate function of the continuous linear functional $p \in X^*$ is the indicator of the singleton $\{p\}$

$$p^* = \psi_{\{p\}}$$

because p can be regarded as the support function of the singleton $\{p\}$:

$$\sigma_{\{p\}}(x) = \langle p, x \rangle \quad \text{for all} \quad x \in X$$

■

PROPOSITION 7

A proper lower semicontinuous, positively homogeneous convex function $\sigma: X^* \to R \cup \{+\infty\}$ is the support function of the subset

$$K := \{x \in X \mid \forall p \in X^*, \langle p, x \rangle \leqslant \sigma(p)\}$$

▲

Proof. Since σ is positively homogeneous, we observe that $\sigma^*(x)=0$ when $x \in K$ and $\sigma^*(x) = +\infty$ when $x \notin K$. Therefore, σ^* is the indicator of K, and $\sigma = \sigma^{**}$ is its support function. ■

We now compute the conjugate functions of the functions $x \to (1/\alpha)\|x\|^\alpha$.

PROPOSITION 8

a. *The conjugate function of $x \to \|x\|$ is the indicator of the unit ball B^* of the dual*

$$\|\cdot\|^*(p) = \begin{cases} 0 & \text{if } \|p\|_* \leqslant 1 \\ +\infty & \text{if } \|p\|_* > 1 \end{cases} \tag{12}$$

b. *Let $\phi: R \to R \cup \{+\infty\}$ be a proper lower semicontinuous, even convex function and ϕ^* its conjugate. Then*

$$\phi(\|\cdot\|)^* = \phi^*(\|\cdot\|_*) \tag{13}$$

c. *In particular, if $\alpha > 1$, then by setting $\alpha^* = \alpha/(\alpha - 1)$, we obtain*

$$\left(\frac{1}{\alpha}\|\cdot\|^\alpha\right)^* = \frac{1}{\alpha^*}\|\cdot\|_*^{\alpha^*} \tag{14}$$ ▲

Proof. **a.** When $\alpha = 1$, we observe that $\|x\| = \sigma_{B^*}(x)$ is the support function of the unit ball B^* of the dual. Hence, $\|\cdot\|_* = \psi_{B^*}$ is the indication of B^*.

b. We check that

$$\begin{aligned}
\phi(\|\cdot\|)^*(p) &:= \sup_{x \in X} [\langle p, x\rangle - \phi(\|x\|)] \\
&= \sup_{\lambda \geqslant 0} \sup_{\substack{x \in X \\ \|x\| = \lambda}} [\langle p, x\rangle - \phi(\|x\|)] \\
&= \sup_{\lambda \geqslant 0} [\lambda\|p\|_* - \phi(\lambda)] \\
&= \sup_{\lambda \in R} [\lambda\|p\|_* - \phi(\lambda)] \qquad \text{(because } \phi \text{ is even)} \\
&= \phi^*(\|p\|_*)
\end{aligned}$$

c. Hölder's inequality states that

$$\text{for all } a, b \in R_+, \quad ab \leqslant \frac{1}{\alpha} a^\alpha + \frac{1}{\alpha^*} b^{\alpha^*}$$

The equality holds true when $\alpha^\alpha = b^{\alpha^*}$. If we set $\phi(\lambda) = (1/\alpha)|\lambda|^\alpha$, this shows that $\phi^*(\mu) = (1/\alpha^*)|\mu|^{\alpha^*}$. Therefore, the last statement follows from the second. ■

We now present elementary properties of conjugate functions.

PROPOSITION 9

If $V \leqslant W$, *then* $W^* \leqslant V^*$.
Let us set $W(x) = V(x - x_0) + \langle p_0, x\rangle + \alpha$. *Then*

$$W^*(p) = V^*(p - p_0) + \langle p, x_0\rangle - (\alpha + \langle p_0, x_0\rangle) \tag{15}$$ ▲

Proof. The first statement is obvious. We verify the second:

$$\begin{aligned}\sup_{x\in X}[\langle p, x\rangle - W(x)] &= \sup_{x\in X}[\langle p - p_0, x\rangle - V(x - x_0)] - \alpha \\ &= \sup_{x\in X}[\langle p - p_0, x - x_0\rangle - V(x - x_0)] + \langle p, x_0\rangle - (\alpha + \langle p_0, x_0\rangle) \\ &= V^*(p - p_0) + \langle p, x_0\rangle - (\alpha + \langle p_0, x_0\rangle)\end{aligned}$$ ■

PROPOSITION 10

Let $U: X \times Y \to R \cup \{+\infty\}$ *be a proper function and* $A \in \mathscr{L}(X, Y)$. *Let* $V(y) := \inf_{x\in X} U(x, y - Ax)$. *Then for all* $q \in Y^*$,

$$V^*(q) = U^*(A^*q, q) \tag{16}$$ ▲

Proof. Indeed

$$\begin{aligned}V^*(q) &:= \sup_{y\in Y}[\langle q, y\rangle - \inf_{x\in X}(U(x, y - Ax))] \\ &= \sup_{x\in X}\sup_{y\in Y}[\langle q, y - Ax\rangle + \langle A^*q, x\rangle - U(x, y - Ax)] \\ &= \sup_{x\in X}\sup_{z\in Y}[\langle A^*q, x\rangle + \langle q, z\rangle - U(x, z)] \\ &= U^*(A^*q, q)\end{aligned}$$ ■

In particular, we obtain the following formulas.

a. Let $U: X \to R \cup \{+\infty\}$ be a proper function and $A \in \mathscr{L}(X, Y)$. Set $V(y) := \inf_{Ax=y} U(x)$. Then

$$V^*(q) = U^*(A^*q) \tag{17}$$

b. Let U_1 and U_2 be two proper functions from X to $R \cup \{+\infty\}$. Set $V(y) := \inf_{x\in X}(U_1(x) + U_2(y - x))$. Then

$$V^*(q) = U_1^*(q) + U_2^*(q) \tag{18}$$

c. Let $U: X \to R \cup \{+\infty\}$, $W: Y \to R \cup \{+\infty\}$ be proper functions and $A \in \mathscr{L}(X, Y)$. Set $V(y) := \inf_{x\in X}(U(x) + W(y - Ax))$. Then

$$V^*(q) = U^*(A^*q) + W^*(q) \tag{19}$$ ■

We now turn our attention to computing the conjugate functions of functions of the form VA, V_1+V_2, $V+WA$ and, more generally, functions of the form $x \to L(x, Ax)$, where

$$\begin{cases} L \text{ is a proper lower semicontinuous, convex function} \\ \text{from } X \times Y \text{ to } R \cup \{+\infty\}. \end{cases} \tag{20}$$

We introduce the lower semicontinuous, convex function defined by

$$V(x) := L(x, Ax) \tag{21}$$

whose domain is nonempty if and only if $0 \in ((A \oplus -1)\text{Dom } L)$. Let L^* denote the conjugate function of L defined on $X^* \times Y^*$ by

$$L^*(p, q) := \sup_{(x,y) \in X \times Y} [\langle p, x\rangle + \langle q, y\rangle - L(x, y)] \tag{22}$$

Hence, the domain of the function $q \to L^*(p - A^*q, q)$ is nonempty if and only if p belongs to $(1 \oplus A^*)\text{Dom } L^*$.

THEOREM 11

Let L satisfy assumption (20) *and V be defined by* (21). *We posit assumption*

$$0 \in \text{Int}((A \oplus -1)\text{Dom } L) \tag{23}$$

Then if $p_0 \in (1 \oplus A^*)\text{Dom } L^*$, *there exists* $\bar{q} \in Y^*$ *such that*

$$V^*(p_0) = L^*(p_0 - A^*\bar{q}, \bar{q}) = \inf_{q \in Y^*} L^*(p_0 - A^*q, q) \tag{24}$$ ▲

Before proving this theorem, we mention explicitely the following particular cases.

COROLLARY 12

a. *Let* $W: Y \to R \cup \{+\infty\}$ *be a proper lower semicontinuous, convex function and* $A \in \mathscr{L}(X, Y)$. *If*

$$0 \in \text{Int}(\text{Im } A - \text{Dom } W) \tag{25}$$

then for all $p \in A^*$ Dom W^*, *there exists* $\bar{q} \in Y^*$ *satisfying* $A^*\bar{q} = p$ *and*

$$(WA)^*(p) = W^*(\bar{q}) = \inf_{A^*q = p} W^*(q) \tag{26}$$

b. *Let* $W_1: X \to R \cup \{+\infty\}$ *and* $W_2: X \to R \cup \{+\infty\}$ *be two proper lower semicontinuous, convex functions such that*

$$0 \in \text{Int}(\text{Dom } W_1 - \text{Dom } W_2) \tag{27}$$

Then for all $p \in \text{Dom } W_1^* + \text{Dom } W_2^*$, *there exists* $\bar{q} \in Y^*$ *such that*

$$(W_1 + W_2)^*(p) = W_1^*(\bar{q}) + W_2^*(p - \bar{q}) = \inf_{q \in X^*} (W_1^*(q) + W_2^*(p-q)) \tag{28}$$

c. *Let* $U: X \to R \cup \{+\infty\}$ *and* $W: Y \to R \cup \{+\infty\}$ *be two proper lower semi-continuous, convex functions and* $A \in \mathscr{L}(X, Y)$ *satisfying*

$$0 \in \text{Int}(A \text{ Dom } U - \text{Dom } W) \tag{29}$$

For all $p \in \text{Dom } U^* + A^* \text{ Dom } W^*$, *there exists* $\bar{q} \in Y^*$ *such that*

$$(U + WA)^*(p) = U^*(p - A^*\bar{q}) + W^*(\bar{q}) = \inf_{q \in Y^*} (U^*(p - A^*q) + W^*(q)) \tag{30}$$ ▲

First proof of Theorem 11 (*self-contained*). Let p_0 belong to $(1 \oplus A^*)\text{Dom } L^*$.
a. We introduce the map ψ from $\text{Dom } L^* \subset X^* \times Y^*$ to $R \times X^*$ defined by

$$\psi(p, q) := (L^*(p + p_0, q), p + A^*q) \tag{31}$$

and we consider the set

$$\psi(\text{Dom } L^* - (p_0, 0)) + R_+ \times \{0\} \subset R \times X^* \tag{32}$$

It is easy to check that this subset is convex. We shall deduce from assumption (23) that it is closed. Indeed, let us consider a sequence of elements (v_n, r_n) belonging to this set and converging to (v_*, r_*) in $R \times X^*$. There exist elements $p_n \in X^*$ and $q_n \in Y^*$ such that

$$v_n \geq L^*(p_n + p_0, q_n), \qquad r_n = p_n + A^*q_n$$

and there exists a ball of radius $\gamma > 0$ contained in $(A \oplus -1)\text{Dom } L$ by assumption (23). Therefore, for all $z \in Y$, there exists $(x, y) \in \text{Dom } L$ such that $\gamma z/\|z\| = y - Ax$. Consequently,

$$\begin{aligned}
\frac{\gamma}{\|z\|} \langle q_n, z \rangle &= \langle q_n, y \rangle - \langle A^*q_n, x \rangle \\
&= \langle q_n, y \rangle + \langle p_n, x \rangle - \langle r_n, x \rangle \\
&\leq L^*(p_n + p_0, q_n) + L(x, y) - \langle r_n, x \rangle - \langle p_0, x \rangle \\
&\leq v_n + L(x, y) - \langle r_n, x \rangle - \langle p_0, x \rangle
\end{aligned}$$

Since the sequences v_n and $\langle r_n, x \rangle$ are convergent and thus bounded, we have proved that

$$\forall z \in Y, \qquad \sup_{n \geq 0} \langle q_n, z \rangle < +\infty \tag{33}$$

The boundedness theorem implies that the sequence of elements $q_n \in Y^*$ lies in a weakly relatively compact subset. Therefore, a subsequence q_n' converges weakly to some q_* in Y^*, and, consequently, $q_n' = r_n' - A^*q_n'$ converges weakly to $p_* := r_* - A^*q_*$. Since the function L^* is lower semicontinuous for the weak topology of $X^* \times Y^*$—as a pointwise supremum of the weakly continuous affine functions

$$(p, q) \to \langle p, x\rangle + \langle q, y\rangle - L(x, y)$$

we deduce that

$$L^*(p_* + p_0, q_*) \leqslant \liminf_{n\to\infty} L^*(p_n' + p_0, q_n') \leqslant \lim_{n\to\infty} v_n' = v_*$$

Thus we have proved that

$$v_* \geqslant L^*(p_* + p_0, q_*), \qquad r_* = p_* + A^*q_*,$$

that is, (v_*, r_*) belongs to

$$\psi(\text{Dom } L^* - (p_0, 0)) + R_+ \times \{0\}$$

b. Now we shall prove that

$$(V^*(p_0), 0) \in \psi(\text{Dom } L^* - (p_0, 0)) + R_+ \times \{0\} \tag{34}$$

Indeed, this statement implies the theorem because there exist $\bar{q} \in Y^*$ such that

$$V^*(p_0) \geqslant L^*(p_0 - A^*\bar{q}, \bar{q}) \tag{35}$$

But inequalities

$$\langle p_0, x\rangle = \langle p_0 - A^*q, x\rangle + \langle q, Ax\rangle \leqslant L^*(p_0 - A^*q, q) + L(x, Ax)$$

imply that $V^*(p_0) \leqslant \inf_{q \in Y^*} L^*(p_0 - A^*q, q)$.
Hence, equation (24) follows from (34). Assume that inclusion (34) is false. Since the subset $\psi(\text{Dom } L^* - (p_0, 0)) + R_+ \times \{0\}$ is convex and closed, the separation theorem implies the existence of a pair $(\alpha, -x) \in R \times X$ and $\varepsilon > 0$ such that

$$\alpha V^*(p_0) \leqslant \inf_{(p,q) \in X^* \times Y^*} (\alpha L^*(p + p_0, q) + \langle p + A^*q, -x\rangle) + \inf_{\theta > 0} \alpha\theta - \varepsilon$$

We deduce that α is not negative and $\inf_{\theta>0} \alpha\theta = 0$. Further, since p_0 belongs to

$(1\oplus A^*)\operatorname{Dom} L^*$, there exists $(\tilde{p}, \tilde{q}) \in \operatorname{Dom} L^* - (p_0, 0)$ such that $\tilde{p}+A^*\tilde{q}=0$. This implies that $\alpha>0$, because if $\alpha=0$, we would deduce from the preceding inequality that $0\leqslant -\varepsilon$. Dividing by $\alpha>0$ and setting $\bar{x}:=x/\alpha$, $\eta=\varepsilon/\alpha$, we would obtain

$$
\begin{aligned}
V^*(p_0) &\leqslant \inf_{p,q}\,[L^*(p+p_0, q)-\langle p+p_0, \bar{x}\rangle - \langle q, A\bar{x}\rangle] + \langle p_0, \bar{x}\rangle - \eta \\
&= -L(\bar{x}, A\bar{x}) + \langle p_0, \bar{x}\rangle - \eta \\
&\leqslant V^*(p_0) - \eta
\end{aligned}
$$

a contradiction. Hence, statement (34) holds true, and our theorem ensues. ■

Second proof of Theorem 11.

a. We temporarily set $W(p):=\inf_{q\in Y^*} L^*(p+A^*q,q)$. Proposition 10 implies that

$$
W^*(x)=L(x, Ax)=V(x) \tag{36}
$$

b. If we prove that the convex function W is also lower semicontinuous, then theorem 2 implies that $V^*(p)=W^{**}(p)=W(p)$, that is, the second statement of the theorem holds true. Since $W(p)=\inf_{q\in Y^*} U(q)$ where $U(q):=L^*(p-A^*q, q)$, it suffices to prove that U is inf compact. Indeed, let $K:=\{q\in Y^*/U(q)\leqslant W(p)+1\}$, which is then a weakly compact set. Then $W(p)=\min_{q\in K} L^*(p-A^*q, q)$. Since L^* is lower semicontinuous, corollary 3.22 implies that W is lower semicontinuous.

c. Therefore, the second statement of theorem 3 follows from the following proposition. ■

PROPOSITION 13

If $0 \in \operatorname{Int}((A\oplus -1)\operatorname{Dom} L)$, *then the functions* $q\to L^*(p-A^*q, q)$ *are inf compact.* ▲

Proof. We begin by estimating the conjugate function of U. If

$$
z_0:=y_0-Ax_0 \in -(A\oplus -1)\operatorname{Dom} L
$$

then

$$
U^*(z_0)\leqslant L(x_0, y_0)-\langle p, x_0\rangle
$$

Indeed,

$$
\begin{aligned}
U^*(z_0) &= \sup_{q\in Y^*}\,[\langle q, z_0\rangle - L^*(p-A^*q, q)] \\
&= \sup_{q\in Y^*}\,\inf_{x\in X}\,\inf_{y\in Y}\,[\langle q, z_0\rangle - \langle p-A^*q, x\rangle - \langle q, y\rangle + L(x, y)]
\end{aligned}
$$

$$\leqslant \sup_{q \in Y^*} [\langle q, z_0 + Ax_0 - y_0 \rangle - \langle p, x_0 \rangle + L(x_0, y_0)]$$
$$= L(x_0, y_0) - \langle p, x_0 \rangle$$

Therefore, assumption (23) implies that there exists $\gamma > 0$ such that for all $z \in Y$, there exists $(x, y) \in \text{Dom } L$ satisfying $\gamma z/\|z\| = y - Ax$ and thus, $U^*(\gamma z/\|z\|) < +\infty$.

Let $K_\lambda := \{q | U(q) \leqslant \lambda\}$ be a level set of U. Hence, for all $z \in Y$,

$$\sup_{q \in K_\lambda} \left\langle q, \frac{\gamma z}{\|z\|} \right\rangle \leqslant \sup_{q \in K_\lambda} U(q) + U^* \left(\frac{\gamma z}{\|z\|} \right) < +\infty$$

Then K_λ is weakly bounded and thus weakly relatively compact. ■

Remark

Formula (1.5.29) on support functions is a consequence of formula (30). Indeed, if $K := \{x \in L / Ax \in M\}$, then its indicator can be written

$$\psi_K(x) = \psi_L(x) + \psi_M(Ax)$$

Therefore, if $0 \in \text{Int } (AL - M) = \text{Int}(A \text{ Dom } \psi_L - \text{Dom } \psi_M)$, for all $p \in b(K)$, there exists $\bar{q} \in Y^*$ such that

$$\sigma_K(p) = \sigma_L(p - A^*\bar{q}) + \sigma_M(\bar{q}) = \inf_{q \in Y^*} (\sigma_L(p - A^*q) + \sigma_M(q)) \tag{37}$$ ■

It will be convenient to investigate further partial conjugate functions of bi-convex functions

Let X and Y be two Banach spaces and

(38) $V : X \times Y \to R \cup \{+\infty\}$ a proper lower semicontinuous, convex function

We introduce the partial conjugate U defined on $X \times Y^*$ by

$$U(x, q) := \sup_{y \in Y} [\langle q, y \rangle - V(x, y)] \tag{39}$$

which is equal to $-\infty$ on $K := \{x \in X | \forall y \in Y, V(x, y) = +\infty\}$. This is obviously a function satisfying

$$\begin{cases} \textbf{i.} \quad \forall x \in X, q \to U(x, q) \text{ is lower semicontinuous and convex.} \\ \textbf{ii.} \quad \forall q \in Y^*, x \to U(x, q) \text{ is concave.} \end{cases} \tag{40}$$

Therefore, by theorem 2, we can write

$$V(x, y) = \sup_{q \in Y^*} [\langle q, y \rangle - U(x, q)] \tag{41}$$

and

(42) $$V^*(p, q) = \sup_{x \in X} [\langle p, x\rangle + U(x, q)]$$

because

$$V^*(p, q) = \sup_{x \in X} [\langle p, x\rangle + \sup_{y \in Y} (\langle q, y\rangle - V(x, y))]$$

If we assume that

(43) $$\forall q \in Y^*, \qquad x \to U(x, q) \text{ is upper semicontinuous}$$

then we can also write that

(44) $$U(x, q) = \inf_{p \in X^*} [V^*(p, q) - \langle p, x\rangle]$$

because by (40, ii) and theorem 2, $x \to -U(x, q)$ is the conjugate function of $p \to V^*(p, q)$.

Remark

Conversely, starting with a concave-convex function U, defined on $X \to Y^*$, we can define a "biconvex" function V on $X \times Y$ by (41). Assumption (43) implies that V is lower semicontinuous.

The next result will be often used in the following proofs. ■

THEOREM 14

We posit assumption (38). *The two following conditions are equivalent*:

(45) $$(p, q) \in \partial V(x, y)$$

and

(46) $$p \in \partial_x(-U)(x, q) \qquad \text{and} \qquad y \in \partial_q U(x, q)$$

where $\partial_x(-U)$ denotes the subdifferential of the convex function $x \to -U(x, q)$ and $\partial_q U$ the subdifferential of the convex function $q \to U(x, q)$. ▲

Proof. **a.** We begin by proving that (46) implies (45). We can write (46) in the form

$$U(x, q) + V(x, y) = \langle q, y\rangle$$

and

$$V^*(p, q) - U(x, q) = \langle p, x \rangle$$

By adding these equalities, we obtain

$$V^*(p, q) + V(x, y) = \langle p, x \rangle + \langle q, y \rangle \tag{47}$$

that is, $(p, q) \in \partial V(x, y)$.

b. Conversely, assume (45), that is, (47). Since we always have $\langle q, y \rangle \leq V(x, y) + U(x, q)$, we obtain $V^*(p, q) - U(x, q) \leq \langle p, x \rangle$, which by (42) implies that $p \in \partial_x(-U)(x, y)$. By using equality $V^*(p, q) = \langle p, x \rangle + U(x, q)$, we obtain $V(x, y) + U(x, q) = \langle q, y \rangle$, that is, $y \in \partial_q U(x, q)$. ■

We set

$$\bar{\partial}_x U(x, q) := -\partial_x(-U)(x, q) \tag{48}$$

and call it the *superdifferential* of the concave function $x \to U(x, p)$. We say that $(x_0, q_0) \in X \times Y^*$ is a *saddle point* of the function defined by

$$(x, q) \to U(x, q) - \langle p_0, x \rangle - \langle q, y_0 \rangle \tag{49}$$

if for all $(x, q) \in X \times Y^*$,

$$\begin{cases} U(x, q_0) - \langle p_0, x \rangle - \langle q_0, y_0 \rangle \leq U(x_0, q_0) - \langle p_0, x_0 \rangle - \langle q_0, y_0 \rangle \\ \qquad\qquad \leq U(x_0, q) - \langle p_0, x_0 \rangle - \langle q, y_0 \rangle \end{cases} \tag{50}$$

The following characterization in proposition 15 is obvious.

PROPOSITION 15

We posit assumption (38). *The following conditions are equivalent*:

$$p_0 \in -\partial_x(-U)(x_0, q_0) \quad \textit{and} \quad y_0 \in \partial_q U(x_0, q_0) \tag{51}$$

and

(52) (x_0, q_0) *is a saddle point of the function defined by* (49)

In particular, $(\bar{x}, \bar{q})$ *is a saddle point of* U *if and only if* $(0, 0) \in \bar{\partial}_x U(\bar{x}, \bar{p}) \times \partial_q U(\bar{x}, \bar{p})$. ▲

PROPOSITION 16

We posit assumption (38). *The set-valued map* $(x, q) \to \partial_x(-U)(x, q) \times \partial_q U(x, q)$ *is monotone.* ▲

Proof. This follows obviously from theorem 14. Indeed, let $(p_i, y_i) \in \partial_x(-U)(x_i, q_i) \times \partial_p U(x_i, q_i)$ $(i=1, 2)$. Then

$$\begin{aligned}&\langle (p_1, y_1)-(p_2, y_2), (x_1, q_1)-(x_2, q_2)\rangle\\&=\langle p_1-p_2, x_1-x_2\rangle+\langle q_1-q_2, y_1-y_2\rangle\\&=\langle (p_1, q_1)-(p_2, q_2), (x_1, x_2)-(y_1, y_2)\rangle \geqslant 0\end{aligned}$$

because $(x, y) \to \partial V(x, y)$ is monotone. ■

5. THE SUBDIFFERENTIAL OF THE MARGINAL FUNCTION AND LAGRANGE MULTIPLIERS

We consider a family of minimization problems

$$W(y) := \inf_{x \in X} V(x, y) \tag{1}$$

where

$$\begin{cases} V \text{ is a proper lower semicontinuous, convex function from} \\ X \times Y \text{ to } R \cup \{+\infty\}. \end{cases} \tag{2}$$

Fix $\bar{y}$ and assume that $\bar{x} \in X$ achieves the minimum of $x \to V(x, \bar{y})$. We can characterize the subdifferential of the marginal function W at $\bar{y}$ in the following way.

PROPOSITION 1

Let $\bar{y} \in Y$ and $\bar{x} \in X$ such that $W(\bar{y}) = V(\bar{x}, \bar{y})$. Then the following conditions are equivalent:

$$\bar{q} \in \partial W(\bar{y}) \tag{3}$$

and

$$(0, \bar{q}) \in \partial V(\bar{x}, \bar{y}) \tag{4}$$

▲

Proof. We observe that

$$\begin{aligned}W^*(q)&=\sup_{y \in Y}\left[\langle q, y\rangle - \inf_{x \in X} V(x, y)\right]\\&=\sup_{x \in X}\sup_{y \in Y}\left[\langle 0, x\rangle + \langle q, y\rangle - V(x, y)\right] = V^*(0, q)\end{aligned}$$

Hence, $\bar{q} \in \partial W(\bar{y})$ if and only if $W(\bar{y}) + W^*(\bar{q}) = \langle \bar{q}, \bar{y} \rangle$, that is, if and only if

$$\langle 0, \bar{x} \rangle + \langle \bar{q}, \bar{y} \rangle - V(\bar{x}, \bar{y}) = \langle \bar{q}, \bar{y} \rangle - W(\bar{y}) = W^*(\bar{q}) = V^*(0, \bar{q})$$

that is, if and only if $(0, \bar{q}) \in \partial V(\bar{x}, \bar{y})$. ■

We consider a more specific problem.
Let

$$(5) \quad \begin{cases} \textbf{i.} & F \text{ be a set-valued map from } X \text{ to } Y \text{ with a closed convex graph} \\ \textbf{ii.} & U \text{ be a proper lower semicontinuous, convex function from} \\ & X \times Y \text{ to } R \cup \{+\infty\} \end{cases}$$

We set

$$(6) \qquad V(x, y) := U(x, y) + \psi_{\text{graph}(F^{-1})}(x, y) = \begin{cases} U(x, y) & \text{if } x \in F^{-1}(y) \\ +\infty & \text{if } x \notin F^{-1}(y) \end{cases}$$

so that the minimization problem (1) can be written in the form

$$(7) \qquad W(y) := \inf_{x \in F^{-1}(y)} U(x, y)$$

THEOREM 2

Let us assume that (5) *holds true. We posit*

$$(8) \qquad 0 \in \operatorname{Int}(\operatorname{Graph}(F) - \operatorname{Dom} U)$$

Let $\bar{y} \in Y$ and $\bar{x} \in F^{-1}(\bar{y})$ be such that $W(\bar{y}) = U(\bar{x}, \bar{y})$. The following statements are equivalent:

$$(9) \qquad \bar{q} \in \partial W(\bar{y})$$

and

$$(10) \quad \exists (\bar{p}, \bar{r}) \in \partial U(\bar{x}, \bar{y}) \quad \textit{such that} \quad \bar{q} \in D(F^{-1})(\bar{y}, \bar{x})^*(\bar{p}) + \bar{r}$$ ▲

Proof. By proposition 1, $\bar{q} \in \partial W(\bar{y})$ if and only if $(0, \bar{q}) \in \partial V(\bar{x}, \bar{y}) = \partial(U + \psi_{\text{graph}(F)}(\bar{x}, \bar{y}))$. Assumption (8) and corollary 4.12. imply that $(0, \bar{q}) \in \partial U(\bar{x}, \bar{y}) + N_{\text{graph}(F)}(\bar{x}, \bar{y})$. Then there exists $(\bar{p}, \bar{r}) \in \partial U(\bar{x}, y)$ such that $(-\bar{p}, \bar{q} - \bar{r}) \in N_{\text{graph}(F)}(\bar{x}, \bar{y})$, that is, such that $-\bar{p} \in DF(\bar{x}, \bar{y})^*(\bar{r} - \bar{q})$. We use the inversion formula

$$\bar{r} - \bar{q} \in DF(\bar{x}, \bar{y})^{*^{-1}}(-\bar{p}) = -D(F^{-1})(\bar{y}, \bar{x})^*(\bar{p})$$ ■

We mention the following corollary.

COROLLARY 3

Let $U: X \to R \cup \{+\infty\}$ be a proper lower semicontinuous, convex function and F a closed convex set-valued map from X to Y. We assume that

$$0 \in \text{Int}(\text{Dom } F - \text{Dom } U) \tag{11}$$

Let $W: Y \to R \cup \{+\infty\}$ be the marginal function defined by

$$W(\bar{y}) = \inf_{x \in F^{-1}(y)} U(x) \tag{12}$$

Let $\bar{x} \in F^{-1}(\bar{y})$ achieve the minimum of U on $F^{-1}(\bar{y})$. The subdifferential of the marginal function W is equal to

$$\partial W(\bar{y}) = D(F^{-1})(\bar{y}, \bar{x})^* \partial U(\bar{x}) \tag{13}$$

▲

Many minimization problems are set in the following form:

$$W(y) := \inf_{\substack{x \in L \\ Ax \in M + y}} U(x, y) \tag{14}$$

where $L \subset X$, $M \subset Y$ are closed convex subsets and $A \in \mathscr{L}(X, Y)$ is a continuous linear operator.

COROLLARY 4

Let $\bar{x} \in L$ satisfy $A\bar{x} \in M + \bar{y}$ and let $W(\bar{y}) = U(\bar{x}, \bar{y})$ be a solution to the minimization problem (14). If we assume that

$$(0, y) \in \text{Int}((1 \times A)L - \{0\} \times M - \text{Dom } U) \tag{15}$$

then the following statements are equivalent:

$$\bar{q} \in \partial W(\bar{y}) \tag{16}$$

and

$$\begin{aligned} &\exists (\bar{p}, \bar{r}) \in \partial U(\bar{x}, \bar{y}) \quad \textit{such that} \quad \bar{q} \in \bar{r} - N_M(A\bar{x} - \bar{y}) \\ &\textit{and} \quad A^*\bar{q} \in \bar{p} + A^*\bar{r} + N_L(\bar{x}) \end{aligned} \tag{17}$$

▲

Proof. We apply theorem 2 when F is the set-valued map defined by

$$F(x) := \begin{cases} Ax - M & \text{when} \quad x \in L \\ \varnothing & \text{when } x \notin L \end{cases} \tag{18}$$

Assumptions (15) imply assumption (8) of theorem 2. We recall that

$$DF(x, y)^*(q)=\begin{cases} A^*q+N_L(x) & \text{when} \quad q \in N_M(Ax-y) \\ \varnothing & \text{when} \quad q \notin N_M(Ax-y) \end{cases} \tag{19}$$

by proposition 2.5. Therefore,

$$D(F^{-1})(y, x)^*(p)=-N_M(Ax-y) \cap A^{*^{-1}}(p+N_L(x)) \tag{20}$$

■

Remark

Since the marginal function W defined by (1) is convex, it is sufficient to prove that it is subdifferentiable at y to establish the existence of $\bar{q}$. The Robinson–Ursescu theorem allows us to find sufficient conditions for W to be subdifferentiable. ■

PROPOSITION 5

Let us assume assumptions (6) *and* (7). *Assume moreover that*

$$\bar{y} \in \operatorname{Int} \operatorname{Im} F \tag{21}$$

Let $\bar{x} \in F^{-1}(\bar{y})$ *be a solution to* $W(\bar{y})=U(\bar{x}, \bar{y})$. *If we assume that*

$$U \text{ is continuous at } (\bar{x}, \bar{y}) \tag{22}$$

then the marginal function W *is continuous at* y *and thus subdifferentiable on the interior of its domain.* ▲

Proof. The Robinson–Ursescu theorem (see theorem 3.3.1.) states that the set-valued map F^{-1} is lower semicontinuous on the interior of $\operatorname{Im}(F)$. Therefore, proposition 3.2.19 implies that the marginal function W is upper semicontinuous at $\bar{y}$ and, thus, W is bounded above on a neighborhood of $\bar{y}$. Therefore, the interior of the domain of W is nonempty. Hence, W is continuous (and thus subdifferentiable) on Int Dom W. ■

We now add supplementary characterizations of the subdifferential $\partial W(\bar{y})$ of the marginal function involving the partial conjugates

$$h(x, q):=\sup_{y \in Y} [\langle q, y\rangle - V(x, y)] \tag{23}$$

and

$$\begin{cases} h^*(p, y):=\sup_{x \in X} [\langle p, x\rangle - V(x, y)] \\ \qquad = \sup_{x \in X} \inf_{q \in Y^*} [\langle p, x\rangle - \langle q, y\rangle + h(x, q)] \end{cases} \tag{24}$$

Then theorem 4.14 and proposition 4.16 imply at once the following characterizations of the subdifferential of W.

PROPOSITION 6
Let $\bar{y} \in Y$ and $\bar{x} \in X$ such that $W(\bar{y}) = V(\bar{x}, \bar{y})$. The following statements are equivalent:

$$\bar{q} \in \partial W(\bar{y}) \tag{25}$$

$$(0, \bar{q}) \in \partial V(\bar{x}, \bar{y}) \tag{26}$$

$$\bar{x} \in \partial_p h^*(0, \bar{y}) \quad \text{and} \quad \bar{q} \in \partial_y(-h^*)(0, \bar{y}) \tag{27}$$

$$0 \in \partial_x(-h)(\bar{x}, \bar{q}) \quad \text{and} \quad \bar{y} \in \partial_q h(\bar{x}, \bar{q}) \tag{28}$$

▲

Statement (27) is quite important, since it gives a way of obtaining both the optimal solution $\bar{x}$ and the subgradients of W in terms of the perturbation y. The regularity of this set-valued map is thus embodied in knowledge about the function h^*. It is traditional to let the function h play an important role through the function ℓ_y defined on $X \times Y^*$ by

$$\ell_y(x, q) = \langle q, y \rangle - h(x, q) \tag{29}$$

called the *Lagrangian of the minimization problem* $W(y)$. ■

PROPOSITION 7
a. *The conjugate function W^* of the marginal function W defined by* (1) *is equal to*

$$W^*(q) = \sup_{x \in X} h(x, q) = V^*(0, q) \tag{30}$$

b. *The marginal function W is lower semicontinuous if and only if for all $y \in Y$,*

$$W(y) = \sup_{q \in Y^*} \inf_{x \in X} \ell_y(x, q) = \inf_{x \in X} \sup_{y \in Y^*} \ell_Y(x, q) \tag{31}$$

▲

Proof. We observe that formula (23) implies

$$W(y) = \inf_{x \in X} \sup_{q \in Y^*} [\langle q, y \rangle - h(x, q)]$$
$$= \inf_{x \in X} \sup_{q \in Y^*} \ell_y(x, q)$$

On the other hand,

$$
\begin{aligned}
W^*(q) &= \sup_y \left[\langle q, y\rangle - \inf_{x \in X} V(x, y)\right] \\
&= \sup_{x \in X} \sup_{y \in Y} \left[\langle q, y\rangle - V(x, y)\right] = \sup_{x \in X} h(x, q) \\
&= \sup_{x \in X} \sup_{y \in Y} \left[\langle 0, x\rangle + \langle q, y\rangle - V(x, y)\right] = V^*(0, q)
\end{aligned}
$$

and

$$
\begin{aligned}
W^{**}(y) &:= \sup_{q \in Y^*} \left[\langle q, y\rangle - W^*(q)\right] \\
&= \sup_{q \in Y^*} \inf_{x \in X} \left[\langle q, y\rangle - h(x, q)\right] = \sup_{q \in Y^*} \inf_{x \in X} \ell_y(x, q)
\end{aligned}
$$

Since the function W is convex, theorem 4.2. states that it is lower semicontinuous if and only if $W = W^*$, that is, if and only if formula (31) holds true for all $y \in Y$. ■

When the function

$$x \to \ell_y(x, q) := \langle q, y\rangle - h(x, q)$$

is simpler to minimize than the function $x \to V(x, y)$, it can be useful to replace the problem $W(y)$ by a problem of the form

$$W_y^{\#}(q) := \inf_{x \in X} \ell_y(x, q) \tag{32}$$

We note that formula (30) implies

$$W_y^{\#}(q) = \langle q, y\rangle - V^*(0, q) = \langle q, y\rangle - W^*(q) \tag{33}$$

and formulas (31) and (32) imply

$$\forall q \in Y^*, \qquad W_y^{\#}(q) \leqslant W(y) \tag{34}$$

Hence, we distinguish the elements $\bar{q} \in Y^*$ (if any) for which

$$W(y) = W_y^{\#}(\bar{q}) := \inf_{x \in X} \ell_y(x, \bar{q}) \tag{35}$$

by calling them *Lagrange multipliers of the minimization problem* (1).

Remark

The problem of finding Lagrange multipliers is called the *dual problem* of the minimization problem $W(y)$. When the function W is lower semicontinuous, this amounts to maximizing the function $q \to W_y^{\#}(q)$. Hence, in this case, we can solve the minimization problem $W(y)$ by

(a) First, finding $\bar{q} \in Y^*$ that maximizes the function $q \to W_y^{\#}(q)$.
(b) Second, minimizing the function $x \to \ell_y(x, \bar{q})$.

The existence of Lagrange multipliers can be proved in the framework of minimax inequalities. For the time being, we emphasize an important—and simple—property of Lagrange multipliers. ■

PROPOSITION 8

The set of Lagrange multipliers of the minimization problem $W(y)$ coincides with the subdifferential $\partial W(y)$ of the marginal function at y. ▲

Proof. Formula (33) implies that

$$\langle q, y\rangle - W^*(q) = \inf_x \ell_y(x, q) = W_y^{\#}(q)$$

Hence, $\bar{q} \in \partial W(y)$ if and only if $\langle \bar{q}, y\rangle = W^*(\bar{q}) + W(y)$; that is, if and only if $W_y^{\#}(\bar{q}) = W(y)$. ■

Remark

Theorem 3.17 implies that there exist Lagrange multipliers when $y \in \text{Int Dom } W$. When Y is a *Hilbert space*, the set of elements y for which there exist Lagrange multipliers is *dense in Dom W* by theorem 3.11.

6. CONVEX OPTIMIZATION PROBLEMS

We consider

$$(1) \quad \begin{cases} \textbf{i.} & \text{two Banach spaces } X \text{ and } Y \\ \textbf{ii.} & \text{two proper lower semicontinuous, convex functions} \\ & U: X \to R \cup \{+\infty\} \quad \text{and} \quad V: Y \to R \cup \{+\infty\} \\ \textbf{iii.} & \text{a continuous linear operator } A \in \mathcal{L}(X, Y) \\ \textbf{iv.} & \text{two parameters } p \in X^* \text{ and } y \in Y \end{cases}$$

We shall study the class of minimization problems

$$(2) \quad W(y) := \inf_{x \in X} [U(x) - \langle p, x\rangle + V(Ax + y)]$$

to which we associate the dual problems

$$W^{\circledast}(p) := \inf_{q \in Y^*} [U^*(-A^*q+p)+V^*(q)-\langle q, y\rangle] \tag{3}$$

We observe that the Fenchel inequalities imply that

$$\forall y \in Y, \quad p \in X^*, \quad W(y)+W^{\circledast}(p) \geq 0 \tag{4}$$

Note that $W(y) < +\infty$ if and only if

$$y \in \text{Dom } V - A \text{ Dom } U \tag{5}$$

and $W^*(p) < +\infty$ if and only if

$$p \in A^* \text{ Dom } V^* + \text{Dom } U^* \tag{6}$$

Inequality (4) and conditions (5) and (6) imply that infima $W(y)$ and $W^{\circledast}(p)$ are finite. Existence of solutions to minimization problems $W(y)$ and $W^{\circledast}(p)$ and the main properties of their solutions follows from assumptions slightly stronger than conditions (5) and (6), which we assume to be satisfied.

THEOREM 1

We posit assumptions (1).

a. *Assume that X is reflexive and that*

$$p \in \text{Int}(A^* \text{ Dom } V^* + \text{Dom } U^*) \tag{7}$$

Then there exists a solution $\bar{x}$ to the problem $W(y)$ and equality $W(y)+W^{\circledast}(p)=0$ holds true.

b. *Assume that*

$$y \in \text{Int}(\text{Dom } V - A \text{ Dom } U) \tag{8}$$

Then there exists a solution $\bar{q}$ to the problem $W^{\circledast}(p)$ and equality $W(y)+W^{\circledast}(p)=0$ holds true.

c. *Assume that both assumptions* (7) *and* (8) *are satisfied. Then $\bar{x}$ is a solution of $W(y)$ and $\bar{q}$ is a solution to $W^{\circledast}(p)$ if and only if they are solutions to the system of inclusions*

$$\begin{cases} p \in \partial U(\bar{x}) + A^*\bar{q} \\ y \in -A\bar{x} + \partial V^*(\bar{q}) \end{cases} \tag{9}$$

d. *Assume that both assumptions* (7) *and* (8) *hold true. The following conditions are equivalent*:

$$(10)\quad \begin{cases} \textbf{i.} & \bar{x} \in X \text{ is a solution to } W(y). \\ \textbf{ii.} & p \in \partial U(\bar{x}) + A^* \partial V(A\bar{x}+y) \\ \textbf{iii.} & \bar{x} \in \partial W^{\circledast}(p) \end{cases}$$

e. *Assume that X is reflexive and that both assumptions* (7) *and* (8) *are satisfied. The following conditions are satisfied.*

$$(11)\quad \begin{cases} \textbf{i.} & \bar{q} \in Y^* \text{ is a solution to } W^{\circledast}(p). \\ \textbf{ii.} & y \in \partial V^*(\bar{q}) - A\partial U^*(-A^*\bar{q}+p) \\ \textbf{iii.} & \bar{q} \in \partial W(y) \end{cases}$$

▲

Proof. **a.** The two statements follow from part **c** of corollary 4.12 to theorem 4.11. For instance, the second statement is obtained by taking in corollary 4.12 the function W to be defined by $W(z)=V(Az+y)$; assumption (8) implies assumption (4.29), and $W^{\circledast}(q)=V^*(q)-\langle q, y\rangle$.
b. Equality $W(y)+W^{\circledast}(p)=0$ implies that solutions $\bar{x}$ and $\bar{q}$ to the minimization problems $W(y)$ and $W^{\circledast}(p)$ satisfy

$$(U(\bar{x})+U^*(-A^*\bar{q}+p)-\langle p-A^*\bar{q}, \bar{x}\rangle)+(V(A\bar{x}+y)+V^*(\bar{q})-\langle \bar{q}, A\bar{x}+y\rangle)=0$$

Since each term on the left-hand side of these equations is nonnegative, they are both equal to zero, which means that inclusions (9) hold true.

Conversely, (9) implies that $W(y)+W^{\circledast}(p)=0$, so that $\bar{x}$ and $\bar{q}$ are solutions to the minimization problems.
c. Theorem 3.5 shows that when X is reflexive, assumption (7) implies that $\bar{x}$ is a solution to $W(y)$ if and only if inclusion (10, **ii**) holds true and assumption (8) implies that $\bar{q}$ is a solution to $W^{\circledast}(p)$ if and only if inclusion (11, **ii**) is satisfied.
d. Proposition 5.1 states that $\bar{q}$ belongs to $\partial W(y)$ if and only if $(0, \bar{q})$ belongs to the subdifferential at $(\bar{x}, \bar{y})$ of the function

$$(x, y) \to U(x) - \langle p, x\rangle + V(Ax+y) = L(x, Ax+y)$$

where $L(x, z) := U(x) - \langle p, x\rangle + V(z)$. Assumption (7) implies that

$$0 \in \text{Int}(\text{Dom } L - \text{Im}(1 \times (A \oplus 1))$$

where $(1 \times (1 \oplus A))$ denotes the continuous linear operator defined by

$$(1 \times (A \oplus 1))(x, y) := (x, Ax+y)$$

Since its transpose is equal to $(1 \oplus A^*) \times 1$, we obtain that $\bar{q}$ belongs to $\partial W(y)$ if and only if there exist $\bar{r} \in \partial U(\bar{x})$ and $\bar{s} \in \partial V(A\bar{x}+y)$ such that $0 = \bar{r} - p + A^*\bar{s}$

and $\bar{q}=\bar{s}$, that is, if and only if $A\bar{x}+y \in \partial V^*(\bar{q})$ and $\bar{x} \in \partial U^*(p-A^*\bar{q})$. By eliminating $\bar{x}$ in these inclusions, we find that $\bar{q}$ belongs to $\partial W(y)$ if and only if $\bar{q}$ solves inclusion (11, **ii**). We use the same arguments for proving the equivalence between statements (10, **ii**) and (10, **iii**). ∎

COROLLARY 2

Let X and Y be Hilbert spaces, $A \in \mathscr{L}(X, Y)$ and $U: X \to R \cup \{+\infty\}$ be a proper lower semicontinuous, convex function. We denote by $J \in \mathscr{L}(Y, Y^*)$ the duality map from Y onto Y^*. We consider the minimization problem

(12)
$$v := \inf_{x \in X} \left[U(x) + \frac{1}{2} \|Ax\|^2 \right]$$

There exists a Lagrange multiplier and the following conditions are equivalent:

(13) $\bar{x} \in X$ *minimizes* $x \to U(x) + \frac{1}{2}\|Ax\|^2$ *on* X.

(14) $\bar{x} \in X$ *is a solution to* $0 \in A^*JA\bar{x} + \partial U(\bar{x})$.

(15) *The solutions* $\bar{q} \in Y^*$ *to* $0 \in \bar{q} - JA\partial U^*(-A^*\bar{q})$ *are the Lagrange multipliers.*

(16) $\bar{x} \in X$ *is related to a Lagrange multiplier* $\bar{q}$ *by the relation* $\bar{q} = JA\bar{x}$ *and* $-A^*\bar{q} \in \partial U(\bar{x})$.

If we assume that

(17)
$$0 \in \operatorname{Int}(\operatorname{Dom} U^* + \operatorname{Im} A^*)$$

then such solutions $\bar{x} \in X$ *and* $\bar{q} \in Y^*$ *do exist.* ▲

Proof. We take $p=0$, $y=0$, and V defined by $V(y)=\frac{1}{2}\|y\|^2$, whose domain is Y. Then $\partial V(y)=Jy$, $V^*(q)=\frac{1}{2}\|q\|_*^2$, and $\partial V^*(q)=J^{-1}$. ∎

COROLLARY 3

Let X and Y be Banach spaces, $K \subset Y$ be a closed convex subset, $U: X \to R \cup \{+\infty\}$ a proper lower semicontinuous, convex function, and $A \in \mathscr{L}(X, Y)$. We consider the following minimization problem:

(18)
$$v := \inf_{Ax \in K} U(x)$$

Let us assume that

(19)
$$0 \in \operatorname{Int}(A \operatorname{Dom} U - K)$$

Then there exists a Lagrange multiplier $\bar{q}$, and the following conditions are equivalent:

(20) $\bar{x} \in K$ *minimizes* $x \to U(x)$ *under the constraint* $Ax \in K$.

(21) $\bar{x} \in X$ *is a solution to* $0 \in \partial U(\bar{x}) + A^* N_K(A\bar{x})$.

(22) *The solutions* $\bar{q} \in Y^*$ *to* $0 \in \partial\sigma_K(\bar{q}) - A\partial U^*(-A^*q)$ *are the Lagrange multipliers.*

(23) $\bar{x}$ *is related to a Lagrange multiplier* $\bar{q}$ *by the relation*

$$\bar{q} \in N_K(A\bar{x}) \quad \text{and} \quad 0 \in \partial U(\bar{x}) + A^*\bar{q}$$

Moreover, if we assume that X is reflexive and that

$$0 \in \operatorname{Int}(\operatorname{Dom} U^* + A^* b(K)) \tag{24}$$

[*where* $b(K)$ *is the barrier cone of* K], *then such solutions* $\bar{x} \in X$ *and* $\bar{q} \in Y^*$ *do exist.* ▲

Proof. We take $V(y) = \psi_K(y)$. Then $\partial\psi_K(y) = N_K(y)$, $V^*(q) = \sigma_K(q)$, and $\operatorname{Dom} V^* = b(K)$. ■

These kinds of results, involving Lagrange multipliers, are known in mathematical economics as decentralization principles. Specifically, let us consider

$$\begin{cases} n \text{ Banach spaces } X_i \\ n \text{ proper lower semicontinuous, convex functions } U_i\colon X \to R \cup \{+\infty\} \\ n \text{ continuous linear operators } A_i \in \mathscr{L}(X_i, Y) \end{cases} \tag{25}$$

The problem

$$v := \inf_{x_i \in X_i} \left(\sum_{i=1}^{n} U_i(x_i) + V\left(\sum_{i=1}^{n} A_i x_i \right) \right) \tag{26}$$

is decentralizable in the following sense: Lagrange multipliers exist provided that

$$0 \in \operatorname{Int}\left(\sum_{i=1}^{n} A_i \operatorname{Dom} U_i - \operatorname{Dom} V \right) \tag{27}$$

If $\bar{q} \in Y^*$ is such a Lagrange multiplier, any optimal solution $\bar{x} \in X$ satisfies

$$\forall i = 1, \ldots, n, \; \bar{x}_i \text{ minimizes } x_i \to U_i(x_i) - \langle \bar{q}, A_i x_i \rangle \tag{28}$$

Optimal solutions do exist when the Banach spaces X_i are reflexive and when

(29) $$\forall x_i \in X_i, \quad \text{there exist } \lambda > 0 \quad \text{and} \quad q \in \text{Dom } V^* \quad \text{such that for all } i=1, \ldots, n, \quad \lambda x_i - A^* q_i \in \text{Dom } U_i$$ ▲

Once the Lagrange multiplier $\bar{q}$ is known, problem (26) is split into n problems (28) of smaller dimension.

7. REGULARITY OF SOLUTIONS TO CONVEX OPTIMIZATION PROBLEMS

We introduce

(1)
- **i.** two *finite dimensional* spaces X and Y
- **ii.** a linear operator A_0 from X to Y
- **iii.** two proper lower semicontinuous, convex functions $U: X \to R \cup \{+\infty\}$ and $V: Y \to R \cup \{+\infty\}$

We take

(2) $$\begin{cases} \textbf{i.} & y_0 \in \text{Int}(\text{Dom } V - A_0 \text{ Dom } U) \\ \textbf{ii.} & p_0 \in \text{Int}(A_0^* \text{ Dom } V^* + \text{Dom } U^*) \end{cases}$$

We recall that the solutions $(x_0, q_0) \in X \times Y^*$ of the optimization problem

(3) $$\begin{cases} \textbf{i.} & U(x_0) + V(A_0 x_0 + y_0) - \langle p_0, x_0 \rangle \\ & \qquad = \min_{x \in X} (U(x) + V(A_0 x + y_0) - \langle p_0, x \rangle) \\ \textbf{ii.} & U^*(-A_0^* q_0 + p_0) + V^*(q_0) - \langle q_0, y_0 \rangle \\ & \qquad = \min_{q \in Y^*} (U^*(-A_0^* q + p_0) + V^*(q) - \langle q, y_0 \rangle) \\ \textbf{iii.} & U(x_0) + U^*(-A_0^* q_0 + p_0) + V(A_0 x_0 + y_0) + V^*(q_0) \\ & \qquad = \langle p_0, x_0 \rangle + \langle q_0, y_0 \rangle \end{cases}$$

are the solutions to the system of inclusions

(4) $$\begin{cases} \textbf{i.} & p_0 \in \partial U(x_0) + A_0^* q_0 \\ \textbf{ii.} & y_0 \in -A_0 x_0 + \partial V^*(q_0) \end{cases}$$

(See theorem 6.1.)

We shall study the behavior of the solutions (x_0, q_0) to this system with respect to the parameters p_0, y_0, and A_0.

For that purpose, let us denote by $F^{-1}(p, y, A)$ the subset of solutions (x, q) to the problem

$$(5)\qquad \begin{cases} \text{i.} & p \in \partial U(x) + A^* q \\ \text{ii.} & y \in -Ax + \partial V^*(q) \end{cases}$$

We use the definition of the *generalized second derivative* of a proper convex function introduced in dealing with nonsmooth analysis (definition 7.4.1).

DEFINITION 1

Let $U: X \to R \cup \{+\infty\}$ be a proper convex function. Assume that U is subdifferentiable at x_0 and let $p_0 \in \partial U(x_0)$ be a subgradient of U at x_0. We shall say that the derivative of the set-valued map $x \to \partial U(x)$

$$\partial^2 U(x_0, p_0) := C\partial U(x_0, p_0)$$

is the second derivative of U at (x_0, p_0). ▲

Then $\partial^2 U(x_0, p_0)$ is a monotone closed convex process from X to X^*. We set

$$d(A, B) := \sup_{x \in A} \inf_{y \in B} \|x - y\|$$

We now make precise what we mean by Lipschitz behavior.

DEFINITION 2

Let F be a proper set-valued map from X to Y and let (x_0, y_0) belong to the graph of F. We say that F is pseudo Lipschitz around (x_0, y_0) if there exists a neighborhood $\mathscr{W}$ of x_0, two neighborhoods $\mathscr{U}$ and $\mathscr{V}$ of y_0, $\mathscr{U} \subset \mathscr{V}$, and a constant $\ell > 0$ such that

i. $\forall x \in \mathscr{W}, \quad F(x) \cap \mathscr{U} \neq \varnothing$

ii. $\forall x_1, x_2 \in \mathscr{W}, \quad d(F(x_1) \cap \mathscr{U}, F(x_2) \cap \mathscr{V}) \leqslant \ell \|x_1 - x_2\|$ ▲

We can now state the regularity theorem.

THEOREM 3

We posit assumptions (1) and (2). Let (x_0, q_0) be a solution to problem (3). We assume that the monotone closed convex process from X to Y to itself defined by

$$\begin{pmatrix} \partial^2 U(x_0, p_0 - A_0^* q_0) & A_0^* \\ -A_0 & \partial^2 V(Ax_0 + y_0, q_0)^{-1} \end{pmatrix}$$

is surjective.

Then

(6) F^{-1} *is pseudo Lipschitz around* $(p_0, y_0, A_0, x_0, q_0)$.

Furthermore, the derivative of F^{-1} *is defined as follows*:

$$(\delta x, \delta q) \in CF^{-1}(p_0, y_0, A_0; x_0, q_0)(\delta p, \delta y, \delta A) \tag{7}$$

if and only if

$$\begin{pmatrix}\delta x\\ \delta q\end{pmatrix} \in \begin{pmatrix}\partial^2 U(x_0, p_0 - A_0^* q_0) & A_0^* \\ -A_0 & \partial^2 V(Ax_0 + y_0, q_0)^{-1}\end{pmatrix}^{-1} \begin{pmatrix}\delta p - \delta A^* \cdot q_0 \\ \delta y + \delta A \cdot x_0\end{pmatrix} \tag{8}$$ ▲

This theorem is a consequence of the inverse function theorem 7.5.2. applied to the map F defined by (5). For simplicity, we set $G(x) := \partial U(x)$ and $H(x) = \partial V(y)$, so that $H^{-1}(q) = \partial V^*(q)$.

Let F be the map from $X \times Y$ to $X \times Y \times L(x, y)$ defined by

$$(p, y, A) \in F(x, q) \tag{9}$$

if and only if

$$\begin{cases} \textbf{i.} & p \in G(x) + A^* q \\ \textbf{ii.} & y \in -Ax + H^{-1}(q) \end{cases} \tag{10}$$

We shall characterize the derivative of F in terms of derivatives of the set-valued maps G and H (or H^{-1}), respectively, in Section 7.5. Our theorem follows then from theorem 7.5.2. and lemma 7.5.9., which we state here.

LEMMA 4

Let x_0, q_0 *be a solution to the system of inclusions*

$$\begin{cases} \textbf{i.} & p_0 \in G(x_0) + A_0^* q_0 \\ \textbf{ii.} & y_0 \in -Ax_0 + H^{-1}(q_0) \end{cases} \tag{11}$$

The following conditions are equivalent:

$$(\delta p, \delta y, \delta A) \in CF(x_0, q_0, y_0, A_0)(\delta x, \delta q) \tag{12}$$

$$\begin{cases} \textbf{i.} & \delta p - \delta A^* \cdot q_0 \in CG(x_0, p_0 - A_0^* q_0)(\delta x) + A_0^* \delta q \\ \textbf{ii.} & \delta y + \delta A \cdot x_0 \in -A_0 \delta x + CH^{-1}(q_0, y_0 + Ax_0)(\delta q) \end{cases} \tag{13}$$ ▲

Example 5

We consider a minimization problem with equality constraints, defined by

$$(14)\quad \begin{cases} \text{i.} & \text{two finite dimensional spaces } X \text{ and } Y \\ \text{ii.} & \text{a linear operator } A_0 \text{ from } X \text{ to } Y \\ \text{iii.} & \text{a lower semicontinuous, convex function } U \text{ from } X \text{ to } R \end{cases}$$

We take

$$(15)\quad \begin{cases} \text{i.} & y_0 \in -\operatorname{Int}(A_0 \operatorname{Dom} U) \\ \text{ii.} & 0 \in \operatorname{Int}(\operatorname{Im} A_0^* + \operatorname{Dom} U^*) \end{cases}$$

Let x_0 be a solution to the minimization problem

$$(16)\quad \begin{cases} \text{i.} & Ax_0 = -y_0 \\ \text{ii.} & U(x_0) = \min\limits_{Ax=-y_0} U(x) \end{cases}$$

and q_0 the associated Lagrange multiplier. Assume that

$$(17)\quad \begin{cases} \text{i.} & A_0 \text{ is surjective.} \\ \text{ii.} & U \text{ is twice continuously differentiable at } x_0 \text{ and } \nabla^2 U(x_0) \\ & \text{is positive definite.} \end{cases}$$

Then F^{-1} is pseudo Lipschitz around $(0, y_0, A_0, x_0, q_0)$. We set

$$(18)\quad \begin{cases} \text{i.} & J(x_0) = (A_0 \nabla^2 U(x_0)^{-1} A_0^*)^{-1} \\ \text{ii.} & A_0^+ = \nabla^2 U(x_0)^{-1} A_0^* J(x_0), \text{ which is a right inverse of } A_0 \\ \text{iii.} & q^* \otimes : A \in \mathscr{L}(x, y) \to A^* q \in X^* \\ \text{iv.} & x \otimes : A \in \mathscr{L}(x, Y) \to Ax \in Y \end{cases}$$

The derivative of the map F^{-1} is given by the formula

$$\begin{pmatrix} \delta x \\ \delta q \end{pmatrix} = \begin{pmatrix} (1 - A_0^+ A_0)\nabla^2 U(x_0)^{-1} & -A_0^+ & -q_0^* \otimes \\ (A_0^+)^* & J(x_0) & x_0 \otimes \end{pmatrix} \begin{matrix} \delta p \\ \delta y \\ \delta A \end{matrix}$$

▲

Proof. We apply theorem 3 to the case when V is the indicator of $\{0\}$. Then $\partial^2 V^*(q_0, A_0 x_0 + y_0) = \partial^2 V^*(q_0, 0)$ is the constant map equal to zero. So inclusion (8) can be written

$$\begin{pmatrix} \delta x \\ \delta q \end{pmatrix} = \begin{pmatrix} \nabla^2 U(x_0) & A_0^* \\ -A_0 & 0 \end{pmatrix}^{-1} \begin{pmatrix} \delta p - \delta A^* q_0 \\ \delta y + \delta A q_0 \end{pmatrix}$$

which can be inverted explicitly. ■

Example 6

We consider the items defined by (14). We set $Y:=R^n$ and we take

$$(19)\quad \begin{cases} \textbf{i.} & y_0 \in -A_0 \text{ Dom } U - R^n_+ \\ \textbf{ii.} & 0 \in \text{Int}\,(A_0^* R^n_+ + \text{Dom } U^*) \end{cases}$$

Let x_0 be a solution to the minimization problem

$$(20)\quad \begin{cases} \textbf{i.} & Ax_0 + y_0 \leqslant 0 \\ \textbf{ii.} & U(x_0) = \min\limits_{Ax+y_0 \leqslant 0} U(x) \end{cases}$$

and let $q_0 \geqslant 0$ be an associated Lagrange multiplier.

We denote by I_1 the set of indexes such that $(A_0x_0 + y_0)_i = 0$. We posit assumption (17) and

$$(21)\quad \forall i \in I_1, \quad q_{0_i} > 0$$

Then F^{-1} is pseudo Lipschitz around $(0, y_0, A_0, x_0, q_0)$. We write

$$(22)\quad \begin{cases} \textbf{i.} & R^n = R^{I_1} \times R^{I_2}, \quad \text{where } I_2 = \{i = 1, \ldots, n \mid i \notin I_1\} \\ \textbf{ii.} & q = (q^1, q^2), \quad A = (A^1, A^2) \\ \textbf{iii.} & J_1(x_0) = (A_0^1 \nabla^2 U(x_0)^{-1} A_0^{1^*})^{-1} \\ \textbf{iv.} & A_0^{1^+} = \nabla^2 U(x_0)^{-1} A_0^{1^*} J_1(x_0) \end{cases}$$

Then the derivative of the set-valued map F^{-1} at $(0, y_0, A_0)$ is given by the inclusion written symbolically

$$\begin{pmatrix} \delta x \\ \delta q_1 \\ \delta q_2 \end{pmatrix} \in \begin{pmatrix} (1 - A_0^{1^+} A_0^1)\nabla^2 U(x_0)^{-1} & -A_0^{1^+} & -q_0^{1^*} \otimes \\ (A_0^{1^+})^* & J_1(x_0) & x_0^1 \otimes \\ 0 & 0 & 0 \end{pmatrix} \begin{pmatrix} \delta p \\ \delta y \\ \delta A_0^1 \end{pmatrix}$$

Proof. We apply theorem 1 to the case when V is the indicator $\psi_{-R^n_+}$ of the cone $-R^n_+$. Then $V^* = \psi_{R^n_+}$ is the indicator of R^n_+ and $\partial V^* = N_{R^n_+}$ is the normal cone to R^n_+. We take

$$-y_0 \in \text{Int}(A_0 \text{ Dom } U + R^n_+) = A_0 \text{ Dom } U + \mathring{R}^n_+$$

(which is the Slater condition). Then (x_0, q_0) is a solution to the inclusion

$$(23)\quad \begin{cases} \textbf{i.} & 0 = U(x_0) + A_0^* q_0 \\ \textbf{ii.} & y_0 \in -A_0 x_0 + N_{R^n_+}(q_0) \end{cases}$$

The latter condition implies that

(24) $$\langle q_0, A_0x_0+y_0\rangle=0$$

Since $A_0x_0+y_0 \in -R^n_+$, we deduce that

(25) $$\text{if } (A_0x_0+y_0)_i<0, \quad \text{then } q_0^i=0$$

Then q_0^2 is equal to zero, and we assume that $q_0^i>0$ for all $i \in I_1$.

By corollary 7.2.12., an element δy of $CN_{R^n_+}(q_0, A_0x_0+y_0)(\delta q)$ is defined by

- **i.** For $i \in I_1$, $\delta y_i=0$ and δq_i is arbitrary.
- **ii.** For $i \in I_2$, $\delta y_i \in R$ and δq_i is equal to zero.

Let us write $R^n=R^{I_1}\times R^{I_2}$ and $q=(q^1, q^2)$. The domain of $\partial^2 V^*(q_0, A_0x_0+y_0)$ is $R^{I_1}\times\{0\}$ and

$$\partial V^*(q_0, A_0x_0+y_0)(\delta q_1, 0)=\{0\}\times R^{I_2}$$

Hence, the matrix of second derivatives can be written symbolically

$$\begin{pmatrix}\delta p\\ \delta y_1\\ \delta y_2\end{pmatrix} \in \begin{pmatrix}\nabla^2 U(x_0) & A_0^{1*} & A_0^{2*}\\ -A_0^1 & 0 & 0\\ -A_0^2 & 0 & R\end{pmatrix}\begin{pmatrix}\delta x\\ \delta q_1\\ 0\end{pmatrix}$$

Then it is surjective if and only if the matrix of linear operators

$$\begin{pmatrix}\nabla^2 U(x_0) & A_0^{1*}\\ -A_0^1 & 0\end{pmatrix}$$

from $X\times R^{I_1}$ to itself is surjective. This is the case by assumption (17). We can even invert the preceding inclusion explicitely and obtain the formula for the derivative of the map F^{-1}. ■

Example 7

Still considering the items defined by (14), we introduce a closed convex subset P of Y.

We take

(26) $$\begin{cases} y_0 \in \operatorname{Int}(P-A_0 \operatorname{Dom} U)\\ 0 \in \operatorname{Int}(A_0^*b(P)+\operatorname{Dom} U^*)\end{cases}$$

where $b(P):=\{q|\sup_{y\in P}\langle q, y\rangle<+\infty\}$ is the barrier cone of P. Let x_0 be a solution to the minimization problem

(27)
$$\begin{cases} \textbf{i.} & Ax_0 \in P - y_0 \\ \textbf{ii.} & U(x_0) = \min\limits_{Ax \in P - y_0} U(x) \end{cases}$$

and q_0 an associated Lagrange multiplier. We posit assumption (17) and the following assumption on P:

(28)
$$\begin{cases} \forall y \in Y, \quad \exists q \text{ solution to} \\ y - J(x_0)q \in C\pi_P(Ax_0 + y_0 + q_0)(y + (1 - J(x_0))q) \end{cases}$$

Then the conclusion of theorem 3 holds true. ▲

Proof. It is sufficient to check the surjectivity of

$$\begin{pmatrix} \nabla^2 U(x_0) & A_0^* \\ -A_0 & CN_P(Ax_0 + y_0, p_0)^{-1} \end{pmatrix}$$ ∎

This method, using the inverse function theorem for set-valued maps, can be used to treat more general convex minimization problems.

8. LAGRANGIANS AND HAMILTONIANS

The convex minimization problems $W(y)$ studied in the two preceding sections are particular cases of minimization problems of the form

(1)
$$v := \inf_{x \in X} L(x, Ax)$$

where X and Y are Banach spaces, A a continuous linear operator from X to Y, and L a proper lower semicontinuous, convex function from $X \times Y$ to $R \cup \{+\infty\}$. Because of their formal analogy with problems arising in the calculus of variations, we may call the function L a Lagrangian. In the calculus of variations, as we shall see, spaces X and Y are spaces of functions or distributions and A is a differential operator. The following results, which use the transpose A^* of A, cannot be applied, because several differential operators cannot be transposed explicitly (depending on the spaces on which they are defined). These operators possess only explicit "formal transposes" related to A by a Green formula instead of the usual relation $\langle A^*q, x\rangle = \langle q, Ax\rangle$. Still, it is worth beginning with the simpler case when A is an abstract continuous linear operator, because it embodies all the main ideas that we shall adapt in the case of calculus of variations (see Chapter 8).

This being said, we associate with problem (1) its *dual problem*

$$v^* := \inf_{q \in Y^*} L^*(-A^*q, q) \tag{2}$$

We observe that the function $\mathscr{A}$ defined on $X \times Y^*$ by

$$\mathscr{A}(x, q) := L(x, Ax) + L^*(-A^*q, q) \tag{3}$$

is always nonnegative, because

$$L(x, Ax) + L^*(-A^*q, q) \geq \langle -A^*q, x\rangle + \langle q, Ax\rangle = 0$$

Hence,

$$v + v^* = \inf_{(x,q) \in X \times Y^*} \mathscr{A}(x, q) \geq 0 \tag{4}$$

We say that an element $\bar{q} \in Y^*$ is a *Lagrange multiplier* if

$$\begin{cases} \textbf{i.} \quad v + v^* = 0 \\ \textbf{ii.} \quad v^* = L^*(-A^*\bar{q}, \bar{q}) \end{cases} \tag{5}$$

THEOREM 1

Let us assume that L is a proper lower semicontinuous, convex function from $X \times Y$ to $R \cup \{+\infty\}$ satisfying

$$0 \in \operatorname{Int}((A \oplus -1)\operatorname{Dom} L) \tag{6}$$

Then there exists a Lagrange multiplier $\bar{q}$ of the minimization problem (1), *and the two following conditions are equivalent:*

$$\bar{x} \in X \quad \textit{minimizes } x \to L(x, Ax) \tag{7}$$

and

$$\bar{x} \in X \quad \textit{is a solution to the inclusion} \quad 0 \in (1 \oplus A^*)\partial L(\bar{x}, A\bar{x}) \tag{8}$$ ▲

Proof. **a.** Let V be the function defined by $V(x) := L(x, Ax)$. Hence, $v = \inf_{x \in X} V(x) = -V^*(0)$. By part (b) of theorem 4.11., there exists $\bar{q} \in Y^*$ such that

$$V^*(0) = L^*(-A^*\bar{q}, \bar{q}) = \inf_{q \in Y^*} L^*(-A^*q, q) = v^*$$

This means that $\bar{q}$ is a Lagrange multiplier.

b. We know that $\bar{x} \in X$ minimizes V on X if and only if $0 \in \partial V(\bar{x}) = (1 \oplus A^*)\partial L(\bar{x}, A\bar{x})$, by theorem 3.7. ■

We now mention a list of useful—and simple—equivalent statements. We introduce the *Hamiltonian* of the problem v, which is the function H defined on $X \times Y^*$ by

$$H(x, q) := \sup_{y \in Y} [\langle q, y \rangle - L(x, y)] \tag{9}$$

PROPOSITION 2
The following statements are equivalent:

(10) $\bar{x} \in X$ *minimizes* $x \to L(x, Ax)$ *on* X, *and* $\bar{q} \in Y^*$ *is a Lagrange multiplier*

$$\mathscr{A}(\bar{x}, \bar{q}) = 0 (= \min_{(x,q) \in X \times Y^*} \mathscr{A}(x, q)) \tag{11}$$

$$(-A^*\bar{q}, \bar{q}) \in \partial L(\bar{x}, A\bar{x}) \tag{12}$$

$$A^*\bar{q} \in \partial_x H(\bar{x}, \bar{q}), \quad \text{and} \quad A\bar{x} \in \partial_q H(\bar{x}, \bar{q}) \tag{13}$$

(14) $\bar{x}$ *is a solution to the inclusion* $0 \in (1 \oplus A^*)\partial L(\bar{x}, A\bar{x})$

(15) $\bar{q}$ *is a solution to the inclusion* $0 \in (-A \oplus 1)\partial L^*(-A^*\bar{q}, \bar{q})$ ▲

Proof. The implications (10)⇒(11)⇒(12)⇒(10) are obvious. Equivalence between statements (12) and (13) follows from theorem 4.14. Statement (12) is equivalent to (14) by eliminating $\bar{q}$ and equivalent to (15) by eliminating $\bar{x}$ in the relation $(\bar{x}, A\bar{x}) \in \partial L^*(-A^*\bar{q}, \bar{q})$. ■

Remark
When L is Gâteaux differentiable at $(\bar{x}, A\bar{x})$, the inclusion $0 \in (1 \oplus A^*)\partial L(\bar{x}, A\bar{x})$ becomes the equation

$$\nabla_x L(\bar{x}, A\bar{x}) + A^*\nabla_y L(\bar{x}, A\bar{x}) = 0 \tag{16}$$

If $A = d/dt$ (formally), so that $Ax = \dot{x}$, we recognize the Euler–Lagrange equation from the calculus of variations. For this reason, we shall call the inclusion $0 \in (1 \oplus A^*)\partial L(\bar{x}, A\bar{x})$ the *Euler–Lagrange inclusion.* When the Hamiltonian H is Gâteaux differentiable at $(\bar{x}, \bar{q})$, inclusion (13) can be written

$$A^*\bar{q} = \nabla_x H(\bar{x}, \bar{q}) \quad \text{and} \quad A\bar{x} = \nabla_q(\bar{x}, \bar{q}) \tag{17}$$

We recognize the *Hamilton equations*; therefore, relations

$$A^*\bar{q} \in \partial_x H(\bar{x}, \bar{q}) \quad \text{and} \quad A\bar{x} \in \partial_q H(\bar{x}, \bar{q}) \tag{18}$$

will be called the *Hamilton inclusions.* When the dual function L^* is Gâteaux differentiable at $(-A^*\bar{q}, \bar{q})$, inclusion (15) becomes

(19) $$A\nabla_p L^*(-A^*\bar{q}, \bar{q}) - \nabla_q L^*(-A^*\bar{q}, \bar{q}) = 0$$

We shall call the relation

(20) $$0 \in (-A \oplus 1)\partial L^*(-A^*\bar{q}, \bar{q})$$

the *dual Euler–Lagrange inclusions.* ■

We observe that a Lagrange multiplier of the dual problem v^* is a solution to the minimization problem v. By applying the preceding theorem 1 to the dual problem [where X is replaced by Y^*, Y by X^*, A by A^*, L by $(q, p) \rightarrow L^*(p, q)$, etc.], we obtain

THEOREM 3

Let us assume that X is a reflexive Banach space and that

(21) $$0 \in \operatorname{Int}((1 \oplus A^*)\operatorname{Dom} L^*)$$

Then $v = v^*$, and there exists $\bar{x} \in X$ such that

(22) $$v = L(\bar{x}, A\bar{x}) = \min_{x \in X} L(x, Ax)$$ ▲

When both assumptions are satisfied, we obtain the existence of both an optimal solution and a Lagrange multiplier.

THEOREM 4

Let us assume that X is a reflexive Banach space and that

(23) $$\begin{cases} \textbf{i.} & 0 \in \operatorname{Int}((A \oplus -1)\operatorname{Dom} L) \\ \textbf{ii.} & 0 \in \operatorname{Int}((1 \oplus A^*)\operatorname{Dom} L^*) \end{cases}$$

Then there exist $\bar{x} \in X$ and $\bar{p} \in Y^*$ satisfying the equivalent conditions (10)–(15). ▲

Remark

We have used the terminology **Lagrange multiplier** in two different instances. However, the terminology is consistent when we regard the minimization problem v defined by (1) as the particular case $v = W(0)$ of the family of minimization problems

(24) $$W(y) := \inf_{x \in X} L(x, Ax + y)$$

So if we set $V(x, y):=L(x, Ax+y)$, we observe that

$$h(x, q):=\sup_{y \in Y} [\langle q, y\rangle - V(x, y)] = H(x, q) - \langle q, Ax\rangle$$

and that

$$\ell_y(x, q):=\langle q, y\rangle - U(x, q) = \langle q, y+Ax\rangle - h(x, q)$$

There is, at this point, a slight terminology ambiguity concerning the word *Lagrangian*, which designates both the function $L(x, Ax+y)$ we want to minimize (terminology coming from the calculus of variations) and the function $\ell_y(x, q)$ (terminology used in optimization). The functional of the dual problem is then equal to

$$W_y^{\#}(q):=\inf_{x \in X} \ell_y(x, q) = \langle q, y\rangle - L^*(-A^*q, q)$$

Hence, for $y=0$, $\bar{q}$ is a Lagrange multiplier in the sense that

$$W(0) = W_0^{\#}(\bar{q})$$

if and only if

$$v = -L^*(-A^*\bar{q}, \bar{q})$$

that is, if and only if $\bar{q}$ is a Lagrange multiplier according to definition (5). ■

We now proceed with the case of convex Hamiltonians. Theorem 4 states that when the Lagrangian L is convex and the assumptions (23) hold true, the solutions $\bar{x}$ to the minimization problem

$$v = \inf_{x \in X} L(x, Ax) \tag{1}$$

are the solutions to the Euler–Lagrange inclusion (14) and, by duality, the solutions $\bar{q}$ to the dual minimization problem

$$v^* = \inf_{q \in Y^*} L^*(-A^*q, q) \tag{2}$$

are related to the solutions $\bar{x}$ to the Euler–Lagrange inclusion (14) through the formula $(\bar{x}, A\bar{x}) \in \partial L^*(-A^*\bar{q}, \bar{q})$.

We can still use such variational principles (allowing us to solve inclusions by solving minimization problems) for solving the Euler–Lagrange inclusion when L is no longer convex but concave–convex. In this case, the Hamiltonian

H is a convex function. Therefore, from now on, we assume that

$$\begin{cases} H: X \times Y^* \to R \cup \{+\infty\} \text{ is a proper lower} \\ \text{semicontinuous, convex function} \end{cases} \tag{25}$$

to which we associate the Lagrangian $L: X \times Y \to \bar{\mathbb{R}}$ (note that the value $-\infty$ is allowed) defined by

$$L(x, y) := \sup_{q \in Y^*} [\langle q, y \rangle - H(x, q)] \tag{26}$$

Such a Lagrangian L is *concave with respect to x, convex and lower semicontinuous with respect to y*. If L is assumed to be upper semicontinuous with respect to x, then the function H is its Hamiltonian in the sense that

$$H(x, q) = \sup_{y \in Y} [\langle q, y \rangle - L(x, q)] \tag{27}$$

In this case, the *Euler–Lagrange inclusion* takes the form

$$0 \in -\partial_x(-L)(\bar{x}, A\bar{x}) + A^*\partial_y L(\bar{x}, A\bar{x}) \tag{28}$$

and the *Hamiltonian inclusion* can be written

$$(A^*\bar{q}, A\bar{x}) \in \partial H(\bar{x}, \bar{q}) \tag{29}$$

When the Hamiltonian is convex and lower semicontinuous, these equations are equivalent, because theorem 4.14 implies that the Hamiltonian inclusions (29) are equivalent to the pair

$$\bar{q} \in \partial_y L(\bar{x}, A\bar{x}) \quad \text{and} \quad A^*\bar{q} \in \partial_x(-L)(\bar{x}, A\bar{x}) \tag{30}$$

Eliminating $\bar{q}$ from (30) yields the Euler–Lagrange inclusion (28).

Note that the function $\mathscr{B}$ defined on $X \times Y^*$ by

$$\mathscr{B}(x, q) := H(x, q) - \langle q, Ax \rangle + H^*(A^*q, Ax) - \langle A^*q, x \rangle \tag{31}$$

is always nonnegative because

$$H(x, q) + H^*(A^*q, Ax) \geq \langle A^*q, x \rangle + \langle q, Ax \rangle$$

We obtain a result analogous to proposition 2.

PROPOSITION 5

The following statements are equivalent:

$$\mathscr{B}(\bar{x}, \bar{q}) = 0 (= \min_{(x,q) \in X \times Y^*} \mathscr{B}(x, q)) \tag{32}$$

(33) $\bar{x}$ *is a solution to the Euler–Lagrange inclusion*

$$0 \in -\partial_x(-L)(\bar{x}, A\bar{x}) + A^*\partial_y L(\bar{x}, A\bar{x})$$

(34) *The pair* $(\bar{x}, \bar{q})$ *is a solution to the Hamiltonian inclusion*

$$(A^*\bar{q}, A\bar{x}) \in \partial H(\bar{x}, \bar{q}).$$ ▲

Proof. Conditions (32) and (34) are obviously equivalent, and we already mentioned that statements (33) and (34) are equivalent. ■

We shall now prove that we can solve these equations by solving either the abstract *least action principle*

$$(35) \qquad w := \inf_{(x,q) \in X \times Y^*} (H(x, q) - \langle q, Ax \rangle)$$

or the *dual least action principle*

$$(36) \qquad w^* := \inf_{(x,q) \in X \in Y^*} (H^*(A^*q, Ax) - \langle A^*q, x \rangle)$$

THEOREM 6

We assume that the Hamiltonian is proper, convex, and lower semicontinuous. Any minimizer $(\bar{x}, \bar{q})$ *of the least action principle* (35) *solves the Hamiltonian inclusion* (34). *If we assume that*

$$(37) \qquad 0 \in \operatorname{Int}(\operatorname{Im}(A^* \times A) - \operatorname{Dom} H^*)$$

one of the minimizers to the dual least action principle (36) *also solves this Hamiltonian inclusion.* ▲

Proof. **a.** It is easy to check that if $(\bar{x}, \bar{q})$ is a solution to the minimization problem (35), then

$$\forall (x, q) \in X \times Y^*, \qquad \langle A^*\bar{q}, x \rangle + \langle q, A\bar{x} \rangle \leq D_+H(\bar{x}, \bar{q})(x, q)$$

Therefore, $(A^*\bar{q}, A\bar{x})$ belongs to $\partial H(\bar{x}, \bar{q})$.

b. Let $(\tilde{x}, \tilde{q})$ be a solution to the dual least action principle. We easily deduce that,

$$\forall (x, q) \in X \times Y^*, \qquad \langle A^*\tilde{q}, x \rangle + \langle q, A\tilde{x} \rangle \leq D_+H^*(A^*\tilde{q}, A\tilde{x})(A^*q, Ax)$$

We apply corollary 3.6 and obtain

$$(A^*\tilde{x}, A\tilde{q}) \in (A \times A^*)\partial H^*(A^*\tilde{q}, A\tilde{x})$$

Then there exists $(\bar{x}, \bar{q})$ such that

$$(\bar{x}, \bar{q}) \in \partial H^*(A^*\tilde{q}, A\tilde{x}), \qquad A^*\tilde{q} = A^*\bar{q}, \qquad A\tilde{x} = A\bar{x}$$

that is, such that

$$(\bar{x}, \bar{q}) \in \partial H^*(A^*\bar{q}, A\bar{x})$$

Since $(\bar{x}, \bar{q})$ is also a solution to the minimization problem (36), the last statement ensues. ■

Unfortunately, neither the least action principle nor its dual are convex minimization problems, because we add to the convex functions $H(x, q)$ and $H^*(A^*q, Ax)$ the bilinear function $-\langle q, Ax\rangle$. But the least action principles do retain enough structure to allow us to solve the Hamiltonian inclusions (34) even when the Hamiltonian is convex. Since H and H^* have dual properties, we shall be able to choose among the least action principle and its dual the one that has a solution.

In Chapter 8, these ideas will be developed, and methods will be given for solving the Hamiltonian inclusions by finding critical points (not necessarily minimizers) of the dual least action principle.

CHAPTER 5

A General Variational Principle

This chapter introduces Ekeland's ε-variational principle (corollary 3.2), which is an important tool for nonlinear analysis, as we shall see in subsequent chapters.

We have tried to motivate its use as much as possible, by providing an appealing proof and giving early and powerful applications. Our proof relies on the notion of dissipative dynamical system.

Along the way, we prove Caristi's fixed-point theorem and Baillon's nonlinear mean ergodic theorem.

Among applications, we give a strengthened version of the Ambrosetti–Rabinowitz "mountain pass" theorem. This is the simplest (and most powerful) method known for finding critical points that are not local maxima or minima. Its use is illustrated in the last chapter on Hamiltonian systems, as will the global inverse function theorem we give in example 5.8.

The last sections give applications to the geometry of Banach spaces and to optimization theory. These are very active fields of research, and many related questions are still open.

1. WALKING ON COMPLETE METRIC SPACES

Let us start somewhere in a complete metric space X and walk around. At each point $x \in X$, the rules of the game specify the set $F(x)$ of points that can be reached from x in one step. Starting from, say, $x_0 \in X$, we reach $x_1 \in F(x_0)$ at the first step, then $x_2 \in F(x_1)$ at the second step, then $x_3 \in F(x_2)$ at the third step, and so on. The question is whether we shall end up somewhere.

We formally introduce definition 1.

DEFINITION 1

Let X be a complete metric space. A dynamical system *on X is a set-valued map $F: X \to X$ with $F(x) \neq \varnothing$ for all x. Any infinite sequence $x_{\dagger} = x_0, x_1, \ldots, x_n, \ldots$ such that*

$$x_{n+1} \in F(x_n), \qquad \textit{for all } n \tag{1}$$

is called a motion *starting at* x_0. *The set*

$$\mathcal{T}(x_\dagger):=\{x_n | n \in N\} \tag{2}$$

is the trajectory *of this motion. The set*

$$\mathcal{C}(x)=\cup\{C(x_\dagger) | x_0=x\}$$

is the forward cone *of the point x in X.* ▲

Here the integer n must be understood to denote successive times, or steps; it runs from zero (initial time) to infinity. In general, starting from any point x in X, several motions are possible unless F is single valued. For instance, if $F(x)$ contains two different points, say $y \in F(x)$ and $z \in F(x)$ with $y \neq z$, then we can start two different motions from x, say $y_\dagger$ and $z_\dagger$, to wit

$$y_0=x, \quad y_1=y, \quad y_2 \in F(y), \quad \text{and so on, inductively} \tag{3}$$

$$z_0=x, \quad z_1=z, \quad z_2 \in F(z), \quad \text{and so on, inductively} \tag{4}$$

A trajectory is the set of points through which an individual motion will run. Note that knowing the trajectory does not give us full information about the motion, since we are not told when an individual point will be reached. The forward cone of a point x is the set of all points in X that can be reached in a finite number of steps if we start at x. It has the obvious properties

$$\begin{cases} \textbf{i.} \quad x \in \mathcal{C}(x) & \text{(reflexivity)} \\ \textbf{ii.} \quad [y \in \mathcal{C}(x) \quad \text{and} \quad z \in \mathcal{C}(y)] \Rightarrow z \in \mathcal{C}(x) & \text{(transitivity)} \end{cases} \tag{5}$$

Indeed, if $y \in \mathcal{C}(x)$, there is a motion $x_\dagger$ and an integer n such that $x_0=x$ and $x_n=y$. Since $z \in \mathcal{C}(y)$, there is a motion $y_\dagger$ and an integer k such that $y_0=y$ and $y_k=z$. Setting $z_p=x_p$ for $p \leqslant n$ and $z_p=y_{p-n-1}$ for $p \geqslant n+1$, we obtain a sequence $z_\dagger$ such that

$$z_0=x_0, \qquad z_{k+n+1}=z, \qquad \text{and} \qquad z_p \in F(z_{p-1}) \qquad \text{for all } p$$

So $z_\dagger$ is a motion, and $z \in \mathcal{C}(x)$.

We are interested in finding motions that converge. Let us start with a simple criterion. It is stressed that all the arguments rely heavily on the fact that the metric space X is complete.

PROPOSITION 2

Assume there exists a nonnegative function U: $X \to \mathbb{R} \cup \{+\infty\}$ *such that*

$$\forall x \in X, \qquad \exists y \in F(x): U(y)+d(x, y) \leqslant U(x) \tag{6}$$

Then for all points x where $U(x) < +\infty$, there is a motion $x_{\dagger}$ starting at x that converges to a limit point $\bar{x}$:

$$x_0 = x \quad \text{and} \quad x_n \to \bar{x} \tag{7}$$

If the graph of F is closed, then $\bar{x}$ is a fixed point. ▲

Proof. Construct by induction, using assumption (6) at each step, a motion $x_{\dagger}$ such that

$$x_0 = x \tag{8}$$

$$d(x_n, x_{n+1}) \leqslant U(x_n) - U(x_{n+1}), \qquad \text{for all } n \tag{9}$$

This implies that $U(x_n)$ is a decreasing sequence of real numbers. Since all its terms are positive, this sequence has to converge. Adding up both sides of inequality (9) from $n = N$ to $n = M - 1$, we have

$$\sum_{n=N}^{M-1} d(x_n, x_{n+1}) \leqslant U(x_N) - U(x_M) \tag{10}$$

Using the triangle inequality, this yields

$$d(x_N, x_M) \leqslant U(x_N) - U(x_M) \tag{11}$$

Since the sequence $U(x_n)$ converges, the right-hand side can be made smaller than any prescribed $\varepsilon > 0$ by choosing $N < M$ large enough. This proves that $x_{\dagger}$ is a Cauchy sequence, and since the space is complete, it converges.

Assume now that the graph of F is closed.

We know $x_{n+1} \in F(x_n)$, since $x_{\dagger}$ is a motion. It follows that the pair (x_n, x_{n+1}) belongs to the graph of F. Since (x_n, x_{n+1}) converges to $(\bar{x}, \bar{x})$, the latter pair must also belong to the graph of F. Hence, the result. ■

Of course, when F is single-valued, $F(x) = \{f(x)\}$ for all x, and the statement of proposition 2 can be simplified:

PROPOSITION 3

Let $f: X \to X$ be single-valued. Assume there is a positive function $U: X \to R \cup \{+\infty\}$ such that

$$\forall x \in X, \qquad U(f(x)) + d(x, f(x)) \leqslant U(x) \tag{12}$$

Then for all points x where $U(x) < +\infty$, the sequence $x_{\dagger}$ defined recursively by

$$x_0 = x, \quad x_{n+1} = f(x_n) = f^n(x_0) \tag{13}$$

converges to some limit $\bar{x}$. If f has a closed graph (for instance, if f is everywhere finite and continuous), then $\bar{x}$ is a fixed point

$$f(\bar{x})=\bar{x} \tag{14}$$ ▲

No proof is needed, since this is just a transposition of proposition 2.

Criterion (6) is easy to understand. Think of the space X as the map of a mountain range, with $U(x)$ as the altitude of point x. Then inequality (9) implies that $U(x_{n+1}) \leqslant U(x_n)$, which means going down. Moreover, we have

$$\frac{U(x_n)-U(x_{n+1})}{d(x_n, x_{n+1})} \geqslant 1$$

which means losing height at some fixed rate. It is intuitively clear that we cannot go on like this indefinitely (unless, of course, we fall into a bottomless pit; this is why we have to assume $U(x) \geqslant 0$ everywhere). We have to end up somewhere.

Criterion (6) is useful in so far as we can construct a function U. When F has nice continuity properties, we know even the smallest function U satisfying property (6).

PROPOSITION 4

Let us associate with a strict set-valued map F from X to X the nonnegative function U_F from X to $\mathbb{R}_+ \cup \{+\infty\}$ defined by

$$\forall x \in X, \qquad U_F(x) := \inf_{\{x\dagger / x_0 = x\}} \sum_{n=0}^{\infty} d(x_n, x_{n+1}) \tag{15}$$

If F is upper semicontinuous with compact values, then U_F satisfies property (6)

$$\forall x \in X, \qquad \exists y \in F(x):\ U_F(x) \geqslant U_F(y) + d(x, y) \tag{16}$$

Moreover, if U is any other function satisfying property (6), *we have*

$$\forall x \in X, \qquad U_F(x) \leqslant U(x) \tag{17}$$ ▲

Proof. Inequality (17) is obvious. Let us associate with any $\varepsilon > 0$ the function U_ε defined by

$$U_\varepsilon(x) := \inf\left\{\sum_{n=0}^{\infty} d(x_n, x_{n+1}) | x_0 = x \text{ and } x_{n+1} \in B(F(x_n), \varepsilon)\right\} \tag{18}$$

and the function U_0 defined by

$$U_0(x) := \lim_{\varepsilon \to 0} U_\varepsilon(x) = \sup_{\varepsilon > 0} U_\varepsilon(x)$$

These properties are well defined, for if $\varepsilon_2 \leqslant \varepsilon_1$, then $U_{\varepsilon_1} \leqslant U_{\varepsilon_2} \leqslant U_0 \leqslant U_F$.

Next, for each $\varepsilon > 0$, select $x_\varepsilon \in B(F(x), \varepsilon)$ such that

$$U_\varepsilon(x_\varepsilon) + d(x, x_0) \leqslant U_\varepsilon(x) + \varepsilon \tag{19}$$

Since $F(x)$ is compact, we can select a sequence $x_k \in B(F(x), 1/k)$ converging to some $\bar{x} \in F(x)$. Also, since F is upper semicontinuous, for any $\delta > 0$ there exists $k_0 > 1/\delta$ such that for all $k \geqslant k_0$,

$$B\left(F(x_k), \frac{1}{k}\right) \subset B(F(\bar{x}), \delta)$$

Thus for $k \geqslant k_0$,

$$U_\delta(\bar{x}) \leqslant d(\bar{x}, x_k) + U_{1/k}(x_k) \tag{20}$$

Combining (19) and (20), we have for $k \geqslant k_0$,

$$U_\delta(\bar{x}) - d(\bar{x}, x_k) + d(x, x_k) \leqslant U_{1/k}(x) + 1/k$$

Letting k go to ∞ in this inequality, we obtain

$$U_\delta(\bar{x}) + d(x, \bar{x}) \leqslant U_0(x)$$

and finally, since δ is arbitrary, we find that for some $\bar{x} \in F(x)$

$$U_0(\bar{x}) + d(x, \bar{x}) \leqslant U_0(x) \tag{21}$$

Hence, U_0 satisfies property (6) and thus, by the first part of the proposition, $U_F \leqslant U_0$. Since $U_0 \leqslant U_F$, we deduce that $U_0 = U_F$. ■

DEFINITION 5
A set-valued map F is called a contraction *if it is Lipschitz with constant* $\lambda \in]0, 1[$. *It is* nonexpansive *if* $\lambda = 1$. ▲

THEOREM 6
A (set-valued) contraction from a complete metric space X to itself has a fixed point. ▲

Proof. Since F is Lipschitz—and thus upper semicontinuous—with compact values, the function U_F defined by (15) satisfies property (6). The function U_F is also finite. To see this, we select a motion such that

$$d(x_{n+1}, x_n) = d(x_n, F(x_n))$$

which is possible because the values of F are compact. Since F is Lipschitz, we observe that

$$U_F(x) \leqslant \sum_{n=0}^{\infty} d(x_{n+1}, x_n) \leqslant \sum_{n=0}^{\infty} \lambda^n d(x_0, x_1) = \frac{1}{1-\lambda} d(x_0, x_1) < +\infty$$

Our theorem follows from proposition 2. ■

When F is simple valued, this is the well-known Banach contraction principle.

THEOREM 7

Any single-valued contraction from a complete metric space to itself has a unique fixed point. ▲

Proof. Uniqueness is a special feature here. Assume there are two fixed points $\bar{x}$ and $\bar{y}$, and apply the Lipschitz condition. We have

$$d(\bar{x}, \bar{y}) = d(f(\bar{x}), f(\bar{y})) \leqslant \lambda d(\bar{x}, \bar{y}) \tag{22}$$

which is absurd since $\lambda < 1$; therefore, $\bar{x} = \bar{y}$. ■

We now go back to the general, set-valued case, and we strengthen condition (6) to bring it closer to the single-valued case.

DEFINITION 8

A dynamical system $F: X \to X$ is called dissipative *with respect to a function U: $X \to \mathbb{R} \cup \{+\infty\}$, positive and not identically $+\infty$, if*

$$\forall x \in X, \qquad \forall y \in F(x), \qquad U(y) + d(x, y) \leqslant U(x) \tag{23}$$ ▲

In our mountaineering analogy, this means that there is no way to go but down. The terminology comes from physics: We can think of x as describing the state of a physical system, with $U(x)$ its energy. Condition (23) tells us that whichever way the system evolves, it is going to lose energy at some fixed rate. It is then natural to think that the system will evolve toward a stable equilibrium. It is this idea that is expressed in the following proposition.

PROPOSITION 9

Assume F is dissipative. Then for any x where $U(x) < +\infty$, there is a motion $x_\dagger$ and a point $\bar{x}$ such that

$$x_0 = x \quad \text{and} \quad \{\bar{x}\} = \bigcap_{n \in N} \overline{\mathscr{C}(x_n)} \tag{24}$$ ▲

Proof. For any y where $U(y) < +\infty$, define

$$V(y) = \inf\{U(z) \mid z \in \mathscr{C}(y)\} \tag{25}$$

Since U is positive and $\mathscr{C}(y)$ nonempty, $V(y)$ is some real number. Now take any $x \in X$ and $y \in \mathscr{C}(x)$. By definition of the forward cone $\mathscr{C}(x)$, there is a sequence $x_\dagger$ and an integer k such that $x_0 = x$ and $x_k = y$. Writing assumption (23) at each step $n=0$ to $n=k-1$ and summing up, we have

$$\forall y \in \mathscr{C}(x), \qquad d(x, y) \leqslant U(x) - U(y) \tag{26}$$

By definition of V, this yields

$$\forall y \in \mathscr{C}(x), \qquad d(x, y) \leqslant U(x) - V(x) \tag{27}$$

and, hence, for any $x \in X$

$$\text{diameter } \mathscr{C}(x) \leqslant 2(U(x) - V(x)) \tag{28}$$

We now consider a sequence (not a motion) $y_\dagger$ in X defined inductively as follows. Start at $y_0 = x$, and pick $y_{n+1} \in \mathscr{C}(y_n)$ such that

$$U(y_{n+1}) \leqslant V(y_n) + 2^{-n} \tag{29}$$

This is always possible by definition of V. By relation (5), we have

$$\mathscr{C}(y_{n+1}) \subset \mathscr{C}(y_n) \tag{30}$$

This implies $V(y_{n+1}) \geqslant V(y_n)$. Let us estimate $V(y_{n+1})$. We have by relation (29)

$$V(y_{n+1}) \leqslant U(y_{n+1}) \leqslant V(y_n) + 2^{-n} \leqslant V(y_{n+1}) + 2^{-n} \tag{31}$$

Hence, $U(y_{n+1}) - V(y_{n+1}) \leqslant 2^{-n}$. Setting $x = y_{n+1}$ in formula (28), we obtain

$$\text{diameter } \mathscr{C}(y_{n+1}) \leqslant 2^{1-n} \tag{32}$$

It follows that $\overline{\mathscr{C}(y_n)}$, $n \in N$, is a nested (decreasing) sequence of closed sets whose diameters go to zero. Since X is a complete metric space, their intersection is a singleton

$$\exists \bar{x} \in X: \bigcap_{n \in N} \overline{\mathscr{C}(y_n)} = \{\bar{x}\} \tag{33}$$

Since $y_{n+1} \in \mathscr{C}(y_n)$, there is a motion $x_\dagger^n$ and an integer k_n such that $x_0^n = y_n$ and $x_{k_n}^n = y_{n+1}$. Using motion $x_\dagger^0$ to go from $y_0 = x$ to y_1 and then motion $x_\dagger^1$ to go from y_1 to y_2, and so on inductively, define a motion $x_\dagger$ that will eventually carry us through all the points y_n. Clearly, for any k, we can find n and m such that

$$x_k \in \mathscr{C}(y_n) \qquad \text{and} \qquad y_m \in \mathscr{C}(x_k) \tag{34}$$

Hence, we have

$$\mathscr{C}(y_m) \subset \mathscr{C}(x_k) \subset \mathscr{C}(y_n) \tag{35}$$

and the desired result (24) follows from (33). ■

We have proved that the diameter of $\mathscr{C}(x_n)$ goes to zero, which conveys the idea of stability: For any $\varepsilon > 0$, there is an integer N such that if a motion $x'_\dagger$ branches off from $x_\dagger$ at a later time than N, it will stay within a distance ε of $\bar{x}$.

We derive from proposition 9 several results pertaining to the existence of points $\bar{x}$ where $F(\bar{x}) = \{\bar{x}\}$. Such points will be called *invariant*. They are dead ends where the system is stuck. This is stronger than the fixed-point property, $\bar{x} \in F(\bar{x})$, except, of course, in the single-valued case.

COROLLARY 10

Assume F is dissipative and lower semicontinuous. Then there is an invariant point $\bar{x}$ where

$$F(\bar{x}) = \{\bar{x}\} \tag{36}$$

▲

Proof. Start from a point x where $U(x) < +\infty$, and define $\bar{x}$ as in proposition 9. Take any $\bar{z}$ in $F(\bar{x})$. Since F is lower semicontinuous, a sequence $z_n \in F(x_n)$ can be found converging to $\bar{z}$. We have

$$z_k \in \mathscr{C}(x_n), \qquad \text{for all } k \geq n \tag{37}$$

and, hence, letting $k \to \infty$

$$\bar{z} \in \overline{\mathscr{C}(x_n)}, \qquad \text{for all } n \tag{38}$$

So $\bar{z}$ belongs to the intersection of all $\overline{\mathscr{C}(x_n)}$. By formula (24), we have $\bar{z} = \bar{x}$. ■

COROLLARY 11

Assume F is dissipative and all forward cones are closed. Then whenever $U(x)$ is finite, the forward cone of x contains an invariant point $\bar{x}$.

$$U(x) < +\infty \Rightarrow \exists \bar{x} \in \mathscr{C}(x): F(\bar{x}) = \{\bar{x}\} \tag{39}$$

▲

Proof. Start from a point x where $U(x) < +\infty$, and define $\bar{x}$ as in proposition 9. Since all the $\mathscr{C}(x_n)$ are closed, condition (24) becomes

$$\{\bar{x}\} = \bigcap_{n \in N} \mathscr{C}(x_n) \tag{40}$$

Since $\bar{x} \in \mathscr{C}(x_n)$, we have $F(\bar{x}) \subset \mathscr{C}(x_n)$ for all n. Hence,

$$F(\bar{x}) \subset \bigcap_{n \in N} \mathscr{C}(x_n) \tag{41}$$

The result follows immediately from (40) and (41). ■

The question naturally arises, how do we construct dissipative dynamical systems; there is a standard way to do this.

PROPOSITION 12

Let $U: X \to R \cup \{+\infty\}$ be a positive function, not identically $+\infty$. The dynamical system G defined by

$$G(x) = \{y \mid U(y) + d(x, y) \leq U(x)\} \tag{42}$$

is dissipative with respect to U and has the property that

$$G(x) = \mathscr{C}(x), \qquad \text{for all } x \tag{43}$$

▲

Proof. $G(x)$ is not empty, since it certainly contains x itself. Thus formula (42) does define a dynamical system, and it is obviously dissipative with respect to U. Indeed, G could even be defined as the largest dynamical system that is dissipative with respect to U.

Relation (43) is obvious when $U(x) = +\infty$, both sets being equal to X. Now assume $U(x)$ is finite. We claim that $G(y) \subset G(x)$ whenever $y \in G(x)$. Indeed, if $z \in G(y)$, we have

$$U(y) + d(x, y) \leq U(x) \tag{44}$$

$$U(z) + d(y, z) \leq U(y) \tag{45}$$

Since $U(x)$ is finite, so are $U(y)$ and $U(z)$. Adding them up and using the triangle inequality, we obtain

$$U(z) + d(x, z) \leq U(x) \tag{46}$$

which means that $z \in G(x)$, as desired.

It follows by inducation that $\mathscr{C}(x) \subset G(x)$. Since the converse inclusion holds by definition, both sets coincide. ■

A dynamical system F is dissipative with respect to U if and only if $F(x) \subset G(x)$ for all x. Of course, any point $\bar{x}$ that is G invariant will also be F invariant

$$\left.\begin{aligned} \varnothing \neq F(\bar{x}) \subset G(\bar{x}) \\ G(\bar{x}) = \{\bar{x}\} \end{aligned}\right\} \Rightarrow F(\bar{x}) = \{\bar{x}\} \tag{47}$$

This simple remark leads us to a remarkable result.

PROPOSITION 13

Let F be a dynamical system, dissipative with respect to some lower semicontinuous positive function $U \not\equiv +\infty$. Then F has an invariant point $\bar{x}$

$$F(\bar{x}) = \{\bar{x}\} \tag{48}$$ ▲

Proof. We associate with U a dynamical system G as in proposition 9. As previously noted, it is enough to show that G has an invariant point.

Since U is lower semicontinuous, the set $G(x)$ defined by (42) is closed and so is the forward cone $\mathscr{C}(x)$ because of equation (43). Applying corollary 11, we see that G has an invariant point. ▲

This last result is particularly striking, because it places no continuity requirement on the mapping F itself. Let us rephrase it in the single-valued case to obtain Caristi's original statement, which was the starting point of this investigation.

THEOREM 14

Let $f: X \to X$ be a (single-valued) map and $U: X \to \mathbb{R}$ a (finite) lower semicontinuous function such that

$$d(x, f(x)) \leq U(x) - U(f(x)), \qquad \text{for all } x \in X \tag{49}$$

Then f has a fixed point $\bar{x}$

$$f(\bar{x}) = \bar{x} \tag{50}$$ ▲

Note that f is not required to be continuous and there may be several fixed points (take f to be the identity and U any positive constant).

Note also that the proof of corollary 11 does not provide us with an iterative procedure for finding $\bar{x}$. Indeed, if we choose some starting point x, there is no guarantee that the motion $x_\dagger$ defined by $x_n = f^n(x)$ will converge to anything, let alone to $\bar{x}$.

Finally, note that a strictly positive constant k can be introduced on the left-hand sides of formulas (6) and (49) to read $kd(x, y)$, without altering subsequent statements. Indeed, kd is again a distance on X, equivalent to the original one d, and we can use either one indifferently.

2. FIXED POINTS OF NONEXPANSIVE MAPS

The Banach contraction theorem implies the existence of an unique fixed point of a contraction f from a complete metric space K to itself.

When K is a closed, convex bounded subset of a *Hilbert space*, we can relax the contraction assumption and replace it by the assumption that f is nonexpansive.

THEOREM 1

Let K be a nonempty, closed, convex bounded subset K of a Hilbert space and let f be a nonexpansive map from K to K. Then f has a fixed point. Furthermore, the subset of fixed points is closed and convex. ▲

Proof. We shall write the proof in three steps by constructing a sequence of elements $x_t \in K$ satisfying

$$\lim_{t\to\infty} \|x_t - f(x_t)\| = 0 \tag{1}$$

then by showing that the weak cluster points of such a sequence are fixed points of the nonexpansive map and finally, by proving that the set of fixed points is closed and convex.

a. Let $x_0 \in K$ and $t > 1$. We define the single-valued map f_t from K to K by

$$f_t(x) := \frac{1}{t} x_0 + \left(1 - \frac{1}{t}\right) f(x)$$

Since

$$\|f_t(x) - f_t(y)\| \leqslant \left(1 - \frac{1}{t}\right) \|f(x) - f(y)\| \leqslant \left(1 - \frac{1}{t}\right) \|x - y\|$$

the maps f_t are contractions. Then there exist fixed points $x_t \in K$ of the maps f_t

$$x_t = \frac{1}{t} x_0 + \left(1 - \frac{1}{t}\right) f(x_t)$$

Therefore,

$$\|x_t - f(x_t)\| = \frac{1}{t} \|x_0 - f(x_t)\| \leqslant \frac{2}{t} \sup \{\|x\| \mid x \in K\}$$

since K is bounded and $f(x_t) \in K$. Hence,

$$\lim_{t\to\infty} \|x_t - f(x_t)\| = 0$$

b. Since x_t belongs to K, which is weakly relatively compact, there exists a generalized subsequence of elements $x_{t'}$ that converges weakly to some $\bar{x} \in K$.

We set $x_\lambda := (1-\lambda)\bar{x} + \lambda f(\bar{x})$, where $\lambda \in]0, 1[$. Since f is nonexpansive, we deduce that

$$\langle x_\lambda - f(x_\lambda) - (x_{t'} - f(x_{t'})),\ x_\lambda - x_{t'} \rangle \geq \|x_\lambda - x_{t'}\|^2 - \|f(x_\lambda) - f(x_{t'})\|\,\|x_\lambda - x_{t'}\| \geq 0$$

By letting $x_{t'}$ converge to $\bar{x}$, this inequality implies that

$$\langle x_\lambda - f(x_\lambda) - 0, \quad x_\lambda - \bar{x} \rangle \geq 0$$

because $x_{t'} - f(x_{t'})$ converges strongly to zero.

By dividing by $\lambda > 0$, we obtain

$$\langle (1-\lambda)\bar{x} + \lambda f(\bar{x}) - f((1-\lambda)\bar{x} + \lambda f(x)),\ f(x) - \bar{x} \rangle \geq 0$$

By letting λ converge to zero, we deduce that

$$\|f(\bar{x}) - \bar{x}\|^2 \leq 0, \qquad \text{that is, } \bar{x} = f(\bar{x})$$

Hence, there exists at least a fixed point of f.

c. The subset of fixed points is obviously closed, because f is continuous. Let x_0 and x_1 be two fixed points of f, and let us prove that $x_\lambda := (1-\lambda)x_0 + \lambda x_1$ is also a fixed point for $\lambda \in [0, 1]$. We have

$$\|f(x_\lambda) - x_0\| = \|f(x_\lambda) - f(x_0)\| \leq \|x_\lambda - x_0\| = \lambda \|x_0 - x_1\|$$

and

$$\|f(x_\lambda) - x_1\| = \|f(x_\lambda) - f(x_1)\| \leq \|x_\lambda - x_1\| = (1-\lambda)\|x_0 - x_1\|$$

It follows that

$$\|f(x_\lambda) - x_0\| + \|f(x_\lambda) - x_1\| \leq \|x_0 - x_1\| \leq \|f(x_\lambda) - x_0\| + \|f(x_\lambda) - x_1\|$$

that is

$$\|f(x_\lambda) - x_0\| = \lambda\|x_0 - x_1\| \qquad \text{and} \qquad \|f(x_\lambda) - x_1\| = (1-\lambda)\|x_0 - x_1\|$$

Since X is a Hilbert space, we deduce that

$$f(x_\lambda) = (1-\lambda)x_0 + \lambda x_1 = x_\lambda$$

that is, x_λ is a fixed point. ■

It is not necessarily true that the sequences of elements $f^n(x_0)$ converge to some fixed point when f is nonexpansive. But we shall prove that the Cesaro means

(2) $$y_T := \frac{1}{T}\sum_{t=1}^{T} f^t(x_0)$$

converge weakly to a fixed point. For that purpose, it is convenient to introduce the concept and the properties of the *asymptotic center* of bounded sequences of a Hilbert space. It always exists, is unique, and coincides with the weak limit whenever the latter exists.

The sequence x_t is bounded. We associate with it the function ϕ defined on X by

(3) $$\phi(y) := \limsup_{t\to\infty} \|x_t - y\|^2 = \inf_{S\geq 0} \sup_{t\geq S} \|x_t - y\|^2$$

The function ϕ is nonnegative, locally Lipschitz, strictly convex, and satisfies

(4) $$\lim_{\|y\|\to\infty} \phi(y) = \infty$$

DEFINITION 2

The unique point $x_\infty \in X$ that minimizes ϕ on X is called the asymptotic center *of the bounded sequence of elements x_t.* ▲

PROPOSITION 3

The asymptotic center of a bounded sequence of elements $x_t \in X$ belongs to the closed convex hull of its weak cluster points. ▲

Proof. Let x_∞ be the asymptotic center of the bounded sequence of elements x_t and let y_∞ be the projection of x_∞ onto the closed convex hull C of the weak cluster points of the sequence $\{x_t\}$. It is nonempty, because any bounded sequence is weakly relatively compact. By definition of ϕ, there exists a subsequence of elements $x_{t'}$ such that

$$\phi(y_\infty) = \lim_{t'\to\infty} \|x_{t'} - y_\infty\|^2$$

Hence,

$$\limsup_{t'\to\infty} \|x_{t'} - x_\infty\|^2 \leq \limsup_{t\to\infty} \|x_t - x_\infty\|^2 = \phi(x_\infty)$$

We can also find a subsequence $x_{t''}$ that converges to some $z \in C$. Therefore,

$$\limsup_{t'\to\infty} \langle y_\infty - x_\infty, y_\infty - x_{t'}\rangle = \langle y_\infty - x_\infty, y_\infty - z\rangle \leq 0$$

because y_∞ is the projection of x_∞ to C.

From the identity

$$\|x_t - x_\infty\|^2 = \|x_t - y_\infty\|^2 + \|y_\infty - x_\infty\|^2 + 2\langle x_t - y_\infty, y_\infty - x_\infty\rangle$$

we deduce that

$$\phi(x_\infty) \geq \phi(y_\infty) + \|x_\infty - y_\infty\|^2 \geq \phi(y_\infty)$$

From the uniqueness of the minimum of ϕ, it follows that $x_\infty = y_\infty \in C$. The second part of the proposition is an immediate consequence of the first. ■

We now detail the preceding result.

PROPOSITION 4

Let us consider a bounded sequence of elements x_t. We denote by C the closed convex hull of its weak cluster points and by N the subset defined by

$$N := \left\{ y \in X \mid \phi(y) = \lim_{t \to \infty} \|x_t - y\|^2 \right\} \tag{5}$$

which we call the "attractor" of the sequence.

If $N \cap C \neq \varnothing$, this intersection reduces to the asymptotic center x_∞

$$N \cap C = \{x_\infty\} \tag{6}$$

▲

Proof. Let $y \in N \cap C$; we shall prove that $y = x_\infty$, and, for that purpose, we check that $\phi(z) \geq \phi(y)$ for all $z \in X$ and y is the unique minimum of ϕ.

Let w be any weak cluster point of $\{x_t\}$; then w is the weak limit of a subsequence $\{x_{t'}\}$ of $\{x\}$. Passing to the limit as $t' \to \infty$ in the identity

$$\|x_{t'} - z\|^2 = \|x_{t'} - y\|^2 + \|y - z\|^2 + 2\langle x_{t'} - y, y - z\rangle$$

we obtain, since y belongs to the attractor N,

$$\phi(z) \geq \phi(y) + \|y - z\|^2 + 2\langle w - y, y - z\rangle$$

Since this inequality holds for all weak cluster points, it also holds for all $w \in C$. In particular, we can take $w = y$ for $y \in C$. Hence,

$$\phi(z) \geq \phi(y) + \|y - z\|^2 \geq \phi(y)$$

■

As a corollary, we obtain the following sufficient condition for weak convergence.

COROLLARY 5

If the weak cluster points of a bounded sequence of elements x_t belong to its attractor N, then this sequence converges weakly to its asymptotic center. ▲

PROPOSITION 6

Let us consider two bounded sequences of elements $\{x_t\}$ and $\{y_t\}$. Let N be the attractor and C the closed convex hull of the weak cluster points of $\{x_t\}$. If the weak cluster points of $\{y_t\}$ belong to $C \cap N$, then y_t converges weakly to the asymptotic center of the sequences $\{x_t\}$. ▲

Proof. Since the weak cluster points of $\{y_t\}$ belong to $C \cap N$, this set is nonempty and by proposition 4, reduces to x_∞, the asymptotic center of $\{x_t\}$. Since the sequence $\{y_t\}$ is relatively compact and since it has a unique cluster point x_∞, then y_t converges weakly to x_∞ as $t \to \infty$. ■

We are now ready to prove that Cesaro means converges to a fixed point.

THEOREM 7

Let f be a nonexpansive map from a closed, convex bounded subset K of a Hilbert space to itself. For all initial point $x_0 \in K$, the sequence of elements $y_T := (1/T)\sum_{t=1}^{T} f^t(x_0)$, $T \in \mathbb{N}$ converges weakly to the asymptotic center of the sequence $f^t(x_0)$, $t \in \mathbb{N}$, which is a fixed point of f. ▲

Proof. We consider the sequence of Cesaro means

$$y_T = \frac{1}{T}\sum_{t=1}^{T} f^t(x_0), \qquad T = 1, \ldots,$$

It is bounded, since f is nonexpansive. For any $y \in K$, write

$$\|f^t(x_0) - f(y) + f(y) - y\|^2$$
$$= \|f^t(x_0) - f(y)\|^2 + 2\langle f^t(x_0) - f(y),\ f(y) - y\rangle + \|f(y) - y\|^2$$

By adding these inequalities from $t = 1$ to T and dividing by $T > 0$, we obtain

$$\frac{1}{T}\sum_{t=1}^{T} \|f^t(x_0) - y\|^2 = \frac{1}{T}\sum_{t=1}^{T} \|f^t(x_0) - f(y)\|^2 + \|f(y) - y\|^2 + 2\langle y_T - f(y),\ f(y) - y\rangle$$

We now take $y := y_T$. We deduce that

$$\frac{1}{T}\sum_{t=1}^{T} \|f^t(x_0) - y_T\|^2 = \frac{1}{T}\sum_{t=1}^{T} \|f^t(x_0) - f(y_T)\|^2 - \|f(y_T) - y_T\|^2$$

Which we rewrite as

$$\|f(y_T)-y_T\|^2 \leqslant \frac{1}{T}\left[\sum_{t=1}^{T} \|f^t(x_0)-f(y_T)\|^2 - \sum_{t=1}^{T} \|f^t(x_0)-y_T\|^2\right]\Bigg]$$

Since f is nonexpansive,

$$\|f^t(x_0)-f(y_T)\| \leqslant \|f^{t-1}(x_0)-y_T\|$$

and all terms in these sums cancel except the first and last, yielding:

$$\|f(y_T)-y_T\|^2 \leqslant \frac{1}{T}\,[\|x_0-y_T\|^2 - \|f^T(x_0)-y_T\|]^2$$

$$\leqslant \frac{4}{T} \max\{\|x\| \mid x \in K\}$$

So, $\lim_{T\to\infty}\|y_T - f(y_T)\| = 0$.

On the other hand, part (b) of the proof of theorem 1 implies that any weak cluster point of the sequence of elements y_T is a fixed point of f.

But any fixed point $\bar{x}$ of f belongs to the set N associated to the sequence $x_t := f^t(x_0)$, because the inequality

$$\|f^t(x_0)-\bar{x}\| = \|f^t(x_0)-f(\bar{x})\| \leqslant \|f^{t-1}(x_0)-\bar{x}\| \tag{7}$$

shows that the sequence of real numbers $\|f^t(x_0)-\bar{x}\|$ is nondecreasing and bounded below and thus convergent.

So the weak cluster points of $\{y_T\}$ belong to the attractor N. They also belong to the closed convex hull C of the weak cluster points of $\{x_T\}$. Hence, proposition 6 implies that y_T converges weakly to the asymptotic center of the sequence of elements $f^t(x_0)$, which is a fixed point of f. ■

3. THE ε-VARIATIONAL PRINCIPLE

We now address ourselves to the problem of minimizing a lower semicontinuous function U on a complete metric space X. Of course, if the space X is compact, there will always be a minimizer, that is, some point $\bar{x}$ where

$$U(x) \geqslant U(\bar{x}), \qquad \text{for all } x \in X \tag{1}$$

Such will not be the case when no compactness assumption is made. In other words, the infimum of the set of real numbers $\{U(x) | x \in X\}$, denoted by inf U, need not be attained (it may even be $-\infty$). However, by the very definition of an infimum, there will be a sequence $x_\dagger$ in X such that

$$U(x_n) \to \inf U \quad \text{when} \quad n \to \infty \tag{2}$$

Such sequences are called *minimizing*. When $\inf U$ is finite, this implies that for any $\varepsilon > 0$, there is some x_ε such that

$$U(x) \geq U(x_\varepsilon) - \varepsilon, \quad \text{for all } x \in X \tag{3}$$

We shall show that even in the general, noncompact case, there is more to be said. Provided that $\inf U$ is finite, we shall associate with any $\varepsilon > 0$ points that satisfy, in addition to (3), other conditions, to be interpreted later.

THEOREM 1

Let X be a complete metric space, $U: X \to \mathbb{R} \cup \{+\infty\}$ a proper, nonnegative, and lower semicontinuous function, and x_0 in Dom U. *Then there exists y in* Dom U *such that*

$$\begin{cases} \textbf{i.} & U(y) + d(x_0, y) \leq U(x_0) \\ \textbf{ii.} & \forall x \neq y, \quad U(y) < U(x) + d(x, y) \end{cases} \tag{4}$$

▲

COROLLARY 2 (ε-VARIATIONAL PRINCIPLE)

We posit the assumptions of theorem 1. Let there be given $\varepsilon > 0$ and $x_\varepsilon \in X$ such that

$$U(x_\varepsilon) \leq \varepsilon + \inf U \tag{5}$$

Then for any $k > 0$, there is some point $y_\varepsilon \in X$ such that

$$U(y_\varepsilon) \leq U(x_\varepsilon) \tag{6}$$

$$d(x_\varepsilon, y_\varepsilon) \leq \frac{1}{k} \tag{7}$$

$$U(x) > U(y_\varepsilon) - k\varepsilon d(x, y_\varepsilon), \quad \text{for all } x \neq y_\varepsilon \tag{8}$$

▲

Proof of Theorem 1. As in proposition 1.12, we associate with U the dynamic system defined by

$$G(x) = \{y \mid U(y) + d(x, y) \leq U(x)\} \tag{9}$$

Since U is lower semicontinuous, $G(x)$ is a closed subset of X, which coincides with the forward cone $\mathscr{C}(x)$ by proposition 1.12. Since $U(x_0)$ is finite, it follows from corollary 1.11 that its forward cone contains an invariant point y

$$y \in \mathscr{C}(x_0) \qquad \text{and} \qquad G(y)=\{y\}$$

Recall that $\mathscr{C}(x_0)=G(x_0)$. The first condition then becomes

$$U(y)+d(x_0, y) \leqslant U(x_0)$$

We now write that y is G invariant

$$U(x)+d(y, x) \leqslant U(y) \Leftrightarrow y=x$$

This is exactly relation (4, ii). The theorem is proved. ■

Proof of Corollary 2. Replace distance d by $k\varepsilon d$, the function U by $U-\inf U$, and the point x_0 by some x_ε satisfying (5). Denote by y_ε a solution provided by theorem 1.

Relations (6) and (7) then follow from combining (4, **i**) and (5). Relation (8) follows from (4, **ii**). ■

This chapter views condition (8) in several different lights. For the time being, we choose to treat it as a variational principle. In other words, we relate it to the theory of necessary conditions for local minima.

From now on, X will be a Banach space, so that $d(x, y)=\|x-y\|$. Transcribing theorem 1 in this new notation would be a loss of time and space, but it might be useful to draw a picture (see Fig. 1).

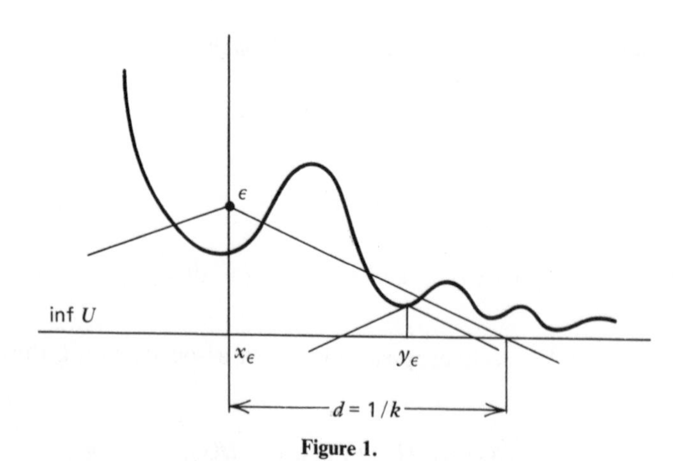

Figure 1.

The set $\mathscr{C}:=\{(x, a) | a+k\varepsilon\|x\|<0\}$ is a cone of revolution in $X \times R$, with vertical axis and vertex at (0, 0), oriented downward. Its angle ω is defined by $\tan \omega = k\varepsilon$.

Translating this cone by a vector $(\bar{x}, \bar{a})$ in $X \times R$ will bring its vertex to $(\bar{x}, \bar{a})$.

The translate of $\mathscr{C}$ can be defined directly as follows:

(10) $$\mathscr{C}+(\bar{x}, \bar{a})=\{(x, a)|(a-\bar{a})+k\varepsilon\|x-\bar{x}\|\leqslant 0\}$$

It is now geometrically clear what condition (8) means: The cone $\mathscr{C}+(y_\varepsilon, U(y_\varepsilon))$ lies entirely under the graph of the function U in $X\times R$ and does not touch it except at its vertex.

The smaller we make $k\varepsilon$, the flatter is the cone $\mathscr{C}$ and the closer $\mathscr{C}+(y_\varepsilon, U(y_\varepsilon))$ to the horizontal hyperplane through $(y_\varepsilon, U(y_\varepsilon))$. The limiting position $k\varepsilon=0$ is not permitted by the statement of theorem 1. Writing $k\varepsilon=0$ in equation (8), leads to the recognition of y_ε as a unique global minimizer for U over X. Such a strong minimizer need not exist under the very weak assumptions that X be complete and U lower semicontinuous and bounded from below. If in some particular case, it can be directly shown that U has a unique global minimizer y_0 over X, then of course this point y_0 will satisfy condition (8) with $k\varepsilon=0$.

The interest of corollary 2 lies in the fact that for any $\varepsilon>0$, there will always be some point x_ε satisfying condition (5) and, hence, some point y_ε satisfying conditions (6), (7), and (8). The last two are somehow complementary, and the choice of $k>0$ allows us to strike a balance between them according to which application we have in view. If k is small, the cone $\mathscr{C}$ will be flat, and the point y_ε will be close to being a global minimizer for U. On the other hand, the right-hand side of inequality (7) will be large, so that we shall have little information on the whereabouts of y_ε. Conversely, if k is chosen large, y_ε will be located close to our initial point x_ε, but the cone $\mathscr{C}$ will be acute, and inequality (8) will give us little information.

The two most important ways to choose k are $k=1$, which means we are losing interest in condition (7), and $k=\varepsilon^{-1/2}$, which means we are keeping $\mathscr{C}$ flat and y_ε close to x_ε (having our cake and eating it). In these two statements, we make X a complete metric space again.

COROLLARY 3

Let X be a complete metric space and $U\colon X\to\mathbb{R}\cup\{+\infty\}$ a proper lower semi-continuous function bounded below. Then for any $\varepsilon>0$, there exists some point y_ε where

(11) $$U(y_\varepsilon)\leqslant\varepsilon+\inf U$$

(12) $$U(x)>U(y_\varepsilon)-\varepsilon d(x, y_\varepsilon), \qquad \textit{for all } x\neq y_\varepsilon$$ ▲

COROLLARY 4

Let X be a complete metric space and $U\colon X\to\mathbb{R}\cup\{+\infty\}$ a lower semicontinuous function bounded below. Let $\varepsilon>0$ and $x_\varepsilon\in X$ be such that

(13) $$U(x_\varepsilon)\leqslant\varepsilon+\inf U$$

Then there exists some point y_ε where

$$U(y_\varepsilon) \leqslant U(x_\varepsilon) \tag{14}$$

$$d(x_\varepsilon, y_\varepsilon) \leqslant \sqrt{\varepsilon} \tag{15}$$

$$U(x) \geqslant U(y_\varepsilon) - \sqrt{\varepsilon}\, d(x, y_\varepsilon), \qquad \text{for all } y \neq x_\varepsilon \tag{16}$$

▲

We now go back to Banach spaces and figure 1. Imagine U to be differentiable at y_ε. The tangent hyperplane $\mathscr{H}$ to U at this point will obviously lie above the cone $\mathscr{C} + (y_\varepsilon, U(y_\varepsilon))$. The flatter the cone $\mathscr{C}$, the more horizontal $\mathscr{H}$. To be precise, if $x^* \in X^*$ is the slope of $\mathscr{H}$, we must have $\|x^*\|_* \leqslant \tan \omega = k\varepsilon$. Let us state and prove this result.

PROPOSITION 5

Assume X is a Banach space and U is finite and Gâteaux differentiable at y_ε. Then condition (8) *of theorem* 1 *implies that*

$$\|U'(y_\varepsilon)\|_* \leqslant k\varepsilon \tag{17}$$

▲

Proof. The left-hand side denotes the norm of $U'(y_\varepsilon)$ as a continuous linear functional on X. Taking any unit vector $y \in X$, $\|y\| = 1$, and setting $x = y_\varepsilon + ty$, with $t > 0$, in formula (8), we have

$$\frac{1}{t}\left[U(y_\varepsilon + ty) - U(y_\varepsilon)\right] > -k\varepsilon, \qquad \text{for all } t > 0 \tag{18}$$

Letting $t \to 0$, we obtain by the definition of $U'(y_\varepsilon)$ as a Gâteaux derivative

$$\langle U'(y_\varepsilon), y\rangle \geqslant -k\varepsilon, \qquad \text{for all } y \in X \quad \text{with } \|y\| = 1 \tag{19}$$

Since $-y$ is also a unit vector, we have

$$-\langle U'(y_\varepsilon), y\rangle \geqslant -k\varepsilon, \qquad \text{for all } y \in X \quad \text{with } \|y\| = 1 \tag{20}$$

From both inequalities, it follows that

$$|\langle U'(y_\varepsilon), y\rangle| \leqslant k\varepsilon, \qquad \text{for all } y \in X \quad \text{with } \|y\| = 1 \tag{21}$$

This is precisely formula (17). ■

We now have the interpretation we were looking for: *condition* (8) *as a variational principle.* If $k\varepsilon = 0$, we have $U'(y_\varepsilon) = 0$, the familiar first-order necessary condition for a local minimum. But it has already been pointed out, and indeed will presently be illustrated by an example, that the assumptions of corollary 2 by themselves do not warrant that such a point y_ε exist. However, for any $k\varepsilon > 0$, we are assured that there is some point y_ε where condition (8), and hence (17), is satisfied. In other words, if $U'(y_\varepsilon)$ can not actually be made zero, it can at least be made arbitrarily small.

This point of view is illustrated by the following results.

COROLLARY 6

Assume X is a Banach space, $U: X \to \mathbb{R}$ is lower semicontinuous, Gâteaux differentiable, and bounded below. Let $\varepsilon > 0$ and $x_\varepsilon \in X$ be given, with

$$U(x_\varepsilon) \leqslant \varepsilon + \inf U \tag{22}$$

Then there exists some point y_ε where

$$U(y_\varepsilon) \leqslant U(x_\varepsilon) \tag{23}$$

$$\|x_\varepsilon - y_\varepsilon\| \leqslant \sqrt{\varepsilon} \tag{24}$$

$$\|U'(y_\varepsilon)\|_* \leqslant \sqrt{\varepsilon} \tag{25}$$

▲

COROLLARY 7

Assume X is a Banach space, $U: X \to \mathbb{R}$ is lower semicontinuous, Gâteaux differentiable, and bounded from below. Then there is a sequence y_n such that when $n \to \infty$

$$U(y_n) \to \inf U \tag{26}$$

$$U'(y_n) \to 0 \qquad \textit{in } X^* \tag{27}$$

▲

The latter result is particularly interesting. *It asserts the existence of minimizing sequences of a particular kind*: Not only do they minimize the function U, but they also satisfy the first-order necessary conditions, up to any desired approximation. To relation (26), which defines all minimizing sequences, we have added condition (27), and our result is that both can be satisfied at once. We illustrate these results with a few examples.

Example 8

Let X be a Banach space, and $U: X \to R$ a lower semicontinuous, Gâteaux-differentiable function. Assume that for some constant $a > 0$ and c, we have

$$U(x) \geqslant a\|x\| + c, \qquad \text{for all } x \tag{28}$$

Then $U'(X)$ is a dense subset of the ball aB^* in X^*.

If condition (28) is strengthened to

$$U(x) \geqslant \phi(\|x\|), \qquad \text{for all } x \tag{29}$$

where $\phi: [0, +\infty) \to R$ is continuous and $t^{-1}\phi(t) \to +\infty$ when $t \to \infty$, then $U'(X)$ is dense in X^* for its norm topology. ▲

To see this, note first that condition (29) implies that for all $a > 0$, some c will be found such that inequality (28) holds. So $U'(X)$ will be dense in all the balls around the origin in X^* and, hence, in X^* itself.

Now, if condition (28) holds, take any $p \in X^*$ such that $\|p\|_* < a$. In other words, the point p belongs to aB^*. We are going to show that there is a sequence p_n in $U'(X)$ that converges to p, which is easy. Simply define a new function $V: X \to R$ by

$$V(x) = U(x) - \langle p, x \rangle, \qquad \text{for all } x. \tag{30}$$

Clearly, V is lower semicontinuous and Gâteaux differentiable. Since

$$V(x) \geqslant U(x) - \|p\|_* \|x\| \geqslant (a - \|p\|_*)\|x\| + c \tag{31}$$

the function V is bounded from below by c. Applying corollary 7, we obtain a sequence y_n in X such that

$$V'(y_n) = U'(y_n) - p \to 0 \text{ in } X^* \tag{32}$$

Setting $p_n = U'(y_n)$, we obtain the desired result. ■

Example 9

We now wish to present a practical situation where there is no exact minimizer. Let X be $W_0^{1,4}(0, 1)$, the Sobolev space of all continuous functions x on the interval $(0, 1)$ such that its distributional derivative $\dot{x} = dx/dt$ belongs to $L^4(0, 1)$ and $x(0) = x(1) = 0$. Let $U: X \to R$ be defined by

$$U(x) = \int_0^1 \{(\dot{x}(t)^2 - 1)^2 + x(t)^2\} dt \tag{33}$$

This is the situation of example 3.4 in chapter 1. We have seen that there is no minimizer.

Corollary 7 will provide us with a substitute. Set

$$X(t) = \int_0^t x(s) ds$$

and write the function U as follows:

$$U(x) = \int_0^1 (\dot{x}(t)^2 - 1)^2 dt - \int_0^1 X(t)\dot{x}(t) dt \tag{34}$$

It is now apparent that U is Fréchet differentiable and its derivative is the linear map on $W_0^{1,4}(0, 1)$ defined by

$$\langle U'(x), y \rangle = \int_0^1 (4\dot{x}(\dot{x}^2 - 1) - 2X)\dot{y}dt \tag{35}$$

Now y can be any function in $L^4(0, 1)$ such that

$$\int_0^1 \dot{y}(t)dt = 0$$

It follows that

$$\begin{aligned} \| U'(x) \|_* &= \sup \left\{ \int_0^1 (4\dot{x}(\dot{x}^2 - 1) - 2X)\dot{y}dt \mid \dot{y} \in L^4, \int_0^1 \dot{y}dt = 0 \right\} \\ &= \inf \{ \| 4\dot{x}(\dot{x}^2 - 1) - 2X - a \|_{L^{4/3}} \mid a \in \mathbb{R} \} \end{aligned} \tag{36}$$

Corollary 7 now yields the following: There is a sequence $y_n \in W_0^{1,4}(0, 1)$ and a sequence $a_n \in \mathbb{R}$ such that

$$\int_0^1 \{(\dot{y}_n^2 - 1)^2 + y_n^2\}dt \to 0 \tag{37}$$

$$\| 4\dot{y}_n(\dot{y}_n^2 - 1) - 2Y_n - a_n \|_{L^{4/3}} \to 0 \tag{38}$$

4. APPLICATIONS TO CONVEX OPTIMIZATION

Let us consider a lower semicontinuous, convex proper function $U: X \to \mathbb{R} \cup \{+\infty\}$, where X is a Banach space. We have proved in theorem 4.3.11 that in the particular case when X is a Hilbert space, U is subdifferentiable on a dense subset of its domain. We now extend this result to Banach spaces.

Let U be bounded below and x_0 any point in X. The ε-variational principle states that there exists a point $x_\varepsilon \in X$ such that

$$\| x_0 - x_\varepsilon \| \leqslant U(x_0) - U(x_\varepsilon) \tag{1}$$

$$\forall x \neq x_\varepsilon, \qquad U(x_\varepsilon) < U(x) + \varepsilon \| x - x_\varepsilon \| \tag{2}$$

Since x_ε minimizes the function

$$x \to U(x) + \varepsilon \| x - x_\varepsilon \|$$

the subdifferential calculus (corollary 4.3.6, in particular) implies that

(3) $$0 \in \partial U(x_\varepsilon) + \varepsilon B$$

Theorem 1 follows immediately.

THEOREM 1

Let $U: X \to \mathbb{R} \cup \{+\infty\}$ be proper, lower semicontinuous, convex, and bounded below. Choose x_0 in Dom U *and $\varepsilon > 0$. Then there exists x_ε in* Dom U *and $p_\varepsilon \in \partial U(x_\varepsilon)$ such that*

(4) $$\|x_0 - x_\varepsilon\| \leqslant U(x_0) - U(x_\varepsilon)$$

(5) $$\|p_\varepsilon\|_\star \leqslant \varepsilon$$ ▲

Changing the norm of X from $\|x\|$ to $k\varepsilon\|x\|$, we obtain a more precise result,

COROLLARY 2

Let $U: X \to \mathbb{R} \cup \{+\infty\}$ be proper, lower semicontinuous, convex, and bounded below. Let $\varepsilon > 0$ and $x_0 \in$ Dom U *be given such that*

(6) $$U(x_0) \leqslant \varepsilon + \inf U$$

Then, for any $k > 0$, there is some point $x_\varepsilon \in$ Dom U *and some $p_\varepsilon \in \partial U(x_\varepsilon)$ such that*

(7) $$U(x_\varepsilon) \leqslant U(x_0)$$

(8) $$\|x_0 - x_\varepsilon\| \leqslant \frac{1}{k}$$

(9) $$\|p_\varepsilon\|_\star \leqslant k\varepsilon$$ ▲

We deduce theorem 3, which is due to Brønsted and Rockafellar.

THEOREM 3

Let X be a Banach space and $U: X \to \mathbb{R} \cup \{+\infty\}$ a convex, lower semicontinuous function. Then the set of points where U is subdifferentiable is dense in Dom U. *More precisely, for any $\bar{x} \in X$ where $U(\bar{x}) < +\infty$, there is a sequence x_k, $k \in N$ such that*

(10) $$\begin{cases} \textbf{i.} & x_k \to \bar{x} \\ \textbf{ii.} & U(x_k) \to U(\bar{x}) \\ \textbf{iii.} & \partial U(x_k) \neq \varnothing, \quad \textit{for all } k \end{cases}$$ ▲

Proof. If $U \equiv +\infty$, Dom U is empty and there is nothing left to prove. If $U \not\equiv +\infty$, there is some $p \in X^*$ and $a \in \mathbb{R}$ such that

$$V(x) = U(x) - \langle p, x \rangle - a > 0, \qquad \text{for all } x \tag{11}$$

Let $\bar{x}$ be any point where U (and hence V) is finite. Apply corollary 2 to V, with $\varepsilon = V(\bar{x}) - \inf V$ (not necessarily small) and k any integer. We obtain some point $x_k \in X$ and $p_k \in \partial V(x_k)$.

$$V(x_k) < V(\bar{x}) \tag{12}$$

$$\|\bar{x} - x_k\| \leqslant k^{-1} \tag{13}$$

$$\|p_k\|_\star \leqslant \varepsilon k \tag{14}$$

This obviously implies that U is also subdifferentiable at x_k, with $\partial U(x_k) = p + \partial V(x_k)$. Conditions (10, **i** and **ii**) have thus been proved, and we are left with condition (10, **ii**).

Expressing V in terms of U in equality (11), we obtain

$$U(x_k) < U(\bar{x}) - \langle p, \bar{x} - x_k \rangle, \qquad \text{for all } k$$

Letting $k \to \infty$, we have lim sup $U(x_k) \leqslant U(\bar{x})$. On the other hand, since U is lower semicontinuous, we have

$$\liminf U(x_k) \geqslant U(\bar{x})$$

So $U(x_k) \to U(\bar{x})$, and the theorem is proved. ∎

We now apply Corollary 2 to convex optimization problems of the form treated in Section 4.8 in situations where there exists a Lagrange multiplier, (given by theorem 4.8.1) but no optimal solution.

PROPOSITION 4

Let X and Y be Banach spaces, $L: X \times Y \to R \cup \{+\infty\}$ a lower semicontinuous, convex function, and $A: X \to Y$ a continuous linear operator. Consider the optimization problem

$$\inf_x L(x, Ax) \tag{15}$$

Assume that $0 \in \operatorname{Int}((A \oplus -I)\operatorname{Dom} L)$, *so that there exists a Lagrange multiplier*

$$-L^*(-A^*\bar{q}, \bar{q}) = \inf_{x \in X} L(x, Ax) \tag{16}$$

Let $\bar{x}_n$, $n \in N$, be a minimizing sequence in problem (15). *Then there is a sequence* (x_n, y_n, p_n, q_n) *such that*

$$\|x_n - \bar{x}_n\| \to 0 \tag{17}$$

$$\|y_n - A\bar{x}_n\| \to 0 \tag{18}$$

$$\|q_n - \bar{q}\|_* \to 0 \tag{19}$$

$$\|p_n + A^*\bar{q}\|_* \to 0 \tag{20}$$

$$(p_n, q_n) \in \partial L(x_n, y_n) \tag{21}$$

▲

Proof. Consider the function U on $X \times Y$ defined by

$$U(x, y) = L(x, y) + \langle x, A^*\bar{q}\rangle - \langle y, \bar{q}\rangle + L^*(-A^*\bar{q}, \bar{q}) \tag{22}$$

This function is always positive. By assumption (16), its infimum is zero, and the sequence $(\bar{x}_n, A\bar{x}_n)$ is minimizing. Setting $U(\bar{x}_n, A\bar{x}_n) = \varepsilon_n \to 0$ and applying corollary 2 to the point $(\bar{x}_n, A\bar{x}_n)$, with $\sqrt{\varepsilon_n} = 1/k$, we obtain points $(x_n, y_n) \in X \times Y$ and $(p'_n, q'_n) \in X^* \times Y^*$, where

$$\|x_n - \bar{x}_n\| \leqslant \sqrt{\varepsilon_n} \quad \text{and} \quad \|y_n - A\bar{x}_n\| \leqslant \sqrt{\varepsilon_n} \tag{23}$$

$$(p'_n, q'_n) \in \partial U(x_n, y_n) \tag{24}$$

$$\|p'_n\|_* \leqslant \sqrt{\varepsilon_n} \quad \text{and} \quad \|q'_n\|_* \leqslant \sqrt{\varepsilon_n} \tag{25}$$

Setting $p_n = -A^*\bar{q} + p'_n$ and $q_n = \bar{q} + q'_n$, these relations yield the desired result. ■

If there is an optimal solution $\bar{x}$ to problem (15), proposition 4 will hold with $(x_n, y_n, p_n, q_n) = (\bar{x}, A\bar{x}, -A^*\bar{q}, \bar{q})$ for all n. More generally, in the absence of any optimal solution, it may well happen that relations (17)–(21) imply some kind of convergence of the sequence (x_n, y_n) to some limit $(\bar{x}, A\bar{x})$, which will be regarded as a "weak," or "generalized," solution to problem (15).

This kind of result was first observed by Temam in the particular case of Plateau's problem. We now describe this example.

Example 5

Let Ω be a bounded, open subset of R^N and let x_0 be some function in the Sobolev space $W^{1,1}(\Omega)$. We want to minimize

$$\int_\Omega \left(1 + \sum_{i=1}^{N} \frac{\partial x^2}{\partial \omega_i}\right)^{1/2} d\omega \tag{26}$$

subject to

$$x - x_0 \in W_0^{1,1}(\Omega) \tag{27}$$

This is a simple version of Plateau's problem: Integral (26) is the area of the hypersurface S in R^{N+1} described in parametric form by

$$S = \{(\omega, x(\omega)) \mid \omega \in \Omega\} \tag{28}$$

and condition (27) means that x and x_0 coincide on the boundary $\partial\Omega$ of Ω. In other words, we are seeking a hypersurface S (if any), admitting a representation of type (28), with prescribed boundary $\partial S = \{(\omega, x_0(\omega)) \mid \omega \in \partial\Omega\}$ and having minimum area. Soap bubbles apparently solve this problem, but mathematicians do not.

The set $x_0 + W_0^{1,1}(\Omega)$ is a closed affine subspace of $W^{1,1}(\Omega)$, Integral (26) defines a function $U(x)$, which is positive and continuous by Krasnoselskii's theorem. It is also Frêchet differentiable, and its derivative at x is the continuous linear map on $W_0^{1,1}(\Omega)$ defined by

$$W_0^{1,1} \ni y \to \int_\Omega \left(1 + \sum_{i=1}^N \frac{\partial x^2}{\partial \omega_i}\right)^{-1/2} \sum_{i=1}^N \frac{\partial x}{\partial \omega_i} \frac{\partial y}{\partial \omega_i} \cdot d\omega \tag{29}$$

By definition this functional is an element of $W^{-1,\infty}(\Omega)$, which we can write as

$$U'(x) = -\sum_{i=1}^N \frac{\partial}{\partial \omega_i} \left[\frac{\partial x}{\partial \omega_i} \left(1 + \sum_{j=1}^N \frac{\partial x^2}{\partial \omega_j}\right)^{-1/2}\right] \tag{30}$$

where the derivatives are to be understood in the sense of distributions. Finding a minimizer for integral (26) under contraint (27) therefore means finding a weak solution for the boundary-value problem

$$\sum_{i=1}^N \frac{\partial}{\partial \omega_i} \left[\frac{\partial x}{\partial \omega_i} \left(1 + \sum_{j=1}^N \frac{\partial x^2}{\partial \omega_j}\right)^{-1/2}\right] = 0 \tag{31}$$

$$x - x_0 \in W_0^{1,1}(\Omega) \tag{32}$$

Unfortunately, if no further assumptions are made on the shape of Ω, this problem need not have a solution. A sufficient condition for a solution x to exist for all x_0 is that the domain Ω be convex (more generally, that the mean curvature of the boundary $\partial\Omega$ be positive). This is very surprising at first glance, since we could always build a wire to the prescription $\{(\omega, x_0(\omega)) \mid \omega \in \partial\Omega\}$, dip it in soapy water, and note the shape S of the bubble. The point is that even though such a "physical" solution will exist in general, it will not admit a parametric representation of the form (28)—except, of course, when Ω is convex.

In the general case, when equations (31) and (32) have no solution, we apply the ε-variational principle to get a substitute. Set $z=x-x_0$, so that problems (26) and (27) becomes

$$\inf U(z+x_0) \tag{33}$$

$$z \in W_0^{1,1}(\Omega) \tag{34}$$

It is well known (Poincaré's inequality) that $\|\text{grad } x\|_{L^1}$ and $\|x\|_{W^{1,1}}$ are equivalent norms on $W_0^{1,1}(\Omega)$. The following inequality then holds for some $k>0$ and c:

$$U(z+x_0)>k\|z\|-c \tag{35}$$

We have the situation in Example 3.8: There exists in the ball kB^* of $W^{-1,\infty}$ a dense subset $\mathcal{S}$ such that for all $T\in\mathcal{S}$, the equation $U'(z+x_0)=T$ has a solution. Taking into account the fact that U is strictly convex, this means that the problem

$$\begin{cases} \inf\left\{\langle T, x\rangle + \int_\Omega \left(1+\sum_{i=1}^N \frac{\partial x^2}{\partial\omega_i}\right)^{1/2} dx\right\} \\ x-x_0\in W_0^{1,1}(\Omega) \end{cases} \tag{36}$$

has a unique solution x for all T in $\mathcal{S}$.

We now apply proposition 4, taking

$$X=W^{1,1}(\Omega) \qquad Y=L^1(\Omega)^N$$

$$Ax=\left(\frac{\partial x}{\partial\omega_1},\ldots,\frac{\partial x}{\partial\omega_n}\right)$$

$$L(x,y)=\begin{cases} \int_\Omega \left(1+\sum_{i=1}^N y^i(\omega)^2\right)^{1/2} d\omega & \text{if} \quad x-x_0\in W_0^{1,1}(\Omega) \\ +\infty & \text{otherwise} \end{cases}$$

Starting from a minimizing sequence $\bar{x}_n\in x_0+W_0^{1,1}(\Omega)$, we find a sequence (x_n, y_n, p_n, q_n) such that

$$x_n-\bar{x}_n\to 0 \qquad \text{in } W^{1,1}(\Omega) \tag{37}$$

$$y_n-A\bar{x}_n\to 0 \qquad \text{in } L^1(\Omega)^N \tag{38}$$

$$q_n-\bar{q}\to 0 \qquad \text{in } L^\infty(\Omega)^N \tag{39}$$

(40) $$p_n + A^*\bar{q} \to 0 \qquad \text{in } [W^{1,1}(\Omega)]^*$$

(41) $$q_n^i = y_n^i \left[1 + \sum_{i=1}^{N} (y_n^i)^2\right]^{-1/2}, \qquad 1 \leq i \leq N, \qquad \text{in } L^\infty(\Omega)$$

(42) $$\langle p_n, x \rangle = 0, \qquad \text{for all } x \in W_0^{1,1}(\Omega)$$

A simple computation shows that

(43) $$L^*(-A^*q, q) = \langle -A^*q, x_0 \rangle - \int_\Omega \left[1 - \sum_{i=1}^{N} q^i(\omega)^2\right]^{1/2} d\omega$$

if $\|q(\omega)\| \leq 1$ almost everywhere and $L^*(-A^*q, q) = +\infty$ otherwise. Since $\bar{q}$ is a Lagrange multiplier, it minimizes (43), and it follows that

(44) $$\|\bar{q}(\omega)\| = \left[\sum_{i=1}^{N} \bar{q}^i(\omega)^2\right]^{1/2} < 1 \quad \text{a.e.}$$

Using very refined a priori estimates, due to Ladyjenskaïa and Uraltseva, Temam has associated with any open set $\mathscr{U}$, whose closure is contained in Ω, a constant $c(\mathscr{U}) > 0$ such that

(45) $$\forall \omega \in \mathscr{U}, \qquad \|\bar{q}(\omega)\| \leq 1 - c(\mathscr{U})$$

Condition (39) them implies that $\|q_n(\omega)\|$ is also bounded by $1 - c(\mathscr{U})$ on $\mathscr{U}$. Relation (41) can be reversed

(46) $$y_n^i(\omega) = q_n^i(\omega) \left[1 - \sum_{i=1}^{N} q_n^i(\omega)^2\right]^{-1/2} \quad 1 \leq i \leq N$$

to show that the sequence y_n converges uniformly on $\mathscr{U}$ toward the function

(47) $$\bar{y}(\omega) = \bar{q}(\omega) \left[1 - \sum_{i=1}^{N} \bar{q}^i(\omega)^2\right]^{-1/2}$$

It is known that the operator A has closed range, so that $\bar{y} = A\bar{x}$, where $\bar{x} \in W^{1,1}(\Omega)$ is defined up to an additive constant

(48) $$\frac{\partial \bar{x}}{\partial \omega_i} = \bar{q}^i(\omega) \left[1 - \sum_{i=1}^{N} \bar{q}^i(\omega)^2\right]^{-1/2}$$

(49) $$\bar{q}^i(\omega) = \frac{\partial \bar{x}}{\partial \omega_i} \left[1 + \sum_{i=1}^{N} \frac{\partial \bar{x}}{\partial \omega_i}(\omega)^2\right]^{-1/2}$$

Conditions (40) and (42) imply that $\langle A^*\bar{q}, x\rangle = 0$ for all $x \in W_0^{1,1}(\Omega)$, which means that in the sense of distributions,

$$\text{div } \bar{q} = 0 \qquad \text{in } \mathscr{D}'(\Omega) \tag{50}$$

Substituting (49) into (50), we have

$$\sum_{i=1}^{N} \frac{\partial}{\partial \omega_i} \left[\frac{\partial \bar{x}}{\partial \omega_i} \left(1 + \sum_{i=1}^{N} \frac{\partial \bar{x}}{\partial \omega_i}(\omega)^2 \right)^{-1/2} \right] = 0 \qquad \text{in } \mathscr{D}'(\Omega) \tag{51}$$

This is the so-called equation of minimal hypersurfaces, which any optimal solution of Plateau's problem has to satisfy, together with the boundary condition

$$x - x_0 \in W_0^{1,1}(\Omega) \tag{52}$$

As noted before, there may well be no solution to boundary-value problem (51) and (52) and, hence, no minimizer. What we do show is that there always exist minimizing sequences x_n and a solution $\bar{x}$ to equation (51) such that x_n converges uniformly to $\bar{x}$ on all compact subsets of Ω. The function $\bar{x}$, however, does not necessarily satisfy boundary condition (52) (if it does, it is a minimizer). It is a "weak" or "generalized" solution of Plateau's problem; a more detailed analysis, carried out by Lichnewski, shows that $\bar{x}(\omega) = x_0(\omega)$ at all points ω of the boundary $\partial\Omega$ where the mean curvature is non-negative. A typical situation is illustrated in Figure 1, with $N=1, \Omega = (-1, 0) \cup (0, 1)$ and $x_0(-1) = x_0(1) = 1$, $x_0(0) = 0$. The generalized solution is $\bar{x}(\omega) \equiv 1$; it is shown in the following figure as well as a minimizing sequence x_n converging to it.

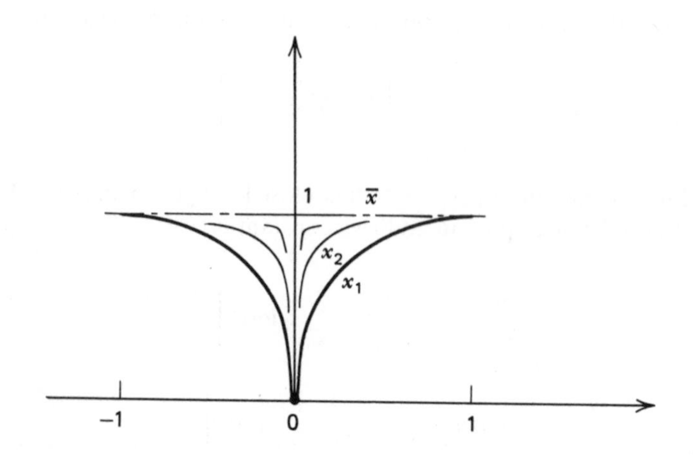

5. CONDITION (C) OF PALAIS AND SMALE

In the preceding section, we have shown the existence of minimizing sequences of a particular type. The question naturally arises whether these sequences converge, thereby proving the existence of an actual minimizer. Some kind of compactness assumption is needed, the weakest being condition (C) of Palais and Smale.

Historically, condition (C) was first stated for C^2 functions on Banach manifolds, where it serves as a substitute for convexity. The differentiability requirement has since been lowered to C^{1+} (locally Lipschitz derivative) and even C^1 for certain questions. It is within this nonlinear framework that condition (C) has proved the most useful.

We shall, however, deal with linear spaces only, because Banach manifolds are outside the scope of this book. The proofs we give will, nevertheless, carry over to the general nonlinear case, since they rely on the ε-variational principle and use only the differentiability structure of the underlying space.

DEFINITION 1

Let X be a Banach space and U: $X \to \mathbb{R}$ a Gâteaux-differentiable function. We say that U satisfies condition (C) on a subspace $\Omega \subset X$ if whenever there is a sequence x_n, $n \in \mathbb{N}$, in Ω, with

$$|U(x_n)| \leqslant constant \tag{1}$$

$$U'(x_n) \to 0 \qquad in\ X^* \tag{2}$$

then in the closure of the set $\{x_n | n \in N\}$, there is some point $\bar{x}$ where $U'(\bar{x})=0$. ■

Note how carefully this condition is worded: It is not stated that there is a convergent subsequence in the sequence x_n. Indeed, if $X=\mathbb{R}$ and $U=0$, then condition (C) is satisfied. For instance, taking the sequence $x_n=n$, we have $U(x_n)=0=U'(x_n)$ for all n, There will be no convergent subsequence, but we can take any of the x_n for $\bar{x}$.

Condition (C) enables us to use Morse theory or Liusternik–Schnirelman theory, that is, to find lower bounds for the number of critical points of U by topological methods (if, for instance, some group acts on X and leaves U invariant). Since our aim is much more modest, we are content with a weakened form of condition (C), which we call (weak C).

DEFINITION 2

Let X be a Banach space and U: $X \to \mathbb{R}$ a Gâteaux-differentiable function. We say that U satisfies condition (weak C) on a subspace $\Omega \subset X$ if whenever there is a *sequence* x_n, $n \in \mathbb{N}$, *in* Ω *with*

$$|U(x_n)| \leqslant \text{constant} \tag{3}$$

$$U'(x_n) \neq 0, \qquad \text{for all } n \tag{4}$$

$$U'(x_n) \to 0 \qquad \text{in } X^* \tag{5}$$

then there is some point $\bar{x} \in X$ *such that*

$$\liminf U(x_n) \leqslant U(\bar{x}) \leqslant \limsup U(x_n) \tag{6}$$

$$U'(\bar{x}) = 0 \tag{7}$$

▲

PROPOSITION 3

If U *is continuous and satisfies condition* (C) *on* Ω, *it satisfies condition* (*weak* C) *on* Ω. *If* X *is reflexive and* U *is convex, lower semicontinuous and* $U(x) \to +\infty$ *when* $\|x\| \to \infty$, *then* U *satisfies condition* (*weak* C) *on* X. ▲

Proof. The first statement is trivial. Its converse, that (weak C) implies (C), is false in general.

We prove the second statement. Take a sequence x_n satisfying (3), (4), and (5). Then take a subsequence y_n such that $\lim U(y_n) = \limsup U(x_n)$. It follows from the assumptions on X and U that y_n contains a weakly convergent subsequence z_n, its limit being called $\bar{x}$. Since U is weakly lower semicontinuous, we have

$$U(\bar{x}) \leqslant \lim U(z_n) = \limsup U(x_n)$$

Fix x in X. Since U is convex, it satisfies inequalities

$$\langle U'(z_n), x - z_n \rangle + U(z_n) \leqslant U(x)$$

Taking the limit, we obtain, because of (5),

$$\limsup U(x_n) = \lim U(z_n) \leqslant U(x)$$

Since x is any point in X, everything follows at once. First $U(\bar{x}) \leqslant U(x)$, so that $\bar{x}$ is a minimizer and $U'(\bar{x}) = 0$. Then $\limsup U(x_n) = \inf U$, so that $U(x_n)$ actually converges to $U(\bar{x})$. ■

Our first existence result concerns minima.

PROPOSITION 4

Let $U: X \to \mathbb{R}$ *be Gâteaux-differentiable, lower semicontinuous, and bounded from below. Assume the restriction of* U' *to straight lines is continuous and* U

satisfies condition (weak C) on X. Then U attains its minimum on X

$$\exists \bar{x} \in X: U(\bar{x}) = \inf U \tag{8}$$

▲

Proof. By corollary 3.7, there is a sequence y_n, $n \in N$, in X such that $U(y_n) \to \inf U$ and $U'(y_n) \to 0$. There are now two cases to consider: Either we can extract a subsequence (denoted by x_n) such that $U'(x_n) \neq 0$ for all n, or $U'(y_n) = 0$ for all but a finite number of n.

In the former case, by condition (weak C), there is some point $\bar{x} \in X$ such that $U'(\bar{x}) = 0$ and

$$U(\bar{x}) \leq \lim U(x_n) = \inf U$$

So $\bar{x}$ is a minimizer, and the result is proved.

In the latter case, set $S = \{x | U'(x) = 0\}$. If $S = X$, then U is constant, and any point is a minimizer. If $S \neq X$, there is some point z where $U'(z) \neq 0$. For fixed n, consider the line segment $t \to tz + (1-t)x_n$, $0 \leq t \leq 1$, from x_n to z, and set

$$\bar{t} = \inf \{t | tz + (1-t)x_n \notin S\}$$

By assumption, the restriction to this line segment of U' is continuous. It follows that we can find $t_1 \leq \bar{t} \leq t_2$ with

$$x_n^1 = t_1 z + (1-t_1)x_n \in S$$

$$x_n^2 = t_2 z + (1-t_2)x_n \notin S$$

$$|U(x_n^2) - U(x_n^1)| \leq n^{-1}$$

$$\|U'(x_n^2)\| \leq n^{-1}$$

Since $t_1 \leq \bar{t}$, we have $tz + (1-t)x_n \in S$ for all $t \leq t_1$. It follows from the mean value theorem that

$$U(x_n^1) = U(x_n)$$

We have found, for each $n \in N$, a point x_n^2 such that $U'(x_n^2) \neq 0$ and $|U(x_n^2) - U(x_n)| \leq n^{-1}$. Letting $n \to \infty$, we have $U(x_n^2) \to \inf U$ and $U'(x_n^2) \to 0$. We are back in the preceding case. ■

If we drop the continuity assumption on U', our argument will give us a critical point $U'(\bar{x}) = 0$, without telling us whether it minimizes U. Note, however, that this will follow automatically when U is convex; as is well known, the result will hold in this case without any differentiability assumption at all on U.

A judicious use of condition (weak C) will enable us to find critical points of various types, not only local minima or maxima, but saddle points as well. There are now many theorems of this kind, the simplest and perhaps the most beautiful being the following, a strengthened version of a result originally due to Ambrosetti and Rabinowitz.

THEOREM 5

Let X be a Banach space and U: $X \to \mathbb{R}$ a continuous and Gâteaux-differentiable function. Assume that U': $X \to X^*$ is strong-to-weak continuous and*

$$\exists \alpha > 0: m(\alpha) := \inf \{U(x) \mid \|x\| = \alpha\} > U(0) \tag{9}$$

$$\exists z \in X: \|z\| > \alpha \quad \text{and} \quad U(z) < m(\alpha) \tag{10}$$

$$U \text{ satisfies condition (weak } C) \text{ on } \{x \mid U(x) \geqslant m(\alpha)\} \tag{11}$$

Then there is a point $\bar{x} \in X$ where

$$U(\bar{x}) \geqslant m(\alpha) \quad \text{and} \quad U'(\bar{x}) = 0 \tag{12}$$

▲

Proof. A path from zero to z is a continuous map c: $[0, 1] \to X$ with $c(0)=0$ and $c(1)=z$. Denote by $\mathscr{C}$ the set of all paths from zero to z, endowed with the distance

$$d(c_1, c_2) = \max \{\|c_1(t) - c_2(t)\| \mid 0 \leqslant t \leqslant 1\}$$

It is a complete metric space. We define a function I: $\mathscr{C} \to R$ as follows:

$$I(c) = \max \{ U(c(t)) \mid 0 \leqslant t \leqslant 1\}$$

The function I is lower semicontinuous, since it can be written $I(c) = \sup_t I_t(c)$, each function $I_t(c) = U(c(t))$ continuous. Note also that it is bounded below by $m(\alpha)$. Indeed, since c continuously joins zero to z, it must cross the ball of radius α somewhere: There is some $t_\alpha \in [0, 1]$ where $\|c(t_\alpha)\| = \alpha$ and, hence,

$$I(c) \geqslant U(c(t_\alpha)) \geqslant m(\alpha)$$

It follows that we can apply corollary 3.2: For any $\varepsilon > 0$, there is some path c_ε where

$$I(c_\varepsilon) \leqslant \inf \{I(c) \mid c \in \mathscr{C}\} + \varepsilon$$
$$\forall c \in \mathscr{C}, \quad I(c) \geqslant I(c_\varepsilon) - \varepsilon d(c, c_\varepsilon)$$

Now let γ be any continuous map from $[0, 1]$ to X such that $\gamma(0) = 0$ and

$\gamma(1)=0$. For any $h \in \mathbb{R}$, we have the inequality

$$I(c_\varepsilon+h\gamma)-I(c_\varepsilon)\geqslant -\varepsilon d(c_\varepsilon+h\gamma, c_\varepsilon)$$

which becomes

$$h^{-1}[I(c_\varepsilon+h\gamma)-I(c_\varepsilon)]\geqslant -\varepsilon \max_t \|\gamma(t)\|$$

On the other hand, we can write

$$I(c_\varepsilon+h\gamma)-I(c_\varepsilon)=\max_t\ U(c_\varepsilon(t)+h\gamma(t))-\max_t\ U(c_\varepsilon(t))$$

$$=\max_t\ \{U(c_\varepsilon)+h\langle U'(c_\varepsilon), \gamma\rangle\}+o(h)-\max_t\ U(c_\varepsilon)$$

with $h^{-1}o(h)\to 0$ when $h\to 0$ (since $[0, 1]$ is compact).

Set $U(c_\varepsilon)=f$ and $\langle U'(c_\varepsilon), \gamma\rangle=g$. Our assumptions will imply that f and g are continuous maps from $[0, 1]$ to X.

Define a function $\Phi\colon C^o([0, 1])\to\mathbb{R}$

$$\Phi(\phi)=\max_t\ |\phi(t)|$$

This function is convex and continuous, hence, subdifferentiable everywhere. The dual of $C^0([0, 1])$ is the space of Radon measures μ on $[0, 1]$, and the subdifferential of Φ is given by

$$\partial\Phi(\phi)=\left\{\mu\geqslant 0\ \middle|\ \int d\mu=1 \text{ and supp } \mu\subset M(\phi)\right\}$$

where $M(\phi)=\{t|\phi(t)=\Phi(\phi)\}$

We now see that

$$-\varepsilon \max_t \|\gamma(t)\|\leqslant \lim_{h\to 0}\ [I(c_\varepsilon+h\gamma)-I(c_\varepsilon)]h^{-1}$$

$$=\lim_{h\to 0}\ [\Phi(f+hg)-\Phi(f)]h^{-1}$$

$$=\max\ \{\langle g, \mu\rangle \mid \mu\in\partial\Phi(f)\}$$

$$=\max\left\{\int \langle U'(c_\varepsilon), \gamma\rangle d\mu \mid \mu\in\partial\Phi(f)\right\}$$

On both sides take the infimum over the set of $\gamma\in C^0([0, 1]; X)$ such that

$$\|\gamma\|\leqslant 1, \qquad \gamma(0)=0=\gamma(1)$$

We obtain

$$-\varepsilon \leqslant \inf_{\gamma} \max_{\mu} \left\{ \int \langle U'(c_\varepsilon), \gamma \rangle d\mu \,\middle|\, \begin{matrix} \mu \in \partial\Phi(f), \ \|\gamma\| \leqslant 1 \\ \gamma(0) = 0 = \gamma(1) \end{matrix} \right\}$$

The set $\partial\Phi(f)$ is weak* compact, so that we can use the inf sup theorem 5.2.7

$$-\varepsilon \leqslant \max_{\mu} \inf_{\gamma} \left\{ \int \langle U'(c_\varepsilon), \gamma \rangle d\mu \,\middle|\, \begin{matrix} \mu \in \partial\Phi(f), \ \|\gamma\| \leqslant 1 \\ \gamma(0) = 0 = \gamma(1) \end{matrix} \right\}$$

$$= \max \left\{ - \int \| U'(c_\varepsilon) \|_* d\mu \,|\, \mu \in \partial\Phi(f) \right\}$$

$$= -\min \left\{ \| U'(c_\varepsilon(t)) \|_* \,|\, t \in M(U o c_\varepsilon) \right\}$$

So there is some t_ε such that

$$U(c_\varepsilon(t_\varepsilon)) = \max_{t} U(c_\varepsilon(t))$$

$$\| U'(c_\varepsilon(t_\varepsilon)) \|_* \leqslant \varepsilon$$

Now set $\varepsilon = n^{-1}$ and $c_\varepsilon(t_\varepsilon) = x_n$. We have proved that $m(\alpha) \leqslant U(x_n) \leqslant \inf I + \varepsilon$ and $U'(x_n) \to 0$. If $U'(x_n) = 0$ for some n, we are done. If not, we apply condition (weak C) to find some point $\bar{x}$ where

$$U'(\bar{x}) = 0 \qquad \text{and} \qquad U(\bar{x}) \geqslant \liminf U(x_n) \geqslant m(\alpha)$$ ■

It was quite easy to picture the underlying geometric situation. Imagine the graph of U as the shape of a mountain range lying over X. Then the origin lies in a closed valley, and it is known that there is some point outside this valley with lower altitude than the surrounding mountains. Clearly then, there must be a mountain pass out of the valley, and this is exactly what our proof is looking for.

The nontrivial critical value in theorem 5 is given by

$$v = \inf_{c \in \mathscr{C}} \max_{0 \leqslant t \leqslant 1} U(c(t))$$

The idea of finding a critical value by an inf max formula of this kind can be extended to other situations, yielding different kinds of critical points.

THEOREM 6

Let X be a Banach space and U: $X \to \mathbb{R}$ a continuous and Gâteaux-differentiable function. Assume that U': $X \to X^*$ is strong-to-weak continuous and X splits into a direct sum $X = X_0 \oplus X_\infty$, with*

(13) $\qquad X_0$ *is finite dimensional*

(14) $$\exists R>0\colon [x_0 \in X_0 \text{ and } \|x_0\|=R] \Rightarrow U(x_0, 0)<0$$

(15) $$x_\infty \in X_\infty \Rightarrow U(0, x_\infty)\geq 0$$

(16) *U satisfies condition (weak C) on* $\{x \mid U(x)\geq 0\}$

Then U has a critical point:

(17) $$\exists \bar{x} \in X\colon U'(\bar{x})=0 \quad \text{and} \quad U(\bar{x})\geq 0$$ ▲

Proof. Set

$$B_0=\{x_0 \in X_0 \mid \|x_0\|\leq R\}$$

$$S_0=\{x_0 \in X_0 \mid \|x_0\|=R\}$$

$$\mathscr{C}=\{\phi \in C^0(B_0, X_\infty) \mid \phi(S_0)=0\}$$

We endow $\mathscr{C}$ with the uniform topology, which turns it into a complete metric space, and we define a function $I\colon \mathscr{C}\to\mathbb{R}$ as follows:

$$I(\phi)=\max\{U(x_0, \phi(x_0)) \mid x_0 \in B_0\}$$

The supremum on the right-hand side is attained, since B_0 is finite dimensional and hence compact. It cannot be attained on the boundary S_0 because of conditions (14) and (15)

$$x_0 \in S_0 \Rightarrow U(x_0, \phi(x_0))=U(x_0, 0)<U(0, \phi(0))$$

The function I is lower semicontinuous and bounded below by zero. It follows that we can apply corollary 2.2. For any $\varepsilon>0$, there is some ϕ_ε such that

$$I(\phi_\varepsilon)\leq \inf\{I(\phi) \mid \phi \in \mathscr{C}\}+\varepsilon$$

$$\forall \phi \in \mathscr{C}, \qquad I(\phi)\geq I(\phi_\varepsilon)-\varepsilon\|\phi-\phi_\varepsilon\|$$

Now pick any γ in $\mathscr{C}$. For any $h \in \mathbb{R}$, we have the inequality

$$I(\phi_\varepsilon+h\gamma)-I(\phi_\varepsilon)\geq -\varepsilon h\|\gamma\|$$

Arguing as in the proof of theorem 5, with $C^0(B_0)$ replacing $C^0([0, 1])$, we find some $x_0^\varepsilon \in B_0$ such that

$$U(x_0^\varepsilon, \phi_\varepsilon(x_0^\varepsilon))=\max\{U(x_0, \phi_\varepsilon(x_0)) \mid x_0 \in B_0\}$$

$$\|U'(x_0^\varepsilon, \phi_\varepsilon(x_0^\varepsilon))\|_*\leq\varepsilon$$

The result follows from condition (weak C). ■

Example 7

Let X be a reflexive Banach space and $A\colon X \to X^*$ a compact linear operator, with $A^* = A$. Let $V\colon X \to \mathbb{R}$ be convex and C^1. Assume that

$$\exists k_0 > 0, \qquad \exists c_0 \in \mathbb{R}: V(x) \geqslant k_0\|x\| - c_0 \qquad \forall x \in X \tag{18}$$

$$\exists k_1 < 2, \qquad \exists c_1 \in \mathbb{R}: \langle V'(x), x\rangle \leqslant k_1 V(x) + c_1 \qquad \forall x \in X \tag{19}$$

The latter condition limits the growth of V at infinity; if $c_1 = 0$, it can be shown to be equivalent to the condition that $V(\lambda x) < \lambda^{k_1} V(x)$ for all $x \in V$ and $\lambda > 1$.

Set

$$U(x) = \frac{1}{2}\langle Ax, x\rangle + V(x)$$

We claim that the function U satisfies condition (weak C) on X. ▲

To prove this, we take two constants a and b and a sequence $x_n \in X$ such that for all $n \in N$

$$a \leqslant U(x_n) = \frac{1}{2}\langle Ax_n, x_n\rangle + V(x_n) \leqslant b$$

$$U'(x_n) = Ax_n + V'(x_n) = p_n \to 0 \quad \text{in} \quad X^*$$

Substituting the latter relation into the first, we obtain

$$a \leqslant \frac{1}{2}\langle p_n - V'(x_n), x_n\rangle + V(x_n) \leqslant b$$

Using the assumptions on V, this yields

$$a \leqslant \frac{1}{2}\langle p_n, x_n\rangle + \left(1 - \frac{1}{2}k_1\right) V(x_n) - c_1$$

$$a \leqslant \frac{1}{2}\langle p_n, x_n\rangle + \left(1 - \frac{1}{2}k_1\right)(k_0\|x_n\| - c_0) - c_1$$

Since $p_n \to 0$, it follows that the sequence x_n is bounded: $\|x_n\| \leqslant$ constant. Since X is reflexive, there is a subsequence x_{n_k}, shortened to x_k, such that

$$x_k \to \bar{x} \quad \text{weakly in } X \tag{20}$$

Since A is a compact operator, Ax_k converges to $A\bar{x}$ strongly in X. It follows that:

(21) $$V'(x_k)=p_k-Ax_k\to A\bar{x}$$

Since V is convex and continuous, it is weakly lower semicontinuous, so that $V(\bar{x})\leqslant\liminf V(x_k)$. Taking limits in the inequalities $V(x)\geqslant V(x_k)+\langle V'(x_k),\ x-x_k\rangle$, we obtain $V(x)\geqslant V(\bar{x})+\langle A\bar{x},\ x-\bar{x}\rangle$, so that $-A\bar{x}=V'(\bar{x})$. This means that $U'(\bar{x})=0$.

Starting from $V(\bar{x})\geqslant V(x_k)+\langle V'(x_k),\ \bar{x}-x_k\rangle$, and letting $k\to\infty$, we have $V(\bar{x})\geqslant\limsup V(x_k)$. We have just proved the converse inequality, so that $V(\bar{x})=\lim V(x_k)$. Since A is compact, the quadratic term $\langle Ax,\ x\rangle$ is weakly continuous on bounded subsets. Finally $U(\bar{x})=\lim U(x_k)$, and condition (weak C) is satisfied. ∎

Example 8

We retain the assumptions and notations of the preceding example. We add the following:

(22) $$\exists\alpha>0\colon \inf\{U(x)\mid \|x\|=\alpha\}>V(0)$$

(23) $$\exists z\in X\colon \|z\|>\alpha \qquad \text{and} \qquad U(z)\leqslant V(0)$$

Then there is some point $\bar{x}\in X$ such that

(24) $$\bar{x}\neq 0 \qquad \text{and} \qquad A\bar{x}+V'(\bar{x})=0$$ ▲

This is the situation in theorem 5. Conditions (9) and (10) are assumed and condition (11) has been proved.

We shall use this result in Chapter 8.

Example 9

Let X be a Hilbert space and $F: X\to\mathbb{R}$ a C^1 functional. Assume F is twice weakly differentiable Gâteaux and there is a constant $k>0$ such that

(25) $$\forall x\in X \qquad \|F''(x)^{\star}F'(x)\|\geqslant k\|F'(x)\|$$

Then F has a critical point on X

(26) $$\exists\bar{x}\in X\colon \quad F'(\bar{x})=0$$ ▲

Note that the assumption is certainly satisfied if F has the C^2 property and $F''(x)\in\mathscr{L}(X,\ X)$ is invertible, with $F''(x)^{-1}$ bounded uniformly

(27) $$\|F''(x)^{-1}\|\leqslant k^{-1} \qquad \forall x\in X$$

To prove our result, assume that there is no critical point, so $F'(x)\neq 0$ for all x. We shall derive a contradiction.

LEMMA 10

The real function $x\to\|F'(x)\|$ is continuous and Gâteaux differentiable everywhere. ▲

Proof. Consider the map

$$\varphi(x)=\|F'(x)\|^2=(F'(x),\,F'(x))$$

It is obviously continuous. Check that it is Gâteaux differentiable.

$$\begin{aligned}\frac{1}{t}[\varphi(x+ty)-\varphi(x)]&=\frac{1}{t}[(F'(x+ty),\,F'(x+ty))-(F'(x),\,F'(x))]\\&=\frac{2}{t}(F'(x+ty)-F'(x),\,F'(x))\\&\quad+\left(\frac{F'(x+ty)-F'(x)}{t},\,F'(x+ty)-F'(x)\right)\end{aligned}$$

Now let $t\to 0$. By assumption, F' is weakly differentiable Gâteaux, so

$$\frac{1}{t}(F'(x+ty)-F'(x),\,F'(x))\to(F''(x)F'(x),\,F'(x))$$

Restrict t to a sequence $t_n\to 0$. Since the sequence $t_n^{-1}[F'(x+t_ny)-F'(x)]$ converges weakly in X, it is bounded (Banach–Steinhaus theorem). Since F' is continuous, $F'(x+t_ny)-F'(x)$ converges strongly to zero in X. So the second term on the right converges to zero, and

$$\lim_{t\to 0}\frac{1}{t}[\varphi(x+ty)-\varphi(x)]=2(F''(x)y,\,F'(x))$$

So φ is Gâteaux differentiable. Since

$$\|F'(x)\|=\sqrt{\varphi(x)}$$

and φ never vanishes, the function $x\to\|F'(x)\|$ is also Gâteaux differentiable, with derivative

$$y\to\left(F''(x)y,\,\frac{F'(x)}{\|F'(x)\|}\right)=\left(y,\,F''(x)^*\,\frac{F'(x)}{\|F'(x)\|}\right)\qquad\blacksquare$$

We now apply corollary 3.6 to the function $\|F'\|$. We obtain a sequence x_n along which the derivatives goes to zero

$$F''(x_n)^* \frac{F'(x_n)}{\|F'(x_n)\|} \to 0 \qquad \text{in } V^*$$

contradicting assumption (25). Hence, the result. ■

6. GENERIC DIFFERENTIABILITY

From now on, U will be any lower semicontinuous function, but X will be a special kind of Banach space.

DEFINITION 1

A Banach space X is smooth if there exists a continuous function $\Phi: X \to \mathbb{R}$ such that

(1) $$\Phi(x) \geqslant 0 \qquad \textit{for all } x \in X$$

(2) $$D = \{x | \Phi(x) > 0\} \textit{ is bounded and nonempty.}$$

(3) $$\Phi \textit{ is Fréchet differentiable on } D.$$ ▲

Combining a translation with a homothety, we can assume that $\Phi(0) > 0$ and D is as small as need be. In the sequel, we shall use the function $\Psi = 1/\Phi$. It is well defined and lower semicontinuous as a mapping from X to $\mathbb{R} \cup \{+\infty\}$ and Fréchet differentiable on D.

PROPOSITION 2

Any Banach space X on which there is an equivalent norm, Fréchet differentiable on $X \setminus \{0\}$ is smooth. If X is reflexive, or if X^* is separable, then X is smooth. The spaces ℓ^1 and ℓ^∞ are not nor is any space that contains one of them. ▲

Proof. Denote by $\|x\|$ this norm on X. Pick any C^∞ function $\alpha: R \to [0, \infty)$ with $\alpha(1) > 0$ and $\alpha(t) = 0$ for $t \leqslant 1/2$ and $t \geqslant 2$. Set $\Phi(x) := \alpha(\|x\|)$.

The second part of the proposition follows from standard renorming theorems due to Kadec, Klee, and Asplund and John and Zizler, to be found in Diestel [1975] Chapter 4, Sections 4 and 9, respectively. ■

On smooth Banach spaces, all lower semicontinuous functions enjoy properties that are related to differentiability; namely, those that follow.

DEFINITION 3

Let $\varepsilon \geqslant 0$ be given. We say that a continuous linear functional $p \in X^*$ is ε supporting to U at the point $x \in X$ if $U(x) < +\infty$ and there is some $\eta > 0$ such that

$$\|x-y\| \leqslant \eta \Rightarrow U(y) \geqslant U(x) + \langle p, y-x \rangle - \varepsilon \|x-y\| \tag{4}$$

The set of all such p is called the ε support of U at x and denoted by $S_\varepsilon U(x)$. If it is nonempty, we say that U is ε supported at x. ▲

The following properties are easy consequences:

$$\begin{cases} \textbf{i.} & S_\varepsilon U(x) \neq \varnothing \Rightarrow U \text{ is lower semicontinuous at } x. \\ \textbf{ii.} & S_\varepsilon U(x) \text{ is a convex subset of } X^*. \\ \textbf{iii.} & S_\varepsilon U(x) + S_\alpha V(x) \subset S_{\varepsilon+\alpha}(U+V)(x) \\ \textbf{iv.} & \alpha \geqslant \varepsilon \Rightarrow S_\alpha U(x) \supset S_\varepsilon U(x) \end{cases} \tag{5}$$

The relationship with Fréchet differentiability is given in the following result.

PROPOSITION 4

The function U is Fréchet differentiable at x if and only if for every $\varepsilon > 0$, both U and $-U$ are ε supported at x. We then have

$$\bigcap_{\varepsilon>0} S_\varepsilon U(x) \bigcap_{\varepsilon>0} -S_\varepsilon(-U)(x) = \{U'(x)\} \tag{6}$$ ▲

Proof. The only if part follows immediately from the definitions. The converse is less obvious.

Set $\varepsilon = n^{-1}$ and assume U and $-U$ are ε supported at x. We find for every n some $\eta_n > 0$ and two continuous linear functionals p_n and q_n such that for

$$\|x-y\| \leqslant \eta_n$$

we have:

$$\begin{cases} \textbf{i.} & U(y) \geqslant U(x) + \langle p_n, y-x \rangle - n^{-1}\|x-y\| \\ \textbf{ii.} & -U(y) \geqslant -U(x) + \langle q_n, y-x \rangle - n^{-1}\|x-y\| \end{cases} \tag{7}$$

Write the first inequality for some $m \geqslant n$, and add it to the second. We obtain

$$\|x-y\| \leqslant \eta_m \Rightarrow \langle p_m + q_n, y-x \rangle \leqslant 2n^{-1}\|x-y\|$$

and hence,

$$\forall n \in \mathbb{N}, \qquad \forall m \geqslant n, \qquad \|p_m + q_n\|_* \leqslant 2n^{-1} \tag{8}$$

Similarly,

$$\forall n \in \mathbb{N}, \qquad \forall m \geqslant n, \qquad \|q_m + p_n\|_* \leqslant 2n^{-1} \tag{9}$$

This proves that both p_n and q_n are Cauchy sequences. Since X^* is complete, they converge to p and q, respectively. Taking limits in the preceding inequalities, we see that $p+q=0$.

We claim that p is the Fréchet derivative of U at x. To see this, take any $\varepsilon>0$. Choose $n \geqslant 3\varepsilon^{-1}$, and take $\eta=\min(\eta_n, \varepsilon_n)$. By the preceding inequalities, we have for $\|y-x\| \leqslant \eta$

$$\langle p_n, y-x\rangle - n^{-1}\|x-y\| \leqslant U(y)-U(x) \leqslant \langle q_n, y-x\rangle + n^{-1}\|x-y\|$$

Letting $m \to \infty$, and taking limits in $\|p_m+q_n\|_*$ and $\|p_n+q_m\|_*$, we obtain

$$\|p+q_n\| \leqslant 2n^{-1} \qquad \text{and} \qquad \|q+p_n\| \leqslant 2n^{-1} \tag{10}$$

Substituting this into the preceding inequality, with $\varepsilon \geqslant 3n^{-1}$, we have for all y such that $\|x-y\| \leqslant \eta$,

$$\langle p, y-x\rangle - \varepsilon\|x-y\| \leqslant U(y)-U(x) \leqslant \langle q, y-x\rangle + \varepsilon\|x-y\|$$

which is precisely the definition of Fréchet differentiability. ■

We now state the main result in theorem 5.

THEOREM 5

Let X be a smooth Banach space and U: $X \to \mathbb{R} \cup \{+\infty\}$ a lower semicontinuous function on X. For any $\varepsilon>0$, the set of points x where U is ε supported is dense in Dom U. ▲

Proof. We are given a point x_0 in Dom U (so that $U(x)<+\infty$), a neighborhood $\mathscr{W}$ of the origin in X, and we seek a point $\bar{x} \in \mathscr{W}+x_0$ where U is locally ε supported.

Since U is lower semicontinuous, we can find a smaller neighborhood of the origin $\mathscr{V} \subset \mathscr{W}$ such that U is bounded from below on $\mathscr{V}+x_0$

$$\exists m\colon \ \forall x \in \mathscr{W}+x_0, \qquad U(x) \geqslant m$$

Take a function Φ: $X \to \mathbb{R}$ as in definition 1, assuming $\Phi(0)=0$ and $D \subset \mathscr{V}$ and define a lower semicontinuous function Ψ: $X \to \mathbb{R} \cup \{+\infty\}$ by

$$\Psi(x)=\frac{1}{\Phi(x-x_0)}$$

Set $F=U+\Psi$. The function F is lower semicontinuous and bounded from below on the Banach space X. Applying the ε-variational principle with $k=\frac{1}{2}$, we obtain some point $\bar{x}$ where

$$\forall x \in X, \qquad F(x) \geqslant F(\bar{x}) - \frac{\varepsilon\|x-\bar{x}\|}{2}$$

This implies that zero is locally $\varepsilon/2$ supporting to F at $\bar{x}$. Let $p=\Phi'(\bar{x})$, so that $-p$ is locally $\varepsilon/2$ supporting to $-\Phi$ at $\bar{x}$. It follows that $0-p$ is locally ε supporting to $F-\Phi=U$ at $\bar{x}$. Moreover,

$$F(\bar{x}) \leqslant F(x_0) + \frac{\varepsilon\|x_0-\bar{x}\|}{2} < +\infty$$

so that

$$\bar{x} \in \text{Dom } F = \text{Dom } U \cap \text{Dom } \Psi \subset \text{Dom } U \cap (\mathscr{V} + x_0) \qquad \blacksquare$$

In the sequel, we shall prove a similar result for $\varepsilon=0$, by slightly strengthening the assumption on X (see proposition 7.7).

We apply the preceding result to the special case of convex functions. We wish to prove the following result.

THEOREM 6

Let X be a smooth Banach space and U: $X \to \mathbb{R} \cup \{+\infty\}$ a convex, lower semicontinuous function. Assume U is finite (and hence continuous) on some open convex subset $\Omega \subset X$. Then U is Fréchet differentiable at all points x of some residual subset of Ω. ▲

Let us begin by noting that since U is continuous at every point $x \in \Omega$, it is also subdifferentiable: $\partial U(x) \neq \varnothing$. This gives us indications on $S_\varepsilon U(x)$ and $S_\varepsilon(-U)(x)$.

LEMMA 7

Let q be a subgradient of U at x. Then q is zero supporting to U at x. Moreover, if p is ε supporting to $-U$ at x, then $\|p+q\| \leqslant \varepsilon$. ▲

Proof. We have by definition

$$\forall y \in X, \qquad U(y) \geqslant U(x) + \langle q, y-x \rangle$$

So q is zero supporting to U. If p is ε supporting to $-U$, we have for some $\eta>0$

$$\|y-x\| < \eta \Rightarrow -U(y) \geqslant -U(x) + \langle p, y-x \rangle - \varepsilon\|y-x\|$$

Adding, we obtain

$$\langle p+q, y-x\rangle \leqslant \varepsilon\|y-x\| \qquad \text{for} \qquad \|y-x\|<\eta$$

Hence the result. ■

Using proposition 4, we see that all we have to do is to find a dense G_δ subset A of Ω such that for all $x \in A$ and $\varepsilon>0$, the function $-U$ is ε supported at x. We proceed to do this by defining a sequence A_n of open dense subsets of Ω, whose intersection will be A.

For each $n \geqslant 1$, define A_n to be the set of all points $x \in \Omega$ such that for some $\delta>0$, we have

$$\left.\begin{array}{lll}\|x-x_1\|<\delta & \text{and} & p_1 \in S_{1/n}(-U)(x_1) \\ \|x-x_2\|<\delta & \text{and} & p_2 \in S_{1/n}(-U)(x_2)\end{array}\right\} \Rightarrow \|p_1-p_2\| \leqslant \frac{8}{n}$$

It is clear from the definition that A_n is an open set.

LEMMA 8
Any point $x \in \Omega$ where $-U$ is $1/n$ supported belongs to A_n. ▲

Proof. Take $x \in \Omega$ and $p \in S_{1/n}(-U)(x)$. By definition, there is some $\eta>0$ such that

$$\|y-x\|<\eta \Rightarrow -U(y) \geqslant -U(x)+\langle p, y-x\rangle - n^{-1}\|y-x\|$$

On the other hand, $\partial U(x)$ is not empty, which means that there is some $q \in X^*$ with

$$\forall y, \quad U(y) \geqslant U(x)+\langle q, y-x\rangle$$

By lemma 7, we have $\|p+q\| \leqslant n^{-1}$, and we rewrite the inequality as follows:

$$\forall y, \quad U(y) \geqslant U(x)-\langle p, y-x\rangle - n^{-1}\|y-x\|$$

Finally, we have proved that

$$\|x-y\|<\eta \Rightarrow |U(y)-U(x)-\langle p, y-x\rangle| \leqslant n^{-1}\|y-x\|$$

Now choose $x' \in \Omega$ such that $\|x'-x\|<\eta/2$. The function U is still continuous, and hence subdifferentiable, at x'. Taking any $q' \in \partial U(x')$, we have

$$\forall y, \quad U(y) \geqslant U(x')+q', y-x'\rangle$$

For any y such that $\|x'-y\|<\eta/2$, we can use both preceding inequalities.

This leads us to

$$\begin{aligned}\langle q', y-x'\rangle &\leqslant [U(y)-U(x)]+[U(x)-U(x')] \\ &\leqslant \langle -p, y-x'\rangle + n^{-1}\|y-x\| + n^{-1}\|y-x'\|\end{aligned}$$

Restricting ourselves to vectors y with $\|x'-x\| \leqslant \|x'-y\| < \eta/2$, we obtain the inequality

$$\langle q'+p, y-x'\rangle \leqslant 3n^{-1}\|y-x'\|$$

which implies

$$\|p+q'\|_* \leqslant 3n^{-1}$$

Now if p' belongs to $S_{1/n}(-U)(x')$, we have $\|p'+q'\| \leqslant n^{-1}$ by lemma 7 and hence, $\|p-p'\| \leqslant 4n^{-1}$.

If $x'' \in \Omega$ is another point where $\|x''-x\| < \eta/2$, and $-U$ is $1/n$ supported, we also have $\|p-p''\| \leqslant 4n^{-1}$ for all $p'' \in S_{1/n}(-U)(x'')$. Hence, $\|p'-p''\| \leqslant 8n^{-1}$, which means x belongs to A_n. ■

By theorem 5, the set A_n is dense in Ω. Thus we obtain a sequence A_n of open and dense subsets of Ω. Define

$$A = \bigcap_{n=1}^{\infty} A_n$$

It is a G_δ subset of Ω and dense by Baire's theorem. We claim it has the desired property, given in lemma 9.

LEMMA 9

The function U is Fréchet differentiable at each point of A. ▲

Proof. Take $x \in A$. All we have to show is that $-U$ is $1/n$ supported at x for all n: Since U is zero supported, differentiability will follow from proposition 4.

By definition, for each n there is some $\delta_n > 0$ such that the set

$$K_n = \{S_{1/n}(-U)(y) \mid \|y-x\| < \delta_n\}$$

has diameter less than $8/n$. Letting $n \to \infty$ and assuming the sequence δ_n to be decreasing, we see that the closures $\overline{K_n}$ build up a nested sequence of closed subsets whose diameters go to zero. Since X^* is complete, their intersection is a singleton

$$\exists \bar{p} \in X^*: \bigcap_{n=1}^{\infty} \bar{K}_n = \{\bar{p}\}$$

This implies that for any n, for all $y \in \Omega$ with $\|x-y\|<\delta_n$, and all $p \in S_{1/n}(-U)(y)$, we have $\|\bar{p}-p\|_* \leqslant 9n^{-1}$. Take any $q \in \partial U(y)$; by lemma 7, we know that $\|q+p\| \leqslant n^{-1}$, which yields

$$\begin{aligned} U(y) &\leqslant U(x) - \langle q, x-y\rangle \\ &\leqslant U(x) + \langle p, x-y\rangle + n^{-1}\|x-y\| \\ &\leqslant U(x) + \langle \bar{p}, x-y\rangle + 10n^{-1}\|x-y\| \end{aligned}$$

By theorem 5, the set of y with $S_{1/n}(-U)(y) \neq \varnothing$ is dense in Ω. Since U is continuous, the preceding inequality extends to all y such that $\|x-y\|<\delta_n$, which means precisely that $-U$ is $10n^{-1}$ supported by $\bar{p}$ at x. The result now follows. ■

7. PERTURBED OPTIMIZATION PROBLEMS

Let $U: X \to \mathbb{R} \cup \{+\infty\}$ be a lower semicontinuous function (not necessarily convex). Assume it is bounded from below

$$\inf_{x} U > -\infty$$

so that the conjugate function $U^*: X^* \to \mathbb{R} \cup \{+\infty\}$ is proper and $0 \in \operatorname{Dom} U^*$. By definition, U^* is convex; if X^* happens to be smooth, the function U^* will be Fréchet differentiable almost everywhere on Dom U^* by theorem 6.6. What does it mean for U itself?

From an extensive analysis by Asplund, we extract the relevant information, given in lemma 1.

LEMMA 1

Assume U^* is Fréchet differentiable at $p \in X^*$ and that the derivative is an element x of X. Then for any sequence $x_n \in X$ such that

$$U(x_n) - \langle p, x_n\rangle \to -U^*(p)$$

we have

$$\|x_n - x\| \to 0$$ ▲

Proof. Define $\gamma^*: [0, +\infty) \to [0, +\infty) \cup \{+\infty\}$ by

$$\gamma^*(t) := \sup\{U^*(q) - U^*(p) - \langle x, q-p\rangle \mid \|q-p\|_* \leqslant t\}$$

(p is fixed). It is easily checked that γ^* is convex and lower semicontinuous.

Since U^* is Fréchet differentiable at p, we have $t^{-1}\gamma^*(t) \to 0$ when $t \to 0$. The conjugate function of γ^* is given by

$$\gamma(s) = \sup\{st - \gamma^*(t) | t \geqslant 0\}$$

It is another convex, lower semicontinuous function on $[0, +\infty)$, and $\gamma(s) > 0$ for all $s \neq 0$. We rewrite the inequality

$$U^*(q) \leqslant U^*(p) + \langle x, q-p \rangle + \gamma^*(\|q-p\|_*) \qquad \text{for all } q \in X^*$$

as follows:

$$U^*(q+p) \leqslant U^*(p) + \langle x, q \rangle + \gamma^*(\|q\|_*) \qquad \text{for all } q \in X^*$$

and take convex conjugates. We obtain

$$U^{**}(y) - \langle p, y \rangle \geqslant -U^*(p) + \gamma(\|y - x\|) \qquad \text{for all } y \in X$$

Since $U^{**} \leqslant U$, it follows that

$$U(y) - \langle p, y \rangle \geqslant -U^*(p) + \gamma(\|y - x\|) \qquad \text{for all } y \in X$$

If x_n is a sequence such that $U(x_n) - \langle p, x_n \rangle$ converges to $-U^*(p)$, the preceding inequality yields $\gamma(\|x_n - x\|) \to 0$, which implies $\|x_n - x\| \to 0$ as desired.

We can now combine this with theorem 6.6 to obtain results for closed sets and lower semicontinuous functionals. We first give a definition.

DEFINITION 2

Let A be a subset of some Banach space X and let x be some point in A. We say that x is a strongly exposed point of A if there exists some $p \in X^*$ such that

$$\left.\begin{array}{l} x_n \in A \quad \text{for all } n \in \mathbb{N}, \quad \text{and} \\ \langle p, x_n \rangle \to \inf\limits_{x \in A} \langle p, x \rangle \end{array}\right\} \Rightarrow x_n \to x$$

Any p with this property is said to expose x. ▲

PROPOSITION 3

Assume X^* is a smooth Banach space and $A \subset X$ is closed and bounded. Then there is a residual subset G of X^* such that all $p \in G$ expose some $x \in A$. ▲

Proof. Take $U(x) = \psi_A(x)$, the indicator function of A. It is lower semicontinuous, since A is closed, and U^* is everywhere finite, and hence continuous, since A is bounded. Now apply theorem 6.6 and lemma 1. ■

PROPOSITION 4

Assume X^* is a smooth Banach space and $A \subset X$ is closed, bounded, and convex. Then A is the closed convex hull of its strongly exposed points. ▲

Proof. Let $E \subset A$ be the set of exposed points, and consider $\overline{\text{co}}\, E$. Obviously $\overline{\text{co}}\, E \subset A$. If $\overline{\text{co}}\, E \neq A$, there is some point $x \in A$, $x \notin \overline{\text{co}}\, E$. Separating x from $\overline{\text{co}}\, E$ by the Hahn–Banach theorem, we obtain some $p \in X^*$ and $\alpha \in \mathbb{R}$ such that

$$\langle p, x\rangle < \alpha < \langle p, y\rangle \qquad \text{for all } y \in \overline{\text{co}}\, E$$

Since A is bounded, $\|y\| \leq m$ for all $y \in A$ say, choosing another $q \in X^*$ will perturb these inequalities into

$$\langle q, x\rangle = \langle p, x\rangle + \langle q-p, x\rangle = a'$$

$$\langle q, y\rangle > a - \|p-q\| m = a'' \qquad \text{for all } y \in \overline{\text{co}}\, E$$

For $\|q-p\|$ small enough, $a' < a''$, so q will still separate x from $\overline{\text{co}}\, E$. By proposition 3, there will be some $\bar{q}$ separating x from $\overline{\text{co}}\, E$ and exposing some point $\bar{x}$ in A. So $\bar{x} \in E$, and

$$\langle \bar{q}, \bar{x}\rangle = \inf_{y \in A} \langle \bar{q}, y\rangle \leq \langle \bar{q}, x\rangle$$

But this contradicts the fact that $\bar{q}$ separates x from $\overline{\text{co}}\, E$. Hence the result. ■

We now turn to functions U on X, with the aim of refining theorem 6.5. If X is a smooth Banach space and $\Phi: X \to \mathbb{R}$ satisfies the requirements of definition 6.1, let us agree to call functions of type Φ^{-1} all functions $\Psi: X \to \mathbb{R} \cup \{+\infty\}$, which can be written as

$$\Psi(x) := c + \langle p, x\rangle + a\Phi^{-1}(m(x - x_0))$$

for some $c \in \mathbb{R}$, $p \in X^*$, $a \in \mathbb{R}$, $m \in \mathbb{R}$, $x_0 \in X$.

THEOREM 5

Let X be a Banach space and $U: X \to \mathbb{R} \cup \{+\infty\}$ a lower semicontinuous function on X. Assume X and X^* are smooth. Then there is a dense subset D of Dom U *such that whenever $\bar{x} \in D$, there is a function Ψ of type Φ^{-1} with*

$$(1) \qquad \begin{cases} \textbf{i.} \ U(x) \geq \Psi(x) & \textit{for all } x \in X \\ \textbf{ii.} \ U(\bar{x}) = \Psi(\bar{x}) \end{cases}$$ ▲

Proof. Define $F = U + \Psi$ as in the proof of Theorem 6.5. Now consider the epigraph of F. It is a closed subset of $X \to \mathbb{R}$, but it is not bounded.

However, it is easy to see that it has strongly exposed points. Indeed, consider its indicator function $\psi_{Ep(F)}(x, a)$; we have

$$\sigma(Ep(F), (p, -b)) = \sup\{\langle p, x\rangle - bF(x) | x \in X\}$$
$$< \infty \text{ provided } b > 0$$

Applying lemma 1 and theorem 6.6, we see that the set of $(p, -b)$ that expose some point (x, a) of $Ep(F)$ is a dense G_δ in $]0, \infty[\times X^*$. Let us choose $(\bar{p}, -\bar{b})$, with $\bar{b} > 0$ and the corresponding $(\bar{x}, \bar{a})$. It is easy to see that $\bar{a} = F(\bar{x})$, so that

$$(2) \qquad \begin{cases} \textbf{i.} & \langle \bar{p}, \bar{x}\rangle - \bar{b}F(\bar{x}) \geqslant \langle \bar{p}, x\rangle - \bar{b}F(x) \\ \textbf{ii.} & F(x) \geqslant F(\bar{x}) + \langle x - \bar{x}, \bar{p}/\bar{b}\rangle \qquad \text{for all } x \in X \end{cases}$$

Now write $U = F - \Psi$ to obtain

$$(3) \qquad U(x) \geqslant \langle x - \bar{x}, \bar{p}/\bar{b}\rangle + U(\bar{x}) + \Psi(\bar{x}) - \Psi(x) \qquad \text{for all } x \in X$$

The right side is a function of type Φ^{-1}, which agrees with U at $x = \bar{x}$. Hence the result. ■

COROLLARY 6

Let $\bar{x}$ be any point in D. Then there is some $\bar{p} \in X^*$ such that for any $\varepsilon > 0$, there is some $\eta > 0$ with

$$(4) \qquad \|x - \bar{x}\| \leqslant \eta \Rightarrow U(x) \geqslant U(\bar{x}) + \langle \bar{p}, x - \bar{x}\rangle - \varepsilon\|x - \bar{x}\| \qquad \blacktriangle$$

Proof. Any function of type Φ^{-1} is Fréchet differentiable on its domain. The result then follows from theorem 5, with $\bar{p} = \Phi'(\bar{x})$ ■

We now turn our attention and efforts to perturbed optimization problems.

We are given two Banach spaces X (parameter space) and V (state space), an open subset $\Omega \subset X$, and a function

$$F: \Omega \times V \to \mathbb{R} \cup \{+\infty\}$$

such that

$$(5) \qquad F \text{ is lower semicontinuous on } \Omega \times V$$

$$(6) \qquad F(x, v) \text{ finite} \Rightarrow F(\cdot, v) \text{ is Gâteaux differentiable at } x.$$

We define

$$U(x) = \inf\{F(x, v) | v \in V\}$$

If this infimum is finite, we ask ourselves whether it is attained. In other words, with every $x \in X$, we associate the optimization problem

$$(\mathscr{P}_x) \quad \inf_{v \in V} F(x, v)$$

A *solution* of $(\mathscr{P}_x)$ is a point $v \in V$ where

$$F(x, v) = U(x)$$

It is a well known fact from optimization theory that the existence of a solution for $(\mathscr{P}_x)$ is closely related to the subdifferentiability of U at x (uniqueness being related to differentiability). Let us give a precise statement.

THEOREM 7

Assume that at some point $\bar{x} \in X$, there is an open neighborhood $\mathscr{N}$ of $\bar{x}$ and a function Ψ: $\mathscr{N} \to \mathbb{R}$ such that

$$\Psi \text{ is } C^1 \tag{7}$$

$$\Psi(\bar{x}) = U(\bar{x}) \tag{8}$$

$$\Psi(x) \leqslant U(x) \qquad \text{for all } x \in \mathscr{N} \tag{9}$$

Then for any minimizing sequence of v_n of $(\mathscr{P}_x)$,

$$F(\bar{x}, v_n) \to U(\bar{x}) \tag{10}$$

there is a sequence $x_n \to \bar{x}$ in $\mathscr{N}$ such that

$$\begin{cases} \textbf{i.} & F(x_n, v_n) \to U(\bar{x}) \text{ in } \mathbb{R} \\ \textbf{ii.} & F'_x(x_n, v_n) \to \Psi'(\bar{x}) \text{ in } X^* \end{cases} \tag{11}$$

▲

Proof. Define $\varepsilon_n = F(\bar{x}, v_n) - \Psi(\bar{x})$, so that $\varepsilon_n \geqslant 0$ and $\varepsilon_n \to 0$. For fixed n, define a function G_n on $\mathscr{N}$ by

$$G_n(x) = F(x, v_n) - \Psi(x)$$

This is a non-negative Gâteaux-differentiable, lower semicontinuous function on $\mathscr{N}$. We can assume n to be so large that the closed ball B_n with center $\bar{x}$ and radius $2\sqrt{\varepsilon_n}$ is contained in $\mathscr{N}$ and apply the ε variational principle to G on B_n, starting from the point $\bar{x}$. We obtain a point x_n such that

$$F(x_n, v_n) - \Psi(x_n) = G_n(x_n) \leq G_n(\bar{x}) = \varepsilon_n$$

$$\|x_n - \bar{x}\| \leq \sqrt{\varepsilon_n}$$

$$\|G_n'(x_n)\|_* = \|F_x'(x_n, v_n) - \Psi'(x_n)\|_* \leq \sqrt{\varepsilon_n}$$

Letting $n \to 0$, we obtain the desired result. ■

COROLLARY 8

Assume moreover that the family $F_x'(\cdot, v_n)$, $n \in \mathbb{N}$, is equicontinuous at $\bar{x}$

$$(12) \qquad \forall \varepsilon > 0, \qquad \exists \eta > 0: \|x - \bar{x}\| \leq \eta \Rightarrow \|F_x'(x, v_n) - F_x'(\bar{x}, v_n)\|_* \leq \varepsilon \qquad \forall n$$

Then, $F'_x(\bar{x}, v_n) \to \Psi'(\bar{x})$. ▲

Proof. Write

$$\|F_x'(\bar{x}, v_n) - \Psi'(\bar{x})\|_* \leq \|F_x'(x_n, v_n) - F_x'(\bar{x}, v_n)\|_* + \|F_x'(x_n, v_n) - \Psi'(\bar{x})\|_*$$

The second term on the right goes to zero by condition (11, ii) of theorem 7, and the first term also goes to zero because $x_n \to \bar{x}$ and the family $F'_x(\cdot, v_n)$ is equicontinuous. ■

Certainly corollary 8 is more transparent than previous statements: It tells us that minimizing sequences v_n at $\bar{x}$ have to be such that $F_x'(\bar{x}, v_n)$ converges strongly in X^*. This is a considerable amount of additional information and can be used in many cases to prove that the sequence v_n itself converges. For the sequel, recall that a map between metric space is *proper* if any sequence whose image converges has a convergent subsequence.

PROPOSITION 9

Assume that for some point $\bar{x} \in \Omega$, there is a function $\Psi: X \to \mathbb{R}$, which is C^1 in some neighborhood of $\bar{x}$ and satisfies

$$(13) \qquad \Psi(\bar{x}) = U(\bar{x}) \quad \text{and} \quad \Psi(x) \leq U(x) \quad \text{for all } x \in \Omega$$

Assume moreover that there is some $\varepsilon > 0$ such that the family

$$\{F_x'(\cdot, v) | F(\bar{x}, v) \leq U(\bar{x}) + \varepsilon\}$$

is equicontinuous at $\bar{x}$. Assume finally that the map $F_x'(\bar{x}, \cdot)$ from V to X^* is proper. Then problem $(\mathscr{P}_{\bar{x}})$ has at least one solution $\bar{x}$. If in addition U is Gâteaux differentiable at $\bar{x}$ and the map $F_x'(\bar{x}, \cdot)$ is one to one, the solution is unique. ▲

Proof. Any sequence v_n such that $F'_x(\bar{x}, v_n)$ converges must have convergent subsequences, $v_{n_k} \to \bar{v}$, say. If we take a minimizing sequence for v_n, we obtain a minimizer for $\bar{v}$, since F is lower semicontinuous.

For the last part, just note that we must have $\Psi'(\bar{x}) = U'(\bar{x})$, and the equation

$$F'_x(\bar{x}, v) = U'(\bar{x})$$

defines v uniquely. ■

For instance, if U is C^1 on some open set, we can take $U = \Psi$ on that set. Only rarely do we come across such regular behavior of U. On the other hand, it is often the case that U is continuous, which motivates the following proposition.

PROPOSITION 10

Assume that X is reflexive and there is an open subset $\Omega \subset X$ on which U is continuous. Assume moreover that whenever we have sequences x_n in Ω and v_n in V such that the sequence

$$(x_n, F(x_n, v_n), F'_x(x_n, v_n))$$

converges in $X \times \mathbb{R} \times X^*$, then the sequence v_n must be compact. Then there is a dense subset $D \subset \Omega$ such that whenever $x \in D$, the problem $(\mathscr{P}_x)$ has a solution.

Proof. Since X is reflexive, then X and X^* are smooth, and we can apply theorem 6.5. Now norms that are Fréchet differentiable away from the origin are in fact C^1 (see Diestel, Chapter 11, Section 2). It follows that function of type Φ^{-1} are C^1 if we build Φ from a differentiable norm. So the conditions in theorem 7 are fulfilled at all points of a dense subset; we conclude as in proposition 9. ■

If we slightly strengthen the assumptions, we obtain uniqueness in a strong way.

PROPOSITION 11

Assume that X is reflexive and there is an open subset $\Omega \subset X$ on which U is continuous. Assume moreover that whenever we have sequences x_n in Ω and v_n in V such that

$$(x_n, F(x_n, v_n), F'_x(x_n, v_n))$$

converges in $X \times \mathbb{R} \times X^*$, then v_n converges in V.

Then there is a dense subset $D \subset \Omega$ such that whenever $x \in D$, the problem $(\mathscr{P}_x)$ has a unique solution v, and all minimizing sequences of $(\mathscr{P}_x)$ converge to v. ▲

Proof. Existence at all points of D follows from proposition 10. Take $x \in D$; if $(\mathscr{P}_x)$ had two solutions v_1 and v_2, we could construct a minimizing sequence by alternatively setting $v_n = v_1$ and $v_n = v_2$. By theorem 7, we would get a sequence $x_n \to x$ with $F'_x(x_n, v_n)$ converging in X^*. But then v_n itself would converge, which is absurd, unless $v_1 = v_2$. So there is only one minimizer. If v_n now is any minimizing sequence, the $F'_x(x_n, v_n)$ will converge, so v_n itself will converge, and the limit must be the minimizer. ■

If we throw in equicontinuity, we can strengthen the result still further. In the following proposition, we keep the notations and assumptions of proposition 11.

PROPOSITION 12

Assume in addition that, for every $x \in \Omega$, there is some $\alpha > 0$ such that the family

$$\{F'_x(\cdot, v) | F(x, v) \leqslant U(x) + \alpha\}$$

is equicontinuous at x. Then the set D of existence and uniqueness contains a dense G_δ of Ω. ▲

Proof. Density is the previous result; to obtain a dense G_δ, we have to use Baire's theorem. To do this, take any $\varepsilon > 0$. We shall say that x has the property P_ε, or simply $P_\varepsilon(x)$, if all minimizing sequences of $(\mathscr{P}_x)$ are eventually confined within some ball of radius $< \varepsilon$.

$$\exists r < \varepsilon\colon [F(x, v_n) \to U(x)] \Rightarrow [\exists N\colon \|F'_x(x, v_n) - F'_x(x, v_m)\|_* \leqslant r \qquad \forall n, m \geqslant N]$$

Set $\Omega_\varepsilon = \{x | P_\varepsilon(x)\}$. Clearly $\Omega_\varepsilon \supset D$, so Ω_ε is dense. On the other hand, $P_\varepsilon(x)$ is easily seen to be equivalent to:

$$\exists r < \varepsilon, \qquad \exists \delta > 0\colon \begin{bmatrix} F(x, v) - U(x) \leqslant \delta \\ \text{and} \\ F(x, w) - U(x) \leqslant \delta \end{bmatrix} \Rightarrow \|F'_x(x, v) - F'_x(x, w)\|_* \leqslant r$$

In this form, it is clear from the equicontinuity that Ω_ε is open. The set of existence and uniqueness then certainly contains

$$\bigcap_{n \in \mathbb{N}} \Omega_{1/n}$$

which is a dense G_δ by Baire's theorem. ■

Note that in none of these propositions have our assumptions been strong enough to ensure that any individual problem $(\mathscr{P}_x)$ has a solution. We are thus in a position where we know that for most, or at least many, values of the parameter x the problem can be solved, without being able to pinpoint any of them.

Let us conclude with an example. Let A be a closed subset of a Banach space X, let $x \in X$ be given. We seek the point in A closest to x if there is any.

The problem is set up as follows:

$$(\mathscr{P}_x) \quad \inf_{y \in Y} \{\|x-y\| + \psi_A(y)$$

So $V = X$, $F(x, y) = \|x-y\| + \psi_A(y)$, and $U(x)$ obviously is the distance from x to A

$$U(x) = d(x, A)$$

It is known to be continuous everywhere. The function F itself is lower semicontinuous. If X is reflexive and the norm is chosen to be Fréchet differentiable away from the origin, F is Fréchet differentiable at all points $x \neq y$, with derivative

$$F'_x(x, y) =: j(x-y)$$

Set $\Omega = X \setminus A$. If $x \in \Omega$ and $F(x, y) < +\infty$, we must have $y \in A$, so $x \neq y$ and $F(\cdot, y)$ is differentiable at x; therefore, assumption (6) is satisfied.

Suppose we have sequences x_n in Ω and y_n in X such that

$$x_n \to x \qquad \text{in } \Omega$$

$$\|x_n - y_n\| + \psi_A(y_n) \to d(x, A)$$

$$j\,(x_n - y_n) = j\left(\frac{x_n - y_n}{\|x_n - y_n\|}\right) \to p \qquad \text{in } X^*$$

Now j sends the unit sphere $S \subset X$ into the unit sphere $S^* \subset X^*$, and is characterized by $\langle j(z), z\rangle = 1$, all $z \in S$ (see Diestel Chapter 2). Since X is reflexive, its unit ball is weakly compact, so j must be surjective. The derivative $j^*: S^* \to S$ of the dual norm is characterized by $\langle j^*(q), q\rangle = 1$ for all $q \in S^*$. It follows that $j^* = j^{-1}$, and j is a homeomorphism. Thus we have

$$\frac{x_n - y}{\|x_n - y_n\|} \to j^*(p) \qquad \text{in } S$$

Since $x_n \to x \notin A$ and y_n stays in A, it follows that y_n converges to some limit $y \in A$.

All the assumptions of proposition 11 are satisfied, and we obtain a well-known result, stated in proposition 13.

PROPOSITION 13

Assume X is reflexive and the norms of X and X^* are Fréchet differentiable off the origin. Then for any closed subset A of X, there is in $X \setminus A$ a dense set D of points x with a unique projection on A: All sequences $y_n \in A$ such that $\|x-y_n\| \to d(x, A)$ converge to the same limit in A. ▲

If we strengthen the assumptions slightly, we obtain a much better result. Recall that X^* is *uniformly convex* if for every $\varepsilon > 0$, there is some $\delta > 0$ such that

$$p \in S^*, \qquad p' \in S^* \qquad \text{and} \qquad \left\| \frac{p+p'}{2} \right\|_* \geqslant 1-\delta \Rightarrow \|p-p'\|_* \leqslant \varepsilon \tag{14}$$

PROPOSITION 14

Assume in addition that X^* is uniformly convex. Then the subset $D \subset \Omega$ contains a dense G_δ. ▲

Proof. We have to check the additional assumption of proposition 12, namely, that the family $F'_x(\cdot, y)$, for $y \in$, is equicontinuous on Ω.

Fix $x \in \Omega$ and $\varepsilon > 0$, and set

$$F_x(x, y) = j\left(\frac{x-y}{\|x-y\|}\right) = p_y$$

Since X^* is uniformly convex, there is some $\delta > 0$ such that $\|p_y + p\|_* \geqslant 2-\delta$ implies $\|p_y - p\|_* \leqslant \varepsilon$. Take x' in S with $\|x-x'\| \leqslant \delta$, and set $F'_x(x', y) = p'_y$. Then

$$\begin{aligned} \|p_y + p'_y\|_* &\geqslant |\langle p_y + p'_y, x\rangle| \\ &\geqslant |\langle p_y, x\rangle + \langle p'_y, x'\rangle + \langle p'_y, x - x'\rangle| \\ &\geqslant 2-\delta \end{aligned}$$

thereby forcing $\|p_y - p'_y\|_* \leqslant \varepsilon$ independently of y, as desired. ■

CHAPTER 6

Solving Inclusions

This rather long chapter deals with one central theme: solving "inclusions"; that is, when F is a set-valued map from X to Y, finding $\bar{x} \in \text{Dom}(F)$, a solution to

$$y \in F(\bar{x}) \qquad \text{when } y \text{ is given in } Y$$

which offers a wide array of applications.

Curiously enough, mathematical problems arising from game theory motivate important theorems in nonlinear analysis. Relations between game theory and nonlinear analysis are so deep that a detour through game theory does, in fact, save time before presenting nonlinear analysis. Moreover, a deeper insight is gained by building up intuition.

After a short presentation of two-person games in Section 1, we prove in the second section the following theorem.

LOP SIDED MINIMAX THEOREM

Let M be a compact convex subset, N a convex subset, and $f: M \times N \to R$ satisfy

$$\begin{cases} \textbf{i.} \ \ \forall y \in N, & x \to f(x, y) \text{ is convex and lower semicontinuous.} \\ \textbf{ii.} \ \ \forall x \in M, & y \to f(x, y) \text{ is concave.} \end{cases}$$

Then there exists $\bar{x} \in M$ such that

$$\sup_{y \in N} f(\bar{x}, y) = \sup_{y \in N} \inf_{x \in M} f(x, y) \qquad \blacktriangle$$

Observe that this equality implies the minimax equality

$$\inf_{x \in M} \sup_{y \in N} f(x, y) = \sup_{y \in N} \inf_{x \in M} f(x, y)$$

We shall propose a proof where the roles of convexity assumptions, on one hand, and topological assumptions on the other are well separated. We then present a more sophisticated version of the minimax theorem, where the compactness assumptions are dramatically relaxed. In the third section, we drop the convexity assumption of f with respect to x. However, we still prove a minimax equality involving continuous maps between the strategy sets of the two players.

THEOREM

Let M be a compact space, N a convex subset, and $f: M \times N \to R$ satisfy

$$\begin{cases} \textbf{i.} \quad \forall y \in N, & x \to f(x, y) \text{ is lower semicontinuous.} \\ \textbf{ii.} \quad \forall x \in M, & y \to f(x, y) \text{ is concave.} \end{cases}$$

Then there exists $\bar{x} \in M$ such that

$$\sup_{y \in N} f(\bar{x}, y) = \sup_{C \in \mathscr{C}(M,N)} \inf_{x \in M} f(x, C(x)) = \inf_{D \in \mathscr{C}(N,M)} \sup_{y \in N} f(D(y), y) \qquad \blacktriangle$$

This is a less known but very useful equality. It is equivalent to the *Ky Fan inequality*, which plays a crucial role in proving in a very easy way the existence theorems in this chapter.

KY FAN INEQUALITY

Let K be a compact convex subset and $\phi: K \times K \to R$ satisfy

$$\begin{cases} \textbf{i.} \quad \forall y \in K, & x \to \phi(x, y) \text{ is lower semicontinuous.} \\ \textbf{ii.} \quad \forall x \in K, & y \to \phi(x, y) \text{ is concave.} \\ \textbf{iii.} \quad \forall y \in K, & \phi(y, y) \leqslant 0 \end{cases}$$

Then there exists $\bar{x} \in K$ satisfying

$$\sup_{y \in K} \phi(\bar{x}, y) \leqslant 0 \qquad \blacktriangle$$

We proceed by generalizing this inequality to the case where we replace the lower semicontinuity of ϕ with respect to x by the monotonicity condition

$$\forall x, y \in K, \qquad \phi(x, y) + \phi(y, x) \geqslant 0$$

and the lower semicontinuity of ϕ with respect to x for only a very strong topology, called the *finite topology*. Ky Fan's inequality is equivalent to the Brouwer fixed point theorem. Its analytical formulation—unlike the pleasant geometrical form of the Brouwer theorem—makes it more *operational* for proving most of the results in nonlinear analysis. In the fourth section, we investigate the problem of finding a zero of a set-valued map F, solution to the inclusion

$$\exists \bar{x} \in K \qquad \text{such that} \qquad 0 \in F(\bar{x})$$

or a fixed point of F, solution to the inclusion

$$\exists x_* \in K \qquad \text{such that} \qquad x_* \in F(x_*)$$

The common feature is that we shall derive all these results from the Ky Fan inequality, and, therefore, in the final analysis, from the Brouwer fixed point theorem.

All the set-valued maps investigated in this section are upper hemicontinuous maps with closed convex values from a compact subset K of X to a topological vector space Y.

The main theorem of this section states that when $F: K \to X = Y$ satisfies the *tangential condition*

$$\forall x \in K, \qquad F(x) \cap T_K(x) \neq \varnothing$$

then

i. there exists $\bar{x} \in K$, a solution to $0 \in F(\bar{x})$.
ii. $\forall y \in K, \quad \exists \hat{x} \in K$, a solution to $y \in \hat{x} - F(\hat{x})$.

We single out several generalizations and consequences. Among them we obtain the Kakutani fixed point theorem and other fixed point theorems. We follow Poincaré's continuation method for carrying the preceding theorem by homotopy and thus proving an adaptation of the Leray–Schauder theorem to set-valued maps. We shall apply these results to prove the existence of a Walras equilibrium of an exchange economy as well as another concept of equilibrium, which avoids some of the shortcomings of the Walras equilibrium (Section 5).

There are, however, examples of set-valued maps that are not upper semicontinuous; for instance, the subdifferential $x \to \partial U(x)$ of a lower semicontinuous, convex function. Unfortunately, existence theorems in the fourth section do not apply to them. Let us consider an example: K is a weakly compact convex subset of a Hilbert space X, and U is a lower semicontinuous, convex function, related to K by the condition $0 \in \text{Int}(\text{Dom } U - K)$. We know that there exists an element $\bar{x} \in K$ that minimizes U on K, that is a solution to the inclusion

$$(*) \qquad 0 \in \partial U(\bar{x}) + N_K(\bar{x})$$

Therefore, there must be a property of $x \to \partial U(x)$ other than upper hemicontinuity that allows us to solve such an inclusion. This property is monotonicity.

A set-valued map A from X to X is *monotone* if and only if

$$\forall (x, p), \qquad (y, p) \in \text{graph}(A), \qquad \langle p - q, x - y \rangle \geqslant 0.$$

It happens that this algebraic property balances some of the continuity requirements we made in the preceding sections.

Inclusion $(*)$ is a particular case of a class of problems of the form

$$(**) \qquad f \in A(\bar{x}) + \partial V(\bar{x})$$

where A is a monotone map with weakly compact convex values, which we assume to be only finitely upper semicontinuous and where V is a lower semicontinuous, convex function. We shall prove that

$$\operatorname{Int}(\operatorname{Dom} V^* + A \operatorname{Dom} V) \subset \operatorname{Im}(A + \partial V) \subset \operatorname{Dom} V^* + A \operatorname{Dom} V$$

and how more restrictive monotonicity assumptions imply that $A + \partial V$ is surjective.

Subdifferentials of lower semicontinuous, convex functions are actually *maximal monotone*, in the sense that there is no strict monotone extension.

Minty's theorem characterizes maximal monotone maps: They are monotone maps such that $1 + A$ is surjective.

They enjoy many properties. In particular, they can be approximated by Lipschitz single-valued maps A_λ called the **Yosida approximations of** A, which are systematically used when dealing with maximal monotone maps.

We study surjectivity properties of maximal monotone maps and prove that strongly coercive maximal monotone maps are surjective. We also provide conditions under which two maximal monotone set-valued maps A and B satisfy the properties

i. $\operatorname{Int}(\operatorname{Im}(A+B)) = \operatorname{Int}(\operatorname{Im}(A) + \operatorname{Im}(B))$
ii. $\operatorname{cl}(\operatorname{Im}(A+B)) = \operatorname{cl}(\operatorname{Im}(A) + \operatorname{Im}(B))$

We deduce the existence of solutions to inclusions

$$y \in \bar{x} + AB\bar{x}$$

when A and B are maximal monotone maps. We conclude this section with the most important property of maximal monotone maps related to the existence of solutions to differential inclusions

$$x'(t) \in -A(x(t)); \; x(0) = x_0 \text{ is given in } \operatorname{Dom}(A).$$

We prove that there exists a unique solution to the initial value problem, which, amazingly enough, is "lazy."

For almost all t, $-x'(t) = m(A(t))$, the velocity with the smallest norm.

Furthermore, if we denote by $Tx_0 \in \mathscr{C}(0, \infty; X)$ the solution of the preceding initial value problem, we observe that

$$\sup_{t \geq 0} \|T(x_0)(t) - T(x_1)(t)\| \leq \|x_0 - x_1\|$$

1. MAIN CONCEPTS OF GAME THEORY

Let us consider two players, Mike and Nancy. Let M and N be their strategy sets. Our purpose is to *single out pairs* $(x, y) \in M \times N$ *by various optimization methods.* This being said, how can we devise these methods? In this chapter, we suggest deriving them from decision theory, which means providing *decision-makers* mechanisms for selecting elements (called *decisions*) in given subsets (called *decision* sets).

History shows that parlor games provided mathematicians since Blaise Pascal with various problems, so such terms as players (instead of decision makers) and strategies (instead of decisions) were coined quite early, and tradition maintains them. The status of games as an application of mathematical theory is due to von Neumann, who proposed the general framework of conflict and cooperation.

An elementary mechanism that allows Mike and Nancy to choose their respective strategies is obtained by giving them decision rules.

DEFINITION 1

A decision rule for Mike is a set-valued map C_M from N to M. It assigns to each strategy $y \in N$ played by Nancy a strategy $x \in C_M(y)$ that can be implemented by Mike when he knows that Nancy plays y. ▲

Similarly, a decision rule for Nancy is a set-valued map C_N from M to N associating with each strategy $x \in M$ a strategy $y \in C_N(x)$ played by Nancy constrained by the choice $x \in M$ made by Mike.

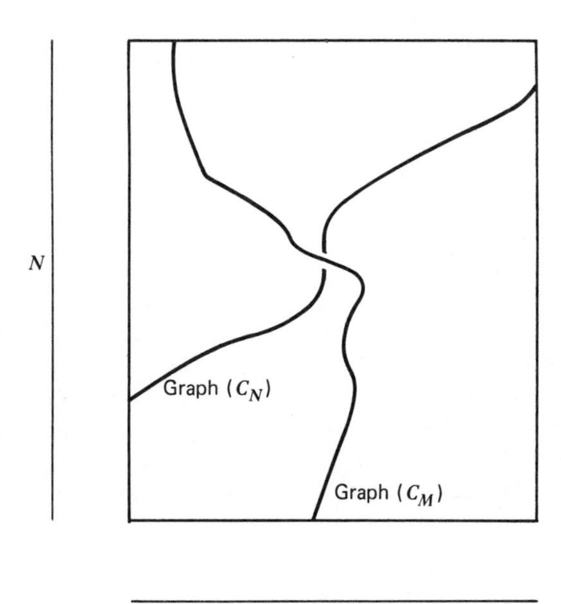

Figure 1. Case of a game with no consistent bistrategies.

Decision rules C_M and C_N being given to Mike and Nancy, we are naturally led to single out pairs of strategies—called bistrategies—that are consistent in the sense that

$$(1) \qquad x \in C_M(y) \quad \text{and} \quad y \in C_N(x)$$

DEFINITION 2

A pair of strategies—or bistrategy—(x, y) satisfying property (1) is said to be a pair of consistent strategies—or a consistent bistrategy—for decision rules C_M and C_N of Mike and Nancy. ▲

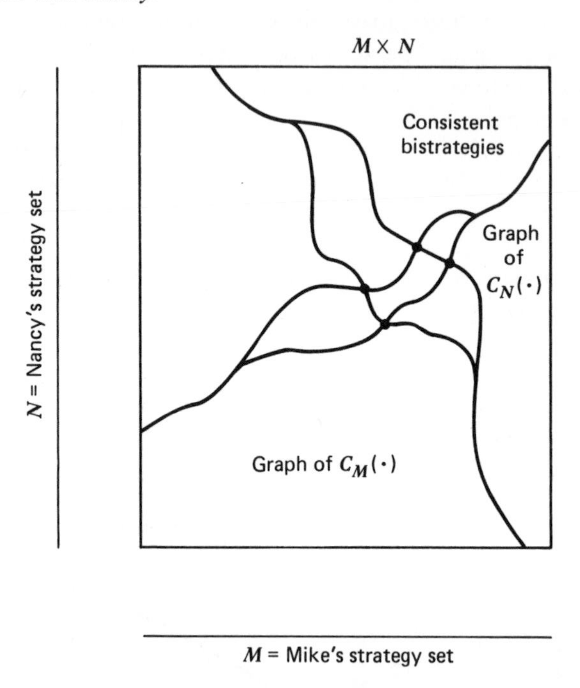

The relevance of such concepts depends on the choice of the decision rules. A trivial example is given by *constant decision rules*. Indeed, we can identify a strategy $x \in M$ of Mike with the constant decision rule $y \in N \rightarrow x \in M$, which describes *stubborn* behavior by Mike, who plays x whatever the strategy played by Nancy is. So, when Mike and Nancy, respectively, play constant decision rules x and y, the associated consistent bistrategy is the pair (x, y). The set of consistent bistrategies may be empty or very large, depending on the properties of C_N and C_M. The problem of finding pairs of consistent strategies amounts to a *fixed point problem*. We denote by $\vec{C}$ the set-valued map from $M \times N$ to itself defined by

$$(2) \qquad \vec{C}(x, y) = C_M(y) \times C_N(x)$$

Inclusions (1) can obviously be written

$$(x, y) \in \vec{C}(x, y) \tag{3}$$

The search for fixed point being a central theme of this book, we shall describe various methods for devising sufficient conditions for the existence of consistent bistrategies. So, decisions rules for Mike and Nancy provide a selection mechanism yielding consistent bistrategies. If it is not sharp enough, in the sense that the set of consistent bistrategies is too large, we may need a further mechanism selecting bistrategies among the consistent ones and so on. Before going further, a first question arises: where do the decision rules come from? How can we construct them?

Games in Normal (or Strategic) Form: Noncooperative Equilibria

The traditional way of presenting game theory is to posit that each player classifies pairs of strategies through a real-valued function. We can think of such a function as a map that associates to each pair of strategies (x, y) its cost, measured by a real number. Since the concept of cost involves the notion of money, which is quite difficult to master in economics, we prefer to call it a loss. So, a player uses a *loss function* $f: M \times N \to R$ to define the *preference preorder on* $M \times N$ as follows:

$$(x_1, y_1) \text{ is preferred to } (x_2, y_2) \text{ if and only if } f(x_1, y_1) \leqslant f(x_2, y_2). \tag{4}$$

Whatever the relevance of this assumption is, we posit from now on that Mike and Nancy select their strategies according to given loss functions $f_M: M \times N \to R$ and $f_N: M \times N \to R$, respectively. We set

$$\vec{f}(x, y) := (f_M(x, y), f_N(x, y)) \in R^2$$

DEFINITION 3
A game in normal (or strategy) form is defined by a map $\vec{f}$ *from* $M \times N \to R^2$, *called the* biloss *map.* ▲

There is a natural way to associate decision rules with a game described by loss functions. Indeed, let f_M be Mike's loss function. If he has the opportunity of knowing a strategy $y \in N$ played by Nancy, he will be tempted to choose a strategy x that minimizes his loss *given Nancy's choice*: In other words, he will choose a strategy in the subset $\bar{C}_M(y)$ defined by

$$\bar{C}_M(y) := \{\bar{x} \in M \mid f_M(\bar{x}, y) = \min_{x \in M} f_M(x, y)\} \tag{5}$$

This set-valued map $\bar{C}_M$ defines a decision rule for Mike. Similarly, Nancy may associate to her loss function f_N the decision rule $\bar{C}_N$ defined by

$$\forall x \in M, \qquad \bar{C}_N(x) := \{\bar{y} \in M \mid f_N(x, \bar{y}) = \min_{y \in N} f_N(x, y)\} \tag{6}$$

DEFINITION 4

The decision rules $\bar{C}_M$ and $\bar{C}_N$ associated to the loss functions f_M and f_N by formulas (5) *and* (6) *are called canonical decision rules. A consistent bistrategy $(\bar{x}, \bar{y})$ for the optimal decision rules is called a noncooperative equilibrium of the game* ▲

Therefore, a pair $(\bar{x}, \bar{y})$ is a noncooperative equilibrium if

$$\bar{x} \in \bar{C}_M(\bar{y}) \qquad \text{and} \qquad \bar{y} \in \bar{C}_N(\bar{x}) \tag{7}$$

Formulas (8) and (9) provide an equivalent definition.

PROPOSITION 5

A pair of strategies $(\bar{x}, \bar{y})$ is a noncooperative equilibrium if and only if

$$f_M(\bar{x}, \bar{y}) = \min_{x \in M} f(x, \bar{y}) \tag{8}$$

and

$$f_N(\bar{x}, \bar{y}) = \min_{y \in M} f_N(\bar{x}, y) \tag{9}$$

▲

So, a noncooperative equilibrium is a situation where each player optimizes his or her own criterion, assuming that the partner's choice is fixed. In other words, this is a situation of *individual stability*.

Pareto Optima and Conservative Strategies

Can we accept the concept of a noncooperative equilibrium as the only reasonable concept of solution? Not necessarily; in particular, not when we assume that the players can communicate and exchange informations. When they do so, they may see that there exist other pairs of strategies that satisfy

$$f_M(x, y) < f_M(\bar{x}, \bar{y}) \qquad \text{and} \qquad f_N(x, y) < f_N(\bar{x}, \bar{y}) \tag{10}$$

that is, yield both Mike and Nancy (strictly) smaller losses than those assigned by the noncooperative equilibrium. This, when it happens, reveals a lack of collective stability, because both players can find better strategies.

DEFINITION 6

A pair of strategies (x_*, y_*) is said to be Pareto optimal if there is no other pair $(x, y) \in M \times N$ such that $f_M(x, y) < f_M(x_*, y_*)$ and $f_N(x, y) < f_N(x_*, y_*)$. ▲

Does there exist noncooperative equilibrium that are not Pareto optimal? Unfortunately, there are many examples of this situation. No general theorem asserting the existence of noncooperative equilibria that are also Pareto optimal is known to the authors.

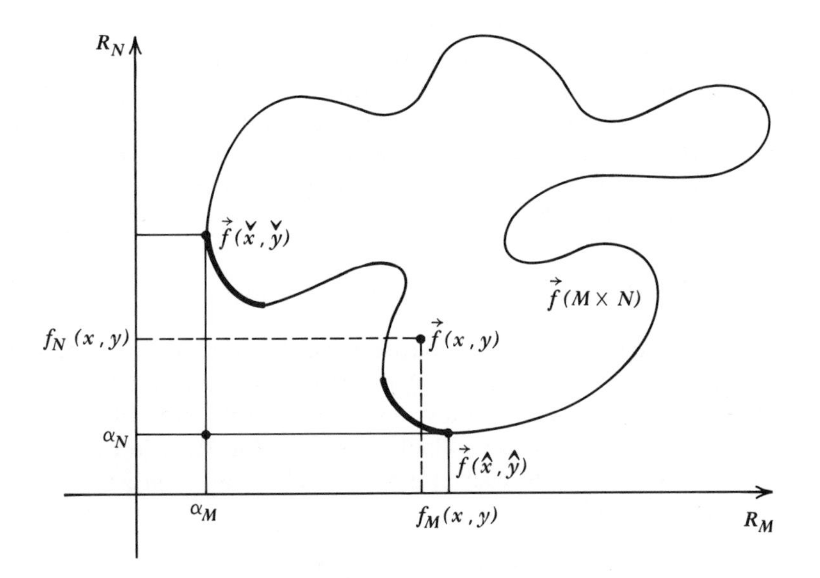

The preceding diagram represents the subset $\vec{f}(M \times N) \subset R^n$ of bilosses yielded by all the pairs of strategies. We also represent by a thick line the bilosses of Pareto optima. We see that selecting Pareto optima is not a sharp selection procedure. Assume, for instance, that there exists a pair $(\check{x}, \check{y})$ that achieves the minimum of Mike's loss function

$$f_M(\check{x}, \check{y}) = \inf_{\substack{x \in M \\ y \in N}} f_M(x, y) =: \alpha_M \tag{11}$$

It is clear that $(\check{x}, \check{y})$ is a Pareto minimum. For Nancy to agree to this situation would probably mean that her only aim in life is to please Mike. Also, any pair $(\hat{x}, \hat{y})$ of strategies that minimizes f_N on $M \times N$ is a Pareto optimum

$$f_N(\hat{x}, \hat{y}) = \inf_{\substack{x \in M \\ y \in N}} f_N(x, y) =: \alpha_N \tag{12}$$

We note that if a pair $(\tilde{x}, \tilde{y})$ minimizes both f_M and f_N on $M \times N$, then it is the best candidate for a concept of solution because

$$f_M(\tilde{x}, \tilde{y}) = \alpha_M, \qquad f_N(\tilde{x}, \tilde{y}) = \alpha_N$$

However, such a situation is naturally quite exceptional. This is why we call the vector

$$\vec{\alpha} := (\alpha_M, \alpha_N) \tag{13}$$

the *shadow minimum* (or *virtual minimum*) of the game. We also note intuitively that neither the pair $(\check{x}, \check{y})$ [defined by (11)] nor the pair $(\hat{x}, \hat{y})$ [defined by (12)] is a realistic choice in the framework of game theory: If it is reasonable for players to agree to choose pairs of strategies that do not yield both players smaller losses, it is not obvious that one of them will agree to give the other the entire benefit (see the preceding figure). Actually, cooperative game theory provides mechanisms of selecting among Pareto optima.

Conservative Strategies and Values

The case when Nancy's behavior is to please Mike without taking into account her own interest leads to strategies $(\check{x}, \check{y})$ defined by (11). Assume that Nancy exhibits the opposite behavior. Her only aim is to hurt Mike, and Mike knows it. (Actually, whenever Nancy behaves kindly, it suffices for Mike to believe that she is nasty.) So, he assigns to each strategy $x \in M$ the *worst loss* $f_M^\#(x)$ (read f sharp sub M) defined by

$$f_M^\#(x) := \sup_{y \in N} f_M(x, y) \tag{14}$$

and by doing so, looks for strategies $x^\# \in M$ that minimize $f_M^\#$ over M

$$f_M^\#(x^\#) = \inf_{x \in M} f_M^\#(x) \tag{15}$$

We say that $x^\#$ is a *conservative strategy* for Mike. We set

$$v_M^\# := \inf_{x \in M} \sup_{y \in N} f_M(x, y) = \inf_{x \in M} f_M^\#(x) \tag{16}$$

and call it Mike's *threat value*. Indeed, he can always reject a pair of strategies (x, y) that satisfies

$$f_M(x, y) > v_M^\# \tag{17}$$

since any conservative strategy $x^\# \in M$ yields Mike a loss $f_M(x^\#, y)$ (strictly) smaller than $f_M(x, y)$. So without any agreement with his opponent, he can always threaten to implement his conservative strategies. Similarly, Nancy's *threat value* is defined by

$$v_N^{\#} := \inf_{y \in N} \sup_{x \in M} f_N(x, y) = \inf_{y \in M} f_N^{\#}(y) \tag{18}$$

We say that the vector

$$\vec{v}^{\#} = (v_M^{\#}, v_N^{\#}) \tag{19}$$

is the *threat vector.*

So, pairs of strategies worth considering are those satisfying $\vec{f}(x, y) \leqslant \vec{v}^{\#}$.

The Duopoly

We present the basic example of duopoly, where both players are producers. The loss functions are the cost functions, and they depend on the production of the two players. This game and the concept of noncooperative equilibrium was introduced by the French economist Cournot in 1838.

We suppose that Mike and Nancy produce a given good, assumed to be homogeneous. We denote by $x \in R_+$ and $y \in R_+$ the quantities of this good produced by Mike and Nancy, respectively. We assume that the price

$$p(x, y) := \alpha - \beta(x + y) \tag{20}$$

is an affine function of the total production $x + y (\alpha \geqslant 0, \beta > 0)$ and cost functions c and d of each producer are also affine functions of the individual productions

$$c(x) = \gamma x + \delta, \qquad d(y) = \gamma y + \delta, \qquad \gamma > 0, \qquad \delta \geqslant 0 \tag{21}$$

The net cost for Mike is

$$f_M(x, y) := \gamma x + \delta - p(x, y)x = \beta x \left(x + y + \frac{\gamma - \alpha}{\beta} \right) + \delta$$

and the net cost for Nancy is

$$f_N(x, y) = \gamma y + \delta - p(x + y)y = \beta y \left(x + y + \frac{\gamma - \alpha}{\beta} \right) + \delta$$

We do not change the game by setting $\beta = 1$ and $\delta = 0$. So by setting $u = \gamma - \alpha$, the duopoly can be regarded as a two-person game, where

$$M := N := [0, u] \tag{22}$$

and the loss functions are

$$\text{(23)} \qquad \begin{cases} \textbf{i.} & f_M(x, y) := x(x+y-u) \\ \textbf{ii.} & f_N(x, y) := y(x+y-u) \end{cases}$$

The biloss map is defined by

$$\text{(24)} \qquad \vec{f}(x, y) = (x(x+y-u), \ y(x+y-u))$$

which maps the upper triangle

$$\text{(25)} \qquad T_+ := \{(x, y) \in [0, u] \times [0, u] \| x+y \geqslant u\}$$

onto the square $S_+ := [0, u^2] \times [0, u^2]$, the diagonal

$$\text{(26)} \qquad T_0 := \{(x, y) \in [0, u] \times [0, u] \| x+y = u\}$$

onto $\{0\}$, and the lower triangle

$$\text{(27)} \qquad T_- := \{(x, y) \in [0, u] \times [0, u] \| x+y \leqslant u\}$$

onto the triangle

$$\text{(28)} \qquad S_- := \left\{(f, g) \in \left[-\frac{u^2}{4}, 0\right] \times \left[-\frac{u^2}{4}, 0\right] \middle\| f+g \geqslant -\frac{u^2}{4}\right\}$$

We observe that the subset

$$\text{(29)} \qquad P := \left\{(x, y) \in [0, u] \times [0, u] \| x+y = \frac{u}{2}\right\}$$

is mapped onto the subset

$$\vec{f}(P) = \left\{(f, g) \in \left[-\frac{u^2}{4}, 0\right] \times \left[-\frac{u^2}{4}, 0\right] \middle\| f+g = -\frac{u^2}{4}\right\}$$

Therefore,

(30) the subset P defined by (29) is the set of Pareto strategies,

The bistrategy

$$\text{(31)} \qquad \left(x_P = \frac{u}{4}, \ y_P = \frac{u}{4}\right)$$

is Pareto optimal; if the producers agree to cooperate, a reasonable compromise is the pair (x_P, y_P).

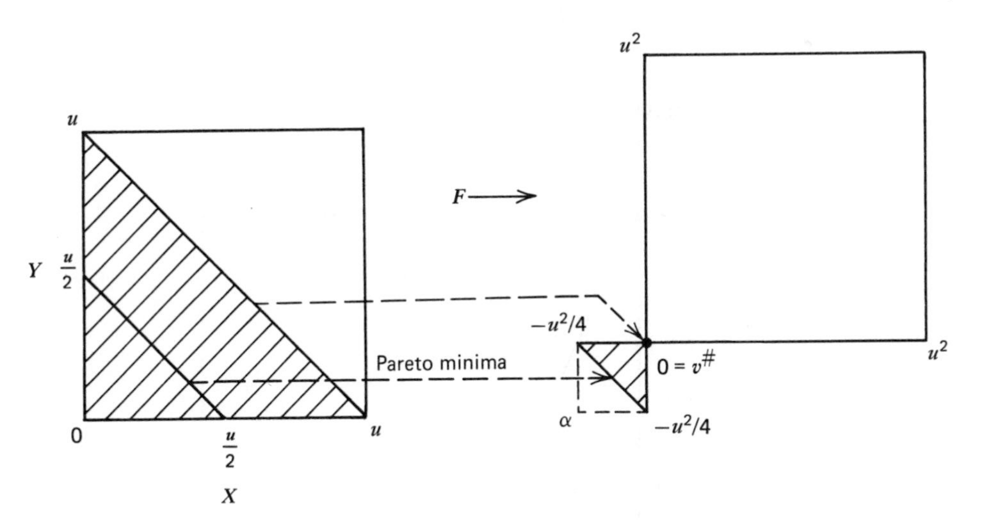

It is clear that

$$f_M^{\#}(x) := \sup_{o \leqslant y \leqslant u} x(x+y-u) = x^2 \tag{32}$$

achieves its minimum at $x=0$ and symetrically, $f_N^{\#}(y)=y^2$ achieves its minimum at $y=0$. Therefore, the *conservative bistrategy* is (0, 0), that is, both players produce nothing. The *threat vector* $\vec{v}^{\#}$ is equal to (0, 0). We note that the shadow minimum is equal to $(-u^2/4, -u^2/4)$.

The concept of noncooperative equilibrium in the case of the duopoly was introduced by Cournot in 1838. Let y be Nancy's production. Then Mike will implement a production x that minimizes over $[0, u]$ his net cost function $x \to x(x+y-u)$; the minimum is achieved at the point

$$x = \bar{C}_M(y) = \tfrac{1}{2}(u-y) \tag{33}$$

and is equal to

$$f_M^{b}(y) = \inf_x [x(x+y-u)] = -\frac{(u-y)^2}{4} \tag{34}$$

So, the map $\bar{C}_M$ defined by (33) is *Mike's canonical decision rule*, and similarly, the map $\bar{C}_N$ defined by

$$\bar{C}_N(x) = \tfrac{1}{2}(u-x) \tag{35}$$

is *Nancy's canonical decision rule*. Therefore, the noncooperative equilibrium of the duopoly is the fixed point of the map $(x, y)\to(C_M(y), C_N(x))$, that is, the point

$$\left(x_C=\frac{u}{3},\ y_C=\frac{u}{3}\right) \tag{36}$$

which yields to each player a cost equal to $-u^2/9$. We note that the *noncooperative equilibrium is not Pareto optimal.*

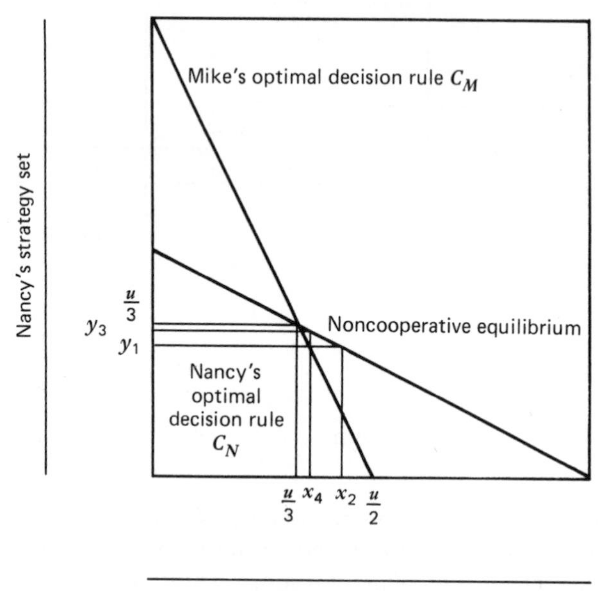

Actually, we observe that the noncooperative equilibrium can be reached by implementing the following algorithm. When Nancy produces y_{2n-1} at the odd period $2n-1$, Mike produces $x_{2n}=\bar{C}_M(y_{2n-1})$ at the even period $2n$, and then Nancy produces $y_{2n+1}=\bar{C}_N(x_{2n})$ at the odd period $2n+1$, and so on, the players taking turn in responding to each other.

Since the sequences (x_{2n}) and (y_{2n+1}) are, respectively, the even and odd subsequences of the sequence of elements z_k defined by

$$2z_{k+1}+z_k=u$$

multiplying each equation by $(-1)^{k+1}2^k$ and summing them, we obtain

$$z_n=\frac{u}{2}\frac{1+2^{-n}}{1+2^{-1}}+(-1)^{n+1}2^{-n-1}z_1$$

So z_n and, consequently, both x_{2n} and y_{2n+1} converges to $u/3$. ∎

We can associate another game with the duopoly, where instead of choosing strategies, the producers choose decision rules. Let us consider Mike's point of view. He may decide to play an affine decision rule of the form

$$C_M^a(y) := a(u-y) \quad \text{where} \quad a \in [0, 1] \tag{37}$$

This means that he produces nothing whenever Nancy produces the maximal production u and he decides to produce au whenever Nancy produces nothing. When Nancy implements an affine decision rule of the form

$$C_N^b(x) := b(u-x) \quad \text{where} \quad b \in [0, 1] \tag{38}$$

the associated consistent bistrategy is

$$\left(\frac{a(1-b)u}{1-ab}, \frac{b(1-a)u}{1-ab}\right) \tag{39}$$

which yields the following costs:

$$\begin{cases} \textbf{i.} \quad g_M(a, b) := -\dfrac{a(1-a)(1-b)^2u^2}{(1-ab)^2} \\ \textbf{ii.} \quad g_N(a, b) := -\dfrac{b(1-b)(1-a)^2u^2}{(1-ab)^2} \end{cases} \tag{40}$$

Therefore, we have a new game, whose strategies *are the slopes of the decision rules.*

For instance, if Nancy plays slope b, Mike will play the slope $a=\sigma_M(b)$ that minimizes $a \to g_M(a, b)$. We find

$$a = \sigma_M(b) := \frac{1}{2-b} \tag{41}$$

For instance, if Nancy plays her canonical decision rule $\bar{C}_N$, which corresponds to the slope $b=1/2$, and if Mike knows (or guesses) her move, he will play the slope $\sigma(1/2)=2/3$. The associated decision rule $C_M^{2/3}$ is called Mike's Stackelberg decision rule. The consistent strategies for the decision rules $C_M^{2/3}$ and $C_N^{1/2}=\bar{C}_N$ are equal to

$$x_S = \frac{u}{2}, \quad y_S = \frac{u}{4} \tag{42}$$

They form the so-called *Stackelberg equilibrium for Mike.* The associated costs are given by

$$f_M(x_S, y_S) = -\frac{u^2}{8}, \qquad f_N(x_S, y_S) = -\frac{u^2}{16} \tag{43}$$

By implementing a Stackelberg equilibrium, Mike is then in a better situation than in the original noncooperative equilibrium ($-u^2/8$ instead of $-u^2/9$).

Mike's advantage lies in the fact that he knows that Nancy will play her canonical decision rule. What would happen if Nancy followed the same reasoning? She would also implement her Stackelberg decision rule $C_N^{2/3}$. Consistent bistrategies for the Stackelberg decision rules are equal to

(44) $$x_D=\frac{2u}{5}, \qquad y_D=\frac{2u}{5}$$

They form the so-called *Stackelberg disequilibrium.*

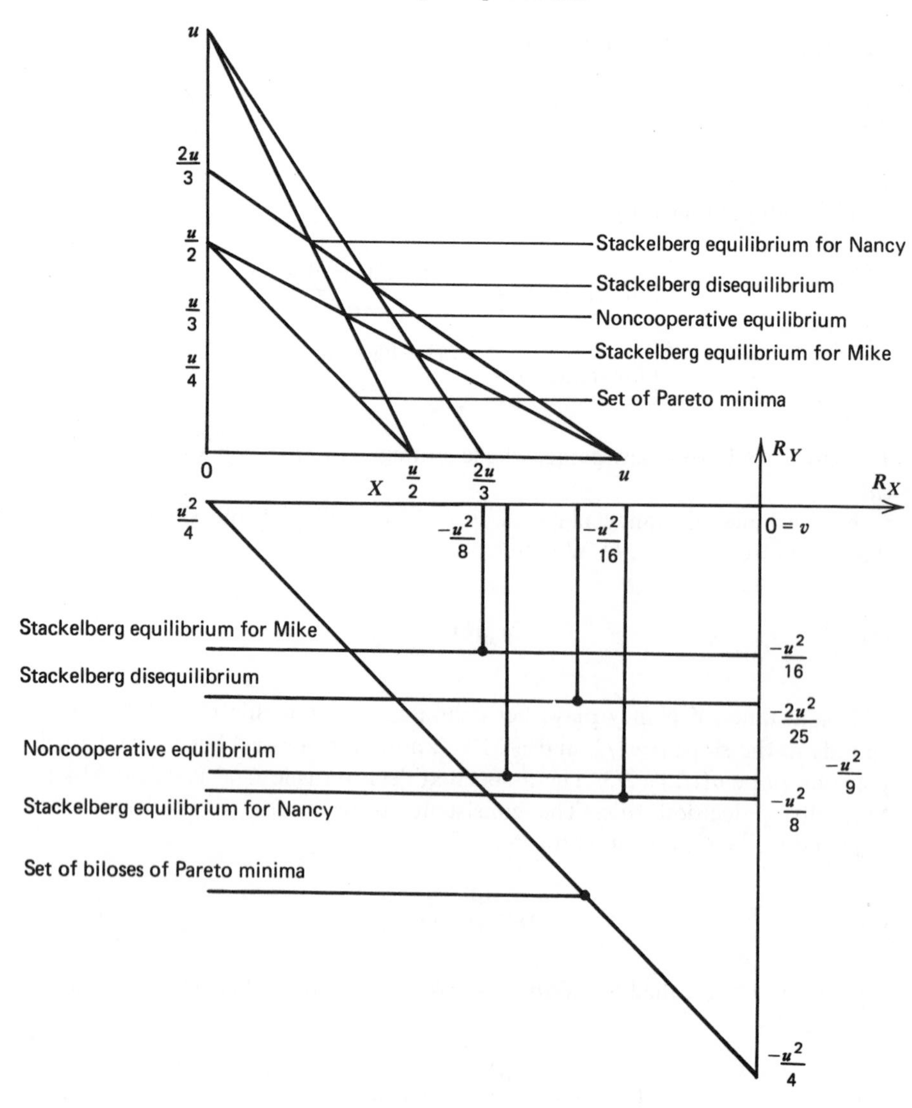

The producer's cost will equal $-2u^2/25$, and both Nancy and Mike are in a worse situation than in the noncooperative equilibrium case. For both players, the Stackelberg decision rules yields a smaller loss than the canonical decision rule, but both players would be better off both choosing the canonical decision rule than both choosing the Stackelberg decision rule.

TABLE 1

Mike \ Nancy	Canonical	Stackelberg
Canonical	$\left(-\frac{u^2}{9}, -\frac{u^2}{9}\right)$	$\left(-\frac{u^2}{16}, \frac{u^2}{8}\right)$
Stackelberg	$\left(-\frac{u^2}{8}, -\frac{u^2}{16}\right)$	$\left(-\frac{2u^2}{25}, -\frac{2u^2}{25}\right)$

The noncooperative equilibrium $(\bar{a}, \bar{b})$ for the game played on decision rules is the solution to the problem

$$\bar{a}=\sigma(\bar{b}) \quad \text{and} \quad \bar{b}=\sigma(\bar{a}) \tag{45}$$

which is $\bar{a}=1, \bar{b}=1$.

The associated consistent bistrategies (x, y) are those defined by equation $x+y=u$. We note that for such strategies, the costs are equal to zero. This noncooperative equilibrium can be achieved by implementing the following algorithm. Nancy begins by playing the slope 1/2 of her canonical decision rule, then Mike implements his Stackelberg decision rule $2/3=\sigma(1/2)$, then Nancy implements the slope $\sigma(2/3)=\sigma^2(1/2)$, and so on. It is easy to show that the sequence of slopes is equal to $1-1/n$ since

$$\sigma\left(1-\frac{1}{n}\right)=\frac{1}{2-1+1/n}=1-\frac{1}{n+1}$$

and they obviously converge to the slope one. Therefore, Mike will implement the slopes $1-1/3, 1-1/5, \ldots, 1-1/(2n+1)$ during the successive even period, while Nancy implements the slopes $1/2, 1-1/4, 1-1/6, \ldots, 1-1/2n$ during the odd periods. During the even periods, the consistent strategies will be

$$x_{2n}=\frac{u}{2}, \qquad y_{2n}=\frac{(2n-2)u}{2(2n-1)}$$

and during the odd periods, the consistent strategies will be

$$x_{2n+1} = \frac{(2n-1)u}{4n}, \qquad y_{2n+1} = \frac{u}{2}$$

They converge to the bistrategy $(u/2, u/2)$.

The analysis of this simple game emphasizes the conceptual difficulties of game theory. The bistrategies we described—the conservative solution $(0, 0)$, the Pareto minimum $(u/4, u/4)$, the noncooperative equilibrium $(u/3, u/3)$, the Stackelberg equilibria $(u/4, u/2)$ and $(u/2, u/4)$, the Stackelberg disequilibrium $(2u/5, 2u/5)$, and the bistrategy $(u/2, u/2)$ just mentioned—each have their own interest. This variety of reasonable concepts indicates the need for richer structures. We shall now turn our attention to the mathematical concepts of the theory.

2. TWO-PERSON ZERO-SUM GAMES: THE MINIMAX THEOREM

We now consider the important class of two-person games that satisfy the condition

$$\text{(1)} \qquad \forall x \in M, \quad y \in N, \quad f_M(x, y) + f_N(x, y) = 0$$

So, Nancy's loss is Mike's gain and vice versa. We observe that every pair of strategies is Pareto optimal, so that this concept is no longer interesting: Indeed, $\vec{f}(M \times N)$ is contained in the second bisector.

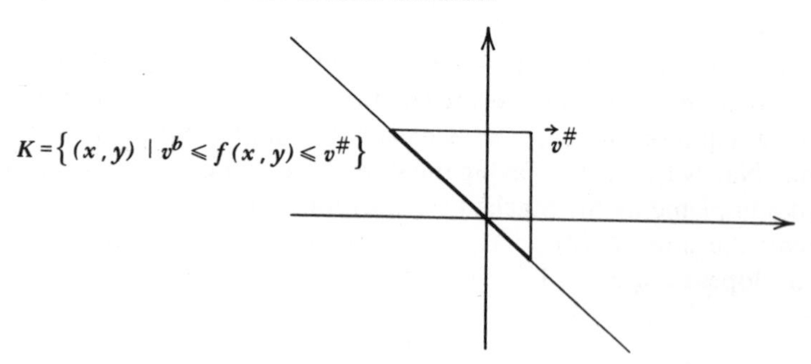

For simplicity, we set

$$\text{(2)} \qquad f_M(x, y) := f(x, y), \qquad f_N(x, y) := -f(x, y)$$

So if we set

$$\text{(3)} \qquad f^\#(x) := \sup_{y \in N} f(x, y), \qquad v^\# := \inf_{x \in M} \sup_{y \in N} f(x, y)$$

(read f-sharp and v-sharp) and

$$f^b(y) := \inf_{x \in M} f(x, y), \qquad v^b := \sup_{y \in N} \inf_{x \in M} f(x, y) \tag{4}$$

(read f flat and v flat) and

$$M^\# := \{x^\# \in M \mid f^\#(x^\#) = v^\#\}, \qquad N^b := \{y^b \in N \mid f^b(y^b) = v^b\} \tag{5}$$

we have

$$f_M^\#(x) = f^\#(x), \qquad f_M^\#(y) = -f^b(y), \qquad v_M^\# = v^\#, \qquad v_N^\# = -v^b \tag{6}$$

We observe that $M^\#$ and N^b are subsets of Nancy and Mike's conservative strategies, respectively.

Since $f^b(y) \leqslant f^\#(x)$ for all $x \in M$, $y \in N$, we deduce that

$$v^b \leqslant v^\# \tag{7}$$

or equivalently,

$$\vec{v}^\# := (v^\#, -v^b) \text{ lies above the second bisector.} \tag{8}$$

We say that the interval $(v^\#, v^b)$ is the *duality gap* of f. The set of pairs of strategies (x, y) satisfying $\vec{f}(x, y) \leqslant \vec{v}^\#$ is

$$K := \{(x, y) \in M \times N \mid v^b \leqslant f(x, y) \leqslant v^\#\} \tag{9}$$

It contains the set $M^\# \times N^b$. There are situations where v^b is strictly less than $v^\#$.

Example

Consider the finite game $M = \{1, 2\}$, $N = \{1, 2, 3\}$, where f is described by the following matrix.

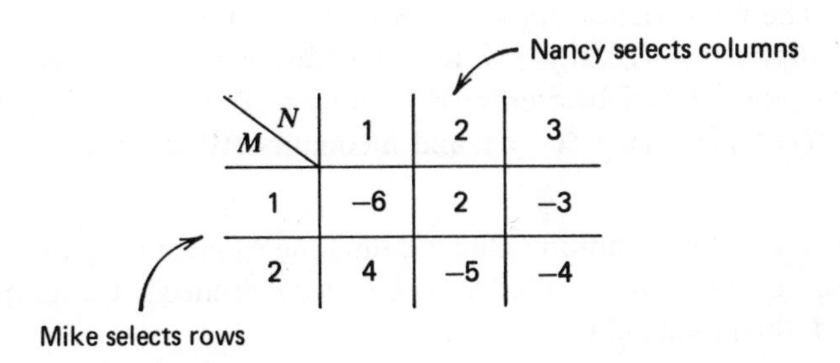

M \ N	1	2	3
1	–6	2	–3
2	4	–5	–4

The entries of this matrix represent Mike's losses. So Mike's biggest losses are 2 and 4, respectively, and thus Mike's conservative strategy is the first row and $v^{\#}=2$. Nancy's least gains are, respectively, -6, -5, and -4 and thus her conservative strategy is the third column and $v^{b}=-4$. The pairs (1, 2), (1, 3), and (2, 3) belong to the subset K. Let us try to play that game for ourselves. First let Mike implement its conservative strategy (first row). He expects Nancy to choose the second column. But the conservative strategy for Nancy is the third column, and she expects Mike to choose the second row. But if Mike is informed of this choice (or guesses it), then he would do better selecting his second row (with a loss of -4) instead of the first one. Similarly, if Mike chooses his conservative strategy, then Nancy would do better playing her second row (with a gain of two) instead of the third. This "wheels-within-wheels" situation illustrates the lack of noncooperative equilibrium; indeed, the canonical decision rule $C^{\#}:=\bar{C}_M$ for Mike is defined by

$$C^{\#}(1)=\{1\}, \qquad C^{\#}(2)=\{2\}, \qquad C^{\#}(3)=\{2\}$$

and the canonical decision rule $C^{b}:=C_N^{\#}$ for Nancy, by

$$C_b(1)=\{2\}, \qquad C^b(2)=\{1\}$$

So the decision rule $\vec{C}:=(C^b, C^{\#})$ has no fixed point, as can be directly checked. The absence of noncooperative equilibria when $v^b<v^{\#}$ is actually a general fact: The following result shows that its existence requires very stringent conditions.

PROPOSITION 1

The following conditions are equivalent:

$$(10)\quad \begin{cases} \textbf{i.} & (\bar{x}, \bar{y}) \textit{ is a noncooperative equilibrium.} \\ \textbf{ii.} & \forall (x, y) \in M \times N, \qquad f(\bar{x}, y) \leqslant f(\bar{x}, \bar{y}) \leqslant f(x, \bar{y}) \\ \textbf{iii.} & v^{\#}=v^b \textit{ and } \bar{x} \in M^{\#} \textit{ and } \bar{y} \in N^b \textit{ are conservative strategies.} \end{cases}$$

▲

Proof. The equivalence between properties (10, i) and (10, ii) is obvious, as well as implication (10, **ii**)⇒(10, **iii**). The converse is easy: Let v denote the common value $v^{\#}=v^b$, $\bar{x}$ belong to $M^{\#}$, and $\bar{y} \in N^b$ be conservative strategies. Then $v=f^b(\bar{y}) \leqslant f(\bar{x}, \bar{y}) \leqslant f^{\#}(\bar{x})=v$ and inequality (10, **ii**) ensues. ■

When $v^b=v^{\#}$, this common value is called the *value* of the *game*, and a noncooperative equilibrium is called a *saddle point*. Indeed, the graph of such a function f then looks like a saddle.

There are examples where saddle points do exist.

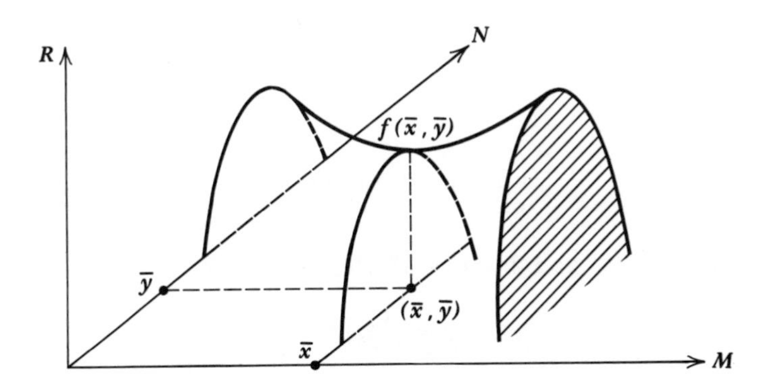

Example
Consider the finite game $M=\{1, 2\}$, $N=\{1, 2, 3\}$, where f is described by the following matrix.

M \ N	1	2	3
1	−2	−1	−4
2	1	0	−6

We observe that $v=-1$ and the pair of conservative strategies (1, 2) is a non-cooperative equilibrium. ■

To find sufficient conditions for the equality $v^{\#}=v^{b}$, we shall introduce another value, $v^{\natural}$ (v-natural) falling within the duality gap and prove successively that $v^{\natural}=v^{\#}$ and $v^{\natural}=v^{b}$.

Let $\mathscr{S}$ denote the set of *finite subsets* K of N. We set

$$v_K^{\#} := \inf_{x \in M} \sup_{y \in K} f(x, y) \tag{11}$$

$$v^{\natural} := \sup_{K \in \mathscr{S}} v_K^{\#} = \sup_{K \in \mathscr{S}} \inf_{x \in M} \sup_{y \in K} f(x, y) \tag{12}$$

(*read* v natural)

Since each point $y \in N$ can be identified with the subset $\{y\} \in \mathscr{S}$, we observe that $v^\sharp_{\{y\}} = f^\flat(y)$ and thus

$$v^\flat = \sup_{y \in N} v^\sharp_{\{y\}} \leqslant \sup_{K \in \mathscr{S}} v^\sharp_K = v^\natural$$

Also since,

$$\sup_{y \in K} f(x, y) \leqslant \sup_{y \in N} f(x, y),$$

we deduce that $v^\sharp_K \leqslant v^\sharp$ and thus $v^\natural \leqslant v^\sharp$. Hence

$$v^\flat \leqslant v^\natural \leqslant v^\sharp \tag{13}$$

We now use convexity assumptions to prove that $v^\flat = v^\natural$.

THEOREM 2

Let us assume that M and N are convex subsets of vector spaces and that

$$\begin{cases} \textbf{i.} \quad \forall y \in N, \quad x \to f(x, y) \text{ is convex.} \\ \textbf{ii.} \quad \forall x \in M, \quad y \to f(x, y) \text{ is concave.} \end{cases} \tag{14}$$

Then $v^\flat = v^\natural$. ▲

Proof. We set $\Sigma^n := \{\lambda \in R^n_+, \sum_{i=1}^n \lambda_i = 1\}$. We associate with any $K := \{y_1, \ldots, y_n\}$ the map F_K from M to R^n defined by

$$F_K(x) := (f(x, y_1), \ldots, f(x, y_n)) \tag{15}$$

and we set

$$w_K := \sup_{\lambda \in \Sigma^n} \inf_{x \in M} \langle \lambda, F_K(x) \rangle \tag{16}$$

We shall prove successively that

$$\begin{cases} \textbf{i.} \quad F_K(M) + R^n_+ \text{ is a convex subset (lemma 3).} \\ \textbf{ii.} \quad \forall K \in \mathscr{S}, \quad v^\sharp_K \leqslant w_K \quad \text{(lemma 4)} \\ \textbf{iii.} \quad \forall K \in \mathscr{S}, \quad w_K \leqslant v^\flat \quad \text{(lemma 5)} \end{cases} \tag{17}$$

Consequently, inequalities

$$v^\natural := \sup_{K \in \mathscr{S}} v^\sharp_K \leqslant \sup_{K \in \mathscr{S}} w_K \leqslant v^\flat \leqslant v^\natural \tag{18}$$

hold true and prove our theorem.

LEMMA 3
If M is convex and in the functions $x \to f(x, y)$ are convex, then the subset $F_K(M) + R^n_+$ is convex. ▲

We set $\Sigma^n := \{\lambda \in R^n_+ | \sum_{i=1}^n \lambda_i = 1\}$, which is convex and compact, and

(19) $$w_K := \sup_{\lambda \in \Sigma^n} \inf_{x \in M} \langle \lambda, F_K(x) \rangle$$

LEMMA 4
If M is convex and the functions $x \to f(x, y)$ are convex, then for all finite subsets K, we have

(20) $$v_K^\# \leqslant w_K$$ ▲

Proof. Let $\varepsilon > 0$, $\mathbb{1} := (1, \ldots, 1)$. We claim that

(21) $$(w_K + \varepsilon)\mathbb{1} \in F_K(M) + R^n_+$$

Assume the contrary, we can use the separation theorem in finite dimensional spaces, because $F_K(M) + R^n_+$ is convex by lemma 3. There exists $\lambda \in R^n$, $\lambda \neq 0$, such that

$$\sum_{i=1}^n \lambda_i(w_K + \varepsilon) = \langle \lambda, (w_K + \varepsilon) 1 \rangle \leqslant \inf_{v \in F_K(N) + R^n_+} \langle \lambda, v \rangle = \inf_{x \in M} \langle \lambda, F_K(x) \rangle + \inf_{u \in R^n_+} \langle \lambda, u \rangle$$

Therefore, $\inf_{u \in R^n_+} \langle \lambda, u \rangle$ is bounded below; consequently, $\lambda \in R^n_+$ and $\inf_{u \in R^n_+} \langle \lambda, u \rangle = 0$. Since $\lambda \neq 0$, then $\sum_{i=1}^n \lambda_i > 0$. We set $\bar{\lambda} = \lambda / \sum_{i=1}^n \lambda_i \in \Sigma^n$ and obtain

$$w_K + \varepsilon \leqslant \inf_{x \in M} \langle \bar{\lambda}, F_K(x) \rangle \leqslant \sup_{\lambda \in \Sigma^n} \inf_{x \in M} \langle \lambda, F_K(x) \rangle = w_K$$

which is impossible.

Hence, there exist $x_\varepsilon \in M$ and $u_\varepsilon \in R^n_+$ such that $(w_K + \varepsilon)\mathbb{1} = F_K(x_\varepsilon) + u_\varepsilon$. By the very definition (15) of F_K, we deduce that

$$\forall i = 1, \ldots, n, \qquad f(x_\varepsilon, y_i) \leqslant w_K + \varepsilon$$

Therefore,

(22) $$v_K^\# \leqslant \max_{i=1,\ldots,n} f(x_\varepsilon, y_i) \leqslant w_K + \varepsilon$$

By letting ε converge to zero, we deduce our lemma. ■

LEMMA 5

Let N be convex and the functions $y \to f(x, y)$ concave. Then for all finite subset K, we have $w_K \leq v^b$. ▲

Proof. We associate with any $\lambda \in \Sigma^n$ the point $y_\lambda := \sum_{i=1}^n \lambda_i y_i$ that belongs to N (by convexity). The concavity of the functions $y \to f(x, y)$ implies that $\forall x \in M, \sum_{i=1}^n \lambda_i f(x, y_i) \leq f(x, y_\lambda)$. Consequently,

$$\inf_{x \in M} \sum_{i=1}^n \lambda_i f(x, y_i) \leq \inf_{x \in M} f(x, y_\lambda) \leq \sup_{y \in N} \inf_{x \in M} f(x, y) := v^b$$

Hence, by taking the supremum on Σ^n, we obtain $w_K \leq v^b$. ■

THEOREM 6

Let us assume that

$$\text{(23)} \qquad \begin{cases} \textbf{i.} & \exists y_0 \in N \text{ such that } x \to f(x, y_0) \text{ is inf compact.}\dagger \\ \textbf{ii.} & \forall y \in N, \quad x \to f(x, y) \text{ is lower semicontinuous.} \end{cases}$$

Then $v^\natural = v^\#$, and there exists $\bar{x} \in K$ such that

$$\text{(24)} \qquad \sup_{y \in N} f(\bar{x}, y) = v^\natural$$ ▲

Proof. We introduce the subsets S_y defined by

$$\text{(25)} \qquad S_y := \{x \in M \mid f(x, y) \leq v^\natural\}$$

They have the *finite intersection property*: For each finite subset $K = \{y_0, \ldots, y_n\} \subset N$ containing y_0, we have $\bigcap_{y \in K} S_y \neq \varnothing$. Indeed, the function $f_K^\#$ defined by $f_K^\#(x) = \max_{y \in K} f(x, y)$ is lower semicontinuous [assumption (23, **ii**)] and inf-compact [assumption (23, **i**)] and $y_0 \in K$. So, it achieves its minimum at a point that belongs to $\bigcap_{y \in K} S_y$.

Furthermore, the subsets S_y are closed [assumption (23, **ii**)] and S_{y_0} is compact [assumption (23, **i**)]. Hence, the intersection $\bigcap_{y \in N} S_y$ is nonempty. Any element $x \in \bigcap_{y \in N} S_y$ satisfies

$$\text{(26)} \qquad \sup_{y \in N} f(x, y) \leq v^\natural$$

Therefore, $v^\# \leq v^\natural$. Since the other inequality holds, the theorem ensues. ■

†We recall that a function f is inf-compact if for all $\lambda \in R$, the lower level sets $\{x \mid f(x) \leq \lambda\}$ are relatively compact.

Putting theorems 2 and 6 together, we obtain the lopsided minimax theorem.

THEOREM 7 (LOPSIDED MINIMAX)
Let M and N be convex subsets of vector spaces, M being supplied with a topology. We assume that

$$(27)\qquad \begin{cases} \textbf{i.} & \forall y \in N, \quad x \to f(x, y) \text{ is convex and lower semicontinuous.} \\ \textbf{ii.} & \exists y_0 \in N \text{ such that } x \to f(x, y_0) \text{ is inf compact.} \end{cases}$$

and

$$(28)\qquad \forall x \in M, \quad y \to f(x, y) \quad \text{is concave}$$

Then f has a value ($v^b = v^{\#}$), *and there exists* $\bar{x} \in M$ *such that* $\sup_{y \in N} f(\bar{x}, y) = v^b$. ▲

We deduce von Neumann's minimax theorem.

THEOREM 8 (MINIMAX)
Let M and N be convex subsets of vector spaces, supplied with topologies. We assume that

$$(29)\qquad \begin{cases} \textbf{i.} & \forall y \in N, \quad x \to f(x, y) \text{ is convex and lower semicontinuous.} \\ \textbf{ii.} & \exists y_0 \in N \quad \text{such that} \quad x \to f(x, y_0) \text{ is inf compact} \end{cases}$$

and

$$(30)\qquad \begin{cases} \textbf{i.} & \forall x \in M, \quad y \to f(x, y) \text{ is concave and upper semicontinuous.} \\ \textbf{ii.} & \exists x_0 \in M \quad \text{such that} \quad y \to f(x, y_0) \text{ is sup compact.} \end{cases}$$

Then there exists a saddle point $(\bar{x}, \bar{y}) \in M \times N$. ▲

We state a corollary to theorem 6 that uses the conjugate functions f_y^* from X^* to $]-\infty, +\infty]$ defined by

$$(31)\qquad f_y^*(p) := \sup_{x \in M} [\langle p, x\rangle - f(x, y)]$$

We set

$$(32)\qquad \operatorname{Dom} f_y^* := \{p \in X^* \mid f_y^*(p) < +\infty\}.$$

COROLLARY 9

We assume that M is a weakly closed subset of a reflexive Banach space X. We posit the following assumptions:

$$(33)\quad \begin{cases} \text{i.} & \forall y \in N, \quad x \to f(x, y) \text{ is lower semicontinuous.} \\ \text{ii.} & \exists y_0 \in N \text{ such that } 0 \in \operatorname{Int}(\operatorname{Dom} f_{y_0}^*) \text{ (for the strong topology of the dual } X^*\text{)} \end{cases}$$

Then $v^{\natural} = v^{\#}$, and there exists $\bar{x} \in K$ such that $\sup_{y \in N} f(\bar{x}, y) = v^{\#}$. ▲

Proof. It suffices to prove that assumption (33) implies that the function $x \to f(x, y_0)$ is inf compact. There exists $\eta > 0$ such that $\eta B \subset \operatorname{Dom} f_{y_0}^*$. Let $S_\lambda := \{x \in M \mid f(x, y_0) \leqslant \lambda\}$ be a lower level set of $x \to f(x, y_0)$. We prove that it is *bounded.* Indeed, for all $p \in X^*$, we deduce from definition (31) of $f_y^*(p)$ that

$$\eta < \left\langle \frac{p}{\|p\|}, x \right\rangle \leqslant f(x, y_0) + f_{y_0}^*(\eta p/\|p\|)$$

and thus

$$\sup_{x \in S_\lambda} \langle p, x \rangle \leqslant \frac{\|p\|}{\eta} (\lambda + f_{y_0}^*(\eta p/\|p\|)) < +\infty$$

The uniform boundedness theorem implies that S_λ is bounded and thus relatively compact in X. Since f is lower semicontinuous, it is actually weakly compact in X and therefore in M. We then apply theorem 6. ■

Remark

Let X and X^* be two paired vector spaces. Corollary 9 remains true when X is supplied with the weak topology $\sigma(X, X^*)$ and X^* with the Mackey topology $\tau(X^*, X)$, since the neighborhoods for the Mackey topology are the polar subsets of weakly compact subsets of X. ■

The compactness assumption we made in Corollary 9 happens to be too strong for many problems. We shall relax it when M is a subset of a Banach space. We consider two Banach spaces X and Y and a function $\tilde{f}$ from $X \times Y$ to $\bar{R} := [-\infty, +\infty]$. We set

$$(34)\qquad M := \{x \in X \mid \forall y \in Y, \tilde{f}(x, y) < +\infty\}$$

and

$$(35)\qquad N := \{y \in Y \mid \forall x \in X, \tilde{f}(x, y) > -\infty\}$$

We shall say that $M \times N$ is the *domain of* $\tilde{f}$. We assume that M and N are *nonempty* and set

(36) $$f \text{ is the restriction of } \tilde{f} \text{ to } M \times N$$

Thus, f maps $M \times N$ to R.

The following theorem and its proof will be very useful in many applications. For that purpose, we replace the covering $\mathscr{S}$ of N by finite subsets by a covering $\mathscr{A}$ of N, which we can choose as a parameter, satisfying

(37) $$\text{when } K \text{ and } L \text{ belong to } \mathscr{A}, \text{ then } K \cup L \text{ belong to } \mathscr{A}.$$

We associate with such a convering the number

(38) $$v^{\natural}(\mathscr{A}) := \sup_{K \in \mathscr{A}} \inf_{x \in M} \sup_{y \in K} f(x, y)$$

We point out that when $\mathscr{A} \subset \mathscr{B}$,

(39) $$v^{\natural} := v^{\natural}(\mathscr{S}) \leqslant v^{\natural}(\mathscr{A}) \leqslant v^{\natural}(\mathscr{B})$$

and that

$$v_b \leqslant v^{\natural}(\mathscr{A}) \leqslant v^{\natural}(\mathscr{U}) = v^{\#}$$

where $\mathscr{U}$ denotes the covering of N by *all* the subsets K of N.

THEOREM 10

Let X be a reflexive Banach space and $\tilde{f}$ be a function from $X \times Y$ to $\bar{R}$ whose domain $M \times N$ is nonempty.

(40) $$\textit{Let } \mathscr{A} \textit{ be a } \text{countable} \textit{ covering of } N.$$

We assume that

(41) $$0 \in \operatorname{Int}\left(\bigcup_{y \in N} \operatorname{Dom} \tilde{f}_y^*\right) \qquad (\textit{for the strong topology})$$

and that

(42) $$\begin{cases} \forall y \in N, & x \to \tilde{f}(x, y) \textit{ is lower semicontinuous on } X \\ \textit{for the weak topology.} \end{cases}$$

Then there exists $\bar{x} \in M$ such that

(43) $$\sup_{y \in N} f(\bar{x}, y) = v^{\natural}(\mathscr{A}) = v^{\#}$$ ▲

Among all the corollaries we can form from these results, we state only the following very useful theorem.

THEOREM 11 (RELAXED LOPSIDED MINIMAX THEOREM)
Let X and Y be reflexive Banach spaces and $\tilde{f}$ be a function from $X \times Y$ to $\bar{R}$ whose domain $M \times N$ is nonempty. We assume that

$$(44) \qquad \forall y \in N, \qquad x \to \tilde{f}(x, y) \text{ is convex, lower semicontinuous}$$

that

$$(45) \qquad \forall x \in M, \qquad y \to \tilde{f}(x, y) \text{ is concave, upper semicontinuous}$$

that

$$(46) \qquad 0 \in \operatorname{Int}\left(\bigcup_{y \in N} \operatorname{Dom} \tilde{f}_y^* \right) \qquad \text{(for the strong topology)}$$

Then f has a value $v := v^b = v^\#$, and there exists $\bar{x} \in M$ such that $\sup_{y \in N} f(\bar{x}, y) = v$. ▲

Proof. We set $\tilde{f}^b(y) := \inf_{x \in X} \tilde{f}(x, y)$ and observe that

$$N = \{ y \in Y \mid \tilde{f}^b(y) > -\infty \}$$

is the domain of $\tilde{f}^b$. We introduce the following countable covering $\mathscr{A}$ of N by the subsets

$$(47) \qquad K_p := \{ y \in N \mid \|y\| \leqslant p \quad \text{and} \quad \tilde{f}^b(y) \geqslant -p \}$$

Since $\tilde{f}^b$ is concave and upper semicontinuous by (45), the subsets K_p are closed, convex, and bounded. Since Y is reflexive, they are *weakly compact*. We introduce

$$(48) \qquad v^\natural(\mathscr{A}) := \sup_{p \geqslant 1} \inf_{x \in M} \sup_{y \in K_p} f(x, y)$$

Theorem 10 implies that there exists $\bar{x} \in M$ such that

$$\sup_{y \in N} f(\bar{x}, y) = v^\natural(\mathscr{A}) = v^\#$$

Now we can apply the lopsided minimax theorem 7 to $-f$, which yields

$$\inf_{x \in M} \sup_{y \in K_p} f(x, y) = \sup_{y \in K_p} \inf_{x \in M} f(x, y)$$

Indeed, K_p is convex compact, M is convex, $y \to f(x, y)$ is concave and upper semicontinuous for all $x \in M$, and $x \to f(x, y)$ is convex for all $y \in K_p$. Therefore,

$$v^\natural(\mathscr{A}) = \sup_{p \geq 1} \sup_{y \in K_p} \inf_{x \in M} f(x, y) = \sup_{y \in N} \inf_{x \in M} f(x, y) = v^b$$

for $N = \bigcup_{p=1}^{\infty} K_p$. ■

It remains to prove theorem 10. For that purpose, we need lemma 12.

LEMMA 12

Assume that $v^\natural(\mathscr{A})$ is finite. Then there exists a generalized sequence of elements x_μ of M satisfying

$$\forall K \in \mathscr{A}, \quad \exists \mu_K \quad \textit{such that} \quad \limsup_{\mu \geq \mu_K} \left(\sup_{y \in K} f(x_\mu, y) \right) \leq v^\natural(\mathscr{A}) \tag{49}$$

When $\mathscr{A}$ is a countable covering, this sequence is a usual one (i.e., *countable*). ▲

Proof. By the very definition of $v^\natural(\mathscr{A})$, inequalities

$$\inf_{x \in M} \sup_{y \in K} f(x, y) \leq v^\natural(\mathscr{A}) \tag{50}$$

hold true for all subsets $K \in \mathscr{A}$. Therefore, we can associate with any $n \in N$ an element $x_{K,n} \in M$ such that

$$\sup_{y \in K} f(x_{K,n}, y) \leq v^\natural(\mathscr{A}) + \frac{1}{n} \tag{51}$$

we set $\mathscr{M} := \mathscr{A} \times N$, preordered by the relation

$$(K_1, n_1) \preccurlyeq (K_2, n_2) \quad \text{if and only if} \quad K_1 \subset K_2 \quad \text{and} \quad n_1 \leq n_2 \tag{52}$$

which is filtered (or directed): Any pair $((K_1, n_1), (K_2, n_2))$ has an upper bound $(K_1 \cup K_2, \max(n_1, n_2))$. It is countable whenever $\mathscr{A}$ is countable. Therefore, the map $(K, n) \in \mathscr{M} \to x_{K,n} \in M$ is a generalized sequence. By taking $L \supset K$ and $m \geq n$, we obtain

$$\sup_{y \in K} f(x_{L,m}, y) \leq \sup_{y \in L} f(x_{L,m}, y) \leq v^\natural(\mathscr{A}) + \frac{1}{m} \leq v^\natural(\mathscr{A}) + \frac{1}{n}$$

Hence, if we fix $K_0 \in \mathscr{A}$ and $K \supset K_0$, we obtain

$$\sup_{(L,m) \succcurlyeq (K,n)} \left(\sup_{y \in K_0} f(x_{L,m}, y) \right) \leq v^\natural(\mathscr{A}) + \frac{1}{n}$$

Consequently,

$$\limsup_{(K,n)\geqslant(K_0,1)} \left[\sup_{y\in K_0} f(x_{K,n}, y)\right]$$

$$= \inf_{(K,n)\geqslant(K_0,1)} \sup_{(L,m)\geqslant(K,n)} \left[\sup_{y\in K_0} f(x_{L,m}, y)\right] \leqslant v^{\natural}(\mathscr{A})$$ ■

Proof of Theorem 10. Let N be the union of an increasing sequence of subsets K_n and $\mathscr{A}$ the family of the K_n. We set

$$\text{(53)} \qquad v^{\natural}(\mathscr{A}) := \sup_{n\geqslant 0} \inf_{x\in M} \sup_{y\in K_n} f(x, y)$$

By lemma 1, there exists a countable sequence of elements $x_n \in M$ satisfying

$$\text{(54)} \qquad \forall y \in N, \qquad \exists n(y) \quad \text{such that} \quad \limsup_{n\geqslant n(y)} f(x_n, y) \leqslant v^{\natural}(\mathscr{A})$$

By assumption (41), there exists $\eta > 0$ such that

$$\text{(55)} \qquad \eta B \subset \bigcup_{y\in N} \operatorname{Dom} f_y^*$$

So for any $p \in X^*$, there exists $y_p \in N$ such that $\tilde{f}_{y_p}(\eta p/\|p\|) < +\infty$. By (54), we deduce that there exists $n_p \geqslant n(y_p)$ such that

$$\text{(56)} \qquad \forall n \geqslant n_{p'} \quad f(x_n, y) \leqslant v^{\natural}(\mathscr{A}) + 1$$

Therefore,

$$\forall n \geqslant n_p, \qquad \left\langle \frac{\eta p}{\|p\|}, x_n \right\rangle \leqslant f(x_n, y_p) + \tilde{f}_y\left(\frac{\eta p}{\|p\|}\right)$$

$$\leqslant v^{\natural}(\mathscr{A}) + 1 + \tilde{f}^*_{y_p}\left(\frac{\eta p}{\|p\|}\right) < +\infty$$

Since the sequence is countable, we deduce that

$$\sup_{n\geqslant n_p} \langle p, x_n \rangle < +\infty$$

Therefore, the uniform boundness theorem implies that the sequence x_n is bounded and thus weakly relatively compact. A subsequence $x_{n'}$ converges weakly to some $x \in X$, and the lower semicontinuity of f with respect to x implies that

(57) $$\forall y \in N, \quad \tilde{f}(x, y) \leqslant \liminf_{x_n \to x} \tilde{f}(x_{n'}, y) \leqslant v^{\natural}(\mathscr{A})$$

Then x belongs to M, and the theorem ensues. ■

3. THE KY FAN INEQUALITY

Let M and N be the strategy sets of Mike and Nancy and f be Nancy's gain function. She can use it to assign to each decision rule $C_N: M \to N$ a gain defined by

(1) $$f^b(C_N) := \inf_{x \in M} f(x, C_N(x))$$

This represents the worst *gain* she can expect using decision rule C_N and assuming that Mike's behavior is noncooperative. Note that this definition is consistent with the definition of the worst gain yielded by a strategy $\bar{y}$, regarded as a constant decision rule $x \to \bar{y}$

$$f^b(\bar{y}) := \inf_{x \in M} f(x, \bar{y}) = \inf_{x \in M} f(x, \bar{y}(x))$$

Consequently, if $\mathscr{C}_N$ is a set of decision rules containing N,

(2) $$v^b := \sup_{y \in N} \inf_{x \in M} f(x, y) \leqslant \sup_{C_N \in \mathscr{C}_N} f^b(C_N) \leqslant \inf_{x \in M} \sup_{y \in N} f(x, y) := v^\#$$

PROPOSITION 1
If Nancy is allowed to use all possible decision rules, then

(3) $$\sup_{C_N \in N^M} f^b(C_N) = v^\#$$ ▲

Proof. Indeed, we can associate with every $\varepsilon > 0$ and every $x \in M$ a strategy $C_N^\varepsilon(x) \in N$ such that

$$v^\# \leqslant \sup_{y \in N} f(x, y) \leqslant f(x, C_N^\varepsilon(x)) + \varepsilon$$

From this, we deduce that

$$v^\# := \inf_{x \in M} \sup_{y \in N} f(x, y) \leqslant f^b(C_N^\varepsilon(x)) + \varepsilon \leqslant \sup_{C_N \in N^M} f^b(C_N) + \varepsilon$$

Since this inequality holds for every $\varepsilon > 0$, we obtain the inequality $v^\# \leqslant \sup_{C_N \in N^M} f^b(C_N)$, which, together with inequaity (2), proves the proposition. ■

We show that with additional hypotheses, equality (3) remains true when we require Nancy to *use only continuous decision rules* (in order to describe a *stable* behavior for her).

THEOREM 2

Let M be a topological space, N a convex subset of a topological vector space, and f a function from $M \times N$ to R. Let us suppose that

$$(4)\qquad \begin{cases} \text{i.} & \exists y_0 \in N \quad \text{such that} \quad x \to f(x, y_0) \text{ is inf compact} \\ \text{ii.} & \forall y \in N, \quad x \to f(x, y) \text{ is lower semicontinuous} \end{cases}$$

and that

$$(5)\qquad \forall x \in M, \quad y \to f(x, y) \text{ is concave}$$

If $\mathscr{C}(M, N)$ denotes the set of continuous maps from M to N, the following equality holds true:

$$(6)\qquad \sup_{C_N \in \mathscr{C}(M,N)} \inf_{x \in M} f(x, C_N(x)) = v^{\#}$$

▲

Proof. We already know from (2) that

$$(7)\qquad \sup_{C_N \in \mathscr{C}(M,N)} \inf_{x \in M} f(x, C_N(x)) \leq v^{\#}$$

Thus, we must establish the opposite inequality. First of all, we can associate with every $\varepsilon > 0$ a map C_N^{ε} from M to N, not necessarily continuous, that satisfies inequality (3). Moreover, since the functions $x \to f(x, y)$ are lower semicontinuous, there exist open neighborhoods $\mathscr{N}(x)$ such that

$$(8)\qquad \forall x' \in \mathscr{N}(x), \quad f(x, C_N^{\varepsilon}(x)) \leq f(x', C_N^{\varepsilon}(x)) + \varepsilon$$

We introduce the subset

$$M_0 := \{x \in M \mid f(x, y_0) \leq v^{\#}\}$$

which is compact, since $x \to f(x, y_0)$ is lower semicontinuous and inf compact. Therefore, it can be covered by n neighborhoods $\mathscr{N}_i := \mathscr{N}(x_i)$. Let $\mathscr{N}_0 := M \setminus M_0$, which is an open set. Then $M \subset \bigcup_{i=0}^{n} \mathscr{N}_i$. Let $\{p_i\}_{i=0,\ldots,n}$ be a continuous partition of unit subordinate to this finite convering. We introduce the function $\bar{C}_N^{\varepsilon}$ from M to N defined by

$$(9)\qquad \bar{C}_N^{\varepsilon}(x) := p_0(x) y_0 + \sum_{i=1}^{n} p_i(x) C_N^{\varepsilon}(x_i)$$

This map is continuous. Furthermore, since the function $y \to f(x, y)$ is concave and $p_i(x) \geq 0$ for $i = 0, \ldots, n$ and $\sum_{i=0}^{n} p_i(x) = 1$, we obtain

$$f(x, \bar{C}_N^\varepsilon(x)) \geq p_0(x) f(x, y_0) + \sum_{i=1}^{n} p_i(x) f(x, C_N^\varepsilon(x_i)) \tag{10}$$

Now, if $p_0(x) > 0$, then $x \in \mathcal{N}_0$ and consequently, $f(x, y_0) \geq v^\# > v^\# - \varepsilon$. If $p_i(x) > 0$, then $x \in \mathcal{N}_i$ and consequently

$$f(x, C_N^\varepsilon(x_i)) \geq f(x_i, C_N^\varepsilon(x_i)) - \varepsilon \geq v^\# - \varepsilon$$

[inequalities (7) and (8)]. Since $\sum_{i=0}^{n} p_i(x) = 1$, we deduce from (10) that

$$f(x, \bar{C}_N^\varepsilon(x)) \geq \sum_{i=0}^{n} p_i(x)(v^\# - \varepsilon) = v^\# - \varepsilon$$

Hence,

$$v^\# - \varepsilon \leq f^b(\bar{C}_N^\varepsilon(x)) \leq \sup_{C_N \in \mathscr{C}(M,N)} \inf_{x \in M} f(x, C_N(x))$$

Letting ε converge to zero, we conclude the proof. ■

We shall prove an equivalent formulation of the Brouwer fixed point theorem, known as the Ky Fan inequality. Since we have proved the Brouwer fixed point theorem for compact convex subsets of R^n, we begin by proving the Ky Fan inequality for compact convex subsets of R^n. We shall extend it to any compact convex subset (see theorem 5).

LEMMA 3

Let K be a compact convex subset of R^n and let ϕ be a real-valued function defined on $K \times K$ satisfying

$$\begin{cases} \textbf{i.} \quad \forall y \in K, & x \to \phi(x, y) \text{ is lower semicontinuous} \\ \textbf{ii.} \quad \forall x \in K, & y \to \phi(x, y) \text{ is concave} \end{cases} \tag{11}$$

Then there exists $\bar{x} \in K$ satisfying

$$\sup_{y \in K} \phi(\bar{x}, y) \leq \sup_{y \in K} \phi(y, y) \tag{12}$$ ▲

Proof. Since the functions $x \to \phi(x, y)$ are lower semicontinuous and the functions $y \to \phi(x, y)$ are concave, theorem 2 and theorem 2.6 imply the existence of $\bar{x} \in K$ satisfying

$$v^\# = \sup_{y \in K} \phi(\bar{x}, y) = \sup_{C \in \mathscr{C}(K,K)} \inf_{x \in K} \phi(x, C(x))$$

Since K is compact and convex and C is a continuous map, there exists a fixed point $x_c \in K$ of C. Hence,

$$\inf_{x \in K} \phi(x, C(x)) \leqslant \phi(x_c, C(x_c)) = \phi(x_c, x_c) \leqslant \sup_{y \in K} \phi(y, y).$$

Lemma 3 is proved. ■

Remark

Conversely, assume that Ky Fan's inequality holds true. We associate with any $C \in \mathscr{C}(K, K)$ the function ϕ defined on $K \times K$ by

$$\phi(x, y) = \langle C(x) - x, y - x \rangle \tag{13}$$

This function satisfies the assumptions of Ky Fan's inequality. Hence, there exists $\bar{x} \in K$ such that $\sup_{y \in K} \langle C(\bar{x}) - \bar{x}, y - \bar{x} \rangle \leqslant 0$. By taking $y = C(\bar{x}) \in K$, we get $\|C(\bar{x}) - \bar{x}\|^2 \leqslant 0$, that is, $\bar{x}$ is a fixed point of C.

This proves that the Ky Fan inequality in finite dimensional space is equivalent to the Brouwer fixed point theorem. ■

We proceed to prove the other inequality mentioned earlier. Let M and N be the strategy sets of Mike and Nancy and let f be Mike's loss function. Mike assigns to each decision rule C_M: $N \to M$ the worst loss

$$f^{\#}(C_M) = \sup_{y \in N} f(C_M(y), y) \tag{14}$$

If $\mathscr{C}_M$ is a set of decision rules containing the set M of constant decision rules, we have

$$v^b \leqslant \inf_{C_M \in \mathscr{C}_M} f^{\#}(C_M) \leqslant v^{\#} \tag{15}$$

THEOREM 4

Let M be a topological space, N a convex subset of a topological vector space, and f a function from $M \times N$ to R. Let us suppose that

$$\begin{cases} \textbf{i.} & \exists y_0 \in N \quad \text{such that} \quad x \to f(x, y_0) \text{ is inf compact} \\ \textbf{ii.} & \forall y \in N, \quad x \to f(x, y) \text{ is lower semicontinuous} \end{cases} \tag{16}$$

and that

$$\forall x \in M, \qquad y \to f(x, y) \text{ is concave} \tag{17}$$

If $\mathscr{C}(N, M)$ denotes the set of continuous maps from N to M, there exists $\bar{x} \in M$ such that

$$\sup_{y \in N} f(\bar{x}, y) = \inf_{C_M \in \mathscr{C}(N,M)} \sup_{y \in M} f(C_M(y), y) \tag{18}$$

▲

Proof. Since inequality $\inf_{C_M \in \mathscr{C}(N,M)} f^{\#}(C_M) \leqslant v^{\#}$ holds true, we have to prove the existence of $\bar{x} \in M$ such that for all continuous maps C_M from N to M, we have the following inequality:

$$\sup_{y \in N} f(\bar{x}, y) \leqslant \sup_{y \in N} f(C_M(y), y) \tag{19}$$

By theorem 2.6., we know that there exists $\bar{x} \in M$ such that

$$\sup_{y \in N} f(\bar{x}, y) = v^{\natural} := \sup_{K \in \mathscr{S}} \inf_{x \in M} \sup_{y \in K} f(x, y) \tag{20}$$

where $\mathscr{S}$ is the family of finite subsets of N. Since N is convex, $\text{co}(K) \subset N$ and $C_M(\text{co}(K)) \subset M$. We set $K := \{y_1, \ldots, y_n\}$ and $\Sigma^n := \{\lambda \in R_+^n | \sum_{i=1}^n \lambda_i = 1\}$. We can write

$$\begin{aligned}
\inf_{x \in M} \max_{i=1,\ldots,n} f(x, y_i) &= \inf_{x \in M} \sup_{\lambda \in \Sigma^n} \sum_{i=1}^n \lambda_i f(x, y_i) \\
&\leqslant \inf_{x \in C_M(\text{co}(K))} \sup_{\lambda \in \Sigma^n} \sum_{i=1}^n \lambda_i f(x, y_i) \\
&= \inf_{\mu \in \Sigma^n} \sup_{\lambda \in \Sigma^n} \sum_{i=1}^n \lambda_i f\left(C_M\left(\sum_{j=1}^n \mu_j y_j\right), y_i\right)
\end{aligned}$$

We set

$$\phi(\mu, \lambda) := \sum_{i=1}^n \lambda_i f\left(C_M\left(\sum_{j=1}^n \mu_j y_j\right), y_i\right)$$

It maps $\Sigma^n \times \Sigma^n$ to R, is lower semicontinuous with respect to μ (because C_M is continuous from N to M), and is affine with respect to λ. Hence, Ky Fan's inequality in finite dimensional spaces (lemma 3) implies that

$$\inf_{\mu \in \Sigma^n} \sup_{\lambda \in \Sigma^n} \phi(\mu, \lambda) \leqslant \sup_{\lambda \in \Sigma^n} \phi(\lambda, \lambda)$$

Since the functions $y \to f(x, y)$ are concave, we obtain

$$\begin{aligned}
\phi(\lambda, \lambda) &= \sum_{i=1}^n \lambda_i f\left(C_M\left(\sum_{j=1}^n \lambda_j y_j\right), y_i\right) \\
&\leqslant f\left(C_M\left(\sum_{j=1}^n \lambda_j y_j\right), \sum_{i=1}^n \lambda_i y_i\right) \leqslant f^{\#}(C_M)
\end{aligned}$$

Thus, we have proved that for all $K \in \mathscr{S}$ and for all $C_M \in \mathscr{C}(N, M)$,

$$\inf_{x \in M} \sup_{y \in K} f(x, y) \leqslant f^{\#}(C_M) \tag{21}$$

Hence,

(22) $$v^{\natural} \leqslant \inf_{C_M \in \mathscr{C}(N,M)} f^{\#}(C_M)$$

and our theorem ensues. ■

As a consequence, we obtain the Ky Fan inequality for compact convex subsets of any topological vector space.

THEOREM 5 (KY FAN)

Let K be compact convex in a topological vector space and $\phi: K \times K \to R$ be a function satisfying

(23) $$\begin{cases} \text{i.} & \forall y \in K, \quad x \to \phi(x, y) \text{ is lower semicontinuous.} \\ \text{ii.} & \forall x \in K, \quad y \to \phi(x, y) \text{ is concave.} \end{cases}$$

Then there exists $\bar{x} \in K$ satisfying

(24) $$\sup_{y \in K} \phi(\bar{x}, y) \leqslant \sup_{y \in K} \phi(y, y)$$ ▲

Proof. We take $M=K$, $N=K$, and $f(x, y)=\phi(x, y)$. Since the identity is a continuous map from K to K, we deduce that there exists $\bar{x} \in K$ such that

$$\sup_{y \in K} \phi(\bar{x}, y) = \inf_{C \in \mathscr{C}(K,K)} \sup_{y \in K} \phi(C(y), y) \leqslant \sup_{y \in K} \phi(y, y)$$ ■

Let us consider theorems 2 and 4, stating that

(25) $$\sup_{C_N \in \mathscr{C}(M,N)} \inf_{x \in M} f(x, C_N(x)) = \inf_{C_M \in \mathscr{C}(N,M)} \sup_{y \in N} f(C_M(y), y)$$

In both theorems, *the topology on N can be chosen as a parameter*; also, the stronger the topology on N, the larger the set $\mathscr{C}(N, M)$ the smaller the set $\mathscr{C}(M, N)$, and consequently, the stronger equality (25). We can design a topology on N, which is not (necessarily) a vector space topology, that is stronger than any vector space topology and for which theorems 2 and 4 hold true.

Let N be a convex subset of a vector space Y. We associate with any finite set $K=\{y_1, \ldots, y_n\}$ of N the affine map β_K from Σ^n to N defined by

(26) $$\forall \lambda \in \Sigma^n, \quad \beta_K(\lambda) = \sum_{i=1}^{n} \lambda_i y_i$$

DEFINITION 6

The finite topology on a convex subset N is the strongest topology for which the maps β_K are continuous when K ranges over the family $\mathscr{S}$ of finite subsets of N. ▲

So, a map C from N, supplied with the finite topology, to a topological space M is continuous if and only if

$$\text{(27)} \qquad \forall K \in \mathscr{S}, \qquad \text{the maps } C\beta_K \text{ from } \Sigma^n \text{ to } M \text{ are continuous}$$

Also, any map C from a topological space M to N of the form

$$\text{(28)} \qquad C(x) := \beta_K \vec{p}(x) = \sum_{i=1}^{n} p_i(x) y_i$$

where $\vec{p}$ is a continuous map from M to Σ^n, is continuous from M to N supplied with the finite topology.

PROPOSITION 7

The finite topology on a convex subset N of Y is stronger than any vector space topology. Any affine map C from N to a vector space X is continuous when both N and X are supplied with the finite topology. ▲

Proof. **a.** Suppose that Y is supplied with a vector space topology. Let us take $C := I$ to be the canonical injection from N to Y. Since for all $K \in \mathscr{S}$, the map $I\beta_K\colon \lambda \in \Sigma^n \to \sum_{i=1}^n \lambda_i y_i$ is obviously continuous, we deduce that I is continuous, that is, the finite topology is stronger than the restriction of the vector space topology to N.

b. Let C be any affine map from N to X. We have to prove that for any $K := \{y_1, \ldots, y_n\}$, the map

$$C\beta_K\colon \lambda \in \Sigma^n \to C\beta_K(\lambda) = C\left(\sum_{i=1}^{n} \lambda_i y_i\right) = \sum_{i=1}^{n} \lambda_i C(y_i) = \beta_{C(K)}(\lambda)$$

is continuous. But $C\beta_K = \beta_{C(K)}$ is indeed continuous by the very definition of the finite topology on X. ■

THEOREM 8

Let M be a topological space, N a convex subset supplied with the finite topology, and $f\colon M \times N \to R$ a function satisfying

$$\text{(29)} \qquad \begin{cases} \textbf{i.} \quad \exists y_0 \in N & \textit{such that} \quad x \to f(x, y_0) \textit{ is inf compact.} \\ \textbf{ii.} \quad \forall y \in N, & x \to f(x, y) \textit{ is lower semicontinuous.} \end{cases}$$

and

(30) $$\forall x \in M, \quad y \to f(x, y) \text{ is concave.}$$

Then there exists $\bar{x} \in M$ *such that*

(31) $$\begin{cases} \sup_{y \in N} f(\bar{x}, y) = \sup_{C_N \in \mathscr{C}(M,N)} \inf_{x \in M} f(x, C_N(x)) \\ \quad = \inf_{C_M \in \mathscr{C}(N,M)} \sup_{y \in N} f(C_M(y), y) \end{cases}$$ ▲

Proof. Indeed, in the proof of theorem 2, the map $\bar{C}_N^\varepsilon$ defined by (9) can be written $\beta_K \vec{p}$, where $K := (y_0, C_N^\varepsilon(x_1), \ldots, C_N^\varepsilon(x_I))$ and $\vec{p}(x) = (p_0(x), p_1(x), \ldots, p_n(x))$, which is continuous from M to N supplied with the finite topology. In the proof of theorem 4, we needed the continuity of C_M only to infer that for all $\lambda \in \Sigma^n$, $\mu \to \phi(\mu, \lambda)$ is lower semicontinuous. But we can write $\phi(u, \lambda) = \sum_{i=1}^n \lambda_i f(C_M \beta_K(\mu), y_i)$; therefore, $\mu \to \phi(\mu, \lambda)$ is lower semicontinuous whenever C_M is continuous for the finite topology on N. ■

We now prove an extension of the Ky Fan inequality in which the assumption of lower semicontinuity with respect to x is relaxed.

THEOREM 9 (KY FAN'S INEQUALITY FOR MONOTONE FUNCTIONS)
Let $K \subset X$ *be a convex subset of a topological vector space and* $\phi: K \times K \to R$ *be a function satisfying*

(32) $$\begin{cases} \textbf{i.} & \forall y \in K, \quad x \to \phi(x, y) \text{ is lower semicontinuous for the finite topology.} \\ \textbf{ii.} & \forall x \in K, \quad y \to \phi(x, y) \text{ is concave and upper semicontinuous.} \end{cases}$$

We also assume that

(33) $$\exists y_0 \in K \quad \text{such that} \quad x \to \phi(x, y_0) \text{ is inf compact}$$

and that ϕ *is* **monotone** *in the sense that*

(34) $$\begin{cases} \textbf{i.} & \forall y \in K, \quad \phi(y, y) \leqslant 0 \\ \textbf{ii.} & \forall x, y \in K, \quad \phi(x, y) + \phi(y, x) \geqslant 0 \end{cases}$$

Then there exists $\bar{x} \in K$ *such that*

(35) $$\sup_{y \in K} \phi(\bar{x}, y) \leqslant 0$$ ▲

Proof. Since we have the inequality

(36) $$v^\natural \leqslant \sup_{\{y_1, \ldots, y_n\} \in \mathscr{S}} \inf_{x \in \text{co}\{y_1, \ldots, y_n\}} \max_{y \in \text{co}\{y_1, \ldots, y_n\}} \phi(x, y)$$

and since the Ky Fan inequality in finite dimensional spaces implies that

(37) $$\inf_{x \in \text{co}\{y_1,\ldots,y_n\}} \sup_{y \in \text{co}\{y_1,\ldots,y_n\}} \phi(x, y) \leqslant \sup_{y \in K} \phi(y, y) \leqslant 0$$

we deduce that

(38) $$v^{\natural} \leqslant 0$$

Lemma 2.12 implies the existence of a generalized sequence of elements $x_\mu \in K$ satisfying

(39) $$\forall y \in K, \quad \exists \mu(y) \text{ such that } \limsup_{\mu \geqslant \mu(y)} f(x_\mu, y) \leqslant v^{\natural} \leqslant 0$$

By the compactness assumption (33), we infer that the subsequence remains in a compact subset and consequently, a subsequence (again denoted by) x_μ converges to some $\bar{x} \in K$. We shall prove that $\bar{x}$ solves inequality (35). If not,

(40) $$\text{there exists } y \in K \text{ such that } 0 < \phi(\bar{x}, y)$$

Since the function $t \to \phi(\bar{x} + t(y - \bar{x}), y)$ is lower semicontinuous by assumption (31, **i**), we deduce from (39) that there exists $\bar{t} \in]0, 1[$ such that

(41) $$0 < \phi(\bar{x} + \bar{t}(y - \bar{x}), y)$$

We prove now that

(42) $$0 \leqslant \phi(\bar{x} + \bar{t}(y - \bar{x}), \bar{x})$$

Indeed, the monotonicity (33, **ii**) of ϕ implies that by setting $z = \bar{x} + \bar{t}(y - \bar{x})$,

$$0 \leqslant \limsup_{\mu \geqslant \mu(z)} (\phi(x_\mu, z) + \phi(z, x_\mu))$$

$$\leqslant \limsup_{\mu \geqslant \mu(z)} \phi(x_\mu, z) + \limsup_{\mu \geqslant \mu(z)} \phi(z, x_\mu)$$

but $\limsup_{\mu \geqslant \mu(z)} \phi(x_\mu, z) \leqslant v^{\natural} \leqslant 0$ by (39) and $\limsup_{\mu \geqslant \mu(z)} \phi(z, \bar{x}_\mu) \leqslant \phi(z, x)$ because $y \to \phi(x, y)$ is upper semicontinuous [(32, **ii**)], hence, (42) holds true. The concavity of the function $z \to \phi(\bar{x} + \bar{t}(y - \bar{x}), z)$ and inequalities (41) and (42) imply that

(43) $$0 < \phi(\bar{x} + \bar{t}(y - \bar{x}), \bar{x} + t(y - \bar{x}))$$

which contradicts assumption (34, **i**). ■

As a first application of Ky Fan's inequality, we prove the existence of noncooperative equilibria in n-person games.

Let us consider n players. When i is a player, we denote by $\hat{i}=\{j\in\{1,\ldots,n\}|j\neq i\}$ the set of the $n-1$ other players. Each player must choose a strategy x_i in a strategy set M_i according to a given mechanism. An example of such a mechanism occurs when each player chooses a strategy through a decision rule C_i that constrains his or her choice when the $n-1$ other players $j\in\hat{i}$ have implemented their strategy $x_{\hat{i}}\in \ldots_{j\neq i}M_j$. For simplicity, we set

$$\text{(44)} \qquad \forall i, \quad M_{\hat{i}}:=\prod_{j\neq i} M_j, \quad x_{\hat{i}}=\{x_j\}_{j\in\hat{i}}, \quad \vec{M}:=\prod_{i=1}^{n} M_i$$

A decision rule for the ith player is a set-valued map C_i from $M_{\hat{i}}$ to M_i.

DEFINITION 10

A multistrategy $\vec{x}=(x_1,\ldots,x_n)\in\vec{M}$ is consistent with respect to decision rules C_i if and only if

$$\text{(45)} \qquad \forall i=1,\ldots,n, \quad x_i\in C_i(x_{\hat{i}}) \qquad \blacktriangle$$

If we set

$$\forall \vec{x}\in\vec{M}, \quad \vec{C}(\vec{x})=\prod_{i=1}^{n} C_i(x_{\hat{i}})$$

the subset of consistent multistrategies is the subset of fixed points $\vec{x}$ of the set-valued map $\vec{C}$

$$\vec{x}\in\vec{C}(\vec{x})$$

This problem can be solved by various available fixed point theorems. We consider here the particular case where decision rules are canonical decision rules constructed from loss functions as follows. We describe the game by

$$\text{(46)} \qquad n \text{ loss functions } f_i \text{ from } \vec{M} \text{ to } R$$

that associate to each player i and to every multistrategy $\vec{x}\in M$ a loss $f_i(\vec{x})\in R$. From the point of view of player i, we can write a multistrategy $\vec{x}$ in the form $(x_i, x_{\hat{i}})\in M_i\times M_{\hat{i}}$, since each player can choose $x_i\in M_i$ but has no control over the choice of $x_{\hat{i}}$. So, we set

$$\text{(47)} \qquad f_i(\vec{x}):=f_i(x_i, x_{\hat{i}})$$

We associate with $x_{\hat{i}}\in M_{\hat{i}}$ the subset

$$\text{(48)} \qquad \bar{C}_i(x_{\hat{i}}):=\{x_i\in M_i|\, f_i(x_i, x_{\hat{i}})=\inf_{y_i\in M_i} f_i(y_i, x_{\hat{i}})\}$$

of the strategies x_i that minimizes this loss function $y_i \to f_i(y_i, x_{\hat{i}})$ when the other players implement $x_{\hat{i}}$.

DEFINITION 11
A n-person game in normal (or strategic) form is defined by n loss functions f_i defined on $\vec{M}$. The maps $\bar{C}_i$ from $M_{\hat{i}}$ to M_i defined by (7) are called canonical decision rules. *The multistrategies $\vec{x}$ that are consistent with these canonical decision rules are called* noncooperative equilibria (*or* Nash equilibria) *of the game. We set*

$$\text{(49)} \qquad \forall \vec{x}, \vec{y} \in \vec{M}, \qquad \phi(\vec{x}, \vec{y}) := \sum_{i=1}^{n} (f_i(x_i, x_{\hat{i}}) - f_i(y_i, x_{\hat{i}})) \qquad \blacktriangle$$

PROPOSITION 12
The following statements are equivalent:

$$\text{(50)} \qquad \begin{cases} \textbf{i.} & \vec{x} \in \vec{M} \textit{ is a noncooperative equilibrium.} \\ \textbf{ii.} & \forall i = 1, 2, \dots, n, \quad f_i(x_i, x_{\hat{i}}) = \min_{y_i \in M_i} f_i(y_i, x_{\hat{i}}) \\ \textbf{iii.} & \vec{x} \in M \textit{ satisfies } \sup_{\vec{y} \in \vec{M}} \phi(\vec{x}, \vec{y}) = 0. \end{cases} \qquad \blacktriangle$$

Proof. **a.** Conditions (50, **i** and **ii**) are obviously equivalent by the very definition of canonical decision rules.
b. We prove that (50, **i**) implies (50, **iii**). Let $\vec{y} := (y_1, \dots, y_n) \in \vec{M}$ be given. Since for all $i = 1, \dots, n$, $f_i(x_i, x_{\hat{i}}) - f_i(y_i, x_{\hat{i}}) \leq 0$, by adding these inequalities we obtain $\phi(\vec{x}, \vec{y}) \leq 0$.
c. We prove that (50, **iii**) implies (50, **ii**). Let i be fixed. We choose $\vec{y}$ such that $y_j = x_j$ for all $j \in \hat{i}$ and $y_i \in M_i$. Therefore,

$$f_i(x_i, x_{\hat{i}}) - f_i(y_i, x_{\hat{i}}) = \phi(\vec{x}, \vec{y}) \leq 0 \qquad \blacksquare$$

We then prove the fundamental theorem of noncooperative game theory.

THEOREM 13 (NASH)
Suppose that for all $i = 1, \dots, n$,

$$\text{(51)} \qquad \begin{cases} \textbf{i.} & \textit{the strategy set } M_i \textit{ is convex and compact.} \\ \textbf{ii.} & \textit{the loss function } f_i \textit{ is continuous, and for every } x_{\hat{i}} \in M_{\hat{i}}, \\ & y_i \to f(y_i, x_{\hat{i}}) \textit{ is convex.} \end{cases}$$

Then there exists a noncooperative equilibrium. $\blacktriangle$

Proof. The set $\vec{M} := \prod_{i=1}^{n} M_i$ is convex and compact and the function $\vec{\phi}$ defined on $\vec{M} \times \vec{M}$ by (49) satisfies

$$\text{(52)} \qquad \begin{cases} \textbf{i.} \ \forall \vec{y} \in \vec{M}, & \vec{x} \to \phi(\vec{x}, \vec{y}) \text{ is lower semicontinuous.} \\ \textbf{ii.} \ \forall \vec{x} \in \vec{M}, & y \to \phi(\vec{x}, \vec{y}) \text{ is concave.} \end{cases}$$

Hence, Ky Fan's inequality implies the existence of $\vec{x} \in \vec{M}$ satisfying

$$\sup_{\vec{y} \in \vec{M}} \phi(\vec{x}, \vec{y}) \leqslant \sup_{\vec{y} \in \vec{M}} \phi(\vec{y}, \vec{y}) = 0. \tag{53}$$

By proposition 12, $\vec{x}$ is a noncooperative equilibrium. ■

4. EXISTENCE OF ZEROS OF SET-VALUED MAPS

In a straightforward way, Ky Fan's inequality gives an array of useful sufficient conditions implying the existence of solutions to inclusions of the form

$$\bar{x} \in K \quad \text{and} \quad 0 \in F(\bar{x}) + P \; [\text{or } F(\bar{x}) \cap -P \neq \varnothing]$$

where P is a closed convex cone. The case where $P = \{0\}$ yields results on the existence of zeros of F. Many problems arising in economics and game theory fall in this class. Namely, let Y and Y^* be two paired vector spaces supplied with their weak topologies and $P \subset Y^*$ a closed convex cone and $P^- \subset Y$ be its negative polar cone.

THEOREM 1

Let K be a compact metric space and $F: K \to Y^*$ a strict set-valued map. We assume that

$$\begin{cases} \textbf{i.} & F \text{ is } P\text{-upper hemicontinuous in the sense that } \forall y \in P^- \\ & x \to \sigma(F(x), y) \text{ is upper semicontinuous.} \\ \textbf{ii.} & \forall x \in K, \quad F(x) + P \text{ is closed and convex.} \end{cases} \tag{1}$$

$$\text{there exists a finitely continuous map } C \text{ from } P^- \text{ to } K \text{ such that } \forall y \in P^-, \; \sigma(F(C(y)), y) \geqslant 0. \tag{2}$$

Then there exists $\bar{x} \in K$ such that $0 \in F(\bar{x}) + P$. ▲

Proof. We introduce the function ϕ defined on $K \times P^-$ by

$$\phi(x, y) := -\sigma(F(x), y) \tag{3}$$

This function is lower semicontinuous with respect to x thanks to the assumption (1, **i**), and concave with respect to y. Hence, theorem 3.8 implies the existence of $\bar{x} \in K$ such that, C being finitely continuous,

$$\sup_{y \in P^-} \phi(x, y) \leqslant \sup_{y \in P^-} \phi(C(y), y) \tag{4}$$

By assumption (2), $\phi(C(y), y) \leqslant 0$ for all $y \in P^-$. Hence, we have proved that

$$\forall y \in P^-, \qquad 0 \leqslant \sigma(F(\bar{x}), y) \tag{5}$$

Since $\sigma(P^-, y) = 0$ when $y \in P^-$ and $\sigma(P^-, y) = +\infty$ when $y \notin P^-$, we deduce that

$$\forall y \in Y, \qquad 0 \leqslant \sigma(F(\bar{x}) + P, y) \tag{6}$$

The subset $F(\bar{x}) + P$ being closed and convex by assumption (1, **ii**), it follows that $0 \in F(\bar{x}) + P$. ∎

The problem, then, amounts to finding finitely continuous maps from p^- to K satisfying assumption (2). When they do exist, we can think of taking a retraction to K, a subset of Y: A retraction from Y to a subset $K \subset Y$ is a map r from Y to K such that $r(y) = y$ for all $y \in K$. For instance, if K is a closed convex subset of a Hilbert space, the projection $r := \pi_K$ is a continuous retraction. So by taking $P = \{0\}$ and C a retraction onto K, we obtain the following corollary.

COROLLARY 2

Let K be a compact subset of Y that has a finitely continuous retraction r. Let F be a strict upper hemicontinuous map from K to Y^* with closed convex images satisfying

$$\forall y \in Y, \qquad \sigma(F(r(y)), y) \geqslant 0 \tag{7}$$

Then there exists $\bar{x} \in K$ such that $0 \in F(\bar{x})$. ▲

For instance, when K is the ball of radius a, we obtain the following consequence, the set-valued extension of theorem 3 of Chapter 2, Section 2.

COROLLARY 3

Let Y be a finite dimensional space and F an upper semicontinuous map from the ball of radius a to the closed convex subsets of Y satisfying

$$\forall x \in Y, \qquad \|x\| = a, \qquad \sigma(F(x), x) \geqslant 0 \tag{8}$$

Then there exists a solution $\bar{x}$ to

$$\|\bar{x}\| \leqslant a \quad \textit{and} \quad 0 \in F(\bar{x}) \tag{9}$$ ▲

Proof. Indeed, take $r(y) = y$ when $\|y\| \leqslant a$ and $r(y) = ay/\|y\|$ when $\|y\| \geqslant a$, so that (8) implies (7). ∎

The following theorem has many applications to economics, as we shall see in Section 5.5. We consider the simplex

$$\Sigma^n := \left\{ x \in R^n_+ \,\middle|\, \sum_{i=1}^{n} x_i = 1 \right\}$$

THEOREM 4

Let F be a strict upper hemicontinuous map from Σ^n to R^n with compact convex values. If

$$\text{(10)} \qquad \forall x \in \Sigma^n, \qquad \sigma(F(x), x) \geqslant 0$$

then there exists $\bar{x} \in \Sigma^n$ such that $0 \in F(\bar{x}) - R^n_+$. ▲

Proof. We take $Y = Y^* = R^n$, $P = -R^n_+$, $P^- = R^n_+$, $K = \Sigma^n$, and C to be the map defined on K by $C(y) := y / \sum_{i=1}^n y_i$. Assumptions of theorem 1 are obviously satisfied, and the existence of a solution $\bar{x} \in \Sigma^n$ to $0 \in F(\bar{x}) - R^n_+$ ensues. ■

Remark

More generally, we can assume that

$$\text{(11)} \qquad K \text{ is a weakly compact subset of } Y^* \text{ such that } 0 \notin K$$

and take

$$\text{(12)} \qquad P = K^-$$

because we can prove that $P^- = \bigcup_{\lambda > 0} \lambda K$ is spanned by K. It can also be proved that P^- is spanned by a weakly compact subset that does not contain the origin if and only if the interior of P (for the Mackey topology) is nonempty. ■

The following generalization, although somewhat technical, may be very useful.

THEOREM 5

We posit assumptions **(i)** *and* **(ii)** *of theorem* 1. *We assume that*

(13) $\forall \varepsilon > 0$, *there exists a finitely continuous map C_ε from P^- to K such that $\forall y \in P^-$,* $\qquad \sigma(F(C_\varepsilon(y)), y) \geqslant -\varepsilon$

Then there exists $\bar{x} \in K$ such that $0 \in F(\bar{x}) + P$. ▲

Proof. The proof is the same as the proof of theorem 1, where we replace (4) by the following consequence of theorem 3.8:

$$\sup_{y \in P^-} \phi(\bar{x}, y) \leqslant \inf_{\varepsilon > 0} \sup_{y \in P^-} \phi(C_\varepsilon(y), y)$$

Assumption (10) implies that

$$\inf_{\varepsilon>0} \sup_{Y \in P^-} \phi(C_\varepsilon(y), y) \leq \inf_{\varepsilon>0} \varepsilon = 0. \qquad \blacksquare$$

Remark

We can also relax the assumption that K is compact. For instance, it is sufficient to assume that there exists $y_0 \in P^-$ such that

(14) $$K_0 := \{x \in K \mid \sigma(F(x), y_0) \geq 0\} \text{ is compact} \qquad \blacksquare$$

Remark

We can imagine applying an analogous proof for upper hemicontinuous maps F with closed convex values from a compact convex subset $K \subset X$ to X^* by applying theorem 3.8 to the function ϕ defined by

(15) $$\phi(x, p) := -\sigma(F(x), p)$$

We then need the existence of a finitely continuous map C from X^* to K such that

(16) $$\forall p \in X^*, \qquad \sigma(F(C(p), p)) \geq 0$$

One natural example of such a map C would be a selection of the subdifferential $\partial\sigma_K(-p)$ of the support function of K (which is continuous—hence subdifferentiable—, because K is compact). Since a finitely continuous selection does not necessarily exist, we have to devise another strategy for proving that assumption (16) for $\partial\sigma_K$ implies the existence of zeros of F. We recall that x belongs to $\partial\sigma_K(-p)$ if and only if $-p$ belongs to the normal cone $N_K(x)$ to K at x. Hence, condition (16) for $\partial\sigma_K$ can be written

(17) $$\forall x \in K, \qquad \forall p \in N_K(x), \qquad \sigma(F(x), -p) \geq 0$$

This condition—which will play an important role—is called the *normal condition*. By using the tangent cone $T_K(x)$ to K at x, which is the negative polar cone of the normal cone, we obtain a dual version of (17), which we call the *tangential condition*. We investigate their properties in a more general framework. $\blacksquare$

Let X and Y be two Hausdorff, locally convex vector spaces, A belong to $\mathscr{L}(X, Y)$, $K \subset X$ a closed convex subset, and $F: K \to Y$ a strict set-valued map.

DEFINITION 6

We shall say that F satisfies the tangential condition *with respect to A if*

(18) $$\forall x \in K, \qquad F(x) \cap cl(A T_K(x)) \neq \varnothing$$

and the normal condition *with respect to A if*

$$(19)\qquad \forall x \in K, \quad \forall p \in A^{*^{-1}} N_K(x), \quad \text{then } \sigma(F(x), -p) \geq 0 \qquad \blacktriangle$$

When $X = Y$ and $A = 1$, we omit mentioning with respect to A.

LEMMA 7

The tangential condition (18) *implies the normal condition* (19). *The converse is true when the images $F(x)$ of F are convex and compact.* ▲

Proof. **a.** Let $x \in K$ and $v \in F(x) \cap \mathrm{cl}(AT_K(x))$ be chosen. Hence, $v = \lim_{n\to\infty} Au_n$ where u_n belongs to $T_K(x)$. Let p such that A^*p belongs to $N_K(x)$. Then,

$$\sigma(F(x), -p) \geq \langle -p, v\rangle = \lim_{n\to\infty} \langle -p, Au_n\rangle = \lim_{n\to\infty} \langle -A^*p, u_n\rangle \geq 0$$

because $\langle A^*p, u_n\rangle \leq 0$ for all $u_n \in T_K(x) = N_K(x)^-$.
b. Conversely, assume that $F(x)$ is convex and compact and that $0 \notin F(x) - \mathrm{cl}(AT_K(x))$, that is, $F(x) \cap \mathrm{cl}(AT_K(x)) \neq \varnothing$. The Hahn–Banach separation theorem implies the existence of $p \in Y^*$ and $\varepsilon > 0$ such that

$$\sigma(F(x), -p) \leq \inf_{v \in T_K(x)} \langle -p, Av\rangle - \varepsilon < 0$$

Since $T_K(x)$ is a cone, this implies that $A^*p \in T_K(x)^- = N_K(x)$ and that $\sigma(F(x), -p) \leq -\varepsilon < 0$. So, the normal condition (19) is contradicted. ■

The calculus on tangent cones to convex subsets allows us to check the tangential condition in many instances of closed convex subsets. We point out the following important, although obvious, remark.

PROPOSITION 8

If two set-valued maps F_1 and F_2 satisfy the (*dual*) *tangential condition, so does the set-valued map $\alpha_1 F_1 + \alpha_2 F_2$ for $a_1, \alpha_2 > 0$.* ▲

In particular, we use the fact that when F satisfies the tangential condition, so does the map $x \to F(x) - Ax + y$ when y is given in $A(K)$.

Remark

This is a first property of stability of the tangential condition and, consequently, of its consequences, under perturbations of F. We mention another property of stability under small perturbations:

PROPOSITION 9

Let K be a convex subset with nonempty interior and let F be a set-valued map from K to X with compact graph satisfying the strong internal tangential condition

$$(20)\qquad \forall x \in K, \quad F(x) \subset \mathrm{Int}\, T_K(x)$$

Then there exists $\alpha > 0$ *such that any set-valued map G close to F in the sense that*

(21) $$\text{graph } G \subset \text{graph } F + \alpha(B \times B)$$

also satisfies the strong internal tangential condition. ▲

Proof. Condition (20) implies that the graph of F, which is compact, is contained in the graph of Int T_K, which is open. Then there exists α such that condition (21) holds true. ■

We now prove our basic result—the existence of a zero of F that belongs to K when the tangential condition holds true.

THEOREM 10

Let X and Y be two Hausdorff, locally convex vector spaces, $A \in \mathscr{L}(X, Y)$ *a continuous linear operator,* $K \subset X$ *compact convex, and F an upper hemicontinuous map from K to Y with nonempty closed convex images. We posit the following tangential condition*:

(22) $$\forall x \in K, \qquad F(x) \cap \text{cl}(AT_K(x)) \neq \varnothing$$

Then,

(23) $$\begin{cases} \textbf{i.} & \textit{there exists a zero } \bar{x} \in K \textit{ of } F: 0 \in F(\bar{x}) \\ \textbf{ii.} & \forall y \in A(K), \quad \exists \hat{x} \in K \textit{ such that } A\hat{x} - y \in F(\hat{x}) \end{cases}$$ ▲

Remark

We can regard the second condition as a perturbation result

(24) $$\begin{cases} \text{if there exists a solution } \tilde{x} \in K \text{ to the linear equation} \\ y = A\tilde{x}, \text{ then there exists a solution } \hat{x} \in K \text{ to the} \\ \text{perturbed inclusion } y \in A\hat{x} - F(\hat{x}) \end{cases}$$

which also amounts to saying that

(25) $(A - F)^{-1} \cap K$ is a strict set-valued map from $A(K)$ to K. ■

When $X = Y$ and $A = 1$, we obtain the following consequence, stated in theorem 11.

THEOREM 11

Let X be a Hausdorff, locally convex vector space, $K \subset X$ *a compact convex subset, and F a strict upper hemicontinuous map from K to X with closed convex images satisfying the tangential condition*

(26) $$\forall x \in K, \qquad F(x) \cap T_K(x) \neq \varnothing$$

Then,

$$(27)\qquad \begin{cases} \textbf{i.} & \textit{there exists } \bar{x} \in K, \textit{ solution to } 0 \in F(\bar{x}). \\ \textbf{ii.} & \forall y \in K, \quad \exists \hat{x} \in K, \textit{ solution to } y \in \hat{x} - F(\hat{x}) \end{cases}$$

▲

Remark

Haddad's viability theorem states that under the assumptions of theorem 11, for all $x_0 \in K$, the differential inclusion

$$(28)\qquad x'(t) \in F(x(t)), \qquad x(0) = x_0$$

has a solution $x(.)$ that is viable in the sense that

$$(29)\qquad \forall t \geqslant 0, \qquad x(t) \in K$$

Actually, the tangential condition (26) is also necessary: It holds true whenever problems (28–29) has a solution for all $x_0 \in K$ [see Aubin-Cellina [1984], Chapter 4]. ■

Proof of Theorem 10. **a.** The second conclusion follows from the first: Since the set-valued map G defined by $G(x) := F(x) + y - Ax$ is the sum of the two maps F and $y - A$ that satisy the tangential condition, then G satisfies it and, consequently, has a zero $\hat{x}$ that is a solution to the inclusion $A\hat{x} - y \in F(\hat{x})$.

b. We denote by $\sigma(F(x), q) := \sup_{v \in F(x)} \langle q, v \rangle$ the support function of the closed convex subset $F(x)$. To prove the existence of a zero, we assume the contrary: $\forall x \in K$, $0 \notin F(x)$ and derive a contradiction. Since the subsets $F(x)$ are closed and convex, the separation theorem implies that

$$\forall x \in K, \qquad \exists p \in Y^* \text{ such that } \sigma(F(x), -p) < 0$$

We set

$$(30)\qquad \Delta_p := \{x \in K \mid \sigma(F(x), -p) < 0\}$$

So, the statement that no zero exists takes the form

$$K \subset \bigcup_{p \in Y^*} \Delta_p$$

c. Since F is upper hemicontinuous, the subsets Δ_p are open. Hence, the compact subsets K can be covered by n open subsets Δ_{p_i}. Let $\{a_i\}$ for $i = 1, \ldots, n$ be a continuous partition of unity associated to this covering.

We introduce the function ϕ defined $K \times K$ by

$$(31)\qquad \phi(x, y) = -\sum_{i=1}^{n} a_i(x) \langle A^* p_i, x - y \rangle$$

It is continuous with respect to x, affine with respect to y, and satisfies $\phi(y, y)=0$ for all $y \in K$.

So the assumptions of the Ky Fan inequality (theorem 3.5) hold and, consequently, there exists $x_* \in K$ such that

$$\forall y \in K, \qquad \phi(x_*, y)=\langle -A^*p_*, x_*-y\rangle \leqslant 0 \tag{32}$$

where we set $p_*=\sum_{i=1}^n a_i(x_*)p_i$. In other words, A^*p_* belongs to $N_K(x_*)$. The tangential condition (18) implies that the normal condition (19) holds true (by lemma 7). Therefore,

$$\sigma(F(x_*), -p_*)\geqslant 0 \tag{33}$$

d. The latter inequality is impossible: Let I be the set of indices i such that $a_i(x_*)>0$. It is nonempty since $\sum_{i=1}^n a_i(x_*)=1$. If $i \in I$, then $x_* \in \Delta_{p_i}$ and thus, $\sigma(F(x_*), -p_i)<0$. Therefore

$$\sigma(F(x_*), -p_*)=\sigma(F(x_*), -\sum_{i\in I} a_i(x_*)p_i)\leqslant \sum_{i\in I} a_i(x_*)\sigma(F(x_*), -p_i)$$

(by the convexity of support functions). We have proved that $\sigma(F(x_*), -p_*)<0$, which is the contradiction we were looking for. ■

Remark

Theorem 10 remains true even if we assume only that F satisfies the normal condition (19). ■

Remark

We can even let the continuous linear operator A depend on x as in theorem 12.

THEOREM 12

We introduce

(34)
- **i.** *K, a compact convex subset of X*
- **ii.** *F, an upper hemicontinuous map from K to Y with closed convex values*
- **iii.** *$A\colon K\to \mathscr{L}(X, Y)$ a continuous map associating with each $x \in K$ a continuous linear operator from X to Y*

We posit that the tangential condition,

$$\forall x \in K, \qquad F(x)\cap \mathrm{cl}(A(x)T_K(x))\neq\varnothing \tag{35}$$

holds. Then

(36) $\begin{cases} \textbf{i.} & \textit{There exists a zero } \bar{x} \in K \textit{ of } F: 0 \in F(\bar{x}). \\ \textbf{ii.} & \forall y \in K, \textit{ there exists } \hat{x} \in K \textit{ satisfying } A(\hat{x})(\hat{x}-y) \in F(\hat{x}). \end{cases}$ ▲

Proof. **a.** The second statement follows from the first applied to the map G defined by $G(x):=F(x)+A(x)(y-x)$.
b. The proof of the first statement is the same as the proof of theorem 10, where the function ϕ defined by (31) is replaced by the function ϕ defined by

$$\phi(x, y) = -\sum_{i=1}^{n} a_i(x)\langle p_i, A(x)(x-y)\rangle \tag{37}$$ ■

We deduce now from theorem 11 the well known Kakutani fixed point theorem as well as some of its extensions.

THEOREM 13 (KAKUTANI)
Let K be a compact convex subset and G an upper semicontinuous map from K to K with compact convex images. Then there exists a fixed point $x_* \in K$ of G. ▲

Proof. We set $F(x):=G(x)-x \subset K-x \subset T_K(x)$. Hence, $F(.)$ is an upper hemicontinuous map from K to X that satisfies the tangential condition. By theorem 11, there is a zero $x_* \in K$, of F, which is a fixed point of G. ■

The same proof implies the following statement. We say that G is *inward* if

$$\forall x \in K, \qquad G(x) \cap (x + T_K(x)) \neq \varnothing \tag{38}$$

THEOREM 14
Let G be an upper hemicontinuous map from a compact convex subset $K \subset X$ to the closed convex subsets of X. If G is inward, then it has a fixed point $x_* \in K$. ▲

We also mention the following result. We say that G is *outward* if

$$\forall x \in K, \qquad G(x) \cap (x - T_K(x)) \neq \varnothing \tag{39}$$

THEOREM 15
Let G be an upper hemicontinuous map from a compact convex subset $K \subset X$ to the closed convex subsets of X. If G Is outward, then

(40) $\begin{cases} \textbf{i.} & \textit{It has a fixed point } x_* \in K. \\ \textbf{ii.} & K \subset G(K) \; [\textit{i.e., for all } y \in K, \; \exists x \in K \textit{ such that } y \in G(x)] \end{cases}$ ▲

Proof. It follows from theorem 11 applied to the map F defined on K by $F(x) := x - G(x)$. Then a zero $\bar{x}$ of F is a fixed point of G, and a solution $\hat{x}$ to $y \in \hat{x} - F(\hat{x})$ is a solution $\hat{x}$ to $y \in G(\hat{x})$. ■

We now mention a convenient sufficient condition implying that a set-valued map G is outward. We recall that

$$\partial\sigma_K(p) := \{x \in K \mid \langle p, x\rangle = \sigma_K(p)\} \tag{41}$$

is the support zone of K at p and that $x \in \partial\sigma_K(p)$ if and only if $p \in N_K(x)$.

PROPOSITION 16

Let K be a closed convex subset of X and let G be a set-valued map from K to K satisfying

$$\forall p \in X^*, \quad \forall x \in \partial\sigma_K(p), \qquad G(x) \cap \partial\sigma_K(p) \neq \varnothing \tag{42}$$

Then G is outward. In particular, this is the case of a map G satisfying

$$\forall p \in X^*, \qquad G(\partial\sigma_K(p)) \subset \partial\sigma_K(p) \tag{43}$$ ▲

Proof. Let $x \in \partial\sigma_K(p)$ and $y_0 \in G(x) \cap \partial\sigma_K(p)$ be fixed. Hence, $v = y_0 - x$ belongs to $G(x) - x$ and $-T_K(x)$ because, $\forall p \in N_K(x)$, $\langle p, y_0\rangle - \langle p, x\rangle = \sigma_K(p) - \sigma_K(p) = 0$. ■

We can relax the assumption that K is compact by replacing it with a coerciveness assumption.

THEOREM 17

Let K be a closed convex subset of a finite dimensional space and let F be a strict upper hemicontinuous map with compact convex images satisfying the coerciveness assumption

$$\lim_{\substack{\|x\| \to \infty \\ x \in K}} \sigma(F(x), x) < 0 \tag{44}$$

We posit the tangential condition,

$$\forall x \in K, \quad F(x) \cap T_K(x) \neq \varnothing \tag{45}$$

Then there exists a zero $x \in K$ of F. If we posit the stronger coerciveness assumption,

(46)
$$\lim_{\substack{\|x\|\to\infty \\ x\in K}} \frac{\sigma(F(x), x)}{\|x\|} = -\infty$$

then for all $y \in K$, there exists a solution $\hat{x} \in K$ to the inclusion $y \in \hat{x} - F(\hat{x})$. ▲

Proof. Coerciveness assumption (44) implies that there exist $\varepsilon > 0$ and $a > 0$ such that

$$\sup_{\substack{\|x\|\geq a \\ x\in K}} \sigma(F(x), x) \leq -\varepsilon < 0$$

This implies that $F(x) \subset T_{aB}(x)$. By taking the number a large enough so that $K \cap aB \neq \emptyset$, we know that $T_{K\cap aB}(x) = T_K(x) \cap T_{aB}(x)$. So the tangential condition (45) implies that

(47)
$$\forall x \in K \cap aB, \quad F(x) \cap T_{K\cap aB}(x) \neq \emptyset$$

It suffices to apply theorem 11. To prove the second part of the theorem, we replace F by the map G defined by $G(x) = F(x) + y - x$, which satisfies the coerciveness assumption (44) whenever F satisfies the stronger coerciveness assumption (46). ■

We can deduce the Leray–Schauder theorem on the existence of stationary points from theorem 11 by Poincare's continuation method.

We take $X = R^n$ and K to be a compact convex subset with a nonempty interior, so that the boundary $\partial K = K \cap \complement \operatorname{Int} K$ of K is distinct from K.

THEOREM 18 (LERAY–SCHAUDER)

Consider a compact convex subset $K \subset R^n$ with nonempty interior and a strict upper hemicontinuous set-valued map F from $[0, 1] \times K$ to R^n, with closed convex values. Suppose that for $\lambda = 0$, the set-valued map $x \to F(0, x)$ satisfies the tangential condition,

(48)
$$\forall x \in \partial K, \qquad F(0, x) \cap T_K(x) \neq \emptyset$$

and

(49)
$$\forall x \in \partial K, \qquad \forall \lambda \in [0, 1[, \qquad 0 \notin F(\lambda, x)$$

Then there exists $\bar{x} \in K$ such that $0 \in F(1, \bar{x})$. ▲

Proof. We suppose that the conclusion is false and derive a contradiction. We set $N := \partial K$, a closed subset of K and introduce the subset

(50)
$$M := \{x \in K \mid \exists \lambda \in [0, 1] \text{ satisfying } 0 \in F(\lambda, x)\}$$

The subset M is nonempty because it contains a solution $\tilde{x} \in K$ to the inclusion $0 \in F(0, \tilde{x})$, which exists by theorem 11 [thanks to assumption (48)]. It is closed, since the graph of F is closed. The intersection $M \cap N$ is empty: If $x \in N$ and if $\lambda \in [0, 1[$, assumption (49) implies that $x \notin M$.

We then introduce a continuous function ϕ from K to $[0, 1]$ that is equal to zero on N and to one on M; for instance,

$$\phi(x) := \frac{d(x, N)}{d(x, M) + d(x, N)}$$

We define the set-valued map G on K by

$$G(x) := F(\phi(x), x) \tag{51}$$

Then G is clearly upper hemicontinuous with nonempty closed convex values. It coincides with $F(0, \cdot)$ on $N = \partial K$ and, consequently, satisfies the assumptions of theorem 21. Hence, there exists a solution $\bar{x} \in K$ to $0 \in G(\bar{x}) = F(\phi(\bar{x}), \bar{x})$. But this implies that $\bar{x} \in M$ and, therefore, $\phi(\bar{x}) = 1$, so that $0 \in F(1, \bar{x})$. This is a contradiction. ■

The following consequence in corollary 19 is very useful.

COROLLARY 19

Let K be a compact convex subset of R^n with a nonempty interior and let G and H be two strict upper hemicontinuous maps from K to R^n with closed convex values. We assume that

$$\forall x \in \partial K, \qquad G(x) \cap T_K(x) \neq \varnothing \tag{52}$$

and

$$\forall x \in \partial K, \qquad \forall \mu \geqslant 0, \qquad 0 \notin G(x) + \mu H(x) \tag{53}$$

Then there exists a solution $\bar{x} \in K$ to the inclusion $0 \in H(\bar{x})$. ▲

Proof. We apply the preceding theorem with $F(\lambda, x) := (1-\lambda)G(x) + \lambda H(x)$. Condition (49) can be written in form (53). It follows that $H = F(1, \cdot)$ has a zero in K. ■

COROLLARY 20

Let x_0 belong to the interior of a compact convex subset $K \subset R^n$ and let H be a strict upper hemicontinuous map from K to R^n with closed convex values. We *suppose that*

$$\forall x \in \partial K, \quad \forall \mu > 0, \quad x_0 \notin x + \mu H(x) \tag{54}$$

Then there exists a solution $\bar{x} \in K$ *to* $0 \in H(\bar{x})$. ▲

Proof. We apply corollary 19 with $G(x) = x - x_0$. ■

We proceed by giving a method for selecting a fixed point of a set-valued map F.

THEOREM 21

We assume that

(55) *K is a compact convex subset of a Hilbert space X*

and

(56) $\begin{cases} F \text{ is an upper hemicontinuous set-valued map from } K \text{ to } K \\ \quad \text{with nonempty closed convex values.} \end{cases}$

We consider a function $f : K \times K \to R$ *satisfying*

$$\begin{cases} \textbf{i.} & \forall y \in K, & x \to f(x, y) \text{ is lower semicontinuous.} \\ \textbf{ii.} & \forall x \in K, & y \to f(x, y) \text{ is concave.} \\ \textbf{iii.} & \forall y \in K, & f(y, y) \leqslant 0 \end{cases} \tag{57}$$

Finally, suppose that the set-valued map F and the function f are related by the property

$$\{x \in K \text{ such that } \sup_{y \in F(x)} f(x, y) \leqslant 0\} \text{ is closed.} \tag{58}$$

Then there exists a solution $\bar{x} \in K$ *to the "quasi-variational inequalities"*

$$\begin{cases} \textbf{i.} & \bar{x} \in F(\bar{x}) \\ \textbf{ii.} & \sup_{y \in F(\bar{x})} f(\bar{x}, y) \leqslant 0 \end{cases} \tag{59}$$ ▲

Remark

Assumptions (55) and (56) are those of Kakutani's fixed point theorem, which imply conclusion (59, **i**). Assumptions (55) and (57) are those of the Ky Fan inequality, and assumption (58) is a consistency hypothesis between f and F. ■

Proof. We argue by contradiction: If the conclusion is false, then for all $x \in K$, either $x \notin F(x)$ or $\alpha(x) := \sup_{y \in F(x)} f(x, y)$ is strictly positive. Saying that $x \in F(x)$ implies that there exists $p \in X^*$ such that $\langle p, x\rangle - \sigma(F(x), p) > 0$. We set

$$\text{(60)} \qquad \begin{cases} \textbf{i.} & \Delta_0 := \{x \in K \mid \alpha(x) > 0\} \\ \textbf{ii.} & \Delta_p := \{x \in K \mid \langle p, x\rangle - \sigma(F(x), p) > 0\} \end{cases}$$

Then the negation of the conclusion can be expressed in the form

$$\text{(61)} \qquad K \subset \Delta_0 \cup \bigcup_{p \in X^*} \Delta_p$$

Assumptions (56) and (57, **i**) imply that the sets Δ_0 and Δ_p are open. Since K is compact, it follows that there exist $p_1, \ldots, p_n$ such that

$$\text{(62)} \qquad K \subset \Delta_0 \cup \bigcup_{i=1}^{n} \Delta_{p_i}$$

and there exists a continuous partition of unity $\{a_i\}$ for $i = 0, \ldots, n$ associated to this covering.

We then introduce the function $\phi: K \times K \to R$ defined by

$$\text{(63)} \qquad \phi(x, y) := a_0(x) f(x,y) + \sum_{i=1}^{n} a_i(x)\langle p_i, x - y\rangle$$

This function ϕ is lower semicontinuous with respect to x, concave with respect to y, and satisfies $\phi(y, y) \leqslant 0$ for all $y \in K$, thanks to assumption (57, **iii**). The Ky Fan inequality (see theorem 3.5) implies the existence of $\bar{x} \in K$ satisfying

$$\text{(64)} \qquad \sup_{y \in K} \phi(\bar{x}, y) \leqslant 0$$

We are going to contradict this inequality by proving that there exists $\bar{y} \in K$ such that

$$\text{(65)} \qquad \phi(\bar{x}, \bar{y}) > 0$$

We take

$$\text{(66)} \qquad \begin{cases} \bar{y} \in F(\bar{x}) \text{ arbitrary when } \alpha(\bar{x}) \leqslant 0 \text{ and satisfying} \\ f(\bar{x}, \bar{y}) \geqslant \alpha(\bar{x})/2 \text{ when } \alpha(\bar{x}) > 0 \text{ (which is possible)} \end{cases}$$

Since $\{a_i\}$ for $i = 0, \ldots, n$ is a partition of unity, then $a_i(\bar{x}) > 0$ for at least one index $i=0,\ldots,n$. Inequality (65) will then follow from the following statements:

$$\text{(67)} \qquad \begin{cases} \textbf{i.} & a_0(\bar{x})>0 \text{ implies that } f(\bar{x},\bar{y})>0 \\ \textbf{ii.} & a_i(\bar{x})>0 \text{ implies that } \langle p_i, \bar{x}-\bar{y}\rangle>0 \end{cases}$$

Let us verify these statements. If $a_0(\bar{x})>0$, then $\bar{x}\in\Delta_0$ and, consequently, $\alpha(\bar{x})>0$. Hence, $f(\bar{x},\bar{y})\geqslant\alpha(\bar{x})/2>0$. If $a_i(\bar{x})>0$ for $i\neq 0$, then $\bar{x}\in\Delta_{p_i}$ and, consequently

$$\langle p_i, \bar{x}\rangle>\sigma(F(\bar{x}),p_i)\geqslant\langle p_i,\bar{y}\rangle$$

because $\bar{y}\in F(\bar{x})$. Hence, $\langle p_i,\bar{x}-\bar{y}\rangle>0$. ■

Since the function α defined by $\alpha(x):=\sup_{y\in F(x)} f(x,y)$ is lower semicontinuous when F and f are lower semicontinuous, we obtain the following corollary because assumption (58) is satisfied.

COROLLARY 22

We assume that K is convex and compact, $F: K\to K$ is a strict continuous set-valued map with closed convex values, and f is a lower semicontinuous function defined on $K\times K$, concave with respect to y and satisfying $\sup_{y\in K} f(y,y)\leqslant 0$. *There exists a solution $\bar{x}\in K$ to the quasi-variational inequalities* (59). ▲

We continue the study of n-person noncooperative games begun earlier. A game is defined by n strategy sets M_i and n loss functions f_i defined on the product $\vec{M}:=\prod_{i=1}^n M_i$ of the strategy sets, to which we have associated the canonical decision rules $\bar{C}_i$ defined by

$$\bar{C}_i(x_{\hat{i}}):=\{x_i\in M_i \mid f_i(x_i,x_{\hat{i}})=\inf_{y_i\in M_i} f_i(y_i,x_{\hat{i}})\}$$

This time, we add a concept of feasibility to the game through the n set-valued maps F_i restricting the strategies of the ith player to the subset $F_i(x_{\hat{i}})\subset M_i$ when all the other players have chosen their strategies $x_j\in M_j$, $j\neq i$. In such a game (often called a metagame), the other players influence player j

(a) Indirectly, by restricting j's feasible strategies to $F_j(x_{\hat{j}})$.

(b) Directly, by affecting j's loss function f_j.

The conjunction of those two effects lead to feasible decision rules defined by

$$\text{(68)} \qquad \tilde{C}_i(x_{\hat{i}}):=\{x_i\in F_i(x_{\hat{i}}) \mid f_i(x_i,x_{\hat{i}})=\inf_{y_i\in F_i(x_i)} f_i(y_i,x_{\hat{i}})\}$$

The multistrategies $\vec{x}\in\vec{M}$ that are consistent with these decision rules are called

social equilibria of the metagame. So, they are defined by the conditions

$$(69)\quad \begin{cases} \textbf{i.}\ \ \forall i=1,\ldots,n, & x_i \in F_i(x_{\hat{\imath}}) \quad \text{(feasibility)} \\ \textbf{ii.}\ \ \forall i=1,\ldots,n, & f_i(x_i, x_{\hat{\imath}}) = \inf\limits_{y_i \in F_i(x_{\hat{\imath}})} f_i(y_i, x_{\hat{\imath}}) \end{cases}$$

THEOREM 23

Suppose that for all $i=1,\ldots,n$,

(70)
- **i.** *The strategy set* M_i *is convex and compact.*
- **ii.** *The loss function* f_i *is continuous and for every* $x_{\hat{\imath}} \in M_{\hat{\imath}}$, $y_i \in f_i(y_i, x_{\hat{\imath}})$ *is convex.*
- **iii.** *The feasible map* $F_i\colon M_{\hat{\imath}} \to M_i$ *is continuous with closed convex images.*

Then there exists a social equilibrium $\vec{x}$, *that is, a multistrategy* $\vec{x}$ *satisfying conditions* (69). ▲

Proof. We set

$$(71)\qquad K := \prod_{i=1}^{n} M_i, \qquad F(\vec{x}) := \prod_{i=1}^{n} F_i(x_{\hat{\imath}})$$

and

$$(72)\qquad f(\vec{x}, \vec{y}) := \sum_{i=1}^{n} \left(f_i(x_i, x_{\hat{\imath}}) - f_i(y_i, x_{\hat{\imath}})\right)$$

Then the set K is convex and compact, and the set-valued map F is continuous with closed convex images. The function f is continuous and concave with respect to $\vec{y}$ and satisfies $\sup_{y \in K} f(\vec{y}, \vec{y}) = 0$. Thus corollary 22 implies the existence of a solution to the quasi-variational inequalities

$$\vec{x} \in F(\vec{x}) \qquad \text{and} \qquad \sup_{y \in F(x)} f(\vec{x}, \vec{y}) \leqslant 0$$

The first statement implies that $x_i \in F_i(x_{\hat{\imath}})$ for all $i=1,\ldots,n$. The second statement implies that for all $i=1,\ldots,n$ and for all $y_i \in M_i$, $f_i(x_i, x_{\hat{\imath}}) \leqslant f_i(y_i, x_{\hat{\imath}})$. It suffices to take $\vec{y}$ such that $\vec{y}_i = y_i$ and $\vec{y}_j = x_j$ for all $j \neq i$. Hence $\vec{y} \in F(\vec{x})$ and

$$f(\vec{x}, \vec{y}) = f_i(x_i, x_{\hat{\imath}}) - f_i(y_i, x_{\hat{\imath}}) + \sum_{j \neq i} \left(f_j(x_j, x_{\hat{\jmath}}) - f_j(x_j, x_{\hat{\jmath}})\right) = f_i(x_i, x_{\hat{\imath}}) - f_i(y_i, x_{\hat{\imath}}) \leqslant 0$$

■

In 1929, Knaster, Kuratowski, and Mazurkiewicz provided a new proof of Brouwer's fixed point theorem based on lemma 26. This theorem was generalized by Shapley in view of applications to cooperative game theory. It is a direct consequence of theorem 11.

Let

$$\Sigma^n := \left\{ x \in R^n_+ \,\middle|\, \sum_{i=1}^{n} x_i = 1 \right\}$$

be the $(n-1)$ simplex. We associate with any nonempty subset T of the set $N := \{1, \ldots, n\}$ the face

(73) $$\Sigma^T := \{x \in \Sigma^n | x_i \neq 0 \quad \text{for all } i \in T\}$$

Let $\mathscr{P}(N)$ denote the family of nonempty subsets T of N and let C_T be the characteristic function of the subset T [defined by $C_T(i)=1$ when $i \in T$ and $C_T(i)=0$ when $i \notin T$].

THEOREM 24 (KKMS)

Let $\{F_T\}_{T \in \mathscr{P}(N)}$ be a family of closed (possibly empty) subsets of Σ^n satisfying the property

(74) $$\forall T \in \mathscr{P}(N), \qquad \Sigma^T \subset \bigcup_{R \subset T} F_R$$

Then there exist nonnegative scalars $m(T)$ such that

(75) $$\begin{cases} \textbf{i.} \quad C_N = \sum_{T \in \mathscr{P}(N)} m(T) C_T \\ \textbf{ii.} \quad \bigcap_{m(T)>0} F_T \neq \varnothing \end{cases}$$

▲

We observe that condition (75, **i**) can be written as

(76) $$\forall i \in N, \qquad \sum_{i \in T} m(T) = 1$$

DEFINITION 25

A family $\mathscr{B}$ of subsets $T \in \mathscr{P}(N)$ such that

(77) $$C_N = \sum_T m(T) C_T, \qquad m(T) > 0 \qquad \textit{for all } T \in \mathscr{B}$$

is called a balanced family. ▲

Hence, theorem 24 can be restated as saying that assumption (74) implies the existence of a balanced family $\mathscr{B}$ such that $\bigcap_{T \in \mathscr{B}} F_T \neq \varnothing$. This is a generalization of the famous "Three Poles lemma" or "KKM lemma".

LEMMA 26 (KKM)
Let us consider n closed subsets F_i of Σ^n satisfying

$$\forall x \in \Sigma^n, \qquad x \in \bigcup_{\{i|x_i>0\}} F_i \tag{78}$$

Then

$$\bigcap_{i=1}^{n} F_i \neq \varnothing \tag{79}$$

▲

Proof of Lemma 26. We apply theorem 24 with $F_S := F_i$ when $T = \{i\}$, $i = 1, \ldots, n$, and $F_T := \varnothing$ when $\text{card}(T) \geqslant 2$. Then assumption (78) implies the corresponding assumption (74). Then condition (75, **i**) shows that $m(T) = m(\{i\})$ is equal to one for all i so that (75, **ii**) is reduced to (79). ■

Proof of Theorem 24. We apply theorem 11 to the set-valued map G from Σ^n to R^n defined by

$$G(x) := \frac{C_N}{\text{card}(N)} - \text{co}\left\{\frac{C_T}{\text{card}(T)}\right\}_{F_T \ni x}$$

The values of G are obviously convex and compact, and G is clearly upper semicontinuous. It satisfies the tangential condition,

$$\forall x \in \Sigma^n, \qquad G(x) \cap T_{\Sigma^n}(x) \neq \varnothing$$

Indeed, let T be the subset of indexes i such that $x_i > 0$. By assumption (74), there exists a nonempty subset $R \subset T$ such that x belongs to F_R. Hence,

$$y := \frac{C_N}{\text{card}(N)} - \frac{C_R}{\text{card}(R)}$$

belongs to $G(x)$. It also belongs to $T_{\Sigma^n}(x)$ because $\sum_{i=1}^{n} y_i = 0$ and if $x_i = 0$, i does not belong to T and, consequently, does not belong to R, so that $y_i = 1/\text{card}(N) \geqslant 0$. Therefore, theorem 11 implies the existence of $\bar{x} \in \Sigma^n$ such that $0 \in G(\bar{x})$, that is, such that

$$C_N = \sum_{F_T \ni x} \frac{\lambda(T)\text{card}(N)}{\text{card}(T)} C_T \quad \text{and} \quad \bar{x} \in \bigcap_{\lambda(T)>0} F_T \tag{80}$$

■

Remark
We can prove Brouwer's fixed point theorem from the KKM lemma. Indeed, let f be a continuous map from the simplex Σ^n to itself. We introduce the sub-

sets F_i defined by

$$F_i := \{x \in \Sigma^n | x_i \geqslant f_i(x)\}$$

which are closed because f is continuous. They satisfy assumption (78), otherwise there would be $x \in \Sigma^n$, which belongs to $\bigcap_{\{i | x_i > 0\}} \complement F_i$. In other words, we would have $f_i(x) > x_i$ whenever $x_i > 0$. Since both x and $f(x)$ belong to Σ^n, we obtain the contradiction $1 = \sum_{i=1}^n f_i(x) > \sum_{i=1}^n x_i = 1$. Therefore, there exists $\bar{x}$ that belongs to the intersection of the subsets F_i, that is, that satisfies

$$\forall i = 1, \ldots, n, \qquad \bar{x}_i \geqslant f_i(\bar{x})$$

We cannot have the strict inequality $\bar{x}_i > f_i(\bar{x})$ for at least one i, because such an inequality would imply that $1 > 1$. Hence, $\bar{x}_i = f_i(\bar{x})$ for all $i = 1, \ldots, n$. ∎

Remark

Ky Fan's inequality and, more generally, theorem 3.4 can also be deduced from the KKM lemma.

Therefore, Brouwer's fixed point theorem, Ky Fan's inequality, theorem 11, Kakutani's fixed point theorem, the KKM lemma and KKMS theorem are all equivalent, as indicated by the diagram on the opposite page.

5. WALRAS EQUILIBRIA AND PRICE DECENTRALIZATION

We apply the preceding theorems to find a possible explanation for the role of price systems in decentralizing the behavior of different consumers; that is, knowledge about the price system allows each consumer to make a choice without knowing the global state of the economy and, in particular, without (necessarily) knowing the choice of other consumers.

There is no doubt that Adam Smith is at the origin of what we now call decentralization, that is, the ability of a complex system moved by different actions in pursuit of different objectives to achieve an allocation of scarce resources. He introduced this mysterious and quite paradoxical property in a poetic way. Let us quote the famous passage from *The Wealth of Nations*, published in 1776, two centuries ago.

> Every individual endeavours to employ his capital so that its produce may be of greatest value. He generally neither intends to promote the public interest, nor knows how much he is promoting it. He intends only his own security, only his own gain. And he is in this led by an invisible hand to promote an end which was no part of his intention. By pursuing his own interest, he frequently thus promotes that of society more effectually than when he really intends to promote it.

Adam Smith did not provide a careful statement of what the *invisible hand* manipulates nor, a fortiori, a rigourous argument for its existence. We had to

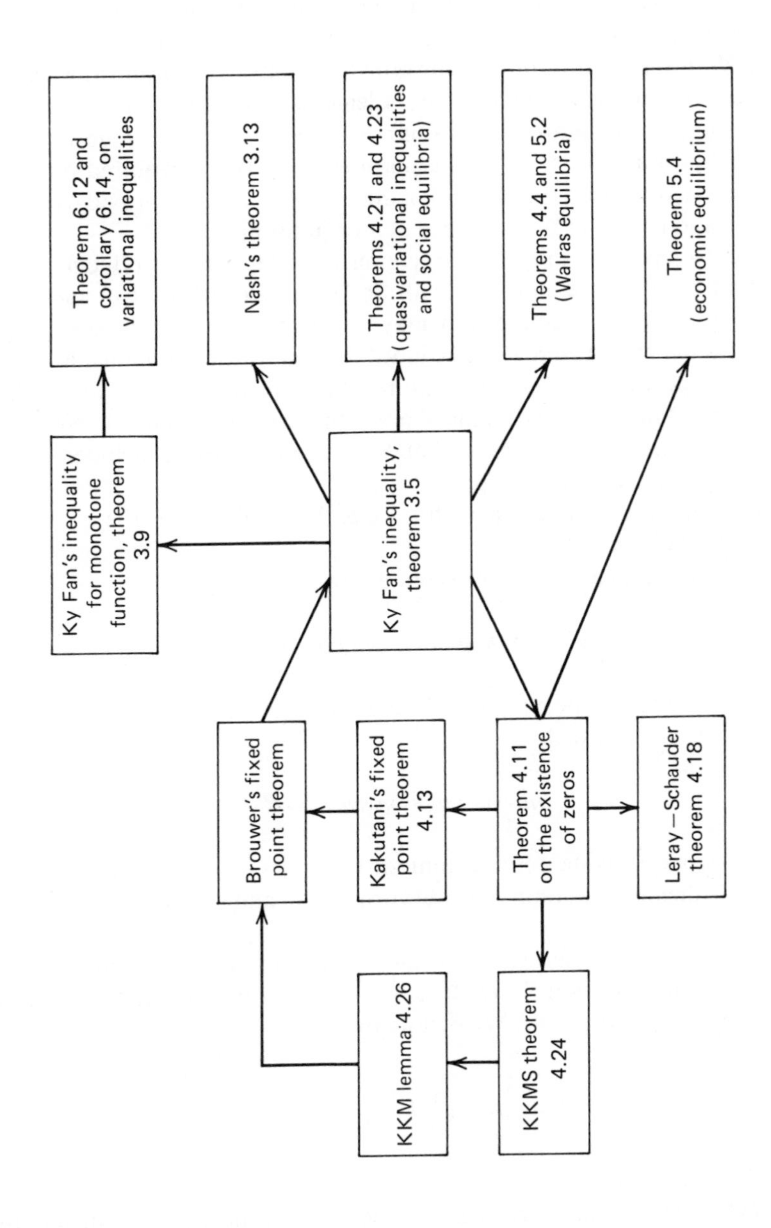
Theorem 6.12 and corollary 6.14, on variational inequalities
Nash's theorem 3.13
Theorems 4.21 and 4.23 (quasivariational inequalities and social equilibria)
Theorems 4.4 and 5.2 (Walras equilibria)
Theorem 5.4 (economic equilibrium)
Ky Fan's inequality for monotone function, theorem 3.9
Ky Fan's inequality, theorem 3.5
Brouwer's fixed point theorem
Kakutani's fixed point theorem 4.13
Theorem 4.11 on the existence of zeros
Leray – Schauder theorem 4.18
KKM lemma 4.26
KKMS theorem 4.24

wait a century for Leon Walras to recognize that price systems are the elements on which the invisible hand acts and that actions by various agents are guided by those price systems, providing enough information to all the agents to guarantee the consistency of their actions with the scarcity of available commodities. Walras presented in 1874 the general equilibrium concept as a solution to a system of nonlinear equations in "Elements d'économie politique pure." An equal number of equations and unknowns led him and his followers quite optimistically to assume that a solution does exist. But it required one more century for Arrow, Debreu, Gale, Nikaido, and others to provide rigorous statements and proofs of the existence of an equilibrium.

In order to represent an exchange economy, we begin by introducing ℓ types of commodities that are endowed with an "unit," so that we can speak of x units of a commodity. A commodity may involve not only its physical properties, but also the place where it is available, the date when it is available, and, in the case of uncertainty about the future, the elementary event that will be realized (for instance, 100 kilograms of bread that will be available in New York in 32 days if there is a truckers' strike). *Services* may also be included as long as units are perfectly well defined.

So, a commodity bundle is a vector $x \in R^l$, which describes the quantity x_h of each commodity $h=1, \dots, \ell$.

The description of an exchange economy begins with

(1) *the subset* $M \subset R^\ell$ of *available commodities*

and continues with the specification of n consumers $i=1, \dots, n$. A first definition of a consumer i begins with

(2) *the consumption set* $L_i \subset R^\ell$

which is interpreted as the set of commodity bundles he or she needs. If $x \in L_i$, then x_h is i's demand of commodity h when $x_h > 0$, and $|x_h|$ is the i's supply of commodity h when $x_h < 0$.

The question now arises, can consumers share an available commodity?

We define an allocation $\vec{x} \in (R^\ell)^n$ as n commodity bundles $x_i \in L_i$ such that their sum $\sum_{i=1}^n x_i$ is available. We denote by

(3) $$K := \left\{ \vec{x} \in \prod_{i=1}^n L_i \,\middle|\, \sum_{i=1}^n x_i \in M \right\}$$

the set of allocations.

The next question that arises is whether we can devise mechanisms that provide allocations. Indeed, when the number of consumers is large, finding an allocation is difficult, because it requires possessing a large amount of informations. Can we find a way of summarizing enough information to allow each consumer to choose his or her consumption in a decentralized manner?

This is possible by using price systems. *A price system* is a linear functional p that associates with any commodity $x \in R^{\ell}$ its value $\langle p, x\rangle \in R$. We denote by $\Sigma^{\ell} := \{p \in R_{+}^{\ell} \mid \sum_{i=1}^{\ell} p_i = 1\}$ the price simplex. We next regard the support function

$$r(p) := \sup_{y \in M} \langle p, y\rangle \tag{4}$$

as the *gross income*, which is the maximum value of the available commodity bundles for the price system p.

The solution proposed by Walras and his followers consists in letting price systems play a crucial role by summarizing enough information about the economic system for the n consumers. A consumer is defined as an automaton associating to every price p and every income r (in monetary units) his or her demand $\delta_i(p, r) \in L_i$, which is the commodity bundle that the consumer buys when the price system is p and income is r. In other words, a consumer i is characterized by the demand function δ_i: $\Sigma^{\ell} \times R \to L_i$, which describes the behavior of consumers. We observe that it is possible to devise other mathematical descriptions—models—of the same behavior, and we shall do so. We also mention that neoclassical economists assume that demand functions are derived by maximizing a **utility function** in compliance with the first sentence in Adam Smith's passage. But this is by no means necessary.

Anyway, assume for the first time that consumers are just demand functions $\delta_i(.,.)$, given independently of the set M (which cannot be known by the consumers).

We need another assumption to define the Walrasian mechanism. If p is the price system, we assume that the gross income $r(p)$ is allocated among consumers in incomes $r_i(p)$

$$\sum_{i=1}^{n} r_i(p) = r(p) \tag{5}$$

We insist on the fact that the model does not provide this allocation of income but assumes that it is given. In summary, the mechanism we are about to describe depends on

(a) the description of each consumer i by its demand function $\delta_i(.,.)$
(b) an allocation $r(p) = \sum_{i=1}^{n} r_i(p)$ of the gross income

Hence, when p is the price on the market, the income of consumer i is $r_i(p)$ and demand is $\delta_i(p, r_i(p))$. The mechanism works if and only if demand balances supply, that is, if and only if there exists a price $\bar{p}$ such that

$$\sum_{i=1}^{n} \delta_i(\bar{p}, r_i(\bar{p})) \in M \tag{6}$$

DEFINITION 1

A price system $\bar{p} \in \sum^{\ell}$ such that (6) *holds true is called a* Walras equilibrium; *we say that "it clears the market."* ▲

Observe that this mechanism is decentralized: The choice of the ith consumer depends only on the price system p (through its income function r_i) and does not require knowledge of the other consumers' choices. This illustrates better Adam Smith's quotation, because by choosing a Walras equilibrium $\bar{p}$, the invisible hand promotes an end, $\sum_{i=1}^{n} \delta_i(\bar{p}, r_i(\bar{p}))$, which was not part of the consumers' intentions.

The task now is to solve problem (6), which requires us to choose among all the sufficient conditions that can be devised those that have an economic interpretation.

It is remarkable that a budgetary constraint on consumers' behavior, known as the **Walras law**, provides such a sufficient condition. The Walras law forbids consumers to spend more than their incomes, that is,

$$\forall i=1, \ldots, n, \qquad \langle p, \delta_i(p, r)\rangle \leqslant r \tag{7}$$

Actually, this law can be less rigorous; it is sufficient to assume that

$$\sum_{i=1}^{n} \langle p, \delta_i(p, r_i)\rangle \leqslant \sum_{i=1}^{n} r_i \tag{8}$$

This latter law, the **collective Walras law**, allows financial transactions among consumers. We insist on the fact that the Walras laws (7) or (8) do not involve the set M of available resources.

We shall prove that Walras law (8) implies the existence of a Walras equilibrium, that is, allows Adam Smith's invisible hand to provide a price system summarizing information on the state of this economy—the subset M and behavior of each consumer described by the demand functions δ_i, and their share $r_i(.)$ of the gross income.

After pioneer work by Wald, von Neumann, Kakutani, and so on, started in the 1930s, the first proof of the existence of a Walras equilibrium was due to Arrow and Debreu in 1954. Further work on this problem was due to McKenzie, Nikaido, Uzawa, and many others.

THEOREM 2

Let us assume that

(9) $\quad M = M_0 - R^{\ell}_{+}$ *is closed and convex, where* M_0 *is compact*

and

(10) *the demand functions* δ_i *are continuous and satisfy the collective Walras law* (8)

and

(11) *the income functions r_i are continuous.*

Then there exists a Walras equilibrium $\bar{p} \in \Sigma^\ell$. ▲

Proof. It is consequence of theorem 4.4 applied to the set-valued map F defined by

$$F(p) := M_0 - \sum_{i=1}^{n} \delta_i(p, r_i(p)) \tag{12}$$

which is obviously upper hemicontinuous with compact convex values and satisfies

$$\sigma(F(p), p) = r(p) - \sum_{i=1}^{n} \langle p, \delta_i(p, r_i(p))\rangle \geqslant r(p) - \sum_{i=1}^{n} r_i(p) \geqslant 0 \tag{13}$$

Hence, there exists a solution $\bar{p} \in \Sigma^\ell$ to the inclusion

$$0 \in F(\bar{p}) - R_+^\ell = M - \sum_{i=1}^{n} \delta_i(\bar{p}, r_i(\bar{p})) \quad ■$$

Remark

We can relax assumption (9) (see for instance Aubin, 1979b, theorem 8.2.2, p. 248). Theorem 1 can also be extended to the case of set-valued demand maps. ■

We are going to propose another model that keeps the essential ideas underlying Adam Smith and Léon Walras's proposals but can take into account the dynamic nature of the behavior of each agent by its *instantaneous demand function* d_i: $L_i \times \Sigma^\ell \to R^\ell$, which sets the variation in consumer's i demand when the price is p and consumption is x.

We assume that

$$(14) \quad \begin{cases} \textbf{i.} & M = M_0 - R_+^\ell \text{ is closed and convex where } M_0 \text{ is compact.} \\ \textbf{ii.} & \forall i = 1, \ldots, n,\ L_i \text{ is closed, convex, and bounded below.} \\ \textbf{iii.} & 0 \in \operatorname{Int}\left(\sum_{i=1}^{n} L_i - M\right) \end{cases}$$

We posit the following assumptions on the instantaneous demand functions d_i:

$$(15) \quad \begin{cases} \textbf{i.} & \forall i = 1, \ldots, n, \text{ the function } d_i : L_i \times \Sigma^\ell \to R^\ell \text{ is continuous.} \\ \textbf{ii.} & \forall x \in L_i, \quad \forall p \in \Sigma^\ell, \quad d_i(x, p) \in T_{L_i}(x) \end{cases}$$

and

$$\text{(16)} \qquad \forall x \in L_i, \quad p \to d_i(x, p) \quad \text{is affine.}$$

We still have a concept of equilibrium associated to this mechanism.

DEFINITION 3

An economic equilibrium is a sequence $(\bar{x}_1, \ldots, \bar{x}_n, \bar{p})$ *of* n *consumptions* $\bar{x}_i$ *and a price system* $\bar{p}$ *such that*

$$\text{(17)} \qquad \begin{cases} \textbf{i.} \ \forall i=1, \ldots, n, & d_i(\bar{x}_i, \bar{p})=0 \\ \textbf{ii.} \ \forall i=1, \ldots, n, & \bar{x}_i \in L_i, \quad \bar{p} \in \Sigma^\ell \\ \textbf{iii.} \ \sum_{i=1}^n \bar{x}_i \in M & \end{cases}$$

▲

In order to keep all the good features of the Walras model, we must check that there are sufficient conditions with an economic interpretation. This is still the case, since we shall prove that equilibria (17) do exist if the instantaneous demand functions d_i satisfy the instantaneous Walras law

$$\text{(18)} \qquad \forall p \in \Sigma^\ell, \quad \forall x_i \in L_i, \quad \langle p, d_i(x_i, p)\rangle \leq 0$$

This is a financial rule that requires that for each price, the value of the rate of change of each consumer is not positive, that is, each consumer *does not spend more than he or she earns* on an *instantaneous exchange of goods.*

As in the Walras model, the instantaneous Walras law does not involve the subset M of available supplies. More generally, we shall assume that the *instantaneous collective Walras law*

$$\text{(19)} \qquad \forall \vec{x} \in \prod_{i=1}^n L_i, \quad \forall p \in \Sigma^\ell, \quad \langle p, \sum_{i=1}^n d_i(x_i, p)\rangle \leq 0$$

holds true.

THEOREM 4

We posit assumptions (14) *on the economy and assumptions* (15) *and* (16) *on the instantaneous demand functions* d_i. *We assume that the instantaneous collective Walras law* (19) *holds true. Then there exists an equilibrium* $(\bar{x}, \bar{p}) \in K \times \Sigma^\ell$ [*satisfying* (17)]. ▲

Proof. It follows from theorem 4.21 applied to the set-valued map D from $K \subset (R^\ell)^n$ *to* $(R^\ell)^n$ defined by

$$\text{(20)} \qquad \begin{cases} D(x) := \{d(x, p)\}_{p \in \Sigma^\ell}, \text{ where} \\ d(x, p) := (d_1(x_1, p), \ldots, d_n(x_n, p)) \in (R^\ell)^n \end{cases}$$

Obviously, assumptions (15) and (16) imply that D is upper hemicontinuous with compact convex images. Furthermore, we deduce from assumptions (14, **i** and **ii**) that K is convex compact and from assumptions (14, **iii**) that the tangent cone to K at x is equal to

$$(21) \qquad T_K(x) := \left\{ v \in \prod_{i=1}^{n} T_{L_i}(x) \,\middle|\, \sum_{i=1}^{n} v_i \in T_M\left(\sum_{i=1}^{n} x_i\right)\right\}$$

It remains to check that

$$(22) \qquad \forall x \in K, \qquad F(x) \cap T_K(x) \neq \varnothing$$

or, equivalently, thanks to assumption (15, **ii**) and formula (21), that

$$(23) \qquad \forall x \in K, \qquad \exists p \in \Sigma^{\ell} \text{ such that } \sum_{i=1}^{n} d_i(x_i, p) \in T_M\left(\sum_{i=1}^{n} x_i\right)$$

Otherwise, there would exist $x \in K$ such that

$$T_M\left(\sum_{i=1}^{n} x_i\right) \cap \left\{\sum_{i=1}^{n} d_i(x_i, q)\right\}_{q \in \Sigma^{\ell}} = \varnothing$$

The separation theorem implies that in this case, there would exist

$$p \in T_M\left(\sum_{i=1}^{n} x_i\right)^{-} =: N_M\left(\sum_{i=1}^{n} x_i\right) \qquad \text{and} \qquad \varepsilon > 0$$

such that

$$(24) \qquad \inf_{q \in \Sigma^{\ell}} \left\langle p, \sum_{i=1}^{n} d_i(x_i, q)\right\rangle \geqslant \varepsilon$$

Since $N_M(\sum_{i=1}^{n} x_i) \subset R_+^{\ell}$ because $M = M_0 - R_+^{\ell}$, we deduce a contradiction to the instantaneous collective Walras law (19) by taking $q := p / \sum_{h=1}^{\ell} p_h$. Inequality (24) implies that

$$\sum_{i=1}^{n} \langle q, d_i(x_i, q)\rangle \geqslant \varepsilon > 0 \qquad ■$$

Walras defined not only the concept of equilibrium but proposed a process known as Walras's *tâtonnement* (*tâtonnement* means tentative process, trial and error, etc.—literally, groping, feeling one's way in the dark). Indeed, a Walras equilibrium is an equilibrium for the "excess demand map" E defined by

$$\forall p, \qquad E(p) := \sum_{i=1}^{n} \delta_i(p, r_i)p)) - M \tag{25}$$

So, the idea was to associate the dynamic system

$$p'(t) \in E(p(t)); \quad p(0) = p_0 \tag{26}$$

and study under which conditions we can prove that $p(t)$ converges to a Walras equilibrium $\bar{p}$, solution to the inclusion $0 \in E(\bar{p})$.

We observe that if $p(t)$ is a price supplied by the Walras tâtonnement process (26) and $p(t)$ is not a Walras equilibrium, it cannot be implemented, because the associated total demand $\sum_{i=1}^{n} \delta_i(p(t), r_i(p(t))$ is not necessarily available.

Hence, this model forbids consumers to transact as long as the price $p(t)$ is not an equilibrium. It is as if there were a superauctioneer calling prices and receiving transactions offers from consumers. If the offers do not match, he or she calls another set of price according to rule (26) but does not allow transactions to take place as long as the offers are not consistent.

Tâtonnement does not provide a model of how prices are actually evolving. The fundamental nature of the Walras world is static, while we live in a dynamical environment where no equilibria have been observed.

By means of the second mechanism, which describes consumers i as automata associating to each price system p and consumption bundle x_i their rate of change x_i, we can regard an equilibrium $(\bar{x}_1, \ldots, \bar{x}_n, \bar{p})$ as an equilibrium for the dynamical system

$$\forall i = 1, \ldots, n, \qquad x_i'(t) = d_i(x_i(t), p(t)); \qquad x_i(0) = x_i^0 \tag{27}$$

When the price $p(t)$ evolves, so do consumptions $x_i(t)$ according to the differential equations (27).

So, a viability problem arises: Does there exist a price function $p(t)$ such that the sum $\sum_{i=1}^{n} x_i(t)$ of the consumptions remains available? In other words, do the trajectories $x_i(.)$ of the n coupled differential equations satisfy the viability condition

$$\forall t \geq 0, \qquad \sum_{i=1}^{n} x_i(t) \in M \tag{28}$$

We observe that this mechanism shares with the Walras model the decentralization property: The price system $p(t)$ summarizes enough information about the economic system to allow each consumer to change her or his own consumption independently of other consumers and ignore the set M of available supplies.

It is possible to prove that under the assumption of theorem 4, we can associate to each initial allocation x_0 a price function $p(t)$ such that trajectories in (27) satisfying viability condition (28) do exist (see Aubin and Cellina, [1984], Chapter 5).

Furthermore, we can show that the price system evolves as a feedback control: It depends on time through the state of the system, in the sense that there exists a set-valued map C from the set of allocations to the set of prices such that

$$\text{(29)} \qquad \text{for almost all } t \geqslant 0, \qquad p(t) \in C(x_1(t), \ldots, x_n(t))$$

In this framework, we see Adam Smith's invisible hand (which, more to the point, should be called invisible brain) setting prices as functions of allocations for the purpose of promoting consumers to respect scarcity constraints. It is in this sense that we may regard such a dynamical system as a *regulation mechanism.* So the feedback condition (29) involves the price system, *but does not influence its variation.* This is the second important difference between dynamical systems such as (26) that require, so to speak, Adam Smith's invisible hand to actively, as in mechanics, set prices to adjust demand to supply. In this model, only consumers are supposed to have the ability to act dynamically according to differential equation (27), and prices "follow" consumption according to relation (29). ∎

6. MONOTONE MAPS

Let K be a closed convex subset of a Hilbert space X (identified with its dual). The problem arises how to transform a set-valued map $A: K \to X$ that does not satisfy the tangential condition,

$$\text{(1)} \qquad \forall x \in K, \qquad -A(x) \cap T_K(x) \neq \emptyset$$

to another set-valued map F that satisfies the condition. The simplest idea is to project the images of $A(x)$ onto the tangent cone $T_K(x)$, that is, to define F by

$$\text{(2)} \qquad F(x) := \pi_{T_{K(x)}}(-A(x))$$

which, by construction, satisfies the strong tangential condition. The zeros of A (or $-A$) are zeros of F. Unfortunately, we cannot apply theorem 4.21 because this map F inherits neither the upper hemicontinuity of A nor the convexity of the images of A. But we observe that the zeros of a set-valued map F are the zeros of the map $m(F): x \to m(F(x))$ associating to x the elements of $F(x)$ with the smallest norm. Since $N_K(x)$, the normal cone to K at x, is the negative polar cone of the tangent cone $T_K(x)$, we know that the orthogonal projections satisfy

(3) $$\pi_{T_{K(x)}}+\pi_{N_{K(x)}}=1$$

(see Fig. 1). This implies the following lemma.

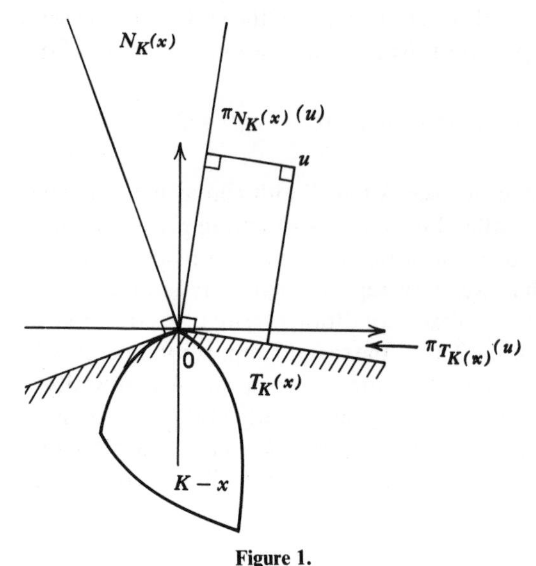

Figure 1.

LEMMA 1

(4) $$m(\pi_{T_{K(x)}}(-A(x)))=-m(A(x)+N_K(x))$$ ▲

(See Fig. 2).

Proof. Set $A:=A(x)$, $T:=T_K(x)$ and $N:=N_K(x)$. The lemma follows from the equality

$$\inf_{y\in A}\|\pi_T(-y)\|=\inf_{y\in A}\|-y-\pi_N(-y)\|=\inf_{y\in A}\inf_{z\in N}\|-y-z\|=\inf_{v\in A+N}\|v\|$$ ■

So, the zeros of the set-valued map $x\to\pi_{T_{K(x)}}(-A(x))$ are the zeros of the set-values map $x\to A(x)+N_K(x)$, which is simpler to handle (see Fig. 3). This justifies the following definition.

DEFINITION 2

Let K be a closed convex subset of X. A "variational inequality" for A on K is an inclusion of the form

(5) $$\begin{cases}\textbf{i.} & \bar{x}\in K\\ \textbf{ii.} & 0\in A(\bar{x})+N_K(\bar{x})\end{cases}$$ ▲

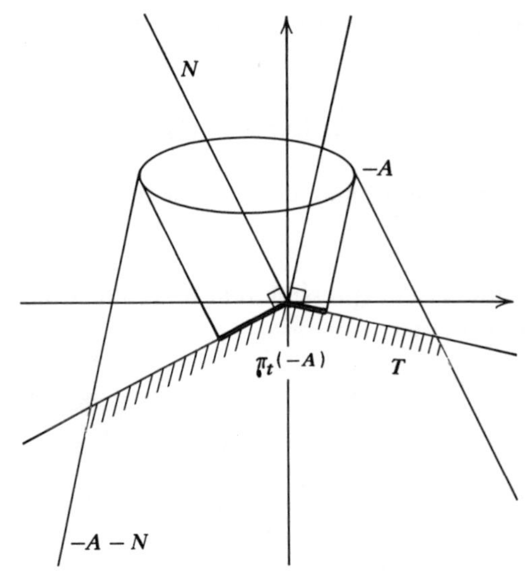

Figure 2.

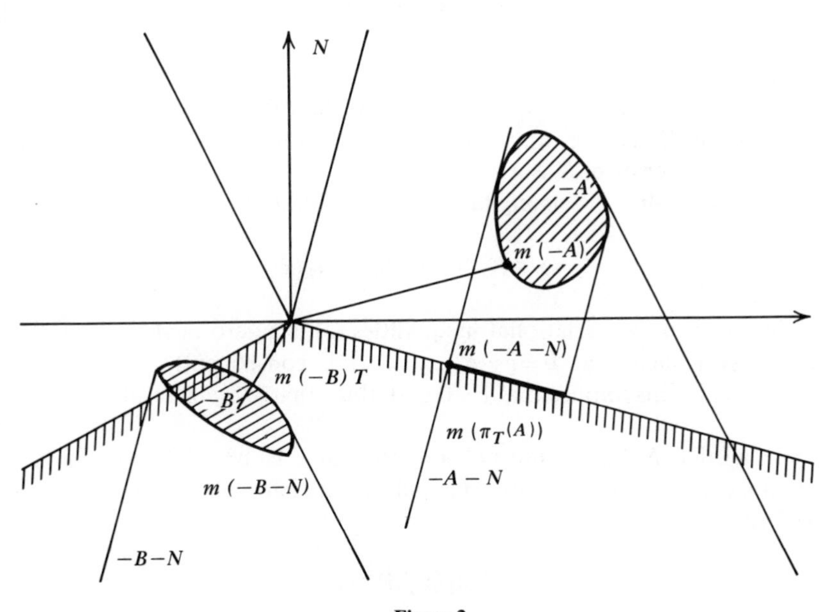

Figure 3.

Example

Optimization with constraints provides examples of variational inequalities, which we saw when U was a lower semicontinuous, convex function such that

$$0 \in \operatorname{Int}(\operatorname{Dom} U - K) \tag{6}$$

because the elements $\bar{x} \in K$ achieving the minimum of U on K are solutions to the inclusion

(7) $$0 \in \partial U(\bar{x}) + N_K(x)$$

that is, a solution to the variational inequality with $A := \partial U$. ■

Remark

Variational inequalities is an expression that was coined because in the single-values case, (5) can be written

(8) $$\begin{cases} \textbf{i.} & \bar{x} \in K \\ \textbf{ii.} & \forall y \in K, \qquad \langle A(\bar{x}), \bar{x} - y \rangle \leqslant 0 \end{cases}$$

We observe that when K is a cone, this system becomes

(9) $$\begin{cases} \textbf{i.} & \bar{x} \in K \\ \textbf{ii.} & A\bar{x} \in -K^{-} \\ \textbf{iii.} & \langle A\bar{x}, \bar{x} \rangle = 0 \end{cases}$$ ■

Remark

Both $x \rightarrow \pi_{T_K(x)} A(x)$ and $x \rightarrow A(x) + N_K(x)$ coincide with $A(x)$ when x belongs to the interior of K. Hence, any solution to the variational inequalities that belongs to the interior of K is a zero of A.

When $-A$ satisfies the strong tangential condition,

(10) $$\forall x \in K, \qquad -A(x) \subset T_K(x)$$

any solution $\bar{x}$ to the variational inequalities (5) is a zero of A. Indeed, there exists $\bar{v} \in A(\bar{x})$ such that $0 = \bar{v} + \pi_{N_K(\bar{x})}(\bar{v})$, and, consequently, $\pi_{N_K(\bar{x})}(\bar{v}) = -\bar{v}$ belong to $T_K(\bar{x}$. This implies that $\bar{v} = 0$ and, thus, that $\bar{x}$ is a zero of A. ■

We recall that $N_K(x)$, the normal cone to K at x, is the subdifferential of the indicator ψ_K. Therefore, variational inequalities are particular cases of inclusions of the form

(11) $$f \in A(\bar{x}) + \partial V(\bar{x})$$

when $V: X \rightarrow]-\infty, +\infty]$ is a proper lower semicontinuous, convex function and A is a set-valued map from the Hilbert space X to itself. We assume once and for all that

(12) $$\begin{cases} \textbf{i.} & \text{Dom } V \subset \text{Dom } A \\ \textbf{ii.} & \forall x \in \text{Dom } A, \qquad A(x) \text{ is convex and weakly compact.} \end{cases}$$

We associate to the function V, the map A, and an element $f \in X$ the function ϕ defined on Dom V by

$$\phi(y) := V(y) + \inf_{u \in A(y)} (V^*(f-u) - \langle f-u, y \rangle) \tag{13}$$

We observe that

$$\forall y \in \operatorname{Dom} V, \qquad \phi(y) \geq 0 \tag{14}$$

since, for all $u \in A(y)$, $V(y) + V^*(f-u) - \langle f-u, y \rangle \geq 0$, thanks to the Fenchel inequality.

We can also characterize the set-valued map A by the function γ defined on Dom $A \times$ Dom A by

$$\gamma(x, y) := \inf_{p \in A(x)} \langle p, x-y \rangle = -\sigma(A(x), y-x) \tag{15}$$

PROPOSITION 3

We posit assumption (12). *The following problems are equivalent*:

$$(16) \quad \begin{cases} \textbf{i.} & \exists \bar{x} \in \operatorname{Dom} V \textit{ such that } f \in A\bar{x} + \partial V(\bar{x}) \\ \textbf{ii.} & \exists \bar{p} \in \operatorname{Dom} V \textit{ such that } f \in \bar{p} + A\partial V^*(\bar{p}) \\ \textbf{iii.} & \exists \bar{x} \in \operatorname{Dom} V \textit{ such that } \forall y \in \operatorname{Dom} V, \\ & \qquad \gamma(\bar{x}, y) - \langle f, \bar{x} - y \rangle + V(\bar{x}) - V(y) \leq 0 \\ \textbf{iv.} & \exists \bar{x} \in \operatorname{Dom} V \textit{ such that } \phi(\bar{x}) = 0 \; (= \min_{y \in \operatorname{Dom} V} \phi(y)) \end{cases}$$

▲

Proof. **a.** Let $\bar{x}$ be a solution to (16, **i**); then there exists $\bar{p} \in \partial V(\bar{x})$ such that $f - \bar{p} \in A\bar{x} \subset A\partial V^*(\bar{p})$. Conversely, let $\bar{p}$ be a solution to (16, **ii**). Then there exists $\bar{x} \in \partial V^*(\bar{p})$ such that $f \in \bar{p} + A\bar{x}$. Since $\bar{p} \in \partial V(\bar{x})$, then $f \in \partial V(\bar{x}) + A\bar{x}$.
b. Let $\bar{x}$ be a solution to (16, **i**). There exists $\bar{u} \in A(\bar{x})$ such that $f \in \partial V(\bar{x}) + \bar{u}$, that is, such that $\forall y \in$ Dom V,

$$\langle \bar{u}, \bar{x} - y \rangle - \langle f, \bar{x} - y \rangle + V(\bar{x}) - V(y) \leq 0$$

By taking the infimum on $A(\bar{x})$, we deduce inequality (16, **iii**).
c. Inequality (16, **iii**) can be written

$$\sup_{y \in \operatorname{Dom} V} \inf_{u \in A(\bar{x})} [V(\bar{x}) - V(y) - \langle f-u, \bar{x} - y \rangle] \leq 0$$

Since Dom V is convex, $A(\bar{x})$ is convex weakly compact, the lopsided minimax theorem 2.7 implies that the left-hand side of this inequality is equal to

$$\inf_{u \in A(\bar{x})} \sup_{y \in \text{Dom } V} [V(\bar{x}) - V(y) - \langle f-u, \bar{x}-y \rangle]$$

$$= \inf_{u \in A(\bar{x})} [V(\bar{x}) + V^*(f-u) - \langle f-u, \bar{x} \rangle] = \phi(\bar{x})$$

Hence, $\phi(\bar{x}) \leqslant 0$, and since we have already observed that $\phi(\bar{x}) \geqslant 0$, we conclude that $\phi(\bar{x}) = 0$.

d. Let $\bar{x} \in \text{Dom } V$ satisfy $\phi(\bar{x}) = 0$. Since $A(\bar{x})$ is weakly compact and V^* is weakly lower semicontinuous, there exists $\bar{u} \in A(\bar{x})$ such that

$$\phi(\bar{x}) = V(\bar{x}) + V^*(f-\bar{u}) - \langle f-\bar{u}, \bar{x} \rangle = 0$$

This is equivalent to saying that $f-\bar{u} \in \partial V(\bar{x})$, that is, $\bar{x}$ solves (16, **i**). ■

The equivalence between (16, **i** and **iv**) allows us to interpret the solutions to problem (11) as solutions to a minimization problem (minimization of the functional ϕ) and provides a variational principle. The equivalence between (16, **i** and **iii**) allows us to solve problem (11) (and, in particular, variational inequalities) by applying minimax inequalities from Section 3 to the function ϕ defined by

$$\phi(x, y) := \gamma(x, y) - \langle f, x-y \rangle + V(x) - V(y) \tag{17}$$

We observe that

$$\begin{cases} \textbf{i.} & \forall x, \quad y \to \phi(x, y) \text{ is concave.} \\ \textbf{ii.} & \forall y, \quad \phi(y, y) = 0 \end{cases} \tag{18}$$

If we want to apply Ky Fan's inequality (theorem 3.5), we have to verify that

$$\forall y, \qquad x \to \phi(x, y) \quad \text{is lower semicontinuous.} \tag{19}$$

To satisfy this assumption, we must require A to be upper semicontinuous from X supplied with the weak topology to X supplied with the strong topology. This is not reasonable (in the case of infinite dimensional spaces), since this requirement is not met by the subdifferential $x \to \partial U(x)$ of lower semicontinuous convex functions. Still, when K is weakly compact, there exists an element $\bar{x} \in K$ achieving the minimum of U on K, that is, a solution to the variational inequalities

$$0 \in \partial U(\bar{x}) + N_K(\bar{x})$$

But the subdifferential ∂U is a *monotone* map, and, consequently, the functions γ and ϕ are monotone in the sense that

$$\gamma(x, y)+\gamma(y, x)\geqslant 0 \qquad (\phi(x, y)+\phi(y, x)\geqslant 0)$$

Therefore, we shall apply theorem 3.9 (Ky Fan's inequality for monotone functions) for solving variational inequalities and problem (11) when A is a monotone map.

DEFINITION 4
We shall say that a set-valued map A from X to X is monotone *if its graph is monotone in the sense that*

(20) $\quad \forall(x, p)\in \text{graph}(A), \quad \forall(y, q)\in \text{graph}(A), \quad \langle p-q, x-y\rangle\geqslant 0.$ ▲

The terminology comes from the monotone maps from R to R; for instance, a nondecreasing map F from R to R is monotone.

Example
The map

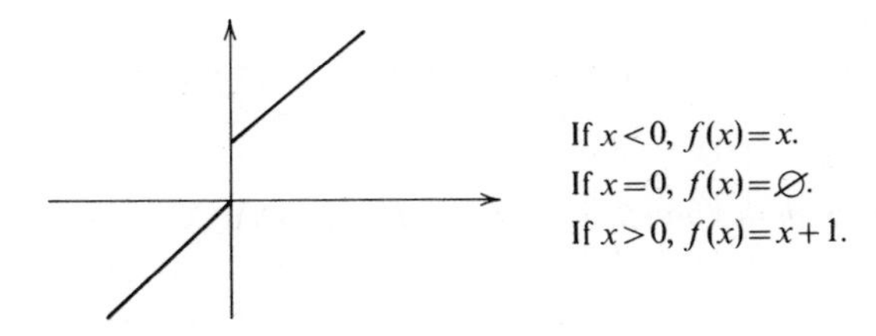

is a monotone map as well as the map

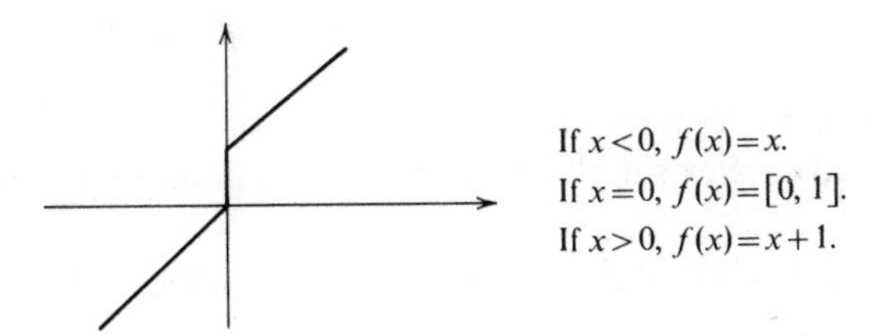

More generally, if f is a nondecreasing map from R to R, the map $x\rightarrow A(x):=[f(x-), f(x+)]\cap R$ is monotone. We give several other examples of monotone maps.

Let f be a nondecreasing function from an interval Dom $f \subset R$ to R. Let $\Omega \in R^n$ be open and $X := L^2(\Omega)$. We define the set-valued map A from X to X by

$$\text{(21)} \qquad \text{for almost all } \omega \in \Omega, \qquad A(x(\cdot))(\omega) := f(x(\omega))$$

It is clear that A is monotone.

Let $V: X \to R \cup \{+\infty\}$ be a proper lower semicontinuous, convex function. Then the set-valued map $x \to \partial V(x)$ is monotone.

Let $U: X \times Y \to R \cup \{-\infty\} \cup \{+\infty\}$ be a proper function satisfying

$$\text{(22)} \qquad \begin{cases} \textbf{i.} \quad \forall y \in Y, & x \to U(x, y) \text{ is concave and upper semicontinuous.} \\ \textbf{ii.} \quad \forall x \in X, & y \to U(x, y) \text{ is convex and lower semicontinuous.} \end{cases}$$

Then the set-valued map

$$\text{(23)} \qquad (x, y) \to \partial_x(-U)(x, y) \times \partial_y U(x, y) \subset X \times Y$$

is monotone. ■

DEFINITION 5

We say that a proper set-valued map F from X to X is nonexpansive if

$$\text{(24)} \qquad \forall (p, x), \qquad (q, y) \in \text{graph}(F), \qquad \|p - q\| \leqslant \|x - y\| \qquad ▲$$

PROPOSITION 6

If F is a proper nonexpansive set-valued map from X to X, then $A := 1 - F$ is monotone. ▲

Proof. Let (x, p) and (y, q) belong to the graph of F. Then

$$\begin{aligned} \langle x - p - (y - q), x - y \rangle &= \|x - y\|^2 - \langle p - q, x - y \rangle \\ &\geqslant \|x - y\|^2 - \|p - q\| \, \|x - y\| \geqslant 0 \end{aligned}$$ ■

The converse is not necessarily true. ■

If K is a closed convex set, the projector π_K is a monotone map, since

$$\langle \pi_K(x) - \pi_K(y), x - y \rangle \geqslant \|\pi_K(x) - \pi_K(y)\|^2 \geqslant 0$$ ■

Since monotonicity is a property bearing on the graph of A, then

$$\text{(25)} \qquad A \text{ is monotone if and only if } A^{-1} \text{ is monotone.}$$

If A and B are monotone and $\lambda > 0$, $\mu > 0$, then $\lambda A + \mu B$ is monotone. If A is monotone, then $\overline{\text{co}}(A)$: $x \to \overline{\text{co}}(A(x))$ is also monotone. If we supply $X \times X$ with the product of the strong topology and the weak topology, we can see that the closure of a monotone graph is still monotone.

We begin by mentioning the following characterization in proposition 7.

PROPOSITION 7

A set-valued map A from X to X is monotone if and only if

$$\forall \lambda > 0, \quad \forall (x, p), \quad \forall (y, q) \in \text{graph}\,(A), \quad \|x-y\| \leqslant \|x-y+\lambda(p-q)\| \tag{26}$$ ▲

Proof. We compute

$$\|x-y+\lambda(p-q)\|^2 = \|x-y\|^2 + \lambda^2\|p-q\|^2 + 2\lambda\langle p-q, x-y\rangle$$

Hence, if A is monotone, inequality (26) ensues; conversely, if (26) holds true, we deduce that

$$\lambda^2\|p-q\|^2 + 2\lambda\langle p-q, x-y\rangle \geqslant 0$$

We obtain monotonicity by dividing by λ and letting λ converge to zero. ■

Remark

We observe that condition (26) does not involve the scalar product; therefore, it can be used on Banach spaces. In this case, maps A satisfying (26) are called *accretive maps.* ■

We associate to the map A its *resolvent*

$$J_\lambda := (1+\lambda A)^{-1} \tag{27}$$

An important consequence of property (26) is given in proposition 8.

PROPOSITION 8

The resolvent J_λ of a monotone map A is a single-valued nonexpansive map from Im $(1+\lambda A)$ *to X.* ▲

Proof. Let $x \in J_\lambda(u)$ and $y \in J_\lambda(v)$. We can write $u \in x + \lambda A(x)$ and $v \in y + \lambda A(y)$. Property (26) implies that

$$\|x-y\| \leqslant \left\|(x-y) + \lambda\left(\frac{u-x}{\lambda} - \frac{v-y}{\lambda}\right)\right\| = \|u-v\|$$

This shows that J_λ is nonexpansive and by taking $u=v$, that J_λ contains a unique point. ■

We shall characterize monotone maps A such that $\mathrm{Im}(I+\lambda A)=X$ for $\lambda>0$; they are the maximal monotone maps, studied in the next section. There are maps that enjoy stronger monotonicity properties, such as maps satisfying

$$\forall(x,p),\quad \forall(y,q)\in\mathrm{graph}(A),\quad \langle p-q,x-y\rangle\geqslant c\|x-y\| \tag{28}$$

or

$$\forall(x,p),\quad \forall(y,q)\in\mathrm{graph}(A),\ \langle p-q,x-y\rangle\geqslant c\|x-y\|^{\alpha} \tag{29}$$

(with $c>0$, $\alpha>1$). We shall describe the monotonicity of A by a proper nonnegative lower semicontinuous, convex function $\beta\colon X\to R_+\cup\{+\infty\}$.

DEFINITION 9

Let $\beta\colon X\to R_+\cup\{+\infty\}$ be a proper lower semicontinuous, convex function. We say that A is β-monotone if

$$\forall(x,p),\quad (y,q)\in\mathrm{graph}(A),\quad \langle p-q,x-y\rangle\geqslant\beta(x-y) \tag{30}$$ ▲

For instance, we can always take

$$\begin{cases}\textbf{i.}\ \ \beta(z):=0\ \ (\textit{and thus } \beta^*=\psi_{\{0\}},\ \mathrm{Dom}\ \beta^*=\{0\})\\ \textbf{ii.}\ \ \beta(z):=\|z\|\ \ (\textit{and thus } \beta^*=\psi_B,\ \mathrm{Dom}\ \beta^*=B)\\ \textbf{iii.}\ \ \beta(z):=\dfrac{1}{\alpha}\|z\|^{\alpha}\left(\textit{and thus } \beta^*=\dfrac{1}{\alpha_*}\|\ \|^{\alpha_*},\ \dfrac{1}{\alpha}+\dfrac{1}{\alpha_*}=1,\ \mathrm{Dom}\ \beta^*=X^*\right)\end{cases} \tag{31}$$

In the following theorem, we shall measure the degree of monotonicity of A by the size of the domain of β^*: the larger Dom β^*, the more monotone A. ■

We now solve the problem

$$f\in A(\bar{x})+\partial V(\bar{x}) \tag{32}$$

when A is a monotone set-valued map with weakly compact convex values. We recall that we have introduced the function γ, defined by

$$\gamma(x,y):=\inf_{p\in A(x)}\langle p,x-y\rangle \tag{33}$$

We observe that the function γ is monotone in the sense that

$$\begin{cases}\textbf{i.}\ \ \forall y\in\mathrm{Dom}\,(A),\quad \gamma(y,y)\leqslant 0\\ \textbf{ii.}\ \ \forall x,y\in\mathrm{Dom}\,(A),\quad \gamma(x,y)+\gamma(y,x)\geqslant 0\end{cases} \tag{34}$$

Indeed,

$$\gamma(x, y)+\gamma(y, x)=\inf_{\substack{p \in A(x) \\ q \in A(y)}} \langle p-q, x-y\rangle \geqslant 0$$

Also, $y \to \gamma(x, y)$ is obviously concave and upper semicontinuous. We shall need the weakest continuity property we can think of.

DEFINITION 10

We say that a set-valued map from a convex subset $M \subset X$ to X^* is finitely upper semicontinuous if it is upper semicontinuous from M supplied with the finite topology to X supplied with the weak topology. ▲

LEMMA 11

If A is finitely upper semicontinuous from M to X, then the functions $x \to \gamma(x, y)$ are lower semicontinuous for the finite topology. ▲

Proof. Let $K:=\{x_1, \ldots, x_n\}$ be a finite subset of M, and set

$$\beta_K: \lambda \in \Sigma^n \to \beta_K(\lambda):=\sum_{i=1}^{n} \lambda_i x_i$$

We have to prove that $\lambda \to \gamma(\beta_K(\lambda), y)$ is continuous.

Let

$$\mathcal{N}:=\left\{p \in X^* \,\middle|\, \max_{i=1,\ldots,n} |\langle p, x_i-y\rangle| \leqslant \frac{\varepsilon}{2}\right\}$$

be a neighborhood of zero for the weak topology. Since A is finitely upper semicontinuous, there exists $\eta>0$ such $\|\lambda-\lambda_0\| \leqslant \eta$ implies that $A(\beta_K(\lambda)) \subset A(\beta_K(\lambda_0))+\mathcal{N}$. Also, η can be chosen so small that

$$\sup_{p \in A(\beta_K(\lambda_0))} \langle p, \beta_K(\lambda)-\beta_K(\lambda_0)\rangle| \leqslant \frac{\varepsilon}{2}$$

Hence, for any $p \in A(\beta_K(\lambda))$, there exists $p_0 \in A(\beta_N(\lambda_0))$ such that $p-p_0 \in \mathcal{N}$, and thus, $\langle p-p_0, \beta_K(\lambda)-y\rangle \leqslant \varepsilon/2$ by the very definition of $\mathcal{N}$. On the other hand, $\langle p_0, \beta_K(\lambda_0)-\beta_K(\lambda)\rangle \leqslant \varepsilon/2$. We deduce that

$$\begin{aligned}
\gamma(\beta_K(\lambda_0), y) &\leqslant \langle p_0, \beta_K(\lambda_0)-y\rangle \\
&=\langle p, \beta_K(\lambda)-y\rangle+\langle p_0-p, \beta_K(\lambda)-y\rangle+\langle p_0, \beta_K(\lambda_0)-\beta_K(\lambda)\rangle \\
&\leqslant \langle p, \beta_K(\lambda)-y\rangle+\varepsilon
\end{aligned}$$

By taking the infimum when p ranges over $A\beta_K(\lambda)$, we deduce that

$$\|\lambda-\lambda_0\|\leqslant\eta\Rightarrow\gamma(\beta_K(\lambda_0), y)\leqslant\gamma(\beta_K(\lambda), y)+\varepsilon$$ ■

We shall now solve inclusion (32) when

(35)
- **i.** V is a proper lower semicontinuous convex function.
- **ii.** A is a monotone finitely upper semicontinuous map with weakly compact convex values.
- **iii.** $\text{Dom } V\subset\text{Dom } A$

We observe that a necessary condition for the existence of a solution to this problem is that

$$f\in\text{Dom } V^*+A\text{ Dom } V \tag{36}$$

Indeed, proposition 3 implies the existence of a solution to the equivalent problem $f\in\bar{p}+A\partial V^*(\bar{p})$; we deduce that $\bar{p}\in\text{Dom }\partial V^*\subset\text{Dom } V^*$ and $\partial V^*(\bar{p})\subset\text{Im }\partial V^*=\text{Dom }\partial V\subset\text{Dom } V$ and thus $f\in\text{Dom } V^*+A\text{ Dom } V$.

We shall prove that this condition is almost sufficient.

THEOREM 12

We posit assumptions (35). *Assume moreover that A is β monotone. Then there exists a solution $\bar{x}$ to the inclusion* (32) *when*

$$f\in\text{Int}(\text{Dom } V^*+A\text{ Dom } V+\text{Dom }\beta^*) \tag{37}$$ ▲

Remark

The size of Dom β^* balances the interiority condition in assumption (37), as the following corollary shows. ■

COROLLARY 13

We posit assumptions (35).

a. *If A is monotone, then*

$$\text{Int}(\text{Dom } V^*+A\text{ Dom } V)\subset\text{Im}(A+\partial V)\subset\text{Dom } V^*+A\text{ Dom } V \tag{38}$$

b. *If there exists $c>0$ such that*

$$\forall(x, p),\quad (y, q)\in\text{graph}(A),\quad \langle p-q, x-y\rangle\geqslant c\|x-y\|$$

then

$$\text{Im}(A+\partial V)=\text{Dom } V^*+A\text{ Dom } V \tag{39}$$

c. *If there exists* $c>0$ *and* $\alpha>1$ *such that*

$$\forall x, p), \quad (y, q) \in \text{graph}(A), \quad \langle p-q, x-y\rangle \geq c\|x-y\|^{\alpha}$$

then

$$\text{Dom } V^* + A \text{ Dom } V = \text{Im } (A+\partial V) = X \tag{40}$$ ▲

We also state the consequence of this theorem for variational inequalities. We recall that

$$b(K) := \{p \in X \mid \sigma_K(p) := \sup_{x \in K} \langle p, x\rangle + \infty\} \tag{41}$$

is the barrier cone of K, whose size measures the lack of boundedness of K.

COROLLARY 14
Let A *be a strict monotone finitely upper semicontinuous map from a closed convex set* $K \subset X$ *to the weakly compact convex subsets of* X. *Assume that* A *is* β *monotone. If*

$$0 \in \text{Int}(b(K) + A(K) + \text{Dom } \beta^*) \tag{42}$$

then there exists a solution $\bar{x} \in K$ *to the variational inequality* $0 \in A(\bar{x}) + N_K(\bar{x})$.▲

Assumption (42) shows how the lack of boundedness of K is compensated by the degree of monotonicity of A. We point out that (42) is satisfied when one of the following instance is satisfied:

$$(43) \quad \begin{cases} \textbf{i.} & K \text{ is bounded} \quad (b(K)=X). \\ \textbf{ii.} & A \text{ is surjective} \quad (A(K)=X). \\ \textbf{iii.} & A \text{ satisfies (28)} \quad (\text{Dom } \beta^*=B) \quad \text{and} \quad A(K) \cap -b(K) \neq \varnothing. \\ \textbf{iv.} & A \text{ satisfies (29)} \quad (\text{Dom } \beta^*=X). \end{cases}$$

COROLLARY 15
We posit assumptions (37) *of theorem* 12. *Then the map* $1+A+\partial V$ *is surjective.* ▲

Proof of Theorem 12. We set $K_n := \{x \in \text{Dom } V \mid V(x) \leq n \text{ and } \|x\| \leq n\}$. The subsets K_n are weakly compact and convex and $\text{Dom } V = \bigcup_{n=1}^{\infty} K_n$.

a. We set

$$\phi(x, y) := \gamma(x, y) - \langle f, x-y\rangle + V(x) - V(y) \tag{44}$$

This function satisfies

$$(45)\quad \begin{cases} \textbf{i.} & \forall y \in \operatorname{Dom} V,\ x \to \phi(x, y) \text{ is lower semicontinuous for the finite topology.} \\ \textbf{ii.} & \forall x \in \operatorname{Dom} V,\ y \to \phi(x, y) \text{ is concave and upper semicontinuous (for the weak topology).} \end{cases}$$

and

$$(46)\quad \begin{cases} \textbf{i.} & \forall y \in \operatorname{Dom} V, \quad \phi(y, y) \leqslant 0 \\ \textbf{ii.} & \forall x, y \in \operatorname{Dom} V, \quad \phi(x, y) + \phi(y, x) \geqslant 0 \end{cases}$$

Since K_n is weakly compact and convex, Ky Fan's inequality for monotone functions (see theorem 3.9) implies that for all $n \geqslant 1$, there exists $x_n \in K_n$ solution to

$$(47)\quad \forall y \in K_n, \qquad \phi(x_n, y) \leqslant 0$$

b. We shall now use assumption (37) to prove that x_n remains in a weakly compact subset of X. For that purpose, thanks to the uniform boundedness theorem, it is sufficient to prove that

$$(48)\quad \forall p \in X^*, \qquad \exists n(p) \text{ such that } \sup_{n \geqslant n(p)} \langle p, x_n \rangle < +\infty$$

By assumption (37), there exist $\eta > 0$, $r \in \operatorname{Dom} \beta^*$, $q \in \operatorname{Dom} V^*$, $y \in \operatorname{Dom} V$, $u \in A(y)$ such that

$$(49)\quad f + \frac{\eta p}{\|p\|} = r + q + u$$

We choose $n(p)$ to be the smallest n such that $y \in K_n$. By taking the duality product with x_n, we obtain

$$\frac{\eta}{\|p\|} \langle p, x_n \rangle = \langle r, x_n - y \rangle + \langle q, x_n \rangle + \langle u, x_n - y \rangle - \langle f, x_n - y \rangle + \langle r + u - f, y \rangle$$

We use Fenchel's inequalities $\langle r, x_n - y \rangle \leqslant \beta(x_n - y) + \beta^*(r)$ and $\langle q, x_n \rangle \leqslant V(x_n) + V^*(q)$. We obtain

$$(50)\quad \frac{\eta}{\|p\|} \langle p, x_n \rangle \leqslant \langle u, x_n - y \rangle + V(x_n) - V(y) - \langle f, x_n - y \rangle + \beta(x_n - y) + \beta^*(r) + V^*(q) + V(y) + \langle r + u - f, y \rangle$$

Since A is β monotone, we deduce that

$$\gamma(x_n, y) - \langle u, x_n - y \rangle = \inf_{p \in A(x_n)} \langle p - u, x_n - y \rangle \geqslant \beta(x_n - y) \tag{51}$$

Therefore, inequality (50) becomes

$$\begin{aligned} \frac{\eta}{\|p\|} \langle p, x_n \rangle \leqslant & (\gamma(x_n, y) - \langle f, x_n - y \rangle + V(x_n) - V(y)) + \beta^*(r) \\ & + V^*(q) + V(y) + \langle r + u - f, y \rangle \end{aligned}$$

Consequently, for all $n \geqslant n(p)$, we deduce from (47) that

$$\langle p, x_n \rangle \leqslant \frac{\|p\|}{\eta} (\beta^*(r) + V^*(q) + V(y) + \langle r + u - f, y \rangle) \tag{52}$$

The right-hand side is finite, because $r \in \operatorname{Dom} \beta^*$, $q \in \operatorname{Dom} V^*$, and $y \in \operatorname{Dom} V$. Hence, the sequence is bounded and thus weakly relatively compact.

c. Therefore, a subsequence of elements $x_{n'}$ converges weakly to some $\bar{x} \in X$. Since V is lower semicontinuous, we deduce from the monotonicity of A and the variational inequalities (47) that

$$\begin{aligned} V(\bar{x}) & \leqslant \liminf_n V(x_n) \\ & \leqslant \liminf_n [(V(y) + \langle f, x_n - y \rangle + \gamma(y, x_n)) - \gamma(y, x_n) - \gamma(x_n, y)] \\ & \leqslant \limsup_n [V(y) + \langle f, x_n - y \rangle + \gamma(y, x_n)] \\ & \leqslant V(y) + \langle f, \bar{x} - y \rangle + \gamma(y, \bar{x}) \end{aligned}$$

Therefore, $\bar{x} \in \operatorname{Dom} V$ and

$$\forall y \in \operatorname{Dom} V, \qquad 0 \leqslant \phi(y, \bar{x}) \tag{53}$$

d. We deduce from properties (45) and (46) that

$$\forall z \in \operatorname{Dom} V, \qquad \phi(\bar{x}, z) \leqslant 0 \tag{54}$$

as in the proof of theorem 3.9.

Indeed, if the conclusion is false, there would exist $z \in \operatorname{Dom} V$ such that $0 < \phi(\bar{x}, z)$ and by (45, **i**), there would exist $\bar{t} \in]0, 1[$ such that

$$0 < \phi(\bar{x} + \bar{t}(z - \bar{x}), z) \tag{55}$$

By taking $y=\bar{x}+\bar{t}(z-\bar{x})$, inequality (53) implies that

$$0 \leqslant \phi(\bar{x}+\bar{t}(z-\bar{x}), \bar{x})$$

Hence, the concavity of ϕ with respect to the second variable yields

$$0<\phi(\bar{x}+\bar{t}(z-\bar{x}), \bar{x}+\bar{t}(z-\bar{x})) \tag{56}$$

a contradiction of (46, **i**). Then proposition 3 implies that the solution $\bar{x}$ of (54) is a solution to problem (32). ■

Remark

We notice that the property

$$A+\partial V \text{ is surjective} \tag{57}$$

provides a decomposition akin to the one furnished by the projection theorem onto closed vector subspaces or convex cones. Indeed, any f can be written

$$\begin{cases} \textbf{i.} & f \in A\bar{x}+\bar{p} \\ \textbf{ii.} & \langle \bar{p}, \bar{x}\rangle = V(\bar{x})+V^*(\bar{p}) \end{cases} \tag{58}$$

since the latter equation expresses that $\bar{p} \in \partial V(\bar{x})$. Actually, when A is the identity and V is the indicator of a closed cone P, V^* is the indicator of the negative polar cone P^- and (58) becomes

$$\begin{cases} \textbf{i.} & f=\bar{x}+\bar{p}, \qquad \bar{x} \in P, \qquad \bar{p} \in P^- \\ \textbf{ii.} & \langle \bar{p}, \bar{x}\rangle = 0 \end{cases} \tag{59}$$

■

7. MAXIMAL MONOTONE MAPS

We continue to assume that X is a Hilbert space identified with its dual.

DEFINITION 1

A monotone set-valued map A is maximal if there is no other monotone set-valued map $\tilde{A}$ whose graph strictly contains the graph of A. ▲

We begin by pointing out the following: A set-valued map is maximal monotone if and only if its inverse A^{-1} is maximal monotone.

Also, the graph of any monotone set-valued map is contained in the graph of a maximal monotone set-valued map by Zorn's lemma, because the union of an increasing family of graphs of a monotone set-valued map is the graph of a set-valued monotone map. Actually, we will use the following equivalent analytical definition of a maximal monotone set-valued map.

PROPOSITION 2
A necessary and sufficient condition for a set-valued map A to be maximal monotone is that the property

$$\forall (y, v) \in \text{graph}(A), \qquad \langle u-v, x-y\rangle \geq 0 \tag{1}$$

be equivalent to

$$u \in A(x) \tag{2}$$ ▲

This provides a useful and manageable way of recognizing that u belongs to $A(x)$.

PROPOSITION 3
Let A be maximal monotone.
a. *Its images $A(x)$ are closed and convex.*
b. *Its graph is weakly-strongly closed in the sense that if x_n converges to x, and if $u_n \in A(x_n)$ converges weakly to u, then $u \in A(x)$.* ▲

Proof. **a.** By the preceding proposition, $A(x)$ is the intersection of the closed half-spaces $\{u \in X \mid \langle u-v, x-y\rangle \geq 0\}$ when (y, v) ranges over the graph of A. Hence, $A(x)$ is closed and convex.
b. Let x_n converge to x and let $u_n \in A(x_n)$ converge weakly to u. Let us choose (y, v) in the graph of A. Inequalities

$$\langle u_n - v, x_n - y\rangle \geq 0$$

imply, by going to the limit, inequalities

$$\langle u-v, x-y\rangle \geq 0$$

Hence, $u \in A(x)$ by proposition 2. ■

PROPSOITION 4
Any finitely continuous monotone single-valued map A from X to X is maximal monotone. ▲

Proof. Let $x \in X$ and $u \in X$ such that

$$\langle u - A(y), x-y\rangle \geq 0 \qquad \text{for all } y \in X \tag{3}$$

To show that A is maximal monotone, we have to check that $u = A(x)$ (proposition 2). For that purpose, we take $y = x - \lambda(z-x)$ where $\lambda \in]0, 1[$ and $z \in X$. Inequality (3) becomes

$$\langle u - A(x-\lambda(z-x)), z-x\rangle \geq 0 \qquad \text{for all } z \in X \tag{4}$$

By letting λ converge to zero and by the continuity of A for the finite topology. we deduce that $\langle u - A(x), z - x\rangle \geqslant 0$ for all $z \in X$, that is, $u = A(x)$. ■

The following theorem provides a very important characterization of maximal monotone maps.

THEOREM 5 (MINTY)

A monotone map A is maximal if and only if $1 + A$ is surjective. ▲

Before proving Minty's theorem, we use it to provide examples of maximal monotone maps.

PROPOSITION 6

a. *Let V be a proper lower semicontinuous, convex function from a Hilbert space X to $R \cup \{+\infty\}$. Then the subdifferential map $\partial V: x \in X \to \partial V(x) \subset X$ is maximal monotone.*

b. *Let U be a proper function from $X \times Y$ to $R \cup \{-\infty\}$ such that*

$$(5) \quad \begin{cases} \text{i. } \forall y \in Y, & x \to U(x, y) \text{ is concave and upper semicontinuous.} \\ \text{ii. } \forall x \in X, & y \to U(x, y) \text{ is convex and lower semicontinuous.} \end{cases}$$

Then the set-valued map $(x, y) \in X \times Y \to \partial_x(-U)(x, y) \times \partial_y U(x, y)$ is maximal monotone. ▲

Proof. The first statement follows from theorem 3.3.11, which states that $1 + \partial V$ is surjective, and the second statement follows from theorem 3.4.14, stating that the graph of ∂U is, up to a canonical isomorphism, the graph of the subdifferential ∂V of the convex function defined on $X \times Y^*$ by

$$V(x, q) = \sup_{y \in Y} [\langle q, y\rangle - U(x, y)]$$ ■

PROPOSITION 7

Let A be a monotone (respectively, maximal monotone) set-valued map from X to X. Let $\mathcal{A}$ be the set-valued map from $L^2(0, T, X)$ to $L^2(0, T; X)$ defined by $(\mathcal{A}x)(t) := A(x(t))$ a.e.

Then $\mathcal{A}$ is a monotone (respectively, maximal monotone) set-valued map. ▲

Proof. We know that $\mathcal{A}$ is monotone. If A is maximal monotone, then $J_1 = (1 + A)^{-1}$ is Lipschitz from X to X. Hence, if $y(\cdot)$ is given in $L^2(0, T, X)$, the function $x(\cdot)$ defined by

$$x(t) = J_1 y(t) \quad \text{a.e.}$$

is obviously measurable. Also,

$$\|x(t)-J_1(0)\|\leqslant\|y(t)\|$$

and thus $x(\cdot)$ belongs to $L^2(0, T; X)$ and is obviously a solution of $y(\cdot)\in(1+A)(x(\cdot))$. ■

Corollary 6.15 also shows that the sum $A+\partial V$ of a monotone finitely upper semicontinuous map with weakly compact images and the subdifferential of a lower semicontinuous, convex function is maximal monotone when $\text{Dom } V \subset \text{Dom } A$.

Proof of Minty's Theorem. **a.** Assume that $1+A$ is surjective. Let $x\in X$ and $u\in X$ satisfy

$$\forall(y, v)\in\text{graph}(A),\qquad \langle u-v, x-y\rangle\geqslant 0 \tag{6}$$

By Proposition 2, we have to prove that $u\in A(x)$. Since $1+A$ is surjective, we can choose y in (6) to be a solution y_0 of the inclusion $u+x\in y_0+A(y_0)$. Let $v_0\in A(y_0)$ such that $u+x=y_0+v_0$. Then

$$\|x-y_0\|^2=\langle x-y_0, x-y_0\rangle=-\langle u-v_0, x-y_0\rangle\leqslant 0$$

because A is monotone. Hence, $x=y_0$, and thus $u=v_0\in A(y_0)=A(x)$.
b. Assume that A is maximal monotone. Let $y\in X$. We have to prove that there exists x such that $y\in x+A(x)$. It is sufficient to choose $y=0$, since this amounts to replacing A by $x\to -y+A(x)$, which is also maximal monotone. By proposition 2, we must prove that there exists $\bar{x}$ satisfying

$$\forall(y, v)\in\text{graph}(A),\qquad \langle -\bar{x}-v, \bar{x}-y\rangle\geqslant 0 \tag{7}$$

For convenience, set $\phi(x; (y, v)):=\langle x+v, x-y\rangle$, or, equivalently,

$$\phi(x; (y, v)):=\|x\|^2+\langle x, v-y\rangle-\langle v, y\rangle$$

We have to prove that there exists $\bar{x}$ such that

$$\forall(y, v)\in\text{graph}(A),\qquad \phi(\bar{x}; (y, v))\leqslant 0 \tag{8}$$

Fix (y_0, v_0) to be any point in the graph of A. If a solution to (8) exists, it certainly belongs to

$$L:=\{x\in\text{Dom}(A)|\phi(x; (y_0, v_0))\leqslant 0\} \tag{9}$$

The map $x \to \phi(x; (y_0, v_0))$ is quadratic; hence, the set $\{x | \phi(x; (y_0, v_0)) \leq 0\}$ is convex, closed, and bounded, hence, weakly compact. Its intersection L with Dom A need not be so, but so is $\overline{\text{co}}(L)$. Also, the maps $x \to \phi(x; (y, v))$ are convex and (strongly) continuous. Therefore, their lower sections are convex and closed, hence, weakly closed. It follows that they are weakly lower semicontinuous.

We use theorem 2.6: Set $\mathscr{S}$ to be the family of all finite subsets $K := \{(y_1, v_1), \ldots, (y_n, v_n)\}$ of graph(A), then there exists $x \in \overline{\text{co}}(L)$ such that

$$
\begin{aligned}
&\sup_{(y,v) \in \text{graph}(A)} \phi(x; (y, v)) \\
&\leq \sup_{K \in \mathscr{S}} \inf_{x \in \overline{\text{co}}(L)} \max_{i=1,\ldots,n} \phi(x; (y_i, v_i)) \\
&\leq \sup_{K \in \mathscr{S}} \inf_{x \in \text{co}(y_1,\ldots,y_n)} \max_{i=1,\ldots,n} \phi(x; (y_i, v_i)) \qquad (10) \\
&\leq \sup_{K \in \mathscr{S}} \inf_{x \in \text{co}(y_1,\ldots,y_n)} \sup_{\mu \in \Sigma^n} \sum_{j=1}^{n} \mu_j \phi(x; (y_j, v_j)) \\
&\leq \sup_{K \in \mathscr{S}} \inf_{\lambda \in \Sigma^n} \sup_{\mu \in \Sigma^n} \phi_K(\lambda, \mu)
\end{aligned}
$$

where we set

$$
\phi_K(\lambda, \mu) := \sum_{j=1}^{n} \mu_j \phi(\beta(\lambda); (y_j, v_j)), \qquad \beta(\lambda) := \sum_{i=1}^{n} \lambda_i y_i \tag{11}
$$

This function is continuous with respect to λ and

$$
\begin{aligned}
\phi_K(\mu, \mu) &= \sum_{i,k=1}^{n} \mu^i \mu^k \langle \beta(\mu) + v_i, y_k - y_i \rangle \\
&= \sum_{i,k=1}^{n} \mu^i \mu^k \langle \beta(\mu), y_k - y_i \rangle + \sum_{i,k=1}^{n} \mu^i \mu^k \langle v_i, y_k - y_i \rangle
\end{aligned}
$$

The first term is zero for reasons of symmetry, while the second can be written

$$
\begin{aligned}
&\frac{1}{2} \sum_{i,k=1}^{n} \mu^i \mu^k \langle v_i, y_k - y_i \rangle + \frac{1}{2} \sum_{j,l=1}^{n} \mu^j \mu^\ell \langle -v_j, y_j - y_\ell \rangle \\
&\frac{1}{2} \sum_{i,k=1}^{n} \mu^i \mu^k \langle v_i - v_k, y_k - y_i \rangle \leq 0
\end{aligned}
$$

since A is monotone. Hence, assumptions of the lopsided minimax theorem are satisfied (see theorem 2.7). Therefore, for all K,

$$
\inf_{\lambda \in \Sigma^n} \sup_{\mu \in \Sigma^n} \phi_K(\lambda, \mu) \leq 0
$$

and thus by (10) inequality (8) is satisfied for $\bar{x}$. We have shown that $\bar{x}$ is a solution of $0 \in \bar{x} + A(\bar{x})$. ■

Suppose that A is maximal monotone. We show that A can be approximated in some sense by single-valued maps A_λ that are also maximal monotone. These maps, called Yosida approximations, play an important role.

THEOREM 8

Let A be maximal monotone. Then for all $\lambda > 0$,

$$\text{the resolvant } J_\lambda = (1 + \lambda A)^{-1} \text{ is a nonexpansive single-valued map from } X \text{ to } X \tag{12}$$

and the map $A_\lambda := (1 - J_\lambda)/\lambda$ satisfies

$$\begin{cases} \textbf{i.} & \forall x \in X, \quad A_\lambda(x) \in A(J_\lambda x). \\ \textbf{ii.} & A_\lambda \text{ is Lipschitz with constant } 1/\lambda \text{ and maximal monotone.} \end{cases} \tag{13}$$

Let $m(A(x))$ denote the element of $A(x)$ with smallest norm. We also have

$$\forall x \in \mathrm{Dom}(A), \qquad \|A_\lambda(x) - m(A(x))\|^2 \leqslant \|m(A(x))\|^2 - \|A_\lambda(x)\|^2 \tag{14}$$

and for all $x \in \mathrm{Dom}(A)$,

$$\begin{cases} \textbf{i.} & J_\lambda x \text{ converges to } x. \\ \textbf{ii.} & A_\lambda x \text{ converges to } m(A(x)). \end{cases} \tag{15}$$

▲

DEFINITION 9

The maps A_λ are called the Yosida approximations *of A.* ▲

Proof. **a.** Let x_i $(i = 1, 2)$ be solutions to the inclusions

$$y_i \in x_i + \lambda A(x_i) \qquad (i = 1, 2) \tag{16}$$

So $y_i = x_i + \lambda v_i$ when $v_i \in A(x_i)$. We obtain

$$\begin{aligned} \|y_1 - y_2\|^2 &= \|x_1 - x_2 + \lambda(v_1 - v_2)\|^2 \\ &= \|x_1 - x_2\|^2 + \lambda^2 \|v_1 - v_2\|^2 + 2\lambda \langle v_1 - v_2, x_1 - x_2 \rangle \\ &\geqslant \|x_1 - x_2\|^2 + \lambda^2 \|v_1 - v_2\|^2 \end{aligned}$$

Hence,

$$\begin{aligned} &\textbf{i.} \quad \|x_1 - x_2\| \leqslant \|y_1 - y_2\| \\ &\textbf{ii.} \quad \|v_1 - v_2\| \leqslant (1/\lambda) \|y_1 - y_2\| \end{aligned} \tag{17}$$

By taking $y_1 = y_2$, (17, **i**) proves the uniqueness of the solution. We note that

$$x_i = J_\lambda y_i \quad \text{and} \quad v_i = A_\lambda(y_i) \tag{18}$$

Hence, inequalities (17) prove that J_λ and A_λ are Lipschitz with constants 1 and $1/\lambda$, respectively.

b. By the very definitions of J_λ and A_λ, we have

$$A_\lambda(y) = \frac{1}{\lambda}(y - J_\lambda(y)) \in A(J_\lambda y) \qquad \text{for all } y \in X \tag{19}$$

Therefore, since $y_i = J_\lambda(y_i) + A_\lambda(y_i)$, we obtain

$$\begin{aligned}
&\langle A_\lambda(y_1) - A_\lambda(y_2), y_1 - y_2 \rangle \\
&\quad = \langle A_\lambda(y_1) - A_\lambda(y_2), J_\lambda(y_1) - J_\lambda(y_2) \rangle + \lambda \|A_\lambda(y_1) - A_\lambda(y_2)\|^2 \\
&\quad \geqslant \lambda \|A_\lambda(y_1) - A_\lambda(y_2)\|^2 \geqslant 0
\end{aligned}$$

Hence, A_λ is monotone (and by proposition 4, maximal monotone).

c. Let $x \in \text{Dom } A$. We compute

$$\begin{aligned}
\|A_\lambda(x) - m(A(x))\|^2 &= \|A_\lambda(x)\|^2 + \|m(A(x))\|^2 - 2\langle A_\lambda(x), m(A(x)) \rangle \\
&= -\|A_\lambda(x)\|^2 + \|m(A(x))\|^2 - 2\langle A_\lambda(x), m(A(x)) - A_\lambda(x) \rangle
\end{aligned}$$

But since A is monotone, $m(A(x)) \in A(x)$, and $A_\lambda(x) \in A(J_\lambda x)$, we obtain

$$\langle A_\lambda(x), m(A(x)) - A_\lambda(x) \rangle = \frac{1}{\lambda} \langle x - J_\lambda(x), m(A(x)) - A_\lambda(x) \rangle \geqslant 0$$

Therefore, we have proved inequality

$$\|A_\lambda(x) - m(A(x))\|^2 \leqslant \|m(A(x))\|^2 - \|A_\lambda(x)\|^2 \tag{20}$$

d. Then when $x \in \text{Dom } A$, we have

$$\|x - J_\lambda(x)\| = \lambda \|A_\lambda(x)\| \leqslant \lambda \|m(A(x))\|$$

Therefore, $J_\lambda(x)$ converges to x when λ converges to zero.

e. Since $J_\lambda(x) = x - \lambda A_\lambda(x)$ and $A_\lambda(x) \in A(J_\lambda x)$, we see that $y = A_\lambda(x)$ is a solution to the equation $y \in A(x - \lambda y)$. Conversely, any such solution y is $A_\lambda(x)$. Indeed, set $z = x - \lambda y$; this equation becomes $x \in z + \lambda A(z)$. Hence, $z = J_\lambda(x)$ and $y = (1/\lambda)(x - J_\lambda(x)) = A_\lambda(x)$

f. This remark implies that

$$A_{\mu+\lambda}(x) = (A_\mu)_\lambda(x)$$

Indeed, $y=A_{\mu+\lambda}(x)$ is a solution to the equation $y\in A(x-\lambda y-\mu y)$; then $y\in A_\mu(x-\lambda y)$. By again applying the preceding remark to the Yosida approximation A_μ, which is maximal monotone, we deduce that $y=(A_\mu)_\lambda(x)$.

g. Now we use inequality (20), replacing A by A_μ. Since $m(A_\mu(x))=A_\mu(x)$, we obtain

$$\|A_{\mu+\lambda}(x)-A_\mu(x)\|^2\leqslant\|A_\mu(x)\|^2-\|A_{\lambda+\mu}(x)\|^2$$

Then the sequence $\|A_\mu(x)\|^2$ is an increasing sequence of real numbers bounded above by $\|m(A(s))\|^2$. Hence, it converges to some real number α. This implies that

$$\lim_{\lambda,\mu\to 0}\|A_{\mu+\lambda}(x)-A_\mu(x)\|^2\leqslant\alpha-\alpha=0$$

Hence, $A_\lambda(x)$ satisfies the Cauchy criterion and converges to some element v in X.

Since $A_\lambda(x)\in A(J_\lambda(x))$ and the graph of A is closed, we deduce that $v\in A(x)$. Also,

$$\|v\|=\lim_{\lambda\to 0}\|A_\lambda(x)\|\leqslant\|m(A(x))\|$$

Since $A(x)$ is closed and convex, the projection of zero to $A(x)$ is unique and, consequently, $v=m(A(x))$. Therefore, $A_\lambda(x)$ converges to $m(A(x))$ for all $x\in\mathrm{Dom}(A)$. ■

We now provide a handy sufficient analytical condition for an open subset to be included in the image of a maximal monotone map. This condition motivates the introduction of a property that is satisfied, for instance, by the subdifferentials of convex functions. When two monotone maps satisfy this property and the sum is maximal monotone, then we obtain the following inclusion:

$$\mathrm{Int}(\mathrm{Im}\,A+\mathrm{Im}\,B)\subset\mathrm{Im}\,A+\mathrm{Im}\,B$$

which allows us to solve inclusions of the form

$$y\in Ax+Bx$$

as well as inclusions of the form

$$y\in x+ABx$$

We recall solving such problems in Section 6, where B was the subdifferential of a lower semicontinuous convex function. We begin with the fundamental result stated in theorem 10.

THEOREM 10

Let A be a maximal monotone map and $K \subset X$ a subset satisfying

$$(21) \quad \forall v \in K, \quad \exists y \in X \quad \textit{and} \quad c \in R \;\Big|\; \inf_{(x,u) \in \mathrm{graph}(A)} \langle u-v, x-y \rangle \geqslant c$$

Then

$$(22) \quad \mathrm{co}(K) \subset \overline{\mathrm{Im}\, A} \quad \textit{and} \quad \mathrm{Int}\, \mathrm{co}(K) \subset \mathrm{Int}(\mathrm{Im}\, A) \quad \blacktriangle$$

Proof. **a.** First, we check that we can replace K by $\mathrm{co}(K)$ in property (21). Indeed, let $v = \sum_{i=1}^{n} \lambda_i v_i$ be a convex combination of elements $v_i \in K$. By assumption (21), there exist y_i and $c_i \in R$ such that for all $(x, u) \in \mathrm{graph}(A)$,

$$\langle u - v_i, x - y_i \rangle \geqslant c_i$$

Let us set $y := \sum_{i=1}^{n} \lambda_i y_i$. We deduce that

$$\langle u, x \rangle - \langle v, x \rangle - \langle u, y \rangle \geqslant \sum_{i=1}^{n} \lambda_i (c_i - \langle v_i, y_i \rangle)$$

Hence,

$$(23) \quad \inf_{(x,u) \in \mathrm{graph}(A)} \langle u - v, x - y \rangle \geqslant \sum_{i=1}^{n} \lambda_i (c_i - \langle v_i, y_i \rangle) + \langle v, y \rangle$$

b. We now assume that K is convex and prove that $K \subset \overline{\mathrm{Im}\, A}$. Indeed, since A is maximal monotone, we can associate with any $\varepsilon > 0$ and any $v \in K$ the solution x_ε to the inclusion $v \in \varepsilon x_\varepsilon + A(x_\varepsilon)$. Let $y \in X$ and $c \in R$ such that property (21) holds true. By taking $(x_\varepsilon, v - \varepsilon x_\varepsilon) \in \mathrm{graph}(A)$, we obtain

$$\langle -\varepsilon x_\varepsilon, x_\varepsilon - y \rangle \geqslant c$$

and thus

$$\varepsilon \|x_\varepsilon\|^2 \leqslant \varepsilon \langle x_\varepsilon, y \rangle - c \leqslant \frac{\varepsilon}{2} \|x_\varepsilon\|^2 + \frac{\varepsilon}{2} \|y\|^2 - c$$

Hence, $\sqrt{\varepsilon} x_\varepsilon$ remains in a bounded set rB. Consequently, $v \in A(x_\varepsilon) + \sqrt{\varepsilon}\sqrt{\varepsilon} x_\varepsilon \subset \mathrm{Im}\, A + \sqrt{\varepsilon}\, rB$ for all $\varepsilon > 0$ which shows that $v \in \overline{\mathrm{Im}\, A}$.

c. We still assume that K is convex and prove that $\mathrm{Int}\, K \subset \mathrm{Int}\ \mathrm{Im}\, A$. Let $v \in \mathrm{Int}\, K$ and $\gamma > 0$ such that $v + \gamma B \subset K$. By assumption, for all $z \in X$, $v + \gamma z / \|z\|$ belongs to K, and there exist $y \in X$ and $c \in R$ such that

$$(24) \quad \forall (x, u) \in \mathrm{graph}(A), \quad \left\langle u - v - \frac{\gamma z}{\|z\|}, x - y \right\rangle \geqslant c$$

Let x_ε be the solution of the inclusion $v \in \varepsilon x_\varepsilon + A(x_\varepsilon)$. By taking $(x_\varepsilon, v-\varepsilon x_\varepsilon) \in \text{graph}(A)$, we obtain

$$\left\langle -\varepsilon x_\varepsilon - \frac{\gamma z}{\|z\|}, x_\varepsilon - y \right\rangle \geq c$$

and thus

$$\frac{\varepsilon}{2}\|x_\varepsilon\|^2 + \frac{\gamma}{\|z\|}\langle z, x_\varepsilon\rangle \leq \frac{\varepsilon}{2}\|y\|^2 + \frac{\gamma}{\|z\|}\langle z, y\rangle - c$$

Therefore, for all $z \in X$,

$$\sup_{\varepsilon} \langle z, x_\varepsilon\rangle < +\infty \tag{25}$$

The uniform boundedness theorem implies that x_ε remains in a weakly compact subset. Then a subsequence $x_{\varepsilon'}$ converges weakly to some x_*. Since $v-\varepsilon' x_{\varepsilon'} \in A(x_{\varepsilon'})$ and $v-\varepsilon' x_{\varepsilon'}$ converges strongly to v, proposition 3 implies that $v \in A(x_*) \subset \text{Im } A$. ■

We deduce the following interesting consequence.

PROPOSITION 11
If A is a maximal monotone map, then

$$\begin{cases} \textbf{i.} & \overline{\text{Dom } A} \textit{ and } \text{Int Dom } A \textit{ are convex.} \\ \textbf{ii.} & \overline{\text{Im } A} \textit{ and } \text{Int Im } A \textit{ are convex.} \end{cases} \tag{26}$$

▲

Proof. We take $K = \text{Im } A$. Since A is monotone, property (21) is obviously satisfied. Then $\text{co}(\text{Im } A) \subset \overline{\text{Im } A}$ and $\text{Int co Im } A \subset \text{Int Im } A$ by theorem 10. This implies that $\overline{\text{Im } A}$ and Int Im A are convex. Since A^{-1} is maximal monotone and $\text{Dom } A = \text{Im } A^{-1}$, we deduce that $\overline{\text{Dom } A}$ and Int Dom A are convex. ■

We also deduce a surjectivity criterion.

THEOREM 12
Let A be a maximal monotone map. We assume that either

$$\text{Dom } A \textit{ is bounded} \tag{27}$$

or that

$$\lim_{\|x\| \to \infty} \frac{\inf_{u \in A(x)} \langle u, x\rangle}{\|x\|} = +\infty \qquad (A \textit{ is strongly coercive.}) \tag{28}$$

Then A is surjective. ▲

Proof. We use theorem 10 with $K=X$. First, we replace A by $\tilde{A}$ defined by $\tilde{A}(x):=A(x+x_0)$ where $x_0 \in \text{Dom } A$, so that $0 \in \text{Dom } \tilde{A}$. The map $\tilde{A}$ is still maximal monotone, its domain is bounded when Dom A is bounded, and $\tilde{A}$ is strongly coercive when A is strongly coercive. Let $v \in X$. We check property (21) when $y=0$.

a. Assume that Dom $A \subset rB$. We take $z \in A(0)$. Then when $(x, u) \in \text{graph}(A)$, we have

$$(29) \qquad \langle u-v, x-0\rangle = \langle u-z, x-0\rangle + \langle z, x\rangle \geq \langle z, x\rangle \geq -\|z\|\,\|x\| \geq -r\|z\|$$

Hence, property (21) is satisfied with $y=0$ and $c=-r\|z\|$.

b. Assume that $\tilde{A}$ is coercive. There exists $r>0$ such that, for all $x \in \text{Dom } \tilde{A}$, $\|x\| \geq r$, for all $u \in \tilde{A}(x)$, $\langle u, x\rangle \geq \|v\|\,\|x\|$. Then when $(x, u) \in \text{graph}(\tilde{A})$, $\|x\| \geq r$, we obtain

$$(30) \qquad \langle u-v, x-0\rangle \geq \|v\|\,\|x\| - \langle v, x\rangle \geq 0$$

By taking into account inequality (29) when $\|x\| \leq r$, we see that property (21) is satisfied when $y=0$ and $c=-r\|z\|$.

c. In both cases, we have $X = \text{Int co}(X) \subset \text{Im } A$, that is, $\tilde{A}$ is surjective or, equivalently, A is surjective. ∎

We now turn our attention to the surjectivity properties of the sum of two maximal monotone maps. For that purpose, we have to provide conditions under which the sum is still a maximal monotone map and to assume that one of these maps satisfies another property, which is satisfied, for instance, by subdifferentials of proper lower semicontinuous, convex functions.

We begin by giving a sufficient condition for the sum of two maximal monotone maps to be maximal monotone. The sum of two monotone maps is monotone, whenever it is proper, that is, whenever $0 \in \text{Dom } A - \text{Dom } B$. We shall prove that a stronger condition,

$$0 \in \text{Int}(\text{Dom } A - \text{Dom } B)$$

implies that $A+B$ is maximal monotone. We have already encountered similar conditions in particular cases. When $A=\partial V$ and $B=\partial W$ are subdifferentials of lower semicontinuous, convex functions V and W, we proved that the weaker condition

$$0 \in \text{Int}(\text{Dom } V - \text{Dom } W)$$

implies that $\partial V + \partial W = \partial(V+W)$ and that thus $\partial V + \partial W$ is maximal monotone.

THEOREM 13

Let A and B be two maximal monotone maps from X to X. If

(31) $$0 \in \operatorname{Int}(\operatorname{Dom} A - \operatorname{Dom} B)$$

then $A+B$ is a maximal monotone map. ▲

Proof. Since $A+B$ is monotone, we must prove that $1+A+B$ is surjective. For that, we shall approximate one of the maps, B, for instance, by its Yosida approximation B_λ, which is maximal monotone and Lipschitz. We shall prove that $A+B_\lambda$ is maximal monotone and thus there exists a solution $x_\lambda \in X$ to the inclusion

(32) $$y \in x_\lambda + Ax_\lambda + B_\lambda x_\lambda$$

In the second step, we shall deduce from assumption (31) that

(33) $$\sup_{\lambda>0} \|B_\lambda x_\lambda\| < +\infty$$

In the third and last step, we shall deduce that the solutions x_λ of (32) do converge to the solution x of

(34) $$y \in x + Ax + Bx$$

a. There exists a solution x_λ to (32). We have to prove that $A+B_\lambda$ is maximal monotone, that is, $1+\mu(A+B_\lambda)$ is surjective for some $\mu>0$. If $y \in X$ is given, we have to find x_λ, solution to

(35) $$x_\lambda = (1+\mu A)^{-1}(y-\mu B_\lambda x_\lambda)$$

so that x_λ is a fixed point of the map $x \to (1+A)^{-1}(y-\mu B_\lambda x)$. Since $(1+\mu A)^{-1}$ is Lipschitz with constant 1 and $x \to y-\mu B_\lambda x_\lambda$ is Lipschitz with constant μ/λ, then $(1+\mu A)^{-1}(y-\mu B_\lambda(\cdot))$ is a contradiction when $\mu<\lambda$. Therefore, there exists a fixed point x_λ of (35) and, consequently, a solution to (32).

b. Let us prove estimate (33). We begin by proving that

(36) $$\sup_{\lambda>0} \|x_\lambda\| < +\infty$$

We choose $z \in \operatorname{Dom} A - \operatorname{Dom} B$. Since $y - x_\lambda \in (A+B_\lambda)(x_\lambda)$ by (32) and $m(Az)+B_\lambda z \in (A+B_\lambda)(z)$, we deduce that

$$\langle y-x_\lambda-m(Az)-B_\lambda z, x_\lambda - z\rangle = \langle y-(z+m(Az)+B_\lambda z), x_\lambda - z\rangle - \|x_\lambda - z\|^2 \geq 0$$

Hence,

$$\|x_\lambda - z\| \leqslant \|y - (z + m(Az) + B_\lambda z)\|$$

Since $\|B_\lambda(z)\| \leqslant \|m(Bz)\|$ by theorem 8, solutions x_λ to (32) remain bounded

$$\|x_\lambda\| \leqslant 2\|z\| + \|y\| + \|m(Az)\| + \|m(Bz)\| \tag{37}$$

By assumption (31), there exists $\gamma > 0$ such that

$$\gamma B \subset \text{Dom } A - \text{Dom } B$$

We can associate to every $p \in X$ elements $u \in \text{Dom } A$ and $v \in \text{Dom } B$ such that $\gamma p/\|p\| = v - u$. Then

$$\frac{\gamma}{\|p\|} \langle p, B_\lambda x_\lambda \rangle = \langle v, B_\lambda x_\lambda \rangle - \langle u, B_\lambda x_\lambda \rangle \tag{38}$$

The monotonicity of B_λ implies that

$$\langle B_\lambda x_\lambda - B_\lambda v, x_\lambda - v \rangle \geqslant 0$$

and thus

$$\langle B_\lambda x_\lambda, v \rangle \leqslant \langle B_\lambda x_\lambda, x_\lambda \rangle - \langle B_\lambda v, x_\lambda - v \rangle \leqslant \langle B_\lambda x_\lambda, x_\lambda \rangle + \|m(Bv)\| \, \|x_\lambda - v\|$$

Hence,

$$\frac{\gamma}{\|p\|} \langle p, B_\lambda x_\lambda \rangle \leqslant \langle B_\lambda x_\lambda, x_\lambda - u \rangle + \|m(Bv)\| \, \|x_\lambda - v\| \tag{39}$$

Since x_λ is a solution to inclusion (32), we can write that $B_\lambda x_\lambda = y - x_\lambda - z_\lambda$ when $z_\lambda \in A(x_\lambda)$. So, inequality (39) becomes

$$\frac{\gamma}{\|p\|} \langle p, B_\lambda x_\lambda \rangle \leqslant \langle y - x_\lambda - z_\lambda, x_\lambda - u \rangle + \|m(Bv)\| \, \|x_\lambda - v\|$$

But the monotonicity of A implies that

$$\langle z_\lambda - m(Au), x_\lambda - u \rangle \geqslant 0$$

and thus

$$-\langle z_\lambda, x_\lambda - u \rangle \leqslant \langle m(Au), x_\lambda - u \rangle \leqslant \|m(Au)\| \, \|x_\lambda - u\|$$

Finally,

$$\langle p, B_\lambda x_\lambda\rangle \leqslant \frac{\|p\|}{\gamma}(\|x_\lambda - u\|(\|y-x_\lambda\| + \|m(Au)\|) + \|m(Bv)\|\;\|x_\lambda - v\|) < +\infty$$

Consequently, for all $p \in X$, $\sup_{\lambda>0}\langle p, B_\lambda x_\lambda\rangle < +\infty$. The uniform boundedncss theorem implies that $B_\lambda x_\lambda$ remains bounded.

c. We prove now that x_λ converges to a solution x to inclusion (34). For that purpose, we prove that the sequence x_λ satisfies the Cauchy criterion. We set

$$z_\lambda := y - x_\lambda - B_\lambda x_\lambda \in A(x_\lambda) \qquad \text{and} \qquad z_\mu := y - x_\mu - B_\mu x_\mu \in A(x_\mu)$$

Since $0 = x_\lambda + z_\lambda + B_\lambda x_\lambda - x_\mu - z_\mu - B_\mu z_\mu$, by taking the scalar product with $x_\lambda - x_\mu$, we obtain

$$\begin{aligned} 0 &= \|x_\lambda - x_\mu\|^2 + \langle z_\lambda - z_\mu, x_\lambda - x_\mu\rangle + \langle B_\lambda x_\lambda - B_\mu x_\mu, x_\lambda - x_\mu\rangle \\ &\geqslant \|x_\lambda - x_\mu\|^2 + \langle B_\lambda x_\lambda - B_\mu x_\mu, x_\lambda - x_\mu\rangle \end{aligned}$$

(because A is monotone). Now, we set $J_\lambda := (1+\lambda B)^{-1}$. By definition of the Yosida approximation,

$$x_\lambda - x_\mu = \lambda B_\lambda x_\lambda - \mu B_\mu x_\mu + J_\lambda x_\lambda - J_\mu x_\mu$$

We recall that $B_\lambda x_\lambda \in B(J_\lambda x_\lambda)$ and $B_\mu x_\mu \in B(J_\mu x_\mu)$. Since B is monotone, we deduce

$$0 \geqslant \|x_\lambda - x_\mu\|^2 + \langle B_\lambda x_\lambda - B_\mu x_\mu, \lambda B_\lambda x_\lambda - \mu B_\mu x_\mu\rangle$$

and, consequently,

$$\|x_\lambda - x_\mu\|^2 \leqslant \|B_\lambda x_\lambda - B_\mu x_\mu\|\;\|\lambda B_\lambda x_\lambda - \mu B_\mu x_\mu\|$$

Since $\|B_\lambda x_\lambda\| \leqslant c$ and $\|B_\mu x_\mu\| \leqslant c$ by (33), we obtain

$$\|x_\lambda - x_\mu\|^2 \leqslant 2c^2(\lambda + \mu) \qquad \text{converges to zero.} \tag{40}$$

Hence, x_λ converges strongly to some x_*. Since $B_\lambda x_\lambda$ is bounded, a subsequence $B_{\lambda'} x_{\lambda'}$ converges weakly to some v_*, and thus

$$z_\lambda = y_\lambda - x_\lambda - B_\lambda x_\lambda$$

converges weakly to some $u_* = y - x_* - v_*$.

Proposition 3 implies that $u_* \in A(x_*)$. Also, $x_\lambda - J_\lambda x_\lambda = \lambda B_\lambda x_\lambda$ implies that $\|x_\lambda - J_\lambda x_\lambda\| \leqslant \lambda \|B_\lambda x_\lambda\| \leqslant \lambda c$ converges to zero, so that $J_\lambda x_\lambda$ converges strongly to x_*. Inclusions $B_{\lambda'}x_{\lambda'} \in B(J_{\lambda'}x_{\lambda'})$ imply that $v_* \in B(x_*)$, because B is maximal monotone. Hence,

$$y = x_* + u_* + v_* \in x_* + A(x_*) + B(x_*)$$ ■

We now introduce a property that we need for studying the image of the sum of two maximal monotone maps.

DEFINITION 14
Let A be a monotone map from X to X. It satisfies the L property if

$$(41) \qquad \begin{cases} \forall w \in \operatorname{Im} A, \quad \forall y \in \operatorname{Dom} A, \quad \exists c := c(y, w) \text{ such that} \\ \inf_{(x,u) \in \operatorname{graph}(A)} \langle u - w, x - y \rangle \geq c. \end{cases}$$ ▲

We begin by giving two examples of monotone maps satisfying the L property.

PROPOSITION 15
The subdifferential of a proper convex function satisfies the L property. ▲

Proof. Let $u \in \partial V(x)$, $v \in \partial V(y)$, and $w \in \partial V(z) \subset \operatorname{Im} \partial V$ be chosen. We deduce from the inequalities

$$V(x) - V(y) \leq \langle u, x - y \rangle, \qquad V(y) - V(z) \leq \langle v, y - z \rangle$$
$$\text{and} \qquad V(z) - V(x) \leq \langle w, z - x \rangle$$

that

$$(42) \qquad \langle u, x - y \rangle + \langle v, y - z \rangle + \langle w, x - x \rangle \geq 0$$

Then

$$\langle u - w, x - y \rangle \leq \langle w - v, y - z \rangle$$

The L property is satisfied by the constant $c := \langle w - v, y - z \rangle$ that depends only on y and w. ■

Remark
Actually, any monotone map satisfying inequality (42) satisfies the L property. ■

Another example is provided by the following proposition.

PROPOSITION 16
Let A be a monotone map such that either

$$(43) \qquad \operatorname{Dom} A \text{ is bounded}$$

or

$$(44)\quad \begin{cases} \text{i.} & \lim_{\|x\|\to\infty} \dfrac{\inf_{u\in A(x)} \langle u, x\rangle}{\|x\|} = +\infty \qquad (A \text{ is strongly coercive}). \\ \text{ii.} & \sup_{(x,u)\in \text{graph}(A)} \dfrac{\|u\|}{\|x\|} \leqslant c \qquad (A \text{ is bounded}). \end{cases}$$

Then A satisfies the L property. ▲

Proof. Let $y \in \text{Dom } A$ and $w \in \text{Im } A$ be fixed.

a. If $\text{Dom } A \subset rB$ is bounded, we take $v \in A(y)$; the monotonicity of A implies that for any $(x, u) \in \text{graph}(A)$,

$$(45)\quad \langle u-w, x-y\rangle \geqslant \langle u-v, x-y\rangle + \langle v-w, x-y\rangle \geqslant 0 - \|v-w\|(\|y\|+r)$$

So the L property holds true with $c := -\|v-w\|(\|y\|+r)$

b. Assume that A is strongly coercive and bounded. There exists $r>0$ so large that for all $\|x\| \geqslant r$, there exists $u \in A(x)$ such that $\langle u, x\rangle \geqslant (\|w\|+c\|y\|)\|x\|$ and $\|u\| \leqslant c\|x\|$. On the other hand, $\langle w, x\rangle \geqslant -\|w\|\,\|x\|$ and $-\langle u, y\rangle \geqslant -\|u\|\,\|y\| \geqslant -c\|y\|\,\|x\|$. Then

$$\langle u-w, x-y\rangle \geqslant (\|w\|+c\|y\|)\|x\| - \|w\|\,\|x\| - c\|x\|\,\|y\| + \langle w, y\rangle = \langle w, y\rangle$$

This inequality, together with inequality (45), implies the L property with

$$c = \min(\langle w, y\rangle, -\|v-w\|(\|y\|+r))$$ ■

The introduction of the L property is justified by the two following results.

THEOREM 17

Let A and B be two monotone maps satisfying

$$(46)\quad \begin{cases} \text{i.} & \text{Dom } A \subset \text{Dom } B \\ \text{ii.} & A+B \text{ is maximal monotone.} \\ \text{iii.} & B \text{ satisfies the L property.} \end{cases}$$

Then

$$(47)\quad \begin{cases} \text{i.} & \text{Int Im}(A+B) = \text{Int}(\text{Im}(A) + \text{Im } B) \\ \text{ii.} & \overline{\text{Im}(A+B)} = \overline{\text{Im } A + \text{Im } B} \end{cases}$$ ▲

Proof. We know that $\text{Im}(A+B) \subset \text{Im } A + \text{Im } B$. We apply theorem 10 to the maximal monotone map $A+B$ and to $K := \text{Im } A + \text{Im } B$. For that purpose, we have to check that

$$(48)\qquad \begin{cases} \forall v \in \operatorname{Im} A + \operatorname{Im} B, \quad \exists y \in X, \quad \exists c \in R \text{ such that} \\ \inf_{(x,u) \in \operatorname{graph}(A+B)} \langle u - v, x - y \rangle \geq c \end{cases}$$

holds true.

Take $v = v_1 + v_2$, where $v_1 \in \operatorname{Im} A$ and $v_2 \in \operatorname{Im} B$. Let $y \in A^{-1}(v_1) \subset \operatorname{Dom} A \subset \operatorname{Dom} B$, Since A is monotone, for all $x \in \operatorname{Dom} A$ and $u_1 \in A(x)$, we have

$$(49)\qquad \langle u_1 - v_1, x - y \rangle \geq 0$$

Since B satisfies the L property, there exists $c \in R$ such that for all $x \in \operatorname{Dom} A \subset \operatorname{Dom} B$ and $u_2 \in B(x)$,

$$(50)\qquad \langle u_2 - v_2, x - y \rangle \geq c$$

By adding these two inequalities, we obtain (48), and thus, $\operatorname{Im} A + \operatorname{Im} B \subset \overline{\operatorname{Im}(A+B)}$ and $\operatorname{Int}(\operatorname{Im} A + \operatorname{Im} B) \subset \operatorname{Im} A + \operatorname{Im} B$ thanks to theorem 10. ■

THEOREM 18

Let A and B be two monotone maps satisfying the L property. If $A+B$ is maximal monotone, then

$$(47)\qquad \begin{cases} \text{i.} \quad \operatorname{Int} \operatorname{Im}(A+B) = \operatorname{Int}(\operatorname{Im} A + \operatorname{Im} B) \\ \text{ii.} \quad \overline{\operatorname{Im}(A+B)} = \overline{\operatorname{Im} A + \operatorname{Im} B} \end{cases}$$

▲

Proof. We proceed in a manner analogous to the proof of theorem 17. We apply theorem 10 for the maximal monotone map $A+B$ with $K = \operatorname{Im} A + \operatorname{Im} B$. We take $y_0 \in \operatorname{Dom} A \cap \operatorname{Dom} B$ and $v = v_1 + v_2$, where $v_1 \in \operatorname{Im} A$ and $v_2 \in \operatorname{Im} B$. Since A and B satisfy the L property, there exist constants c_1 and c_2 such that

$$(51)\qquad \begin{cases} \text{i.} \quad \forall x \in \operatorname{Dom} A \cap \operatorname{Dom} B, \quad \forall u_1 \in A(x), \quad \langle u_1 - v_1, x - y_0 \rangle \geq c_1 \\ \text{ii.} \quad \forall x \in \operatorname{Dom} A \cap \operatorname{Dom} B, \quad \forall u_2 \in B(x), \quad \langle u_2 - v_2, x - y_0 \rangle \geq c_2 \end{cases}$$

By adding these two inequalities, we find that

$$(52)\qquad \begin{cases} \forall v \in \operatorname{Im} A + \operatorname{Im} B, \quad \exists y_0 \in X, \quad \exists c := c_1 + c_2 \in R \text{ such that} \\ \forall (x, u) \in \operatorname{graph}(A+B), \quad \langle u - v, x - y_0 \rangle \geq c \end{cases}$$

Hence, by theorem 10, we obtain the inclusions

$$\operatorname{Im} A + \operatorname{Im} B \subset \overline{\operatorname{Im}(A+B)}$$

$$\operatorname{Int}(\operatorname{Im} A + \operatorname{Im} B) \subset \operatorname{Im}(A+B)$$

from which our theorem ensues. ■

We deduce a result on the surjectivity of maps of the form $1+AB$.

THEOREM 19
Let A and B be two maximal monotone maps satisfying

$$(53)\qquad \begin{cases} \textbf{i.} & 0 \in \operatorname{Int}(\operatorname{Dom} A - \operatorname{Im} B) \\ \textbf{ii.} & y \in \operatorname{Int}(\operatorname{Im} A + \operatorname{Dom} B) \end{cases}$$

Then either the assumptions

$$(54)\qquad A \text{ satisfies the L property and } \operatorname{Im} B \subset \operatorname{Dom} A$$

or the assumption

$$(55)\qquad B \text{ satisfies the L property and } \operatorname{Im} A \subset \operatorname{Dom} B + y$$

imply the existence of a solution x to the inclusion

$$(56)\qquad y \in x + ABx$$

▲

Proof. **a.** We assume (54). Inclusion (56) can be written in the form $y \in B^{-1}u + Au$, where $x \in B^{-1}u$. We have $\operatorname{Dom} B^{-1} = \operatorname{Im} B \subset \operatorname{Dom} A$. Assumption (53, **i**) can be written $0 \in \operatorname{Int}(\operatorname{Dom} A - \operatorname{Dom} B^{-1})$ and theorem 13 implies that $A + B^{-1}$ is maximal monotone. Then theorem 16 states that

$$\operatorname{Int}(\operatorname{Im} A + \operatorname{Dom} B) = \operatorname{Int}(\operatorname{Im} A + \operatorname{Im} B^{-1}) \subset \operatorname{Im}(A + B^{-1})$$

so that assumption (53, **ii**) implies that there exists a solution u to $y \in Au + B^{-1}u$. Hence, there exists $x \in B^{-1}(u)$, which is a solution to the inclusion (56).
b. We assume (55). Inclusion (56) can be written in the form

$$(57)\qquad 0 \in Bx + \tilde{A}(x), \qquad \text{where } \tilde{A}(x) := -A^{-1}(y - x)$$

Then $\tilde{A}$ is a maximal monotone map, $\operatorname{Dom} \tilde{A} = y - \operatorname{Im} A$, and $\operatorname{Im} \tilde{A} = -\operatorname{Dom} A$. We have $\operatorname{Dom} \tilde{A} \subset \operatorname{Dom} B$. Assumption (53, **ii**) implies that $0 \in \operatorname{Int}(\operatorname{Dom} \tilde{A} - \operatorname{Dom} B)$, and theorem 13 implies that $\tilde{A} + B$ is maximal monotone. Theorem 17 states that

$$\operatorname{Int}(\operatorname{Im} B - \operatorname{Dom} A) = \operatorname{Int}(\operatorname{Im} B + \operatorname{Im} \tilde{A}) \subset \operatorname{Im}(\tilde{A} + B)$$

Hence, condition (53, **i**) implies that there exists a solution x to inclusion (57), that is, to inclusion (56). ■

8. EXISTENCE AND UNIQUENESS OF SOLUTIONS TO DIFFERENTIAL INCLUSIONS

THEOREM 1

Let A be a maximal monotone set-valued map from X to X. Consider the initial value problem for the differential inclusion

$$(1) \qquad x' \in -A(x), \qquad x(0)=x_0$$

when x_0 is given in Dom A. *Then there exists a unique solution $x(\cdot)$ defined on $[0, \infty]$, which is the slow solution. Let $m(K)$ denote the projection of zero onto a closed convex set K; a slow solution is a solution to the differential equation*

$$(2) \qquad \text{for almost all } t \geqslant 0, \qquad x'(t) = -m(A(x(t)))$$

Moreover,

$$(3) \qquad t \to \|x'(t)\| \text{ is nonincreasing.}$$

Let $x(\cdot)$ and $y(\cdot)$ be solutions starting at x_0 and y_0. Then

$$(4) \qquad \forall t \geqslant 0, \qquad \|x(t)-y(t)\| \leqslant \|x_0 - y_0\|$$

Finally,

$$(5) \quad \forall t \geqslant 0, \quad x'(t) = \lim_{h \to 0^+} \frac{x(t+h)-x(t)}{h} \quad \text{and } x'(\cdot) \text{ is continuous from the right.}$$

▲

Proof. The proof is based on defining approximate solutions as solutions to ordinary differential equations where A is replaced by the Yosida approximations of A. To show the convergence of a sequence of approximate solutions, we shall prove that they are a Cauchy sequence. Properties (3) and (4) have been pointed out explicitely in the statement of the theorem; property (3) would give a bound on $\|A(x(t))\|$ if we knew we had a solution $x(\cdot)$, yields the same bound on the set of approximate solutions. Property (4), which shows that the solution is unique, also implies that the distance between two approximate solutions is small, that is, that they are a Cauchy sequence.

We shall divide the proof into the following steps:

Step a. We prove (4) [i.e., that the map $x_0 \to x(\cdot)$ is nonexpansive] and derive uniqueness.
Step b. We prove that $t \to \|x'(t)\|$ is nonincreasing.
Step c. We consider the solutions $x_\lambda(\cdot)$ to

$$x'_\lambda(t) = -A_\lambda x_\lambda(t); \quad x_\lambda(0) = x$$

where A_λ is the Yosida approximation of A and prove that $x_\lambda(\cdot)$ is a Cauchy sequence that converges to some $x(\cdot)$ in $\mathscr{C}(0, \infty, X)$.

Step d. We check that $x(t) \in \operatorname{Dom} A$ for all $t \geqslant 0$.

Step e. We prove that $x'(t) \in -Ax(t)$ for almost all $t \geqslant 0$.

Step f. We show that $t \to m(A(x(t)))$ is continuous from the right.

Step g. We conclude by establishing that $x'(t) = -m(A(x(t)))$ [i.e., that $x(\cdot)$ is the slow solution] and $x'(\cdot) = d/dt\, x(t)$ is the derivative from the right.

a. Assume that $x(\cdot)$ and $y(\cdot)$ are solutions of the initial value problems

$$\text{(6)} \qquad \begin{cases} \textbf{i.} & x'(t) \in -Ax(t), \qquad x(0) = x_0 \\ \textbf{ii.} & y'(t) \in -Ay(t), \qquad y(0) = y_0 \end{cases}$$

Therefore, since A is monotone,

$$\frac{d}{dt}\frac{1}{2}\|x(t) - y(t)\|^2 = \langle x'(t) - y'(t), x(t) - y(t)\rangle \leqslant 0$$

and by integrating from zero to t, we deduce that

$$\sup_{t \geqslant 0} \|x(t) - y(t)\| \leqslant \|x_0 - y_0\|$$

In particular, when $x_0 = y_0$, this implies that the solution of (1), if any, is unique.

b. Consider now $y(t) = x(t+h)$ where $h > 0$ and $x(\cdot)$ is a solution of (6, **i**). It satisfies

$$y'(t) \in -Ay(t), \qquad y(0) = x(h)$$

Therefore, (3) implies that

$$\text{(7)} \qquad \|x(t+h) - x(t)\| \leqslant \|x(h) - x(0)\|$$

By dividing by $h > 0$ and letting h tend to zero, we deduce that

$$\|x'(t)\| \leqslant \|x'(0)\| \qquad \text{for all } t \geqslant 0$$

c. We consider solutions x_λ to the approximate problems

$$\text{(8)} \qquad x_\lambda'(t) = -A_\lambda x(t), \qquad x_\lambda(0) = x_0$$

Since A_λ is a Lipschitz map from X into X (see theorem 7.8), it has a unique, continuously differentiable solution $x_\lambda(\cdot)$ defined on $[0, \infty[$. We shall prove that $x_\lambda(\cdot)$ is a Cauchy sequence in $\mathscr{C}(0, \infty; X)$ and check that $x_\lambda(\cdot)$ converges to the solution to the differential inclusion (1).

Since A_λ is also monotone, we deduce from (3) [with $x(\cdot)$ replaced by $x_\lambda(\cdot)$] that

$$\|A_\lambda x_\lambda(t)\| = \|x'_\lambda(t)\| \leqslant \|x'_\lambda(0)\| = \|A_\lambda x_0\| \leqslant \|mA(x_0)\| \tag{9}$$

We compute $\frac{1}{2}\|x_\lambda(t) - x_\mu(t)\|^2 = a(t)$

$$\begin{aligned} a(t) = \frac{1}{2}\|x_\lambda(t) - x_\mu(t)\|^2 &= \int_0^t \frac{d}{dt}\|x_\lambda(\tau) - x_\mu(\tau)\|^2 d\tau \\ &= -\int_0^t \langle A_\lambda x_\lambda(\tau) - A_\mu x_\mu(\tau), x_\lambda(\tau) - x_\mu(\tau)\rangle d\tau \end{aligned}$$

We now use the relation $1 - J_\lambda = A_\lambda$, the fact that $A_\lambda x \in AJ_\lambda(x)$, and the monotonicity of A

$$\begin{aligned} a(t) = &-\int_0^t \langle A_\lambda x_\lambda(\tau) - A_\mu x_\mu(\tau), \lambda A_\lambda x_\lambda(\tau) - \mu A_\mu x_\mu(\tau)\rangle d\tau \\ &-\int_0^t \langle A_\lambda x_\lambda(\tau) - A_\mu x_\mu(\tau), J_\lambda x_\lambda(\tau) - J_\mu x_\mu(\tau)\rangle d\tau \\ \leqslant &-\int_0^t \langle A_\lambda x_\lambda(\tau) - A_\mu x_\mu(\tau), \lambda A_\lambda x_\lambda(\tau) - \mu A_\mu x_\mu(\tau)\rangle d\tau \\ = &\int_0^t \lambda\langle A_\lambda x_\lambda(\tau), A_\mu x_\mu(\tau)\rangle d\tau + \int_0^t \mu\langle A_\mu x_\mu(\tau), A_\lambda x_\lambda(\tau)\rangle d\tau \\ &-\int_0^t \lambda\|A_\lambda x_\lambda(\tau)\|^2 + \mu\|A_\mu x_\mu(\tau)\|^2 d\tau \end{aligned}$$

We note that

$$\lambda\langle A_\lambda x_\lambda(\tau), A_\mu x_\mu(\tau)\rangle \leqslant \lambda\|A_\lambda x_\lambda(\tau)\|\,\|A_\mu x_\mu(\tau)\| \leqslant \lambda\|A_\lambda x_\lambda(\tau)\|^2 + \frac{\lambda}{4}\|A_\mu x_\mu(\tau)\|^2$$

and in the same way

$$\mu\langle A_\mu x_\mu(\tau), A_\lambda x_\lambda(\tau)\rangle \leqslant \mu\|A_\mu x_\mu(\tau)\|^2 + \frac{\mu}{4}\|A_\lambda x_\lambda(\tau)\|^2$$

Therefore, using this and (9), we obtain

$$\begin{aligned} a(t) &\leqslant \frac{1}{4}\int_0^t (\lambda\|A_\mu x_\mu(\tau)\|^2 + \mu\|A_\lambda x_\lambda(\tau)\|^2) d\tau \\ &\leqslant \frac{(\lambda+\mu)}{4} t\|m(A(x_0))\| \end{aligned}$$

Hence, $x_\lambda(\cdot)$ is a Cauchy sequence in $\mathscr{C}(0, \infty; X)$ and thus converges to some continuous function $x(\cdot)$ uniformly over compact intervals.

d. The inequality

$$\|x_\lambda(t) - J_\lambda x_\lambda(t)\| = \lambda \|A_\lambda x_\lambda(t)\| \leqslant \lambda \|m(A(x_0))\|$$

yields that $J_\lambda x_\lambda(\cdot)$ converges to $x(\cdot)$ uniformly.

Also, since $\|A_\lambda x_\lambda(t)\| \leqslant \|m(A(x_0))\|$, there exists a subsequence $A_{\lambda_n} x_{\lambda_n}(t)$ that converges weakly in X to some $v(t)$. But $A_{\lambda_n} x_{\lambda_n}(t)$ belongs to $A(J_{\lambda_n} x_{\lambda_n}(t))$; therefore, we deduce from proposition 7.3 that $v(t) \in A(x(t))$. In particular this implies that $x(t) \in \text{Dom } A$ for all $t \geqslant 0$.

e. We note that $x'_\lambda(t)$ remains in a bounded set of $L^\infty(0, \infty; X)$ and thus a subsequence $x'_{\lambda_n}(\cdot)$ converges weakly to some function, which is equal almost everywhere to $x'(\cdot)$. So, for any $T > 0$, $x'_{\lambda_n}(\cdot)$ converges weakly to $x'(\cdot)$ in $L^2(0, T; X)$, and $x_n(\cdot)$ converges strongly to $x(\cdot)$ in $L^2(0, T; x)$. Proposition 7.7 implies that the map $x(\cdot) \to (\mathscr{A} x)(\cdot) = A(x(\cdot))$ is maximal monotone in $L^2(0, T; X)$; we deduce that $x'(\cdot) \in -Ax(\cdot)$ in $L^2(0, T; X)$. Hence, $x'(t) \in Ax(t)$ almost everywhere.

f. Let $t > t_0$. Since

$$\|x'_\lambda(t)\| = \|A_\lambda x_\lambda(t)\| \leqslant \|m(A(x(t_0)))\|$$

we deduce that

$$\|v(t)\| \leqslant \liminf_{\lambda \to 0} \|x'_\lambda(t)\| \leqslant \|m(A(x(t_0)))\|$$

and the solutions $x_\lambda(\cdot)$ are uniformly Lipschitz. Hence, $x(\cdot)$ is also Lipschitz.

Moreover, $\|m(A(x(t)))\| \leqslant \|v(t)\| \leqslant \|m(A(x(t_0)))\|$. This proves that $t \to \|m(A(x(t)))\|$ is nonincreasing.

g. Let us check that $t \to m(A(x(t)))$ is continuous from the right. Let $t_n \to t_0$ converge to t_{0+}; then $x(t_n)$ converges to $x(t_0)$, and since $\|m(A(x(t_n)))\| \leqslant \|v(t_n)\| \leqslant \|m(Ax(t_0))\|$, we deduce that some subsequence (again denoted) $m(A(x(t_n)))$ converges weakly to some y in X. Proposition 7.3 implies that $y \in A(x(t_0))$. But

$$\|y\| \leqslant \liminf_{t_n \to \infty} \|m(A(x(t_n)))\| \leqslant \|m(A(x(t_0)))\|$$

Hence, $y = m(A(x(t_0)))$, and $m(A(x(t_0))$ is the weak limit of the sequence $m(A(x(t_n)))$. Since

$$\|m(A(x(t_0)))\| = \lim_{n \to \infty} \|m(A(x(t_n)))\|$$

we deduce that $m(A(x(t_n)))$ converges strongly to $m(A(x(t_0)))$ when $t_n \to t_{0+}$. Hence, $t \to m(A(x(t)))$ is continuous from the right.

h. Let N be the subset of $[0, \infty[$ where neither $x(\cdot)$ is differentiable nor $x'(t) \notin -A(x(t))$. Let $t_0 \notin N$.

We deduce from (7) (where zero is replaced by t_0) that

$$\|x(t+h)-x(t)\| \leqslant \|x(t_0+h)-x(t_0)\|$$

But

$$\|x(t_0+h)-x(t_0)\| \leqslant \int_{t_0}^{t_0+h} \|x'(\tau)\|d\tau \leqslant \int_{t_0}^{t_0+h} \|m(A(x(t_0)))\|d\tau \leqslant h\|m(A(x(t_0)))\|$$

(By (9))

Hence,

$$\|x'(t_0)\| = \lim_{h \to 0+} \left\| \frac{x(t_0+h)-x(t_0)}{h} \right\| \leqslant \|m(A(x(t_0)))\|$$

Since $x'(t_0) \in -Ax(t_0)$, we deduce that $x'(t_0) = -m(A(x(t_0)))$. Integrating from t_0 to t_0+h, we deduce that

$$\frac{x(t_0+h)-x(t_0)}{h} = -\frac{1}{h}\int_{t_0}^{t_0+h} m(A(x(\tau)))d\tau$$

Since $m(A(x(\cdot)))$ is continuous from the right, we deduce that

$$\frac{d}{dt}x(t_0) = -mA(x(t_0))$$

■

CHAPTER 7

Nonsmooth Analysis

Nonlinear analysis must provide sufficient conditions for solving inclusions

$$(*) \qquad y \in F(x)$$

when F is a set-valued map from a Banach space X to a Banach space Y.

Our principal objective in this chapter is to prove an inverse function theorem for set-valued maps allowing us to say that when x_0 is a solution to

$$y_0 \in F(x_0)$$

then there exist neighborhoods U of x_0 and V of y_0 such that inclusion (*) has solutions in U whenever y ranges over V. Furthermore, as in the smooth case, we require that the set of solutions $F^{-1}(y) \cap U$ of (*) depends in a Lipschitz manner on the data y. Since the inverse function theorem for usual smooth maps plays such an important role in solving many problems of pure and applied analysis, we can expect an adaptation of the inverse function theorem to be very useful, more than just a generalization that had to be made. Actually, it can be used for solving convex minimization problems and proving the Lipschitz behavior of its solutions when the natural parameters vary. Economists claim that this problem is of utmost importance in their field (marginal theory). We shall take our inspiration from the smooth situation where the sufficient condition is very simply stated: *The derivative at* x_0 *must be surjective.* The question arises, can we define derivatives of set-valued maps such that the surjectivity of the derivative at (x_0, y_0) is sufficient for solving the surjectivity of F around y_0?

The answer to this question is one purpose for this chapter. We now explain how we shall proceed to define derivatives of set-valued maps. We adopt the very first strategy, apparently suggested by Fermat, which defines the graph of the derivative to a smooth function as the tangent to the graph of this function. Therefore, we postpone questions about derivatives until after having tackled the matter of tangent spaces to subsets K of a Banach space X. They do not exist when K is no longer a smooth manifold. However, it is known in convex analysis that we can define in a natural way "tangent cones" to convex sets, which retain enough properties of tangent spaces to be quite useful. This is not enough, because most of the set-valued maps we shall meet have nonconvex graphs.

When K is neither smooth nor convex, there are many ways of defining **tangent cones**, each one being as "natural" as the other. We shall retain only two concepts among the many candidates: the contingent cone and the tangent cone. Namely, they are defined in the following way: Let x_0 belong to K. The contingent cone, defined by

$$T_K(x_0) := \bigcap_{\varepsilon>0} \bigcap_{\alpha>0} \bigcup_{h \in]0,\alpha]} \left(\frac{1}{h}(K - x_0) + \varepsilon B\right)$$

was introduced by Bouligand in the early thirties, and the tangent cone defined by

$$C_K(x_0) := \bigcap_{\varepsilon>0} \bigcup_{\alpha,\beta>0} \bigcap_{\substack{h \in]0,\alpha] \\ x \in B_K(x_0,\beta)}} \left(\frac{1}{h}(K - x) + \varepsilon B\right)$$

was introduced by F. H. Clarke in 1975.

We see at once that the tangent cone $C_K(x)$ is contained in the contingent cone $T_K(x)$. They are both closed, and the tangent cone *is always convex*. We can say that they form a kind of "dipole," in the sense that the tangent cone $C_K(x_0)$ is the Kuratowski lim inf of the contingent cones $T_K(x)$ when $x \to x_0$ (when the space X is finite dimensional). So, several properties of the tangent cone $C_K(x_0)$ at x_0 "diffuse" to generally weaker properties of the contingent cones $T_K(x)$ in a neighborhood of x_0. This dipole collapses to the usual tangent cone of K or the tangent space of K when K is convex and a smooth manifold, respectively. We shall see in the sixth section that the contingent and tangent cones enjoy dual properties.

These are some of the reasons for studying both contingent and tangent cones and treating them as a pair rather than individuals.

Let F be a set-valued map from X to Y and (x_0, y_0) belong to its graph.

We define the *contingent derivative* $DF(x_0, y_0)$ as the *closed process* from X to Y whose graph is the contingent cone to the graph of F

$$v_0 \in DF(x_0, y_0)(u_0) \Leftrightarrow (u_0, v_0) \in T_{\text{graph}(F)}(x_0, y_0)$$

and the *derivative* $CF(x_0, y_0)$ as the *closed convex process* from X to Y whose graph is the tangent cone to the graph of F

$$v_0 \in CF(x_0, y_0)(u_0) \Leftrightarrow (u_0, v_0) \in C_{\text{graph}(F)}(x_0, y_0)$$

They may be different. Consider, for instance, the Lipschitz single-valued map π from $\mathbb{R}$ to $\mathbb{R}$ defined by

$$\pi(x) = 0 \quad \text{when} \quad x \leqslant 0, \qquad \pi(x) = x \quad \text{when} \quad x \geqslant 0$$

Then the contingent derivative of π at (0, 0) is defined by

$$D\pi(0, 0)(u)=0 \quad \text{when } u\leqslant 0, \qquad D\pi(0, 0)(u)=u \quad \text{when} \quad u\geqslant 0$$

and the derivative of π at (0, 0) is defined by

$$C\pi(0, 0)(u)=\varnothing \quad \text{when} \quad u\neq 0, \qquad C\pi(0, 0)(0)=0$$

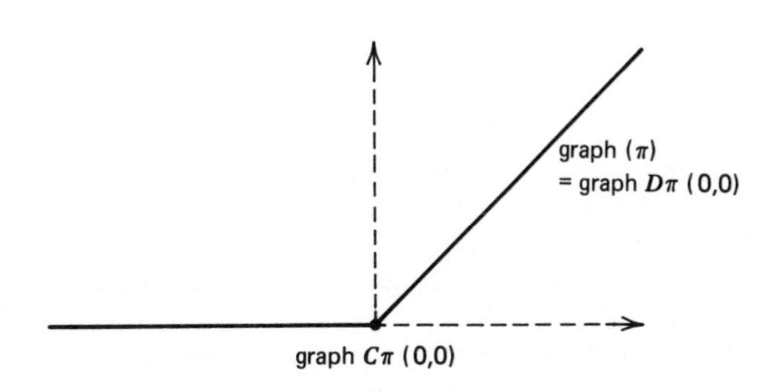

This example shows that the price to pay for having a closed convex process as a derivative is sometimes too high.

These definitions provide intrinsic definitions of derivatives of single-valued maps defined on subsets K that may have an empty interior, as well as formulas for computing them when they are restrictions to K of a smooth map. When F is continuously differentiable on an open neighborhood of K, then the contingent derivatives and derivatives of the restriction $F|_K$ of F to K are the restrictions of the Jacobian ∇F of F to the contingent and tangent cones, respectively

$$\begin{aligned} &\textbf{i.} \quad D(F|_K)(x_0, F(x_0))=\nabla F(x_0)|_{T_K(x_0)} \\ &\textbf{ii.} \quad C(F|_K)(x_0, F(x_0))=\nabla F(x_0)|_{C_K(x_0)} \end{aligned}$$

Also, these concepts of derivatives allow us to compute the inverse of the derivative of a map, in particular, the inverse of the Jacobian of a single-valued map, because we infer immediately from the definitions that

$$\begin{aligned} &\textbf{i.} \quad DF(x_0, y_0)^{-1}=D(F^{-1})(y_0, x_0) \\ &\textbf{ii.} \quad CF(x_0, y_0)^{-1}=C(F^{-1})(y_0, x_0) \end{aligned}$$

Since the derivative is a closed convex process, it is useful to distinguish its transpose $CF(x_0, y_0)^*$, a closed convex process from Y^* to X^*: We shall call it the *codifferential* of F at (x_0, y_0).

For real-valued functions, we can take into account the order relation, which is used in optimization problems (or in the theory of Lyapunov functions, i.e., functions that decrease along the trajectories of a dynamical system).

We associate with a proper function V from X to $\mathbb{R}\cup\{+\infty\}$ the set-valued map $\mathbf{V}_+$ defined by $\mathbf{V}_+(x):=V(x)+\mathbb{R}_+$ when $V(x)<+\infty$, $\mathbf{V}_+(x):=\varnothing$ when $V(x)=+\infty$. We observe that there are numbers $D_+V(x_0)(u_0)$ and $C_+V(x_0)(u_0)$ such that

$$\textbf{i.}\quad D\mathbf{V}_+(x_0, V(x_0))(u_0)=D_+V(x_0)(u_0)+\mathbb{R}_+$$
$$\textbf{ii.}\quad C\mathbf{V}_+(x_0, V(x_0))(u_0)=C_+V(x_0)(u_0)+\mathbb{R}_+$$

where

$$-\infty\leqslant D_+V(x_0)(u_0)\leqslant C_+V(x_0)(u_0)\leqslant+\infty$$

We shall say that the functions $D_+V(x_0)(\cdot)$ and $C_+V(x_0)(\cdot)$ from X to $\{-\infty\}\cup\mathbb{R}\cup\{+\infty\}$ are the *epicontingent derivative* and *epiderivative* of the function V.

In other words, the epigraphs of $D_+V(x_0)(\cdot)$ and $C_+V(x_0)(\cdot)$ are the contingent and tangent cones at the epigraph of V at $(x_0, V(x_0))$.

When the derivative $C_+V(x_0)(\cdot)$ is a proper function from X to $\mathbb{R}\cup\{+\infty\}$, it is convex, positively homogeneous, and lower semicontinuous. This is, then, the support function of the closed convex subset

$$\partial V(x_0):=\{p\in X^*|\forall u\in X,\ \langle p, u\rangle\leqslant C_+V(x_0)(u)\}$$

We shall call this subset the *generalized gradient* introduced by F. H. Clarke in 1975. Indeed, the terminology is justified by the fact that when V is continuously differentiable at x_0, then $\partial V(x_0)=\{\nabla V(x_0)\}$. We also observe that when V is convex, the generalized gradient coincides with the subdifferential $\partial V(x_0)$ of convex analysis.

It is then natural to consider the derivatives of the set-valued map $x\to\partial V(x)$ as candidates for the role of *second derivatives.* Let p_0 belong to $\partial V(x_0)$; the derivative

$$C\partial V(x_0, p_0):=\partial^2 V(x_0, p_0)$$

is a closed convex process from X to X^*, which is monotone when V is convex.

These tangent cones and derivatives enjoy enough properties to make a decent calculus. But the main justification for including this study here is their use in the inverse function theorem. When X and Y are finite dimensional, it has a very simple formulation.

Let F be a set-valued map with a closed graph and let (x_0, y_0) belong to the graph of F. Assume that

(**) *the derivative $CF(x_0, y_0)$ of F at (x_0, y_0) is surjective.*

Then F^{-1} is pseudo Lipschitz around (x_0, y_0) in the sense that there exist a neighborhood W of y_0, two neighborhoods U and V of x_0, $U\subset V$, and a constant

$\ell > 0$ *such that*

$$\begin{aligned} &\text{i. } \forall y \in W, \quad F^{-1}(y) \cap U \neq \varnothing \\ &\text{ii. } \forall y_1, \quad y_2 \in W, \quad \mathrm{d}(F^{-1}(y_1) \cap U, F^{-1}(y_2) \cap V) \leqslant \ell \|y_1 - y_2\| \end{aligned}$$

where $\mathrm{d}(A, B) := \sup_{x \in A} \inf_{y \in B} d(x, y)$.

It is itself a consequence of a more general inverse function theorem, valid in infinite dimensional spaces and involving surjectivity properties of the contingent derivative of F not only at (x_0, y_0), but at all neighboring points.

We conclude this chapter with a section devoted to the calculus of tangent cones, derivatives of set-valued maps, and epiderivatives of real-valued functions.

1. CONTINGENT AND TANGENT CONES

Let K be a nonempty subset of a Banach space X. We denote by εB and $\varepsilon \mathring{B}$ the ball (respectively, open ball) of center zero and radius $\varepsilon > 0$. We set $B_K(x_0, \varepsilon) := K \cap (x_0 + \varepsilon B)$, and the symbol $x \underset{K}{\to} x_0$ denotes the convergence of x to x_0 in K.

DEFINITION 1

We say that the subset

$$(1) \qquad T_K(x) := \bigcap_{\varepsilon > 0} \bigcap_{\alpha > 0} \bigcup_{0 < h \leqslant \alpha} \left(\frac{1}{h} K - x) + \varepsilon B \right)$$

is the contingent cone *to K at x.* ▲

In other words, $v \in T_K(x)$ if and only if

$$(2) \qquad \forall \varepsilon > 0, \quad \forall \alpha > 0, \quad \exists u \in v + \varepsilon B, \quad \exists h \in]0, \alpha] \quad \text{such that } x + hu \in K$$

or, equivalently, $v \in T_K(x)$ if and only if there exist sequences of strictly positive numbers h_n and elements $u_n \in X$ satisfying

$$(3) \qquad \begin{cases} \text{i. } \lim_{n \to \infty} u_n = v. \\ \text{ii. } \lim_{n \to \infty} h_n = 0. \\ \text{iii. } \forall n \geqslant 0, \quad x + h_n u_n \in K. \end{cases}$$

We characterize the contingent cone by using the distance function $d_K(\cdot)$ to K defined by $d_K(x) := \inf \{ \|x - y\| \mid y \in K \}$

$$(4) \qquad v \in T_K(x) \quad \text{if and only if} \quad \liminf_{h \to 0+} \frac{d_K(x + hv)}{h} = 0$$

It is quite obvious that the contingent cone is a *closed cone*, which is trivial when x belongs to the interior of K;

(5) $$\text{when } x \in \operatorname{Int}(K), \qquad \text{then } T_K(x) = X.$$

For all $x \in X$, we have $T_X(x) = X$. We set $T_\phi(x) := \varnothing$. It is convenient to introduce the definition of the lim inf of a family of subsets $F(u)$.

DEFINITION 2

Let U be a metric space, u_0 belong to U, and F be a set-valued map from U to X. We set

(6) $$\liminf_{u \to u_0} F(u) := \bigcap_{\varepsilon > 0} \bigcup_{\eta > 0} \bigcap_{u \in B(u_0, \eta)} (F(u) + \varepsilon B)$$ ▲

We observe that when the images of F are closed,

(7) $$\liminf_{u \to u_0} F(u) \subset F(u_0)$$

and F is lower semicontinuous at u_0 if and only if

(8) $$F(u_0) = \liminf_{u \to u_0} F(u)$$

It is useful to note that v belongs to $\liminf_{u \to u_0} F(u)$ if and only if

(9) $$\forall \varepsilon > 0, \qquad \exists \eta > 0 \quad \text{such that} \quad \sup_{u \in B(u_0, \eta)} d(v, F(u)) \leqslant \varepsilon$$

DEFINITION 3

We say that the subset

(10) $$C_K(x_0) := \liminf_{\substack{h \to 0+ \\ x \to x_0}} \frac{1}{h}(K - x) = \bigcap_{\varepsilon > 0} \bigcup_{\alpha, \beta > 0} \bigcap_{\substack{x \in B_K(x_0, \alpha) \\ h \in]0, \beta]}} \left(\frac{1}{h}(K - x) + \varepsilon B \right)$$

is the tangent cone to K at x_0. ▲

In other words, $v \in C_K(x_0)$ if and only if

(11) $$\begin{cases} \forall \varepsilon > 0, \quad \exists \alpha > 0, \quad \exists \beta > 0 \quad \text{such that} \quad \forall x \in B_K(x_0, \alpha) \\ \forall h \in]0, \beta], \quad \exists u \in v + \varepsilon B \quad \text{satisfying} \quad x + hu \in K \end{cases}$$

or, equivalently, if and only if

(12)
for all sequences of elements $x_n \in X$, $h_n > 0$
converging to x_0 and zero, there exists a sequence
of elements $u_n \in X$ converging to v such that
$x_n + h_n u_n$ belongs to K for all n.

It is also characterized in the following way:

$$v \in C_K(x_0) \quad \text{if and only if} \quad \lim_{\substack{x \to_K x_0 \\ h \to 0+}} \frac{d_K(x+hv)}{h} = 0 \tag{13}$$

We observe that when $x \in \text{Int}(K)$, then $C_K(x) = X$. For all $x \in X$, we have $C_X(x) = X$. We shall set $C_\phi(x) := \varnothing$. Tangent cones enjoy a very attractive property.

PROPOSITION 4
The tangent cone $C_K(x_0)$ to K at x_0 is closed and convex. ▲

Proof. Let v^1 and v^2 belong to $C_K(x_0)$. We take any sequence of elements $(x_n, h_n) \in K \times]0, \infty[$ converging to $(x_0, 0)$. There exists a sequence of elements v_n^1 converging to v^1 such that the elements $y_n := x_n + h_n v_n^1$ belong to K for all n. Since y_n converges to x_0, there exists a sequence of elements v_n^2 converging to v^2 such that $y_n + h_n v_n^2 = x_n + h_n(v_n^1 + v_n^2)$ belongs to K for all n. Since $v_n^1 + v_n^2$ converges to $v^1 + v^2$, we deduce that $v^1 + v^2$ belongs to $C_K(x_0)$. Hence, the tangent cone is convex. ■

We note that

$$C_K(x_0) \subset T_K(x_0) \subset \text{cl}\left(\bigcup_{h>0} \frac{1}{h}(K - x_0)\right)$$

PROPOSITION 5
If K is a convex subset, these three cones coincide

$$C_K(x_0) = T_K(x_0) = \text{cl} \bigcup_{h>0} \frac{1}{h}(K - x_0) \tag{14}$$ ▲

Proof. We have to prove that any $u_0 \in \text{cl} \bigcup_{h>0} (1/h)(K - x_0)$ belongs to $C_K(x_0)$. Let $\varepsilon > 0$ be fixed; there exist $y \in K$ and $\beta > 0$ such that $u_0 - (1/\beta)(y - x_0) \in (\varepsilon/2)B$. Let us take $\alpha := \beta\varepsilon/2$, x in $B_K(x_0, \alpha)$, and $h \in]0, \beta]$. We set $u := (y - x)/\beta$. Then

$$x + hu = \left(1 - \frac{h}{\beta}\right)x + \frac{h}{\beta} y$$

belongs to K, because both x and y belong to K and $h/\beta \leqslant 1$. Also,

$$\|u-u_0\| \leqslant \frac{\|x-x_0\|}{\beta} + \left\| u_0 - \frac{y-x_0}{\beta} \right\| \leqslant \frac{\alpha}{\beta} + \frac{\varepsilon}{2} = \varepsilon$$

Hence, u_0 belongs to $C_K(x_0)$. ∎

These two cones may be different. Consider, for instance, the set K from $\mathbb{R}^2$, which is the graph of the map π from $\mathbb{R}$ defined by

$$\pi(x)=0 \quad \text{when} \quad x \leqslant 0, \qquad \pi(x)=x \quad \text{when} \quad x \geqslant 0$$

Then

$$\text{if } x<0, \qquad C_K(x, 0)=T_K(x, 0)=\mathbb{R} \times \{0\}$$

$$\text{if } x=0, \qquad C_K(0, 0)=\{0, 0\}, \qquad T_K(0, 0)=(-\mathbb{R}_+ \times \{0\}) \cup \{u, u\}_{u \in \mathbb{R}_+}$$

$$\text{if } x>0, \qquad C_K(x, x)=T_K(x, x)=\{u, u\}_{u \in \mathbb{R}}$$

The tangent cone to K at (0, 0) is convex but trivial, whereas the contingent cone to K at (0, 0) is nonconvex but quite large.

We also observe that when K is a smooth manifold (of class C^1), then both the tangent cone and the contingent cone coincide with the usual tangent vector space to K at x of differential geometry.

The contingent and tangent cones are related by the following interesting relation.

PROPOSITION 6

Assume that X is finite dimensional. Then

$$\forall x_0 \in K, \qquad C_K(x_0) \subset \liminf_{x \to_K x_0} T_K(x) \tag{15}$$ ▲

Proof. By definition of the tangent cone, we have

$$C_K(x_0) = \bigcap_{\varepsilon} \bigcup_{\alpha>0} \bigcup_{\beta>0} \bigcap_{x \in B_K(x_0,\alpha)} \bigcap_{h \in]0,\beta]} \left(\frac{1}{h}(K-x) + \varepsilon B \right)$$

Let ε and α be fixed. It is clear that

$$\bigcup_{\beta>0} \bigcap_{x \in B_K(x_0,\alpha)} \bigcap_{h \in]0,\beta]} \left(\frac{1}{h}(K-x) + \varepsilon B \right) \subset \bigcap_{x \in B_K(x_0,\alpha)} \bigcup_{\beta>0} \bigcap_{h \in]0,\beta]} \left(\frac{1}{h}(K-x) + \varepsilon B \right)$$

Since X is finite dimensional, we observe that any v in

$$\bigcup_{\beta>0} \bigcap_{h \in]0,\beta]} \left(\frac{1}{h}(K-x)+\varepsilon B\right) \text{ belongs to } T_K(x)+\varepsilon B.$$

Indeed, there exist β and elements x_h such that

$$v \in \frac{x_h - x}{h} + \varepsilon B \quad \text{for } h \leqslant \beta$$

A subsequence of $(x_h - x)/h$ converges to some w in $T_K(x)$. Hence,

$$C_K(x_0) \subset \bigcap_{\varepsilon>0} \bigcup_{\alpha>0} \bigcap_{x \in B_K(x_0,\alpha)} (T_K(x)+\varepsilon B) = \liminf_{x \underset{K}{\to} x_0} T_K(x)$$

This inclusion is actually an equality. ■

THEOREM 7
Let K be a nonempty weakly closed subset of a Hilbert space. The following inclusions hold true:

$$\text{(16)} \qquad \liminf_{x \underset{K}{\to} x_0} T_K(x) \subset \liminf_{x \underset{K}{\to} x_0} (\overline{\text{co}}\, T_K(x)) \subset C_K(x_0)$$

When X is finite-dimensional, equalities hold true. ▲

Then the set-valued map $x \to T_K(x)$ is lower semicontinuous at x_0 if and only if the contingent cone to K at x_0 coincides with the tangent cone to K at x_0. ■

The proof ensues from the following lemmas.

LEMMA 8
Let $K \subset X$ be a weakly closed subset. We denote by $\pi_K(x)$ the nonempty subset of elements $x \in K$ such that $\|x-y\| = d_K(y)$. We obtain the following:

$$\text{(17)} \quad \forall y \notin K, \qquad \forall x \in \pi_K(y), \qquad \forall v \in \overline{\text{co}}\, T_K(x), \qquad \textit{then } \langle y-x, v\rangle \leqslant 0 \qquad ▲$$

Proof. Let $x \in \pi_K(y)$ and $v \in T_K(x)$. We deduce from the inequalities

$$\|y-x\| - d_K(x+hv) = d_K(y) - d_K(x+hv) \leqslant \|y-x-hv\|$$

that

$$\frac{\langle y-x, v\rangle}{\|y-x\|} = \lim_{h \to 0^+} \frac{\|y-x\| - \|y-x-hv\|}{h} \leqslant \liminf_{h \to 0^+} \frac{d_K(x+hv)}{h} = 0$$

for $y \neq x$, since $u \to \|u\|$ is differentiable at $u \neq 0$. So $\langle y-x, v\rangle \leqslant 0$ for all $v \in T_K(x)$ and, consequently, for all $v \in \overline{\text{co}}\, T_K(x)$. ■

LEMMA 9

For any $y \in X$, we have

$$\liminf_{h\to 0^+} \frac{1}{2h}(d_K(y+hv)^2 - d_K(y)^2) \leqslant d_K(y)d(v, \overline{\text{co}}\, T_K(\pi_K(y))) \tag{18}$$ ▲

Proof. Let us take x in $\pi_K(y)$. We observe that

$$\frac{1}{2h}(d_K(y+hv)^2 - d_K(y)^2) \leqslant \frac{1}{2h}(\|y+hv-x\|^2 - \|y-x\|^2)$$

because $d_K(y) = \|y-x\|$. Therefore,

$$\liminf_{h\to 0^+} \frac{1}{2h}(d_K(y+hv)^2 - d_K(y))^2 \leqslant \langle y-x, v\rangle$$

and for all $w \in \overline{\text{co}}\, T_K(x)$, we deduce from lemma 8 that

$$\liminf_{h\to 0^+} \frac{1}{2h}(d_K(y+hv)^2 - d_K(y)^2) \leqslant \langle y-x, v-w\rangle \leqslant \|y-x\|\,\|v-w\| = d_K(y)\|v-w\|$$

Lemma 9 ensues by taking the infimum when w ranges over $\overline{\text{co}}\, T_K(x)$ and x over $\pi_K(y)$. ■

LEMMA 10

Let us consider the Lipschitz function f defined by $f(t) := \frac{1}{2}d_K(x+tv)^2$. For almost all $t \geqslant 0$, we have

$$f'(t) \leqslant d_K(x+tv)d(v, \overline{\text{co}}\, T_K(\pi_K(x+tv))) \tag{19}$$ ▲

Proof of Theorem 7. Let v_0 belong to $\liminf_{x\to x_0} \overline{\text{co}}\, T_K(x)$. Then, for all $\varepsilon > 0$, there exists $\eta > 0$ such that for all $x \in B_K(x_0, \eta)$, $v_0 \in \overline{\text{co}}\, T_K(x) + \varepsilon B$. Now if x belongs to $B_K(x_0, \alpha)$ and $t \in]0, \beta[$, then $\pi_K(x+tv_0) \subset B_K(x_0, \eta)$ whenever $2\alpha + \beta\|v_0\| \leqslant \eta$. This happens, for instance, when $\alpha := \eta/4$ and $\beta := \eta/2\|v_0\|$. By setting $f(t) := \frac{1}{2}d_K(x+tv_0)^2$, we deduce from lemma 10 that

$$f'(t) \leqslant d_K(x+tv_0)d(v_0, \overline{\text{co}}\, T_K(\pi_K(x+tv_0))) \leqslant \varepsilon d_K(x+tv_0) \leqslant \varepsilon t\|v_0\|$$

because

$$d_K(x+tv_0) \leqslant t\|v_0\|$$

Therefore, for all $x \in B_K(x_0, \alpha)$ and $h \in]0, \beta]$,

$$\frac{1}{h} d_K(x+hv_0)^2 = f(h) - f(0) = \int_0^h f'(t)dt \leqslant \varepsilon\|v_0\| \frac{h^2}{2}$$

and, consequently,

$$\lim_{\substack{x \to_K x_0 \\ h \to 0+}} \frac{d_K(x+hv_0)}{h} = 0$$

This implies that v_0 belongs to the tangent cone $C_K(x_0)$. Then, we obtain

$$\liminf_{x \to_K x_0} T_K(x) \subset \liminf_{x \to_K x_0} \overline{\text{co}}\, T_K(x) \subset C_K(x_0)$$

When X is finite-dimensional, proposition 6 implies that these three cones are equal. ∎

The tangent cone $C_K(x_0)$ being a closed convex cone, it is equal to $C_K(x_0)^{--}$, its negative bipolar cone. Since this duality relation is quite useful, we introduce the following definition.

DEFINITION 11
We shall say that the negative polar cone

$$N_K(x_0) := C_K(x_0)^- \tag{20}$$

to the tangent cone to K at x_0 is the normal cone to K at x_0. ▲

2. CONTINGENT DERIVATIVES AND DERIVATIVES OF A SET-VALUED MAP

We adapt the intuitive definition of a derivative of a function in terms of the tangent to its graph to the case of a set-valued map.

Let F be a proper set-valued map from X to Y and let (x_0, y_0) belong to graph(F). *We denote by $DF(x_0, y_0)$ the set-valued map from X to Y whose graph is the contingent cone $T_{\text{graph}(F)}(x_0, y_0)$ to the graph of F at (x_0, y_0).* In other words,

$$v_0 \in DF(x_0, y_0)(u_0) \quad \text{if and only if} \quad (u_0, v_0) \in T_{\text{graph}(F)}(x_0, y_0) \tag{1}$$

We observe that v_0 belongs to $DF(x_0, y_0)(u_0)$ if and only if

$$\begin{cases} \text{there exist sequences } h_n \to 0^+,\ u_n \to u_0 \text{ and } v_n \to v_0 \\ \text{such that } v_n \in (F(x_0 + h_n u_n) - y_0)/h_n \text{ for all } n \end{cases} \tag{2}$$

DEFINITION 1

We shall say that the set-valued map $DF(x_0, y_0)$ from X to Y is the "contingent derivative" of F at $(x_0, y_0) \in \text{graph}(F)$. ▲

It is a "process," that is, a positively homogeneous set-valued map (since its graph is a cone) with closed graph.

We now give an analytical characterization of $DF(x_0, y_0)$, which justifies that the preceding definition is a reasonable candidate for translating the idea of a derivative as a (suitable) limit of differential quotients

v_0 belongs to $DF(x_0, y_0)(u_0)$ if and only if

$$\liminf_{\substack{h \to 0+ \\ u \to u_0}} d\left(v_0, \frac{F(x_0+hu)-y_0}{h}\right)=0 \tag{3}$$

When F is a single-valued map, we set

$$DF(x_0):=DF(x_0, F(x_0)) \tag{4}$$

since $y_0=F(x_0)$. This formula shows that in this case, v_0 belongs to $DF(x_0)(u_0)$ if and only if

$$\liminf_{\substack{h \to 0+ \\ u \to u_0}} \frac{\|F(x_0+hu)-F(x_0)-hv_0\|}{h}=0 \tag{5}$$

If F is C^1, then $DF(x_0)(u_0)=\nabla F(x_0)u_0$. When the graph of F is convex, we observe that u_0 belongs to $DF(x_0, y_0)(u_0)$ if and only if

$$\liminf_{u \to u_0} \left(\inf_{h>0} d\left(v_0, \frac{F(x_0+hu)-y_0}{h}\right)\right)=0 \tag{6}$$

PROPOSITION 2

Assume that F is Lipschitz on a neighborhood of x_0 (belonging to Int Dom F*). Then v_0 belongs to $DF(x_0, y_0)(u_0)$ if and only if*

$$\liminf_{h \to 0+} d\left(v_0, \frac{F(x_0+hu_0)-y_0}{h}\right)=0 \tag{7}$$

Furthermore, if the dimension of Y is finite, then

$$\text{Dom } DF(x_0, y_0)=X \tag{8}$$ ▲

Proof. **a.** The first statement follows from the fact that

$$F(x_0+hu)-y_0 \subset F(x_0+hu_0)-y_0+\ell h\|u-u_0\|B \tag{9}$$

when both h and $\|u-u_0\|$ are small.

b. Let u_0 belong to X. Then for all $h>0$ small enough,

$$y_0 \in F(x_0) \subset F(x_0+hu_0)+\ell h\|u_0\|B \tag{10}$$

Hence, there exists $v_h \in F(x_0+hu_0)$ such that $(v_h-y_0)/h$ belongs to $\ell\|u_0\|B$, which is compact. A subsequence $(v_{h_n}-y_0)/h_n$ converges to some v_0, which belongs to $DF(x_0, y_0)(u_0)$. ■

We point out that

$$\forall x_0 \in K, \quad \forall y_0 \in F(x_0), \quad DF(x_0, y_0)^{-1}=D(F^{-1})(y_0, x_0) \tag{11}$$

Indeed, to say that $(u_0, v_0) \in T_{\text{graph}(F)}(x_0, y_0)$ amounts to saying that $(v_0, u_0) \in T_{\text{graph}(F^{-1})}(y_0, x_0)$.

Contingent derivatives allow us to differentiate restrictions of a map or a set-valued map to a subset.

PROPOSITION 3
Let F be a single-valued map from an open subset Ω of X to Y of class C^1 and let K be a nonempty subset of Ω containing x_0. Then

$$DF|_K(x_0)(u_0)=\begin{cases}\nabla F(x_0)u_0 & \text{if} \quad u_0 \in T_K(x_0)\\ \varnothing & \text{if} \quad u_0 \notin T_K(x_0)\end{cases} \tag{12}$$

▲

Proof. If F is a C^1 single-valued map at x_0 and u_0 belongs to $T_K(x_0)$, there exist sequences $h_n \to 0^+$ and $u_n \to u_0$ such that $x_0+h_n u_n$ belongs to K. Since

$$F|_K(x_0+h_n u_n)=F(x_0+h_n u_n)=F(x_0)+h_n(\nabla F(x_0)u_n+0(h_n))$$

we deduce that the elements $v_n:=\nabla F(x_0)u_n+0(h_n)$ converge to $\nabla F(x_0)u_0$ and belong to $(F|_K(x_0+h_n u_n)-F|_K(x_0))/h_n$. Therefore,

$$DF|_K(x_0, F(x_0))(u_0)=\nabla F(x_0)u_0$$

■

We follow the same procedure in defining the derivative of a set-valued map from X to Y.

Let (x_0, y_0) belong to the graph of F.

We denote by $CF(x_0, y_0)$ the *closed convex process* from X to Y whose graph is the tangent cone $C_{\text{graph}(F)}(x_0, y_0)$ to the graph of F at (x_0, y_0). Briefly,

$$v_0 \in CF(x_0, y_0)(u_0) \quad \text{if and only if} \quad (u_0, v_0) \in C_{\text{graph}(F)}(x_0, y_0) \tag{13}$$

DEFINITION 4
We shall say that the closed convex process $CF(x_0, y_0)$ from X to Y is the derivative of F at $x_0 \in$ Dom F and $y_0 \in F(x_0)$. ▲

We observe that v_0 belongs to $CF(x_0, y_0)(u_0)$ if and only if

$$(14)\quad \begin{cases} \forall \varepsilon_1, \ \varepsilon_2 > 0, \ \exists \alpha, \beta > 0 \ \text{such that} \ \forall (x, y) \in B_{\text{graph}(F)}(x_0, y_0, \alpha), \\ \forall h \in]0, \beta], \ \exists u \in u_0 + \varepsilon_1 B, \ v \in v_0 + \varepsilon_2 B \ \text{such that} \ v \in (F(x+hu) - y)/h \end{cases}$$

or, equivalently, if and only if

(15) For all sequences of elements $(x_n, y_n, h_n) \in \text{graph}(F) \times]0, \infty[$ converging to $(x_0, y_0, 0)$, there exist sequences of elements u_n converging to u_0 and v_n converging to v_0 such that $y_n + h_n v_n \in F(x_n + h_n u_n)$ for all $n > 0$.

The analytical formula involving "differential quotients" is quite complicated. It is simpler when F is locally Lipschitz; we begin with it.

PROPOSITION 5

Assume that F is Lipschitz on a neighborhood of an element $x_0 \in \text{Int Dom } F$. Then v_0 belongs to $CF(x_0, y_0)(u_0)$ if and only if

$$(16)\qquad \lim_{x \underset{K}{\to} x_0, h \to 0+} d\left(v_0, \frac{F(x+hu_0) - y_0}{h}\right) = 0 \qquad \blacktriangle$$

Remark

We observe that the domain of the derivative of a Lipschitz function is not necessarily the whole space, while the domain of the contingent derivative is the whole space when the dimension of Y is finite. Take, for instance, the map π associating to $x \in \mathbb{R}$, $\pi(x) := 0$ if $x \leq 0$ and $\pi(x) = x$ if $x > 0$. We saw that $C\pi(0, 0)(u) = \varnothing$ when $u \neq 0$ and $C\pi(0, 0)(0) = 0$, whereas $D\pi(0, 0)(u) = \pi(u)$ for all $u \in \mathbb{R}$. ■

For the analytical formula in the general case, we need the following definition.

DEFINITION 6

Let U and V be metric spaces and ϕ be a function from $U \times V$ to $\bar{\mathbb{R}}$. We set

$$(17)\qquad \limsup_{u \to u_0} \inf_{v \to v_0} \phi(u, v) := \sup_{\varepsilon > 0} \inf_{\eta > 0} \sup_{u \in B(u_0, \eta)} \inf_{v \in B(v_0, \varepsilon)} \phi(u, v) \qquad \blacktriangle$$

PROPOSITION 7

Let F be a proper set-valued map from X to Y and let (x_0, y_0) belong to graph(F). *Then v_0 belongs to the derivative $CF(x_0, y_0)(u_0)$ if and only if*

$$(18)\qquad \left\{ \limsup_{\substack{(x,y) \to (x_0, y_0) \\ \text{graph}(F) \\ h \to 0+}} \inf_{u \to u_0} d\left(v_0, \frac{F(x+hu) - y}{h}\right) = 0 \right.$$

▲

Proof of Propositions 5 and 7. Formula (14) can be written

$$\sup_{\varepsilon_1>0}\ \inf_{\alpha,\beta>0}\ \sup_{(x,y)\in B\text{graph}(F)(x_0,y_0;\alpha)}\ \inf_{u\in u_0+\varepsilon_1 B} d\left(v_0, \frac{F(x+hu)-y}{h}\right)=0$$

This proves proposition 7. When F is Lipscitz around x_0,

$$\left|\inf_{u\in u_0+\varepsilon_1 B} d\left(v_0, \frac{F(x+hu)-y}{h}\right)-d\left(v_0, \frac{F(x+hu_0)-y}{h}\right)\right|\leqslant \ell\varepsilon_1$$

and the formulas become

$$\inf_{\alpha,\beta>0}\ \sup_{\substack{(x,y)\in B\text{graph}(F)(x_0,y_0,\alpha)\\ h\in]0,\beta]}} \mathrm{d}\left(v_0, \frac{F(x+hu_0)-y}{h}\right)=0 \qquad \blacksquare$$

When F is single-valued, we set

$$CF(x_0):=CF(x_0, F(x_0)) \tag{19}$$

If F is continuously differentiable at x_0, we have

$$CF(x_0)=\nabla F(x_0) \tag{20}$$

Naturally, the formula for derivatives of inverses is obvious

$$\forall (x_0, y_0)\in \text{graph}(F), \qquad CF(x_0, y_0)^{-1}=C(F^{-1})(y_0, x_0) \tag{21}$$

PROPOSITION 8

Let F be a single-valued map from an open subset Ω of X to Y, continuously differentiable at $x_0\in\Omega$ and let K be a nonempty subset of X containing x_0. Then

$$CF|_K(x_0)u_0=\begin{cases}\nabla F(x_0)u_0 & \text{if } u_0\in C_K(x_0)\\ \varnothing & \text{if } u_0\notin C_K(x_0)\end{cases} \tag{22}$$

▲

Proof. Let $(x_n, h_n)\in K\times]0,\infty[$ converge to $(x_0, 0)$ in $K\times\mathring{\mathbb{R}}_+$. If u_0 belongs to $C_K(x_0)$, there exists a sequence of elements u_n converging to u_0 such that $x_n+h_nu_n$ belongs to K for all n. Then

$$F|_K(x_n+h_nu_n)=F(x_n+h_nu_n)=F(x_n)+h_n(\nabla F(x_n)u_n+O(h_n))$$

Since F is continuously differentiable, the sequence of elements $v_n:=\nabla F(x_n)u_n+o(h_n)$ converges to $\nabla F(x_0)u_0$, and we have $F|_K(x_n)+h_nv_n=F|_K(x+h_nu_n)$ for all n. ■

Since the derivative $CF(x_0, y_0)$ is a closed convex process, it is equal to its bi-transpose $CF(x_0, y_0)^{**}$. This suggests that we introduce the following definition.

DEFINITION 9

We shall say that the transpose $CF(x_0, y_0)^*$ of the derivative of F at $(x_0, y_0) \in$ graph(F) is the codifferential of F at (x_0, y_0). ▲

It is a closed convex process from Y^* to X^* defined by

$$\text{(23)} \qquad \begin{cases} p_0 \in CF(x_0, y_0)^*(q_0) \quad \text{if and only if} \quad \forall u \in X, \\ \forall v \in CF(x_0, y_0)(u), \quad \langle p_0, u\rangle - \langle q_0, v\rangle \leq 0 \end{cases}$$ ■

We mention an example of derivatives of a set-valued map that we shall use later.

PROPOSITION 10

Let X and Y be Banach spaces, A a continuously differentiable operator from an open subset Ω of X to Y, and $L \subset \Omega$, $M \subset Y$ closed subsets of X and Y, respectively. Let F be the set-valued map from X to Y defined by

$$\text{(24)} \qquad F(x) := \begin{cases} A(x) - M & \text{when} \quad x \in L \\ \varnothing & \text{when} \quad x \notin L \end{cases}$$

Let (x_0, y_0) belong to the graph of F. The following conditions are equivalent

$$\text{(25)} \qquad \begin{aligned} &\textbf{i.} \quad v_0 \in CF(x_0, y_0)(u_0). \\ &\textbf{ii.} \quad u_0 \in C_L(x_0) \quad \textit{and} \quad v_0 \in \nabla A(x_0)u_0 - C_M(Ax_0 - y_0). \end{aligned}$$ ▲

Proof. **a.** Let us prove that **(i)** implies **(ii)**. We take sequences $(x_n, z_n, h_n) \in L \times M \times]0, \infty[$ converging to $(x_0, Ax_0 - y_0, 0)$. Then $y_n := A(x_n) - z_n$ converges to y_0, and by **(i)**, there exist sequences u_n and v_n converging to u_0 and v_0 such that $x_n + h_n u_n \in L$ and $A(x_n + h_n u_n) \in M + y_n + h_n v_n$ for all n. This implies that u_0 belongs to $C_L(x_0)$ and that $\nabla A(x_0)u_0 - v_0$ belongs to $C_M(Ax_0 - y_0)$ because $w_n := A(x_n + h_n u_n) - A(x_n) - v_n$ converges to $\nabla A(x_0)u_0 - v_0$ and $z_n + h_n w_n$ belongs to M for all n.

b. Conversely, let us show that **(i)** follows from **(ii)**. We take a sequence $(x_n, y_n, h_n) \in \text{graph}(F) \times]0, \infty[$ converging to $(x_0, y_0, 0)$. There exists a sequence u_n converging to u_0 such that $x_n + h_n u_n$ belongs to L, and since $Ax_n - y_n$ converges to $Ax_0 - y_0$ in M, there exists a sequence of elements w_n converging to $\nabla A(x_0)u_0 - v_0$ and satisfying $Ax_n - y_n + h_n w_n \in M$ for all n. Then the sequence of elements $v_n := \{[A(x_n + h_n u_n) - Ax_n]/h_n\} - w_n$ converges to v_0 and satisfies $y_n + h_n v_n \in F(x_n + h_n u_n)$ for all n. ■

PROPOSITION 11

Let K be a closed convex subset of a Hilbert space X and let p_0 belong to the normal cone $N_K(x_0)$. Let N_K denote the set-valued map $x \to N_K(x)$ and π_K the

Lipschitz single-valued map associating to x *its best approximation* $\pi_K(x) \in K$ *by elements of* K. *Then the two following statements are equivalent*:

$$(26) \qquad \begin{aligned} &\textbf{i.} \quad q_0 \in CN_K(x_0, p_0)(u_0). \\ &\textbf{ii.} \quad u_0 \in C\pi_K(x_0+p_0)(u_0+q_0). \end{aligned}$$

The same result holds when the derivative is replaced by the contingent derivative. ▲

Proof. We recall that p belongs to the normal cone $N_K(x)$ if and only if $x = \pi_K(x+p)$.

a. Assume that q_0 belongs to $CN_K(x_0, p_0)(u_0)$. Let us consider a sequence of elements

$$(y_n, h_n) \in X \times]0, \infty[\quad \text{converging to} \quad (x_0+p_0, 0)$$

We set $x_n := \pi_K(y_n)$, which converges to $x_0 = \pi_K(x_0+p_0)$, and $p_n := y_n - x_n$, which converges to p_0. Then there exist sequences of elements u_n and q_n converging to u_0 and q_0 such that $p_n + h_n q_n$ belongs to $N_K(x_n + h_n u_n)$ for all n; that is, such that

$$\pi_K(y_n) + h_n u_n = \pi_K(y_n + h_n(q_n + u_n)) \qquad \text{for all } n$$

Hence, u_0 belongs to $C\pi_K(x_0+p_0)(u_0+q_0)$.

b. Conversely, assume that u_0 belongs to $C\pi_K(x_0+p_0)(u_0+q_0)$. Let $(x_n, p_n, h_n) \in \text{graph } N_K \times]0, \infty[$ converge to $(x_0, p_0, 0)$. Since $x_n + p_n$ converges to $x_0 + p_0$, there exist sequences of elements u_n and w_n converging to u_0 and $u_0 + q_0$ such that

$$x_n + h_n u_n = \pi_K(x_n + p_n) + h_n u_n = \pi_K(x_n + p_n + h_n w_n) \text{ for all } n$$

Then $q_n := w_n - u_n$ converges to u_0 and we deduce that

$$p_n + h_n q_n \in N_K(x_n + h_n u_n) \text{ for all } n$$

Hence, q_0 belongs to $CN_K(x_0, p_0)(u_0)$. ■

COROLLARY 12

Let us consider the set-valued map associating to $x \in R^n_+$ *the normal cone* $N_{R^n_+}(x)$ *to* R^n_+ *at* x. *Let* p^0 *belong to* $N_{R^n_+}(x^0)$. *Then* q^0 *belongs to* $CN_{R^n_+}(x^0, p^0)(u^0)$ *if and only if*

$$(27) \qquad q_i \in \begin{cases} \{0\} & \text{if } x_i > 0 \quad (\text{and thus } p_i = 0) \\ \varnothing & \text{if } x_i^0 = 0, \ p_i^0 \leqslant 0 \ \text{ and } \ u_i \neq 0 \\ \mathbb{R} & \text{if } x_i^0 = 0, \ p_i^0 < 0 \ \text{ and } \ u_i = 0 \\ \{0\} & \text{if } x_i^0 = 0, \ p_i^0 = 0 \ \text{ and } \ u_i = 0 \end{cases}$$

▲

Proof. We observe that

$$\pi_{R_+^n}(x_1, \ldots, x_n) = (\pi(x_1), \ldots, \pi(x_n))$$

where $\pi(x)=0$ when $x \leqslant 0$ and $\pi(x)=x$ when $x \geqslant 0$. Since $C\pi(x)(u)=0$ when $x<0$, u when $x>0$ and $C\pi(0)(u)=\varnothing$ when $u \neq 0$ and $C\pi(0)(0)=0$, we obtain corollary 12. ■

3. EPICONTINGENT DERIVATIVES AND EPIDERIVATIVES OF REAL-VALUED FUNCTIONS

We can use the concept of contingent derivatives and derivatives for single-valued maps V from Dom $V \subset X$ to $\mathbb{R}$. We obtain, for instance,

$$(1) \qquad v_0 \in DV(x)(u_0) \Leftrightarrow \liminf_{\substack{h \to 0+ \\ u \to u_0}} \left| \frac{V(x+hu) - V(x)}{h} - v_0 \right| = 0$$

In many problems, such as minimization problems, the order relation plays an important role. This is why we associate with a proper function $V: X \to \mathbb{R} \cup \{+\infty\}$ the set-valued map $\mathbf{V}_+$ defined by $\mathbf{V}_+(x) = V(x) + \mathbb{R}_+$ when $V(x) < +\infty$ and $\mathbf{V}_+(x) = \varnothing$ when $V(x) = +\infty$. Its domain is the domain of V, and its graph is the epigraph of V. We consider its contingent derivative $D\mathbf{V}_+(x, V(x))$, whose images are closed half lines. Therefore, for all $u_0 \in X$, $D\mathbf{V}_+(x, V(x)(u_0)$ is either $\mathbb{R}$ or a half line $[v_0, \infty[$ or empty. We set

$$(2) \qquad D_+V(x)(u) := \inf \{v \mid v \in D\mathbf{V}_+(x, V(x))(u)\}$$

It is equal to $-\infty$ if $D\mathbf{V}_+(x, V(x)) = \mathbb{R}$, to v_0 if $D\mathbf{V}_+(x, V(x))(u) = [v_0, \infty[$, and to $+\infty$ if $D\mathbf{V}_+(x, V(x))(u) = \varnothing$.

DEFINITION 1

We shall say that $D_+V(x)(u)$ is the epicontingent derivative of V at x in the direction u. ▲

We begin by computing epicontingent derivatives.

PROPOSITION 2

If V is a proper function from X to $R \cup \{+\infty\}$, then

$$(3) \qquad D_+V(x_0)(u_0) = \liminf_{\substack{h \to 0+ \\ u \to u_0}} \frac{V(x_0+hu) - V(x_0)}{h}$$

The function $u \to D_+V(x_0)(u)$ is positively homogeneous and lower semicontinuous when $D_+V(x_0)(u) > -\infty$ for all $u \in X$. ▲

Proof. Indeed, let $v_0 \in D\underset{\sim}{V}_+(x_0, V(x_0))(u_0)$; then $\forall \varepsilon_1 > 0$, $\varepsilon_2 > 0$, $\forall \alpha > 0$, there exist $u \in u_0 + \varepsilon_2 B$ and $h < \alpha$ such that

$$v_0 \in \frac{\underset{\sim}{V}_+(x_0 + hu) - V(x_0)}{h} + \varepsilon_1 B$$

This implies that

$$v_0 \geqslant \frac{V(x_0 + hu) - V(x_0)}{h} - \varepsilon_1 \geqslant \inf_{h \leqslant \alpha} \inf_{\|u - u_0\| \leqslant \varepsilon_2} \frac{V(x_0 + hu) - V(x_0)}{h} - \varepsilon_1$$

Therefore,

$$v_0 \geqslant \liminf_{\substack{h \to 0+ \\ u \to u_0}} \frac{V(x_0 + hu) - V(x_0)}{h} - \varepsilon_1$$

For the time being, let us set

$$a := \liminf_{\substack{h \to 0+ \\ u \to u_0}} \frac{V(x_0 + hu) - V(x_0)}{h}$$

Thus, we have proved that $a \leqslant D_+ V(x_0)(u_0)$. On the other hand, we know that for any $M > a$,

$$\sup_{\substack{\alpha > 0 \\ \delta > 0}} \inf_{h \leqslant \alpha} \inf_{\|u_0 - u\| \leqslant \delta} \frac{V(x_0 + hu) - V(x_0)}{h} < M$$

that for all $\alpha, \delta > 0$, there exist $h < \alpha$ and $u \in u_0 + \delta B$ such that

$$\frac{V(x_0 + hu) - V(x_0)}{h} \leqslant M$$

Hence, $M \in [\mathbf{V}_+(x_0 + hu) - V(x_0)]/h$, which proves that $a \in D\mathbf{V}_+(x_0, V(x_0))(u_0)$. Since it is smaller than all the other ones, we infer that $a = D_+ V(x_0)(u_0)$. ■

If V is C^1 at x_0, then

$$\forall u_0 \in X, \quad D_+ V(x_0)(u_0) = \langle \nabla V(x_0), u_0 \rangle \tag{4}$$

If V is convex, then

$$\forall u_0 \in X, \quad D_+ V(x_0)(u_0) = \liminf_{u \to u_0} \left(\inf_{h > 0} \frac{V(x_0 + hu) - V(x_0)}{h} \right) \tag{5}$$

We deduce from propositions 2.2 and 2.3 the following statements.

PROPOSITION 3

Let us assume that V is Lipschitz on a neighborhood of $x_0 \in \text{Int Dom } V$. Then

$$(6) \qquad \forall u_0 \in X, \qquad D_+V(x_0)(u_0) = \liminf_{h \to 0^+} \frac{V(x_0 + hu_0) - V(x_0)}{h}$$

and the epicontingent derivative is finite. ▲

PROPOSITION 4

Let V be a proper function from X to $R \cup \{+\infty\}$ and K a subset of X. Let $V|_K$ denote the restriction of V to K (in the sense that $V|_K(x)$ equals $V(x)$ when $x \in K$, ∞ when $x \notin K$).

Then

$$(7) \qquad \forall x_0 \in K, \qquad \forall v_0 \in T_K(x_0), \qquad D_+V(x_0)(u_0) \leqslant D_+V|_K(x_0)(u_0)$$

If V is C^1 at x_0, we have

$$(8) \qquad D_+V|_K(x_0)(u_0) = \begin{cases} \langle \nabla V(x_0), u_0 \rangle & \text{if} \quad u_0 \in T_K(x_0) \\ +\infty & \text{if} \quad u_0 \notin T_K(x_0) \end{cases}$$ ▲

We state the obvious property of the epicontingent derivative at a minimizer.

PROPOSITION 5

Let V be a proper function from a Banach space X to $\mathbb{R} \cup \{+\infty\}$. If $\bar{x} \in \text{Dom } V$ minimizes V on X, then

$$(9) \qquad \forall u \in X, \qquad 0 \leqslant D_+V(\bar{x})(u)$$ ▲

More generally, the ε-variational principle can take the following form.

THEOREM 6

Let V be a proper lower semicontinuous function bounded below from a Banach space X to $\mathbb{R} \cup \{+\infty\}$ and let x_0 belong to $\text{Dom } V$. Then for any $\varepsilon > 0$, there exists $x_\varepsilon \in \text{Dom } V$ satisfying

$$(10) \qquad \begin{cases} \textbf{i.} \quad V(x_\varepsilon) + \varepsilon \|x_\varepsilon - x_0\| \leqslant V(x_0). \\ \textbf{ii.} \quad \forall u \in X, \qquad 0 \leqslant D_+V(x_\varepsilon)(u) + \varepsilon \|u\|. \end{cases}$$ ▲

Proof. By theorem 5.3.1 (the ε-variational principle), there exists $x_\varepsilon \in \text{Dom } V$ satisfying (10, **i**) and $V(x_\varepsilon) = \min_{x \in X}[V(x) + \varepsilon \|x - x_\varepsilon\|]$.

Let $u \in \text{Dom } D_+V(x_\varepsilon)$. Then for any $\eta > 0$, $\delta > 0$, $\alpha > 0$, there exist $h \leqslant \alpha$ and $v \in u + \delta B$ such that

$$\frac{V(x_\varepsilon + hv) - V(x_\varepsilon)}{h} \leqslant D_+V(x_\varepsilon)(u) + \eta$$

Theorem 5.3.1 implies

$$-\varepsilon\delta - \varepsilon\|u\| \leqslant -\varepsilon\|v\| \leqslant \frac{V(x_\varepsilon + hv) - V(x_\varepsilon)}{h}$$

Therefore, we infer that

$$0 \leqslant D_+ V(x_\varepsilon)(u) + \varepsilon\|u\| + \varepsilon\delta + \eta$$

By letting δ and η converge to zero, we obtain the desired inequality. ■

In the same way, we define epiderivatives of functions V from X to $\mathbb{R} \cup \{+\infty\}$. Since images of the derivative $C\underset{\sim}{V}_+(x_0, V(x_0))$ are either R or a half line $[v_0, \infty[$ or empty, we set

$$(11) \qquad C_+ V(x_0)(u_0) := \inf\{v \mid v \in C\underset{\sim}{V}_+(x_0, V(x_0))(u_0)\}$$

It is equal to $-\infty$ when $C\underset{\sim}{V}_+(x_0, V(x_0)) = R$, to v_0 when $C\underset{\sim}{V}_+(x_0, V(x_0))(u_0) = [v_0, \infty[$, and to $+\infty$ when $C\underset{\sim}{V}_+(x_0, V(x_0))(u_0) = \varnothing$.

DEFINITION 7
We shall say that $C_+ V(x_0)(u_0)$ is the epiderivative *of V at x_0 in the direction u_0.* ▲

The epigraph of $u \to C_+ V(x_0)(u)$ is a closed convex cone, because it is the graph of the set-valued map $u \to CV_+(x_0, V(x_0))(u)$, which is a closed convex process. We deduce at once the following important property.

PROPOSITION 8
The epiderivative $u \to C_+ V(x_0)(u)$ is a positively homogeneous lower semicontinuous, convex function when $C_+ V(x_0)(u) > -\infty$ for all $u \in X$. ▲

It is easy to check that the codifferential of $\underset{\sim}{V}_+$ at $(x_0, V(x_0))$ is a closed convex process from $\mathbb{R}$ to X^*, defined by its values $C\underset{\sim}{V}_+(x_0, V(x_0))^*(-1)$ and $C\underset{\sim}{V}_+(x_0, V(x_0))^*(1)$. We observe that $C\underset{\sim}{V}_+(x_0, V(x_0))^*(-1) = \varnothing$ and the support function of $C\underset{\sim}{V}_+(x_0, V(x_0))^*(1)$ equals $C_+ V(x_0)(\cdot)$ when it is not empty. ■

DEFINITION 9
We say that the closed convex subset of X^* defined by

$$(12) \quad \partial V(x_0) := C\underset{\sim}{V}_+(x_0, V(x_0))^*(1) = \{p \in X^* \mid \forall u \in X,\ \langle p, u\rangle \leqslant C_+ V(x_0)(u)\}$$

is the generalized gradient *of V at x_0.* ▲

It is empty whenever there exists a direction u_0 for which $C_+ V(x_0)(u_0) = -\infty$.

If V is continuously differentiable at x_0, then $\forall u_0 \in X$, $C_+V(x_0)(u_0) = \langle \nabla V(x_0), u_0 \rangle$ and, consequently,

$$\partial V(x_0) = \{\nabla V(x_0)\}$$

This justifies the term *generalized gradient*. When V is convex, it coincides with the subdifferential of V at x_0.

PROPOSITION 10

Let us assume that V is Lipschitz on a neighborhood of $x_0 \in \text{Int Dom } V$. Then

$$\text{(13)} \qquad \forall u_0 \in X, \qquad C_+V(x_0)(u_0) = \limsup_{\substack{x \to x_0 \\ h \to 0^+}} \frac{V(x+hu_0) - V(x)}{h}$$

and the epiderivative is finite.
Furthermore, the following properties hold true:

$$\text{(14)} \begin{cases} \textbf{i.} & \forall u_0 \in X, \quad (x, u) \to C_+V(x)(u) \textit{ is upper semicontinuous at } (x_0, u_0). \\ \textbf{ii.} & u \to C_+V(x_0)(u) \textit{ is Lipschitz.} \\ \textbf{iii.} & C_+(-V)(x_0)(u) = C_+V(x_0)(-u). \end{cases}$$

In terms of generalized gradients, these properties become

$$\text{(15)} \qquad \begin{cases} \textbf{i.} & x \to \partial V(x) \textit{ is upper hemicontinuous at } x_0. \\ \textbf{ii.} & \partial V(x_0) \textit{ is (closed convex and) bounded.} \\ \textbf{iii.} & \partial(-V)(x_0) = -\partial V(x_0). \end{cases}$$

▲

Proof. Since V is Lipschitz on a neighborhood of $x_0 \in \text{Int Dom } V$, there exist $\alpha_0 > 0$ and $\ell > 0$ such that for any α, β, η satisfying $\alpha + \beta(\|u_0\| + \eta) \leqslant \alpha_0$, we have: $\forall x \in x_0 + \alpha\beta$, $\forall h \in]0, \beta]$, $\forall u \in u_0 + \eta B$,

$$\text{(16)} \qquad \left| \frac{V(x+hu) - V(x)}{h} \right| \leqslant \ell(\|u_0\| + \eta)$$

a. Let v_0 belong to the derivative $C\underset{\sim}{V}_+(x_0, V(x_0))$ of $\underset{\sim}{V}_+$ at $(x_0, V(x_0))$. This means that for all $\varepsilon, \eta > 0$, there exist $\alpha, \beta > 0$ such that for all $x \in x_0 + \alpha\beta$, $\forall h \in]0, \beta]$, there exists $u \in u_0 + \eta\beta$ such that

$$v_0 \in \frac{\underset{\sim}{V}_+(x+hu) - V(x)}{h} + \varepsilon B$$

Hence,

$$v_0 \geqslant \frac{V(x+hu)-V(x)}{h} - \varepsilon \geqslant \frac{V(x+hu_0)-V(x)}{h} - \varepsilon - \ell\eta$$

(because V is Lipschitz around x_0).

Consequently,

$$v_0 \geqslant \limsup_{\substack{x \to x_0 \\ h \to 0^+}} \frac{V(x+hu_0)-V(x)}{h}$$

and thus

$$C_+V(x_0)(u_0) \geqslant \limsup_{\substack{x \to x_0 \\ h \to 0^+}} \frac{V(x+hu_0)-V(x)}{h}$$

Conversely, let us set

$$a := \limsup_{\substack{x \to x_0 \\ h \to 0^+}} \frac{V(x+hu_0)-V(x)}{h}$$

which is finite by inequality (16). Then we can associate to any $\varepsilon > 0$ constants α, $\beta > 0$ such that

$$a + \varepsilon \geqslant \frac{V(x+hu_0)-V(x)}{h}$$

This implies that a belongs to $CV_+(x_0, V(x_0))$. Hence, formula (13) ensues.

b. The upper semicontinuity of $(x, u) \to C_+V(x)(u)$ at (x_0, u_0) follows at once from formula (13).

Also, inequality (16) implies that

$$C_+V(x_0)(u) \leqslant \ell \|u\| \tag{17}$$

and thus that $u \to C_+V(x_0)(u)$ is Lipschitz. To prove (14, **iii**), we observe that

$$\frac{-V(x+hu_0)-(-V(x))}{h} = \frac{V((x+hu_0)+h(-u_0))-V(x+hu_0)}{h}$$

Since $x+hu_0$ is in a neighborhood of x_0 when x is a neighborhood of x_0 and h is small, we deduce that

$$C_+(-V)(x_0)(u_0) = C_+V(x_0)(-u_0)$$

c. Since $C_+V(x_0)(\cdot)$ is proper, it is the support function of $\partial V(x_0)$. ■

Remark

More generally, we can prove the following formula for epiderivatives of arbitrary functions. For that purpose, it is expedient to use the notation

$$(18)\qquad (x,\lambda)\downarrow x_0 \Leftrightarrow \lambda \geqslant V(x), \qquad x\to x_0 \quad \text{and} \quad \lambda\to V(x_0)$$

and definition 2.6 of lim sup inf.

PROPOSITION 11

Let x_0 belong to the domain of a function V from X to $\mathbb{R}\cup\{+\infty\}$. Then

$$(19)\qquad C_+V(x_0)(u_0) = \limsup_{\substack{(x,\lambda)\downarrow x_0\\ h\to 0^+}} \inf_{u\to u_0} \frac{V(x+hu)-\lambda}{h} \qquad \blacktriangle$$

The proof is left as an exercise.

When V is lower semicontinuous at x_0, formula (19) becomes

$$(20)\qquad C_+V(x_0)(u_0) = \limsup_{\substack{x\to x_0\\ V(x)\to V(x_0)\\ h\to 0^+}} \inf_{u\to u_0} \frac{V(x+hu)-V(x)}{h}$$

It may be useful to use another concept of derivative, easier to manipulate than the epiderivative.

DEFINITION 12

Let V be a proper function from X to $\mathbb{R}\cup\{+\infty\}$ and let x_0 belong to Dom V. *We set*

$$(21)\qquad B_+V(x_0)(u_0) := \limsup_{\substack{(x,\lambda)\downarrow x_0\\ u\to u_0\\ h\to 0^+}} \frac{V(x+hu)-\lambda}{h}$$

We shall say that $B_+V(x_0)(u_0)$ is the strict epiderivative *of V at x_0 in the direction of u_0 and V is strictly epidifferentiable at x_0 if the function $u\to B_+V(x_0)(u)$ is a proper function from X to $\mathbb{R}\cup\{+\infty\}$.* ▲

We always have

$$(22)\qquad \forall u\in X, \qquad D_+V(x_0)(u) \leqslant C_+V(x_0)(u) \leqslant B_+V(x_0)(u)$$

Clearly, a function V that is Lipschitz around x_0 is strictly epidifferentiable at x_0.

The introduction of this concept is justified by the following result, stated in proposition 13. ▲

PROPOSITION 13

Let us assume that the function V is strictly epidifferentiable at $x_0 \in \text{Dom } V$. *Then*

$$\text{Dom } B_+V(x_0) = \text{Int Dom } C_+V(x_0) \tag{23}$$

and

$$\forall u_0 \in \text{Dom } C_+V(x_0), \qquad C_+V(x_0)(u_0) = \liminf_{u \to u_0} B_+V(x_0)(u_0) \tag{24}$$

Furthermore, for any $u_0 \in \text{Int Dom } C_+V(x_0)$,

$$\begin{cases} \textbf{i.} & (x, u) \to C_+V(x)(u) \text{ is upper semicontinuous at } (x_0, u_0) \\ \textbf{ii.} & u \to C_+V(x_0)(u) \text{ is continuous at } u_0. \end{cases} \tag{25}$$

If we assume that $\text{Dom } B_+V(x_0) = X$, *then*

$$\partial(-V)(x_0) = -\partial V(x_0) \tag{26}$$ ▲

Proof. **a.** Let u_0 belong to the domain of $B_+V(x_0)$. Equation (21) implies at once that $\text{Dom } B_+V(x_0)$ is open and $(x, u) \to B_+V(x)(u)$ is upper semicontinuous at (x_0, u_0)

b. Formula (21) implies that

$$B_+V(x_0)(u_0 + u_1) \leq B_+V(x_0)(u_0) + C_+V(x_0)(u_1) \tag{27}$$

We deduce that any u interior to the domain of $C_+V(x_0)$ belongs to the domain of $B_+V(x_0)$. For that purpose, take $u_0 \in \text{Dom } B_+V(x_0)$ and $\lambda > 0$ such that $u - \lambda u_0$ belongs to the domain of $C_+V(x_0)$. Then inequality (27) implies that

$$B_+V(x_0)(u) \leq C_+V(x_0)(u - \lambda u_0) + \lambda B_+V(x_0)(u_0) < +\infty$$

that is, u belongs to the domain of $B_+V(x_0)$. Hence, the domain of $B_+V(x_0)$ coincides with the interior of the domain of $C_+V(x_0)$. Inequality (27) implies also that the epigraph of $B_+V(x_0)$ is dense in the epigraph of $C_+V(x_0)$. Consequently,

$$\liminf_{u \to u_0} B_+V(x_0)(u) \leq C_+V(x_0)(u_0)$$

Since $u \to C_+V(x_0)(u)$ is lower semicontinuous, equality (24) ensues. Furthermore, by letting λ go to zero in the preceding inequality, we obtain

$$\forall u \in \text{Dom } B_+V(x_0), \qquad B_+V(x_0)(u) = C_+V(x_0)(u) \tag{28}$$

Inequality (24) implies that

$$\partial V(x_0)=\{p\in X^*|\forall u\in X, \quad \langle p,u\rangle\leqslant B_+V(x_0)(u)\} \tag{29}$$

Hence, property (26) follows from

$$\forall u\in \text{Dom}\ B_+V(x_0), \qquad B_+V(x_0)(-u)=B_+(-V)(x_0)(u) \tag{30}$$

To prove this statement, we set $v_0:=B_+V(x_0)(-u_0)$; for all $\varepsilon>0$, there exist $\alpha_0, \beta_0, \eta_0>0$ such that, for all $y\in \text{Dom}\ V\cap(x_0+\alpha_0 B)$, $h\in\,]0,\beta]$, $u\in u_0+\eta_0 B$, $[V(y-hu)-V(y)]/h\leqslant v_0+\varepsilon$.

Let us take $\alpha\in\,]0,\alpha_0]$, $\beta\in\,]0,\beta_0]$, and $\eta\in\,]0,\beta_0]$ such that $\alpha+\beta(\|u_0\|+\eta)\leqslant\alpha_0$. Hence, for all $x\in\text{Dom}\ V\cap(x_0+\alpha B)$, $\lambda\in V(x_0)+\alpha B$, satisfying $\lambda\geqslant -V(x)$, $h\in\,]0,\beta]$, $u\in u_0+\eta B$, we have, by setting $y:=x+hu$,

$$\frac{-V(x+hu)-\lambda}{h}\leqslant\frac{V(x)-V(x+hu)}{h}=\frac{V(y-hu)-V(y)}{h}\leqslant v_0+\varepsilon$$

because y belongs to $\text{Dom}\ V\cap(x_0+\alpha_0 B)$. This implies that

$$B_+(-V)(x_0)(u_0)\leqslant v_0:=B_+V(x_0)(-u_0)$$

By exchanging the roles of V and $-V$, we have proved equality (30). ■

PROPOSITION 14

Let P be a closed convex cone. Assume that V is nonincreasing with respect to P in the sense that $V(x+y)\leqslant V(x)$ for all $y\in P$. Let x_0 belong to Dom V. *Then*

$$\forall u_0\in P, \qquad C_+V(x_0)(u_0)\leqslant 0 \tag{31}$$

and

$$\partial V(x_0)\subset P^- \tag{32}$$

If Int $P\neq\varnothing$, *then*

$$\forall u_0\in \text{Int}\ P, \qquad B_+V(x_0)(u_0)\leqslant 0 \tag{33}$$ ▲

Proof. Indeed, for any $x\in x_0+\alpha B$, $\lambda\in V(x_0)+\alpha B$, $\lambda\geqslant V(x)$, $h\in\,]0,\beta[$, $u_0\in P$, we have

$$\frac{V(x+hu_0)-\lambda}{h}\leqslant\frac{V(x+hu_0)-V(x)}{h}\leqslant 0$$

because V is nonincreasing. Hence, $C_+V(x_0)(u_0)\leqslant 0$. If u_0 belongs to the interior of P, there exists $\eta_0>0$ such that $u_0+\eta_0 B\subset P$. Hence, for all $u\in u_0+\eta_0 B$, we would have

$$\frac{V(x+hu)-\lambda}{h} \leqslant \frac{V(x+hu)-V(x)}{h} \leqslant 0$$

and thus $B_+V(x_0)(u_0) \leqslant 0$. ■

We deduced the property of epicontingent derivatives and epiderivatives from the properties of contingent cones and tangent cones. Conversely, we can derive properties of contingent cones and tangent cones from those of epicontingent derivatives and epiderivatives, because we remark that when x_0 belongs to a subset K,

(34) $$D_+\psi_K(x_0, 0)=\psi_{T_K(x_0)} \quad \text{and} \quad C_+\psi_K(x_0, 0)=\psi_{C_K(x_0)}$$

where ψ_L denotes the indicator of the subset L. We also mention the following useful properties.

PROPOSITION 15

a. *If $p \in X^*$ satisfies $\langle p, x_0\rangle = \max_{y \in K} \langle p, y\rangle$, then p belongs to the normal cone $N_K(x_0)$.*

b. *Assume now that X is a Hilbert space. If $y \notin \bar{K}$ and $x \in \pi_{\bar{K}}(y)$ is a projection of y to $\bar{K}$, then $y-x$ belongs to the normal cone $N_K(x)$.* ▲

Proof. **a.** If $p \in X^*$ satisfies $\langle p, x_0\rangle = \max_{y \in K}\langle p, y\rangle$, then $x_0 \in K$ minimizes on K the linear functional $x \to \langle p, x\rangle$, and thus $0 \in \partial(-p|_K)(x_0) \subset -p+N_K(x_0)$ by propositions 4 and 5.

b. Since the function $V: x \to \|y-x\| =: V(x)$ is continuously differentiable at all $x \neq y$ and $x \in \pi_{\bar{K}}(y)$ minimizes V on K, we deduce that

$$0 \in \partial(V|_K)(x) \subset \nabla V(x) + N_K(x) = \frac{x-y}{\|x-y\|} + N_K(x)$$

Hence, $y-x \in N_K(x)$. ■

PROPOSITION 16

Let V be a proper upper semicontinuous function from X to $R \cup \{+\infty\}$. We set

(35) $$K := \{x \in X \mid V(x) \leqslant c\}$$

Let $x_0 \in K$ satisfy $V(x_0)=c$. Then

(36) $$T_K(x) \subset \{v \in X \mid D_+V(x)(v) \leqslant 0\}$$

If we assume that

(37) $$\exists u_0 \in X \quad \text{such that } C_+V(x_0)(u_0) < 0$$

then inclusion

(38) $$\{u \in X \mid C_+V(x_0)(u) \leqslant 0\} \subset C_K(x_0)$$

holds true. ▲

Proof. We first check that if u_0 satisfies $C_+V(x_0)(u_0) < 0$, then u_0 belongs to $C_K(x_0)$. Let us set $v_0 := -C_+V(x_0)(u_0) > 0$. For all $\varepsilon \in]0, v_0[$, there exist $\alpha > 0$ and $\beta > 0$ such that for all $x \in x_0 + \alpha B$, $h \in]0, \beta[$, there exists $u \in u_0 + \varepsilon B$ such that $V(x+hu) \leqslant V(x) + h(-v_0 + \varepsilon)$. Hence, for all $x \in B_K(x_0, \alpha)$, $h \in]0, \beta[$, there exists $u \in u_0 + \varepsilon B$ such that $V(x+hu) \leqslant V(x_0)$, that is, $x + hu \in K$. Therefore, u_0 belongs to $C_K(x_0)$.

Now, if u satisfies $C_+V(x_0)(u) \leqslant 0$, then for all $\lambda \in]0, 1[$, $u_\lambda := (1-\lambda)u + \lambda u_0$ satisfies $C_+V(x_0)(u_\lambda) < 0$ by convexity, and thus $u_\lambda \in C_K(u_0)$. Hence, we deduce that u belongs to $C_K(u_0)$ by letting λ converge to zero. ■

4. GENERALIZED SECOND DERIVATIVES OF REAL-VALUED FUNCTIONS

Let V be a proper function from X to $R \cup \{+\infty\}$. We consider the set-valued map ∂V from X to X^* associating to each $x_0 \in X$ the generalized gradient of V at x_0.

Therefore, if (x_0, p_0) belongs to the graph of ∂V, the derivative $C(\partial V)(x_0, p_0)$ of ∂V at (x_0, p_0) plays the role of a second derivative of V.

DEFINITION 1

We shall say that the derivative

(1) $$\partial^2 V(x_0, p_0) := C(\partial V)(x_0, p_0)$$

of the map ∂V at $(x_0, p_0) \in$ graph (∂V) is the generalized second derivative of V at (x_0, p_0).

Therefore, $\partial^2 V(x_0, p_0)$ is a closed convex process from X to X^*. ▲

It is clear that when V is twice continuously differentiable at x_0, then $p_0 = \nabla V(x_0)$ and $\partial^2 V(x_0, p_0)$ coincides with the Hessian $\nabla^2 V(x_0)$, mapping X to X^*.

PROPOSITION 2

Let V be a proper lower semicontinuous, convex function from X to $R \cup \{+\infty\}$ and V^* its conjugate function. Then $\partial^2 V(x_0, p_0)$ is a monotone closed convex process and

(2) $$\partial^2 V^*(p_0, x_0) = (\partial^2 V(x_0, p_0))^{-1}$$ ▲

Proof. Let $(u^i, q^i)(i=1, 2)$ be two pairs of the graph of $\partial^2 V(x_0, p_0)$. Let h_n converge to 0^+. Then we know that there exist sequences of elements u_n^i and v_n^i converging to u^i and v^i such that $(x_0+h_n u_n^i, p_0+h_n q_n^i)$ belong to the graph of ∂V for $i=1, 2$. Since the graph of ∂V is monotone, we deduce that

$$h_n^2\langle q_n^1-q_n^2, u_n^1-u_n^2\rangle=\langle p_0+h_n q_n^1-(p_0+h_n q_n^2), x_0+h_n u_n^1-(x_0+h_n u_n^2)\rangle\geqslant 0$$

Hence, $\partial^2 V(x_0, p_0)$ is monotone.

Inequality (2) is straightforward, since ∂V^* is the inverse of ∂V.

5. THE INVERSE FUNCTION THEOREM FOR SET-VALUED MAPS

We denote by ρB and ρB° the closed and open balls of radius ρ, respectively. We set

$$\mathrm{d}(A, B):=\sup_{x\in A}\inf_{y\in B}\|x-y\|$$

Note that $\mathrm{d}(A, B)=0$ means that A is contained in the closure of B.

We shall extend the usual inverse function theorem for continuously differentiable single-valued maps to the case of set-valued maps. We need the following definition.

DEFINITION 1

Let F be a proper set-valued map from X to Y and let (x_0, y_0) belong to the graph of F. We say that F is pseudo Lipschitz around (x_0, y_0) if there exist a neighborhood W of x_0, two neighborhoods U and V of y_0, $U\subset V$, and a constant $\ell>0$ such that

$$(1)\quad \begin{cases}\text{i.} & \forall x\in W, \quad F(x)\cap U\neq\varnothing.\\ \text{ii.} & \forall x_1, x_2\in W, \quad \mathrm{d}(F(x_1)\cap U, F(x_2)\cap V)\leqslant\ell\|x_1-x_2\|.\end{cases}$$

▲

We note that if F is single-valued on W, it is pseudo Lipschitz if and only if it is Lipschitz.

THEOREM 2

Let F be a proper set-valued map with closed graph from X to Y and let (x_0, y_0) belong to graph (F). We assume that

(2)

i. *Both X and Y are finite dimensional.*

ii. *The derivative $CF(x_0, y_0)$ of F at (x_0, y_0) is surjective (i.e., $\operatorname{Im} CF(x_0, y_0)=Y$).*

Then F^{-1} is pseudo Lipschitz around (y_0, x_0). ▲

We start with the following lemma.

LEMMA 3

Let us assume that the spaces X and Y are finite dimensional. Let (x_0, y_0) belong to the graph of F. We assume that

$$\text{(3)} \qquad \textit{the derivative } CF(x_0, y_0) \textit{ maps } X \textit{ onto } Y.$$

Then, for all $\alpha>0$, there exist constants $c>0$ and $\eta>0$ such that for all $(x, y) \in \text{graph}(F)$ satisfying

$$\|x-x_0\|+\|y-y_0\|\leqslant\eta$$

and for all $v \in Y$, there exist $u \in X$ and $w \in Y$ satisfying

$$\text{(4)} \qquad v \in DF(x, y)(u)+w, \ \|u\|\leqslant c\|v\|, \quad \textit{and} \quad \|w\|\leqslant\alpha\|v\| \qquad \blacktriangle$$

Proof. Since $CF(x_0, y_0)$ is a closed convex process, proposition 3.3.8 with $x_0=0$ and $y_0=0$ implies the existence of $\gamma>0$ such that

$$\text{(5)} \qquad \gamma B\subset CF(x_0, y_0)(B)$$

Let us introduce the subset

$$\text{(6)} \qquad \mathscr{K}:=(B\times\gamma B)\cap \text{graph } CF(x_0, y_0)$$

Since the spaces X and Y are finite dimensional, the subset $\mathscr{K}$ is a compact subset of the tangent cone $C_{\mathscr{F}}(x_0, y_0)$ to $\mathscr{F}$, the graph of F, at (x_0, y_0), which is the lim inf of the contingent cones $T_{\mathscr{F}}(x, y)$ to the graph of F at points (x, y) converging to (x_0, y_0). Hence, we can associate to every $\alpha>0$ a positive number η such that for all $(u_0, v_0) \in \mathscr{K}$, $(x, y) \in B_{\mathscr{F}}(x_0, y_0, \eta)$, we have $(u_0, v_0) \in T_{\mathscr{F}}(x, y) +\alpha(B\times B)$.

Now, take v in Y. Then $v_0:=\gamma v/\|v\|$ belongs to γB and by (5), there exists $u_0 \in B$ such that (u_0, v_0) belongs to $\mathscr{K}$. Then, for all $(x, y) \in B_{\mathscr{F}}(x_0, y_0; \eta)$, there exist $u_\alpha \in \alpha B$ and $v_\alpha \in \alpha B$ such that $(u_0-u_\alpha, v_0-v_\alpha) \in T_{\mathscr{F}}(x, y)$, that is, such that

$$v_0 \in DF(x, y)(u_0-u_\alpha)+v_\alpha$$

We set $u:=(\|v\|/\gamma)(u_0-u_\alpha)$ and $w=(\|v\|/\gamma)v_\alpha$. Then

$$v \in DF(x, y)(u)+w, \qquad \|u\|\leqslant\frac{1+\alpha}{\gamma}\|v\|, \qquad \text{and} \qquad \|w\|\leqslant\frac{\alpha}{\gamma}\|v\| \qquad \blacksquare$$

Theorem 2 then becomes a consequence of the following general inverse function theorem, valid in all Banach spaces.

THEOREM 4

Let F be a proper set-valued map from a Banach space X to a Banach space Y with closed graph. Let $(x_0, y_0) \in \mathrm{graph}(F)$ be fixed. We assume that there exist constants $\alpha \in [0, 1[$, $\eta > 0$ and $c > 0$ such that for all $(x, y) \in \mathrm{graph}(F)$ satisfying $\|x - x_0\| + \|y - y_0\| \leqslant \eta$, for all $v \in Y$, there exist $u \in X$ and $w \in Y$ such that

$$\text{(7)} \quad \begin{cases} \textbf{i.} & v \in DF(x, y)(u) + w. \\ \textbf{ii.} & \|u\| \leqslant c\|v\| \quad \textit{and} \quad \|w\| \leqslant \alpha\|v\|. \end{cases}$$

Let us set

$$\text{(8)} \quad \begin{cases} r := \dfrac{\eta(1-\alpha)}{3(1+\alpha+c)}, \quad F_0^{-1}(y) := F^{-1}(y) \cap \left(x_0 + \dfrac{c+2\alpha}{1-\alpha} rB\right) \\ \textit{and} \quad F_1^{-1}(y) := F^{-1}(y) \cap \left(x_0 + \dfrac{3(c+2\alpha)}{1-\alpha} rB\right) \end{cases}$$

Then F^{-1} is pseudo Lipschitz around (x_0, y_0). Namely,

$$\text{(9)} \quad \begin{cases} \textbf{i.} & \forall y \in y_0 + rB, \quad F_0^{-1}(y) \neq \varnothing. \\ \textbf{ii.} & \forall y_1, v_2 \in y_0 + r\mathring{B}, \quad d(F_0^{-1}(y_1), F_1^{-1}(y_2)) \leqslant \dfrac{c+2a}{1-\alpha} \|y_1 - y_2\| \end{cases}$$

▲

Proof. Let y_1 and y_2 belong to the open ball $y_0 + r\mathring{B}$. Assume for the time being that there exists x_1 satisfying

$$\text{(10)} \quad x_1 \in F_0^{-1}(y_1) := F^{-1}(y_1) \cap (x_0 + \ell r B), \quad \text{where } \ell := \frac{c+2\alpha}{1-\alpha}$$

(This is possible when we take $y_1 = y_0$ and $x_1 = x_0$!) We associate with any $\rho \in]\|y_1 - y_2\|, 2r[$ the number

$$\varepsilon := \frac{\|y_1 - y_2\|}{\|y_1 - y_2\| + \ell\rho}$$

that satisfies

$$\text{(11)} \quad \frac{3\|y_1 - y_2\|}{2\eta} \leqslant \varepsilon < \frac{1-\alpha}{1+c+\alpha}$$

We apply theorem 1 of Chapter 5, Section 3, to the continuous function V defined on the graph of F by $V(x, y) := \|y_2 - y\|$.

Since it is complete, there exists $(\bar{x}, \bar{y}) \in \text{graph}(F)$ such that

$$(12)\quad \begin{cases} \text{i.} & \|\bar{y}-y_2\|+\varepsilon(\|\bar{x}-x_1\|+\|\bar{y}-y_1\|)\leqslant\|y_1-y_2\|. \\ \text{ii.} & \forall (x, y) \in \text{graph}(F), \qquad \|\bar{y}-y_2\|\leqslant\|y-y_2\|+\varepsilon(\|x-\bar{x}\|+\|y-\bar{y}\|). \end{cases}$$

Inequality (12, **i**) implies that

$$\|\bar{x}-x_1\|+\|\bar{y}-y_1\|\leqslant\frac{1}{\varepsilon}\|y_1-y_2\|\leqslant\frac{2\eta}{3}$$

Therefore,

$$\begin{aligned} \|\bar{x}-x_0\|+\|\bar{y}-y_0\| &\leqslant\frac{2\eta}{3}+\|x_0-x_1\|+\|y_0-y_1\| \\ &\leqslant\frac{2\eta}{3}+\left(\frac{c+2\alpha}{1-\alpha}+1\right)r=\frac{2\eta}{3}+\frac{1+\alpha+c}{1-\alpha}r=\frac{2\eta}{3}+\frac{\eta}{3}=\eta \end{aligned}$$

Consequently, we can use property (7) with $v:=y_2-\bar{y}$. There exist u and w satisfying

$$(13)\quad \begin{cases} \text{i.} & y_2-\bar{y} \in DF(\bar{x}, \bar{y})(u)+w \\ \text{ii.} & \|u\|\leqslant c\|y_2-\bar{y}\| \qquad \text{and} \qquad \|w\|\leqslant\alpha\|y_2-\bar{y}\| \end{cases}$$

By the very definition of the contingent derivative $DF(\bar{x}, \bar{y})$, we can associate to any $\delta>0$ elements $h \in]0, \delta]$, $u_\delta \in \delta B$, and $v_\delta \in \delta B$ such that the pair (x, y) defined by

$$x=\bar{x}+hu+hu_\delta, \qquad y=\bar{y}+h(y_2-\bar{y})-hw-hv_\delta$$

belongs to the graph of F. Using this pair in inequality (12, **ii**), we obtain

$$\begin{aligned} \|y_2-\bar{y}\|\leqslant(1-h)\|y_2-\bar{y}\|+h\|w\|+h\varepsilon(\|u\|+\|y_2-\bar{y}\|+\|w\|) \\ +h((1+\varepsilon)\|v_\delta\|+\varepsilon\|u_\delta\|) \end{aligned}$$

We divide this inequality by $h>0$ and let δ converge to zero. We get

$$\|y_2-\bar{y}\|\leqslant(\varepsilon(c+1)+\alpha(1+\varepsilon))\|y_2-\bar{y}\|$$

Since $\varepsilon<(1-\alpha)/(c+1-\alpha)$, we infer that $y_2=\bar{y}$ and thus that $\bar{x}$ is a solution to the inclusion $y_2 \in F(\bar{x})$. By setting $y_2=\bar{y}$ in inequality (12, **i**), we obtain

$$\|\bar{x}-x_1\|\leqslant\left(\frac{1}{\varepsilon}-1\right)\|y_1-y_2\|=\ell\rho\leqslant2\ell r$$

Therefore, $\bar{x}$ belongs to $F^{-1}(y_2)\cap(x_1+2\ell rB)\subset F_1^{-1}(y_2)$, and thus

$$d(x_1, F_1^{-1}(y_2))\leq\|\bar{x}-x_1\|\leq\left(\frac{1}{\varepsilon}-1\right)\|y_1-y_2\|=\ell\rho$$

By letting ρ converge to $\|y_1-y_2\|$, we deduce that

$$d(x_1, F_1^{-1}(y_2))\leq\ell\|y_1-y_2\| \tag{14}$$

We can always take $(x_1, y_1):=(x_0, y_0)$. We thus have proved

$$\forall y_2\in y_0+r\mathring{B},\qquad x_2\in F_0^{-1}(y_2):=F^{-1}(y_2)\cap\left(x_0+\frac{c+2\alpha}{1-\alpha}rB\right) \tag{15}$$

(because $\|y_2-y_0\|<r$ instead of $2r$).

In other words, the set-valued map F_0^{-1} has nonempty images when y ranges over the open ball $y_0+r\mathring{B}$. Inequality (14) implies that

$$\mathrm{d}(F_0^{-1}(y_1),F_1^{-1}(y_1)):=\sup_{x_1\in F_0^{-1}(y_1)} d(x_1, F_1^{-1}(y_2))\leq\frac{c+2\alpha}{1-\alpha}\|y_1-y_2\| \qquad \blacksquare$$

As a first consequence, we obtain the usual Liusternik theorem.

COROLLARY 5

Let f be a continuously differentiable map from an open subset Ω of a Banach space X to a Banach space Y. Assume that for $x_0\in\Omega$,

$$\nabla f(x_0) \text{ is surjective.} \tag{16}$$

Then there exist neighborhoods U and V of x_0, $U\subset V$, and W of $f(x_0)$ such that for all $y\in W$, there exists a solution $x\in U$ to the equation $f(x)=y$. Furthermore,

$$\forall y_1, y_2\in W,\qquad \mathrm{d}(f^{-1}(y_1)\cap U, f^{-1}(y_2)\cap V)\leq\ell\|y_1-y_2\| \tag{17}$$ ▲

Proof. Let K be a closed neighborhood of x_0 contained in Ω. We apply theorem 4 to the restriction F of f to K. Since $\nabla f(x_0)$ is surjective, there exists a constant $c>0$ such that for all $v\in Y$, there exists a solution u of the equation $\nabla f(x_0)u=v$ satisfying $\|u\|\leq c\|v\|$. Let $\alpha>0$ be given and $\eta>0$ such that $\|\nabla f(x)-\nabla f(x_0)\|\leq\alpha$ when $x\in B(x_0,\eta)\subset\operatorname{Int}K$. Then the assumptions of theorem 4 are satisfied, because $v=\nabla f(x)u+w$ where $\|u\|\leq c\|v\|$ and

$$w:=(\nabla f(x)-\nabla f(x_0))v\qquad\text{is such that}\qquad\|w\|\leq\alpha\|v\| \qquad \blacksquare$$

By taking $\eta=\infty$, we obtain the following corollary.

COROLLARY 6 (NORMAL SOLVABILITY)

Let F be a proper closed set-valued map from a Banach space X to a Banach space Y. We assume that there exists a constant $c>0$ such that for all $(x, y) \in \text{graph}(F)$, for all $v \in Y$, there exists $u \in X$ satisfying $v \in DF(x, y)(u)$ and $\|u\| \leqslant c\|y\|$. Then F maps X onto Y, and its inverse F^{-1} is a Lipschitz set-valued map with Lipschitz constant equal to c. ▲

Let us also mention the following consequence of the proof of theorem 4.

COROLLARY 7

Let F be a proper closed set-valued map from a Banach space X to a Banach space Y. Assume that there exists a constant $c>0$ such that

$$(18) \qquad \begin{cases} \forall (x, y) \in \text{graph}(F), \quad \exists u \in X \quad \textit{satisfying} \\ -y \in DF(x, y)(u) \quad \textit{and} \quad \|u\| \leqslant c\|y\| \end{cases}$$

Then the set $F^{-1}(0)$ of zeros of F is nonempty and

$$(19) \qquad \forall x \in \text{Dom}(F), \quad d(x, F^{-1}(0)) \leqslant c d(0, F(x))$$ ▲

Remark

When $F=f|_K$ is the restriction to a closed subset K of a continuously differentiable single-valued map f, assumption (18) becomes

$$(20) \quad \forall x \in K, \quad \exists u \in T_K(x) \quad \text{such that} \quad -f(x)=\nabla f(x)u \quad \text{and} \quad \|u\| \leqslant c\|f(x)\|$$

We then deduce that there exists a solution $\bar{x} \in K$ to the equation $f(\bar{x})=0$ and

$$(21) \qquad \forall x \in K, \quad d(x, f^{-1}(0)) \leqslant c\|x\|$$

In the book by Aubin and Cellina [1983], it is shown that assumption (20) implies the existence of a trajectory of the implicit differential equation

$$(22) \qquad \begin{cases} \nabla f(x(t))x'(t) = -f(x(t)) \\ x(0)=x_0 \quad \text{given in } K \end{cases}$$

satisfying

$$(23) \qquad \forall t \geqslant 0, \quad x(t) \ \text{belongs to } K$$

and

$$(24) \qquad d(x(t), f^{-1}(0)) \leqslant e^{-ct}\|x(0)\|$$

We observe that the differential equation (22) is the continuous version of the Newton method and inequality (24) implies the convergence of the Newton method. ■

Remark
When the graph of F is compact (i.e., when the domain of F is compact and F is upper semicontinuous with compact values), we need only to assume that

$$(25) \qquad \forall (x, y) \in \text{graph}(F), \qquad \exists u \in X \quad \text{satisfying} \quad -y \in DF(x, y)(u)$$

for deducing that $F^{-1}(0)$ is nonempty.

Indeed, we minimize on the graph of F the function $(x, y) \to \|y\|$ and denote by $(\bar{x}, \bar{y}) \in \text{graph}(F)$ a minimizer. We proceed as in the proof of theorem 4 with $\varepsilon = 0$. ■

The use of set-valued maps abolishes the formal distinction between the inverse function theorem and the implicit function theorem.

Let X, Y, and Z be three Banach spaces and let G be a set-valued map from $X \times Y$ to Z. The implicit function theorem deals with the behavior of the map that associates to any $(y, z) \in Y \times Z$ the set of solutions x to the inclusion $z \in G(x, y)$. This amounts to studying the inverse of the set-valued map F from X to $Y \times Z$ defined by

$$(26) \qquad (y, z) \in F(x) \Leftrightarrow z \in G(x, y)$$

Since the graphs of the set-valued maps F and G coincide as subsets of $X \times Y \times Z$, there are close relations between the derivatives of F and G at (x_0, y_0, z_0), because the graphs of these two derivatives coincide with the tangent cone to the graph of G at (x_0, y_0, z_0). Then we can state the implicit function theorem.

THEOREM 8

Let G be a proper set-valued map with closed graph from $X \times Y$ to Z and let (x_0, y_0, z_0) belong to the graph of G. We assume that

$$(27) \quad \begin{cases} \textbf{i.} & \textit{Both } X, Y, \textit{ and } Z \textit{ are finite dimensional.} \\ \textbf{ii.} & \forall v, \quad w \in Y \times Z, \quad \exists u \in X \quad \textit{such that } w \in CG(x_0, y_0, z_0)(u, v). \end{cases}$$

Then

$$(28) \qquad F^{-1} \textit{ is pseudo Lipschitz around } (x_0, (y_0, z_0)). \qquad \blacktriangle$$

In the case when G is a continuously differentiable function, we obtain the following useful corollary.

COROLLARY 9

Let g be a C^1 function from an open neighborhood of (x_0, y_0) in $X \times Y$ to Z satisfying

(29)
$$\nabla_y g(x_0, y_0) \text{ is surjective from } Y \text{ to } Z$$

Then there exist neighborhoods V_0 and V_1 of y_0: $V_0 \subset V_1$, neighborhoods U of x_0 and W of z_0 and a constant $c > 0$ such that

(30)
$$\forall x \in U, \quad \forall z \in W, \quad \exists y \in V_0 \quad \text{such that} \quad g(x, y) = z$$

and, if we set $F^{-1}(x, z) := \{y \in Y | g(x, y) = z\}$,

(31)
$$\forall x_1, x_2 \in V, \quad z_1, z_2 \in W,$$
$$d(F^{-1}(x_1, z_1) \cap V_0, F^{-1}(x_2, z_2) \cap V_1) \leqslant \ell(\|x_1 - x_2\| + \|z_1 - z_2\|)$$ ▲

Proof. It is analogous to the proof of Corollary 5 and follows from Theorem 4 applied to the set-valued map F from Y to $X \times Z$ defined by

(32)
$$F(y) := \begin{cases} \{(x, z) \in K \times Z | g(x, y) = z\} & \text{when} \quad y \in L \\ \varnothing \quad \text{when} \quad y \in L \end{cases}$$

where K and L are closed neighborhoods of z_0 and y_0 on which g is C^1. The graph of F is closed and assumption (7) is satisfied: let $(u, w) \in X \times Z$ be chosen and define $v \in Y$ as a solution to the equation

(33)
$$\nabla_y g(x_0, y_0) v = w - \nabla_x g(x_0, y_0) u$$

satisfying

(34)
$$\|v\| \leqslant c \|w - \nabla_x g(x_0, y_0) u\|$$

thanks to the Banach open mapping principle. We set $\hat{w} := \nabla g(x, y)(u, v) - \nabla g(x_0, y_0)(u, v)$. We see that

(35)
$$(u, w) \in DF(y, (x, z))(v) + (0, \hat{w})$$

with

(36)
$$\|v\| \leqslant c \max(1, \|\nabla_x g(x_0, y_0)\|)(\|u\| + \|w\|)$$

and, α being given,

(37)
$$\|(0, \hat{w})\| \leqslant \|\nabla g(x, y) - \nabla g(x_0, y_0)\|(\|u\| + \|w\|) \leqslant \alpha(\|u\| + \|w\|)$$

provided that (x, y) remains in a small neighborhood of (x_0, y_0). Hence theorem 4 implies that F^{-1} is pseudo-Lipschitz around $(x_0, (y_0, z_0))$, which is what the conclusion of corollary 9 states. ■

This having been said, it is not always obvious how to obtain "nice" formulas for the derivatives. For instance, let X and Y be Banach spaces, A a continuous linear operator from X to Y, and $G: X \to X^*$ and $H: Y \to Y^*$ set-valued maps.

We consider the set of solutions $x \in X$ to the inclusion

$$p \in G(x) + A^*H(Ax + y) \tag{38}$$

where p is given in X^*. The first idea is to apply the inverse function theorem to the set-valued map E from X to $X^* \times Y$ defined by

$$E(x) = \{p, y) | p \in G(x) + A^*H(Ax + y)\} \tag{39}$$

Unfortunately, there is no nice expression for the tangent cone to the graph of E in $X \times X^* \times Y$. But we can introduce an auxiliary variable $q \in Y^*$ and write inclusion (38) as the equivalent inclusion

$$\begin{cases} \textbf{i.} & p \in G(x) + A^*q. \\ \textbf{ii.} & y \in -Ax + H^{-1}(q). \end{cases} \tag{40}$$

The set of solutions (x, q) to this problem is denoted by $F^{-1}(p, y, A)$, where F is the set-valued map from $X \times Y^*$ to $X^* \times Y \times \mathcal{L}(X, Y)$ defined by

$$(p, y, A) \in F(x, q) \quad \text{if and only if (40) holds} \tag{41}$$

We shall characterize the derivative of F in terms of the derivatives of the set-valued maps G and H(or H^{-1}), respectively.

LEMMA 10

Let x_0, q_0 be a solution to the system of inclusions

$$\begin{cases} \textbf{i.} & p_0 \in G(x_0) + A_0^*q_0. \\ \textbf{ii.} & y_0 \in -Ax_0 + H^{-1}(q_0). \end{cases} \tag{42}$$

The following conditions are equivalent:

$$(\delta p, \delta y, \delta A) \in CF(x_0, q_0; p_0, y_0, A_0)(\delta x, \delta q) \tag{43}$$

$$\begin{cases} \textbf{i.} & \delta p - \delta A^* \cdot q_0 \in CG(x_0, p_0 - A_0^*q_0)(\delta x) + A_0^*\delta q. \\ \textbf{ii.} & \delta y + \delta A \cdot x_0 \in -A_0\delta x + CH^{-1}(q_0, y_0 + Ax_0)(\delta q). \end{cases} \tag{44}$$

▲

Proof. **a.** We prove that (43) implies (44). We choose sequences $(x_n, q_n, p'_n, y'_n, h_n)$ converging to $(x_0, q_0, p_0 - A^*q_0, y_0 + Ax_0, 0)$. By setting $A_n := A_0$, $p_n := p'_n + A_0^*q_n$, and $y_n := y'_n - A_0x_n$, we see that $(x_n, q_n, p_n, y_n, A_n, h_n)$ converges to $(x_0, q_0, p_0, y_0, A_0, 0)$. Therefore by (43), there exist sequences of elements δx_n, δq_n, δp_n, δy_n, and δA_n converging to δx, δq, δp, δy, and δA such that

$$\textbf{i.}\quad p'_n + h_n(\delta p_n - A_0^*\delta q_n - \delta A_n^* q_n + h_n \delta A_n^* \delta q_n) \in G(x_n + h_n \delta x_n).$$
$$\textbf{ii.}\quad y'_n + h_n(\delta y_n + A_0\delta x_n + \delta A_n \cdot x_n + h_n \delta A_n \delta x_n) \in H^{-1}(q_n + h_n \delta q_n).$$

Hence, the system of inclusions (44) holds true.

b. Conversely, let us consider the system of inclusions (44) and prove (43). We choose sequences $(x_n, q_n, y_n, p_n, A_n, h_n)$ converging to $(x_0, q_0, y_0, p_0, A_0, 0)$. Then we know that there exist sequences of elements δx_n, δq_n, u_n, and v_n converging to δx, δq, $\delta p - \delta A^* q_0 - A_0^*\delta q$, and $\delta y + \delta A x_0 + A_0 \delta x$, respectively. We set

$$(45)\quad \begin{cases} \textbf{i.} & \delta A_n := \delta A, \quad \text{which converges to } \delta A. \\ \textbf{ii.} & \delta p_n := u_n + \delta A^* \cdot q_n + A_n^* \delta q_n + h_n \delta A^* \delta q_n, \quad \text{which converges to } \delta p. \\ \textbf{iii.} & \delta y_n := v_n - \delta A x_n - A_n \delta x_n - h_n \delta A x_n, \quad \text{which converges to } \delta x. \end{cases}$$

Hence,

$$(p_n + h_n \delta p_n, y_n + h_n \delta y_n, A_n + h_n \delta A) \in F(x_n + h_n \delta x_n, q_n + h \delta q_n)$$

and, consequently, inclusion (43) holds true. ■

COROLLARY 11

Let X and Y be finite dimensional and let $G: X \to X^*$ and $H: Y \to Y^*$ be set-valued with a closed graph. Let (p_0, y_0, A_0) belong to $X^* \times Y \times \mathscr{L}(X, Y)$. Assume that there exists a solution (x_0, q_0) to the system of inclusions

$$(46)\quad \begin{cases} \textbf{i.} & p_0 \in G(x_0) + A_0^* q_0 \\ \textbf{ii.} & y_0 \in -A_0 x_0 + H^{-1}(q_0) \end{cases}$$

If the matrix of closed convex processes

$$\begin{pmatrix} CG(x_0, p_0 - A_0^* q_0) & A_0^* \\ -A_0 & CH(A_0 x_0 + y_0, q_0)^{-1} \end{pmatrix}$$

is surjective, then there exist neighborhoods U and V of (x_0, q_0), $U \subset V$ and W of (p_0, y_0, A_0) such that the set-valued map

$$(47)\quad (p, y, A) \in W \to F^{-1}(p, y, A) \cap U$$

has nonempty values and is pseudo Lipschitz. Furthermore, the derivative of F^{-1}

is given by

$$(48)\quad \begin{pmatrix}\partial x\\ \delta q\end{pmatrix} \in \begin{pmatrix} CG(x_0, p_0 - A_0^* q_0) & A_0^* \\ -A_0 & CH(Ax_0 + y_0, q_0)^{-1}\end{pmatrix}^{-1} \begin{pmatrix}\delta p - \delta A^* \cdot q_0 \\ \delta y + \delta A \cdot x_0\end{pmatrix} \quad \blacktriangle$$

6. CALCULUS OF CONTINGENT AND TANGENT CONES, DERIVATIVES, AND EPIDERIVATIVES

The applications of nonsmooth analysis to nonlinear analysis to which we have devoted the preceding section motivate the development of a calculus of contingent and tangent cones, contingent derivatives and derivatives of set-valued maps, and epicontingent derivatives and epiderivatives of real-valued functions. In the tables at the end of this book, we summarize this calculus, adding the formulas of convex analysis for the sake of comparison.

PROPOSITION 1

a. *Let $K \subset L \subset X$ be two nonempty subsets. Then*

$$(1)\quad \forall x_0 \in K, \qquad T_K(x_0) \subset T_L(x_0)$$

b. *Let $K := \bigcup_{i \in I} K_i$ be the union of subsets K_i. Then*

$$(2)\quad \forall x_0 \in K, \qquad \bigcup_{i \in I} T_{K_i}(x_0) \subset T_K(x_0)$$

If $I = \{1, \ldots, n\}$ is finite, equality holds true.

c. *Let $K := \bigcap_{i \in I} K_i$ be the intersection of subsets K_i. Then*

$$(3)\quad \forall x_0 \in K, \qquad T_K(x_0) \subset \bigcap_{i \in I} T_{K_i}(x_0)$$

d. *Let $K := \prod_{i \in I} K_i$ be a finite product of subsets K_i. Then*

$$(4)\quad \forall x_0 := (x_0^i)_{i \in I} \in K, \qquad T_K(x_0) \subset \prod_{i=1}^{n} T_{K_i}(x_0^i) \quad \text{and} \quad C_K(x_0) = \prod_{i \in I} C_{K_i}(x_0^i)$$

$\blacktriangle$

PROPOSITION 2

Let X and Y be Banach spaces, A a C^1 map from an open subset Ω of X to Y, and $K \subset \Omega$ a subset of X.

Then

$$(5)\quad \forall x_0 \in K, \qquad \nabla A(x_0) T_K(x_0) \subset T_{A(K)}(A(x_0)) \quad \blacktriangle$$

Proof. Let v_0 belong to $T_K(x_0)$; there exist sequences $h_n \to 0^+$ and $v_n \to v_0$ such that $x_0 + h_n v_n$ belongs to K for all n. Then the sequence of elements $u_n := (A(x_0 + h_n v_n) - A(x_0))/h_n$ converges to $A(x_0)v_0$ and $A(x_0) + h_n u_n$ belongs to $A(K)$ for all n. Hence, $A(x_0)v_0$ belongs to $T_{A(K)}(Ax_0)$. ■

In particular, if $A \in \mathscr{L}(X, Y)$, we obtain the formula

$$\forall x \in K, \quad A T_K(x) \subset T_{A(K)}(A(x)) \tag{6}$$

We now study the contingent cone to the preimage of a set by a smooth map.

PROPOSITION 3

a. *Let X and Y be two Banach spaces, $L \subset X$ and $M \subset Y$ two subsets, and A a C^1 map from an open neighborhood of L to Y. We set*

$$K := \{x \in L \mid A(x) \in M\} = L \cap A^{-1}(M) \tag{7}$$

Then

$$\forall x \in K, \quad T_K(x) \subset T_L(x) \cap \nabla A(x)^{-1} T_M(A(x)) \tag{8}$$

b. *Let X and Y be finite dimensional spaces, A a C^1 map from an open subset $\Omega \subset X$ to Y, and $L \subset \Omega$ and $M \subset Y$ closed subsets of X and Y, respectively.*
We assume that there exists $x_0 \in L \cap A^{-1}(M)$ such that

$$\nabla A(x_0) C_L(x_0) - C_M(Ax_0) = Y \tag{9}$$

Then

$$C_L(x_0) \cap \nabla A(x_0)^{-1} C_M(Ax_0) \subset C_K(x_0) \tag{10}$$ ▲

Proof. **a.** By proposition 1, $T_K(x) \subset T_L(x)$, because $K \subset L$. By proposition 2

$$\nabla A(x) T_K(x) \subset T_{A(K)}(Ax) \subset T_M(Ax)$$

because $A(K) \subset M$. Hence, $T_K(x) \subset \nabla A(x)^{-1} T_M(Ax)$, and, consequently, formula (8) holds true.
b. We introduce the set-valued map F from X to Y defined by

$$F(x) := A(x) - M \quad \text{when} \quad x \in L, \qquad F(x) := \varnothing \quad \text{when} \quad x \notin L \tag{11}$$

We observe that $F^{-1}(0) = K$. We shall prove that there exists a neighborhood U_0 of x_0 in L such that

$$\forall x \in U_0, \quad d(x, F^{-1}(0)) \leqslant \ell\, d_M(Ax) \tag{12}$$

Indeed, we take $y_0=0$ and $x_0 \in F^{-1}(0)$. The inverse function theorem implies that F^{-1} is pseudo Lipschitz around $(0, x_0)$. Then there exist a neighborhood U of x_0, a ball of radius r in Y, and a constant $\ell>0$ such that

$$\forall y \in r\mathring{B}, \qquad \forall x \in F^{-1}(y) \cap U, \qquad d(x, F^{-1}(0)) \leqslant \ell \|y\|$$

We can choose U so small that $\|A(x)-A(x_0)\|<r$ when x ranges over U. Any $x \in L \cap U$ belongs to $F^{-1}(A(x)-\pi_M(A(x)))$, and

$$\|A(x)-\pi_M(A(x))\| \leqslant \|A(x)-A(x_0)\| < r$$

Therefore, we know that for all $x \in U_0 := L \cap U$,

$$d(x, F^{-1}(0)) \leqslant c\|A(x)-\pi_M(A(x))-0\| = d_M(A(x))$$

c. Let u_0 belong to $C_L(x_0) \cap \nabla A(x_0)^{-1} C_M(Ax_0)$. There exist $\alpha>0$ and $\beta>0$ such that $x+hu_0$ belongs to $U_0 := L \cap U$ when $\|x-x_0\| \leqslant \alpha$ and $h \leqslant \beta$. Since $F^{-1}(0) = L \cap A^{-1}(M)$, we deduce from (12)

$$\frac{d(x+hu_0, F^{-1}(0))}{h} \leqslant c\,\frac{d_M(A(x+hu_0))}{h}$$

$$\leqslant \frac{d_M(Ax+h\nabla A(x_0)u_0)}{h} + c\,\frac{\|A(x+hu_0)-A(x)-h\nabla A(x_0)u_0\|}{h}$$

The first term on the right-hand side converge to zero, because $\nabla A(x_0)u_0$ belongs to $C_M(Ax_0)$ and the second converges also to zero, because A is continuously differentiable. Hence, u_0 belongs to $C_{F^{-1}(0)}(u_0)$. ■

COROLLARY 4

Let X and Y be finite dimensional spaces, A a continuously differentiable map from X to Y, and M a closed subset of Y. Let Ax_0 belong to M. If

$$\text{Im } \nabla A(x_0) - C_M(Ax_0) = Y \tag{13}$$

then

$$\nabla A(x_0)^{-1} C_M(Ax_0) \subset C_{A^{-1}(M)}(x_0) \tag{14}$$

▲

COROLLARY 5

Let L and M be two nonempty closed subsets of a finite dimensional space X and let x_0 belong to $L \cap M$. If

$$C_L(x_0) - C_M(x_0) = X \tag{15}$$

then

$$C_L(x_0) \cap C_M(x_0) \subset C_{L \cap M}(x_0) \tag{16}$$ ▲

COROLLARY 6

Let $K_i (i=1, \ldots, n)$ be n nonempty closed subsets of X and let x_0 belong to $\bigcap_{i=1}^n K_i$. We posit the following assumption:

$$\forall v_1, \ldots, v_n \in X, \qquad \bigcap_{i=1}^{n} (C_{K_i}(x_0) - v_i) \neq \emptyset \tag{17}$$

Then

$$\bigcap_{i=1}^{n} C_{K_i}(x_0) \subset C_{\bigcap_{i=1}^n K_i}(x_0) \tag{18}$$ ▲

Proof. Let $\vec{D} \subset X^n$ denote the closed vector space of constant sequences $\vec{x} := (x, \ldots, x)$. Then K is identified with $\vec{K} := \vec{D} \cap \prod_{i=1}^n K_i$. We observe that $C_D(\vec{x}) = \vec{D}$ and $C_{\prod_{i=1}^n K_i}(\vec{x}) = \prod_{i=1}^n C_{K_i}(x)$. Assumption (17) implies that

$$C_D(\vec{x}) - C_{\prod_{i=1}^n K_i}(\vec{x}) = \vec{D} - \prod_{i=1}^{n} C_{K_i}(x) = X^n \tag{19}$$

Therefore, corollary 5 implies that

$$\vec{D} \cap \prod_{i=1}^{n} C_{K_i}(x) \subset C_{\prod_{i=1}^n K_i} \quad (\vec{x}) \tag{20}$$

that is, inclusion (18). ■

We shall derive a calculus of contingent derivatives and derivatives of set-valued maps from the properties of the contingent and tangent cones.

PROPOSITION 7

a. *Let F be a set-valued map from X to Y and let B be a C^1 map from an open neighborhood Ω of* Im $F \subset Y$ *to Z. Then*

$$\forall u_0 \in X, \qquad \nabla B(y_0) \cdot DF(x_0, y_0)(u_0) \subset D(BF)(x_0, By_0)(u_0) \tag{21}$$

If

(22) *F is Lipschitz around $x_0 \in$* Int Dom *F with compact values and* dim *$Y < +\infty$, then*

$$\forall u_0 \in X, \qquad \nabla B(y_0) \cdot DF(x_0, y_0)(u_0) = D(BF)(x_0, By_0)(u_0) \tag{23}$$

b. *Let F be a set-valued map from X to Y and let A be a C^1 map from X_0 to X. Then if Ax_0 belongs to* Dom F,

$$\forall u_0 \in X_0, \qquad D(FA)(x_0, y_0)(u_0) \subset DF(Ax_0, y_0)(\nabla A(x_0)(u_0)) \tag{24}$$

If we assume that either

$$F \text{ is Lipschitz around } Ax_0 \tag{25a}$$

or

$$\nabla A(x_0) \text{ is surjective and } \dim X_0 < +\infty \tag{25b}$$

then

$$\forall u_0 \in X_0, \qquad D(FA)(x_0, y_0)(u_0) = DF(Ax_0, y_0)(\nabla A(x_0)u_0) \tag{26}$$

▲

Proof. **a.** Let $(1 \times B)$ be the map

$$(x, y) \in X \times \Omega \to (x, B(y)) \in Y \times Z$$

The graph of the set-valued map $G := BF$ is related to the graph of F by the relation graph$(G) = (1 \times B)$graph(F). By proposition 2, we know that $(1 \times \nabla B(y_0))T_{\text{graph}(F)}(x_0, y_0)$ is contained in $T_{\text{graph}(G)}(x_0, By_0)$. This implies formula (21).

b. Let w_0 belong to $D(BF)(x_0, By_0)(u_0)$. There exist sequences $h_n \to 0$, $u_n \to u_0$, and $w_n \to w_0$ such that $h_n w_n$ belongs to $B(F(x_0 + h_n u_n)) - B(y_0)$ for all n. Hence, there exists $v_n \in (F(x_0 + h_n u_n) - y_0)/h_n$ such that $h_n w_n = B(y_0 + h_n v_n) - B(y_0)$. Since F is Lipschitz around x_0, v_n belongs to $F(x_0) - y_0 + h_n \|u_n\| B$, which is contained in a compact set, because the values of F are compact and the dimension of Y is finite. Hence, a subsequence (again denoted by) v_n converges to some v_0, which belongs to $DF(x_0, y_0)(u_0)$, and thus the sequence of elements w_n converges to $\nabla B(y_0) \cdot v_0 = w_0$. Therefore,

$$D(BF)(x_0, By_0)(u_0) \subset \nabla B(y_0) DF(x_0, y_0)(u_0)$$

c. Let $A \times 1 : X_0 \times Y \to X \times Y$ be the map defined by $(A \times 1)(x_0, y) = (Ax_0, y)$. The graph of the map $G = FA$ is related to the graph of F by the formula graph $G = (A \times 1)^{-1}$ graph F. By corollary 4, we know that

$$T_{\text{graph}(G)}(x_0, y_0) \subset (\nabla A(x_0) \times 1)^{-1} T_{\text{graph}(F)}(Ax_0, y_0)$$

which implies formula (24).

d. To prove (26), let us pick w_0 in $DF(Ax_0, y_0)(\nabla A(x_0)u_0)$. There exist sequences $h_n \to 0+$, $v_n \to \nabla A(x_0)u_0$ and $w_n \to w_0$ such that $h_n w_n$ belongs to

$F(A(x_0)+h_n v_n)-y_0$. Assume first that F is Lipschitz around x_0; we deduce that

$$F(A(x_0)+h_n v_n)) \subset F(A(x_0+h_n u_0))+\ell\|A(x_0+h_n u_0)-A(x_0)-h_n v_n\|B$$

Hence, there exists a sequence of elements w'_n converging to w_0 such that $h_n w'_n \in F(A(x_0+h_n u_0))-y_0)$ for all n. Thus w_0 belongs to $D(FA)(x_0, y_0)(u_0)$.

Assume now that $\nabla A(x_0)$ is surjective. Then corollary 5.5 implies the existence of a constant $\ell>0$ such that for n large enough, there exists a solution x_n to the equation $A(x_n)=A(x_0)+h_n v_n$ satisfying $\|x_n-x_0\|\leqslant \ell h_n\|v_n\|$. Since $\dim X_0<+\infty$, we deduce that a subsequence of elements $u_n:=(x_n-x_0)/h_n$ converges to some element u_0. Since

$$y_0+h_n w_n \in FA(x_0+h_n u_n) \qquad \text{for all } n$$

then $w_0 \in D(FA)(x_0, y_0)(u_0)$. ■

We now investigate the chain rule formulas.

PROPOSITION 8

Let X_0, X, and Y be three finite-dimensional spaces, F a set-valued map with closed graph from X to Y, and A a continuously differentiable map from X_0 to X. Let x_0 belong to A^{-1} Dom F and $y_0 \in F(x_0)$. We assume that

$$\text{(27)} \qquad \operatorname{Im} \nabla A(x_0)-\operatorname{Dom} CF(Ax_0, y_0)=X$$

Then

$$\text{(28)} \qquad \begin{cases} \text{i.} \quad \forall u_0 \in X_0, & C(FA)(x_0, y_0)(u_0) \supset CF(Ax_0, y_0)(\nabla A(x_0)u_0) \\ \text{ii.} \quad \forall q_0 \in Y^*, & C(FA)(x_0, y_0)^*(q_0) \subset \nabla A(x_0)^* CF(Ax_0, y_0)^*(q_0) \end{cases}$$ ▲

Proof. Since $\operatorname{graph}(FA)=(A\times 1)^{-1}\operatorname{graph}(F)$, we apply the second part of proposition 3, which states that $(\nabla A(x_0)\times 1)^{-1} C_{\operatorname{graph}(F)}(Ax_0, y_0)$ is contained in $C_{\operatorname{graph}(FA)}(x_0, y_0)$; that is formula (28), provided that property

$$\text{(29)} \qquad \operatorname{Im}(\nabla A(x_0)\times 1)-C_{\operatorname{graph}(F)}(Ax_0, y_0)=X\times Y$$

is satisfied. But this property follows from assumption (27).

Assumption (27) also implies that

$$(CF(Ax_0, y_0)\cdot \nabla A(x_0))^* = \nabla A(x_0)^* CF(Ax_0, y_0)^*$$

thanks to proposition 3.3.14. ■

PROPOSITION 9
Let F be a proper set-valued map from X to Y, K a subset of X, and let x_0 belong to $K \cap \text{Dom } F$. Then

$$D(F|_K)(x_0, y_0)(u_0) \subset DF(x_0, y_0)|_{T_K(x_0)}(u_0) \tag{30}$$

If X and Y are finite dimensional, the graph of F is closed, K is closed, and

$$C_K(x_0) - \text{Dom } CF(x_0, y_0) = X \tag{31}$$

then

$$CF(x_0, y_0)|_{C_K(x_0)}(u_0) \subset C(F|_K)(x_0, y_0)(u_0) \tag{32}$$

and for all $q_0 \in Y^*$,

$$C(F|_K)(x_0, y_0)^*(q_0) \subset CF(x_0, y_0)^*(q_0) + N_K(x_0) \tag{33}$$ ▲

Proof. We observe that $\text{graph}(F|_K) = \text{graph}(F) \cap (K \times Y)$. Then

$$T_{\text{graph}(F_K)}(x_0, y_0) \subset T_{\text{graph}(F)}(x_0, y_0) \cap (T_K(x_0) \times Y)$$

from which we deduce formula (30). We observe that assumption (31) implies that

$$C_{\text{graph}(F)}(x_0, y_0) - C_K(x_0) \times Y = X \times Y$$

Therefore, corollary 5 implies that

$$C_{\text{graph}(F)}(x_0, y_0) \cap (C_K(x_0) \times Y) \subset C_{\text{graph}(F|_K)}(x_0, y_0)$$

from which we deduce formula (32). Assumption (31) allows us to deduce formula (33) from formula (32). ■

PROPOSITION 10
a. *Let us consider n set-valued maps F_i from X to Y.*

$$\forall u_0 \in X, \qquad D\left(\bigcup_{i=1}^{n} F_i\right)(x_0, y_0)(u_0) = \bigcup_{i=1}^{n} DF_i(x_0, y_0)(u_0) \tag{34}$$

b. *Let us consider n set-valued maps F_i with closed graph from a finite dimensional space X to a finite dimensional space Y. Let (x_0, y_0) belong to the inter-*

section of the graphs of F_i. *Assume that*

$$
(35)\qquad \begin{cases} \forall (u_i, v_i) \in X \times Y (i=1,\ldots,n), \qquad \exists (u_0, v_0) \in X \times Y \\ \text{such that} \\ v_0 \in CF_i(x_0, y_0)(u_0 + u_i) - v_i \quad \text{for } i=1,\ldots,n \end{cases}
$$

Then

$$
(36)\qquad \forall u_0 \in X, \qquad C\left(\bigcap_{i=1}^{n} F_i\right)(x_0, y_0)(u_0) \supset \bigcap_{i=1}^{n} CF_i(x_0, y_0)(u_0) \qquad \blacktriangle
$$

Proof. We note that graph$(\cup F_i) = \cup$ graph(F_i) and graph$(\cap F_i) = \cap$ graph (F_i) and apply proposition 3 and corollary 6, respectively. ■

We deduce at once a calculus of epicontingent derivatives and epiderivatives of real-valued functions V from the calculus of contingent derivatives and derivatives of the set-valued map $\underset{\sim}{V}_+$ defined by $\underset{\sim}{V}_+(x) := V(x) + R_+$ when $V(x) < +\infty$ and $\underset{\sim}{V}_+(x) := \varnothing$ when $V(x) = \{+\infty\}$.

PROPOSITION 11

a. *Let V be a proper function from Y to $\mathbb{R} \cup \{+\infty\}$ and let A be a C^1 map from X to Y. If Ax_0 belongs to the domain of V, then*

$$
(37)\qquad \forall u_0 \in X, \qquad D_+ V(Ax_0)(\nabla A(x_0)u_0) \leqslant D_+(VA)(x_0)(u_0)
$$

b. *Let X and Y be finite-dimensional spaces, A a C^1 map from X to Y, and V: $Y \to \mathbb{R} \cup \{+\infty\}$ a proper lower semicontinuous function. Let x_0 belong to A^{-1}(Dom V). If we assume that*

$$
(38)\qquad \operatorname{Im} \nabla A(x_0) - \operatorname{Dom} C_+ V(Ax_0) = Y
$$

then

$$
(39)\qquad \forall u_0 \in X, \qquad C_+ V(Ax_0)(\nabla A(x_0)u_0) \geqslant C_+(VA)(x_0)(u_0)
$$

and

$$
(40)\qquad \partial(VA)(x_0) \subset \nabla A(x_0)^* \partial V(Ax_0) \qquad \blacktriangle
$$

We observe that assumption (38) is satisfied when either V is Lipschitz around $A(x_0)$, because in this case Dom $C_+ V(Ax_0) = Y$, or $\nabla A(x_0)$ is surjective.

Proposition 9 implies the following formula for epiderivatives of restrictions.

PROPOSITION 12
Let X be a finite-dimensional space, V a proper lower semicontinuous function from X to $R\cup\{+\infty\}$, and $K\subset X$ a closed subset. Let x_0 belong to $K\cap \mathrm{Dom}\, V$. We assume that

$$\text{(41)} \qquad \mathrm{Dom}\, C_+V(x_0)-C_K(x_0)=X$$

Then

$$\text{(42)} \qquad \forall u \in C_K(x_0), \qquad C_+(V|_K)(x_0)(u)\leqslant C_+V(x_0)(u)$$

and

$$\text{(43)} \qquad \partial(V|_K)(x_0)\subset\partial V(x_0)+N_K(x_0)$$

In particular, these formulas hold true when V is Lipschitz around x_0. ▲

PROPOSITION 13
Let V be a proper function from the product $X\times Y$ of two Banach spaces to $\mathbb{R}\cup\{+\infty\}$. We set

$$\text{(44)} \qquad W(y):=\inf_{x\in X} V(x, y)$$

If $x_y\in X$ minimizes $x\to V(x, y)$ on X, then

$$\text{(45)} \qquad \forall v\in Y, \qquad D_+W(y)(v)\leqslant \inf_{u\in X} D_+V(x_y, y)(u, v)$$ ▲

Proof. Let u and v belong to X and Y, respectively. Then the following inequalities hold true:

$$\frac{W(y+hv)-W(y)}{h}\leqslant\frac{V(x_y+hu, y+hv)-V(x_y, y)}{h}$$

This implies that for all (u_0, v_0) in $X\times Y$,

$$D_+W(y)(v_0)\leqslant D_+V(x_y, y)(u_0, v_0)$$

and, consequently, inequality (45). ■

Remark
When $v=0$, we obtain proposition 3.5 as a consequence. The analogous statement holds true for the supremum of a family of functions. Let

$$\text{(46)} \qquad U(y):=\sup_{x\in X} V(x, y)=V(\tilde{x}_y, y)$$

Then

$$(47) \qquad \forall v \in Y, \quad D_+U(y)(v) \geqslant \sup_{u \in X} D_+V(\tilde{x}_y, y)(u, v) \qquad ■$$

We now provide a formula on the epiderivative of a supremum of a finite number of functions.

PROPOSITION 14

Let us consider n proper lower semicontinuous functions from a finite dimensional space X to $\mathbb{R} \cup \{+\infty\}$. *We set* $U(x) := \max_{i=1,\dots,n} V_i(x)$. *We assume that*

$$(48) \qquad x_0 \text{ belongs to } \bigcap_{i=1}^{n} \operatorname{Int} \operatorname{Dom} V_i$$

Let us set $J(x_0) := \{i = 1, \dots, n \mid V_i(x_0) = U(x_0)\}$. *Assume also that*

$$(49) \qquad \begin{cases} \forall i \in J(x_0), \quad \forall u_i \in X, \quad \text{there exists } u_0 \\ \text{such that } u_0 - u_i \in \operatorname{Dom} C_+V_i(x_0) \quad \text{for all } i \in J(x_0) \end{cases}$$

then

$$(50) \qquad \forall u_0 \in X, \quad C_+U(x_0)(u_0) \leqslant \max_{i \in J(x_0)} C_+V_i(x_0)(u_0)$$

and

$$(51) \qquad \partial U(x_0) \subset \overline{\text{co}} \bigcup_{i \in J(x_0)} \partial V_i(x_0) \qquad ▲$$

Proof. Assumption (48) implies that when $i \notin J(x_0)$, then $(x_0, U(x_0))$ belongs to the interior of $Ep V_i$, so that the tangent cone $C_{Ep(V_i)}(x_0, U(x_0))$ is equal to $X \times R$. Assumption (49) implies that

$$(52) \qquad \begin{cases} \forall (u_i, \lambda_i) \in X \times R (i = 1, \dots, n), \quad \text{there exists} \\ (u_0, \lambda_0) \quad \text{such that} \quad (u_0 - u_i, \lambda_0 - \lambda_i) \\ \text{belongs to } C_{Ep(V_i)}(x_0, U(x_0)) \quad \text{for all } i = 1, \dots, n \end{cases}$$

Indeed, u_0 is given by property (49), and we take

$$\lambda_0 := \max_{i \in J(x_0)} (C_+V_i(x_0)(u_0 - u_i) + \lambda_i)$$

Therefore, we are allowed to use corollary 6, since $Ep\, U = \bigcap_{i=0}^{n} Ep(V_i)$: Then

$$\bigcap_{i \in J(x_0)} C_{Ep(V_i)}(x_0, U(x_0)) = \bigcap_{i=1}^{n} C_{Ep(V_i)}(x_0, U(x_0)) \subset C_{EpU}(x_0, U(x_0))$$

This inclusion implies inequality (50), from which we deduce inclusion (51), because the support function of a union is the supremum of the support functions. ■

We mention the following corollary.

COROLLARY 15
Assume that the n proper lower semicontinuous functions are Lipschitz around x_0. Then

$$\forall u_0 \in X, \qquad C_+U(x_0)(u_0) \leqslant \max_{i \in J(x_0)} C_+V_i(x_0)(u_0) \tag{53}$$

and

$$\partial U(x_0) \subset \overline{\mathrm{co}} \bigcup_{i \in J(x_0)} \partial V_i(x_0) \tag{54}$$ ▲

We now turn our attention to the behavior of epicontingent derivatives and epiderivatives of the sum of two functions. This time, we need the concept of strict epidifferentiability.

PROPOSITION 16
Let V and W be proper functions from X to $\mathbb{R} \cup \{+\infty\}$. Then

$$D_+V(x_0)(u_0) + D_+W(x_0)(u_0) \leqslant D_+(V+W)(x_0)(u_0) \tag{55}$$

Assume that W is strictly epidifferentiable at x_0 and

$$\mathrm{Dom}\, C_+V(x_0) \cap \mathrm{Dom}\, B_+W(x_0) \neq \varnothing \tag{56}$$

Then

$$C_+(V+W)(x_0)(u_0) \leqslant C_+V(x_0)(u_0) + C_+W(x_0)(u_0) \tag{57}$$

and

$$\partial(V+W)(x_0) \subset \partial V(x_0) + \partial W(x_0) \tag{58}$$ ▲

Remark
Assumption (56) is satisfied when W is Lipschitz around x_0.

Proof. The formula for epicontingent derivatives is obvious. We observe that for all u_0 belonging to $\mathrm{Dom}\, C_+V(x_0) \cap \mathrm{Dom}\, B_+W(x_0)$, we have

$$C_+(V+W)(x_0)(u_0) \leqslant C_+V(x_0)(u_0) + B_+W(x_0)(u_0) \tag{59}$$

Now, let $u \in \text{Dom } C_+V(x_0) \cap \text{Dom } C_+W(x_0)$ and $\lambda \in]0, 1[$ be fixed. Since $\text{Dom } B_+W(x_0) = \text{Int Dom } C_+W(x_0)$, we deduce that $(1-\lambda)u + \lambda u_0$ belongs to $\text{Dom } C_+V(x_0) \cap \text{Dom } B_+W(x_0)$, so that formula (59) implies that

$$C_+(V+W)(x_0)((1-\lambda)u+\lambda u_0)$$
$$\leqslant (1-\lambda)C_+V(x_0)(u_0) + \lambda C_+V(x_0)(u_0) + B_+W(x_0)((1-\lambda)u+\lambda u_0)$$

By formula (3-27), we have

$$B_+W(x_0)((1-\lambda)u+\lambda u_0) \leqslant (1-\lambda)C_+W(x_0)(u) + \lambda B_+W(x_0)(u_0)$$

By letting λ converge to 0, we deduce formula (57), which implies formula (58), because the support function of a sum is the sum of support functions. ■

Remark

Observe that assumption (56) implies that

$$\text{(60)} \qquad \text{Dom } C_+V(x_0) - \text{Dom } C_+W(x_0) = X$$ ■

In the same way, without using the inverse function theorem, we can prove that the assumption

$$\text{(61)} \qquad \text{Dom } C_+V(x_0) \cap \nabla A(x_0)^{-1} \text{ Dom } B_+W(x_0) \neq \varnothing$$

implies that

$$\text{(62)} \qquad C_+(V+WA)(x_0)(u_0) \leqslant C_+V(x_0)(u_0) + C_+W(Ax_0)(\nabla A(x_0)u_0)$$

and that

$$\text{(63)} \qquad \partial(V+WA)(x_0) \subset \partial V(x_0) + \nabla A(x_0)^* \partial W(Ax_0)$$

without assuming that the spaces are finite dimensional. ■

CHAPTER 8

Hamiltonian Systems

We now apply the abstract methods of earlier chapters to a specific example of great practical importance: finding periodic solutions to certain differential equations.

The differential equations we are investigating are of Hamiltonian type, that is, they can be written

$$(1)\qquad \begin{aligned} \frac{dq_i}{dt} &= \frac{\partial H}{\partial p_i}(t, q, p) \\ & & i = 1, \ldots, n \\ \frac{dp_i}{dt} &= -\frac{\partial H}{\partial q_i}(t, q, p) \end{aligned}$$

for a suitable choice of the variables $(q, p) \in \mathbb{R}^{2n}$ and the function $H: [0, T] \times \mathbb{R}^{2n} \to \mathbb{R}$.

In the particular case where the equations are autonomous, that is, H is a function of q and p only, the physical or mechanical systems they represent are conservative

$$(2)\qquad \frac{d}{dt} H(q(t), p(t)) = 0$$

The quantity $H(q, p)$, which is preserved throughout the motion, is usually understood as the energy of the system. Since it does not dissipate, the system cannot come to rest, and its motion can be very complicated as $t \to \pm\infty$. This is why it is so interesting to find periodic solutions: They are the only ones whose behavior can be completely described.

Of course, this problem has been studied for a very long time, and giving a survey of the field in anything shorter than a full-length book is out of the question. What we do in this last chapter is to present three selected results, chosen to illustrate the abstract methods described in earlier chapters, particularly Chapter 5.

It may come as a surprise that abstract variational methods are helpful in solving problems in ordinary differential equations. This is the case here because Hamiltonian systems have a special, variational, structure: It has been known since the time of Fermat that the trajectories of equation (1) minimize a certain path integral. This is known as the least action principle, actually a misnomer, since nowadays we know that the trajectories of equation (1) may correspond to critical points other than minima. A precise formulation is given in Section 1.

It turns out, however, that the least action principle is not the appropriate tool for our purpose: The associated functional on the path space is too complicated. For instance, in the cases we consider, it will be unbounded from above and from below. So we introduce another path integral that is much better behaved than the least action principle but yields the same critical paths. This new variational principle, which is in some sense dual to the least action principle, is explained in Section 2. In later sections, we apply the abstract results of Chapters 2 and 5 to obtain existence theorems for periodic solutions. We use the inverse function theorem in Section 3 and the Ambrosetti–Rabinowitz theorem in Section 5. In Section 4, minimization is enough.

1. THE LEAST ACTION PRINCIPLE

We start from a C^2 function $H: \mathbb{R} \times \mathbb{R}^{2n} \to \mathbb{R}$, referred to as the *Hamiltonian.* The integer n is called the *number of degrees of freedom.* It is assumed that the Hamiltonian is time periodic

(1) $$\exists T > 0: H(t+T, x) = H(t, x) \qquad \forall (t, x)$$

We are given a linear operator $J: \mathbb{R}^{2n} \to \mathbb{R}^{2n}$ such that

(2) $$-J = J^* = J^{-1}$$

Note that $J^2 + I = 0$. We can always pick a basis of $\mathbb{R}^{2n}$ where the matrix of J will be

(3) $$J = \begin{pmatrix} 0 & 1 \\ -1 & 0 \end{pmatrix}$$

where 0, 1, and -1 denote $n \times n$ diagonal matrices with 0, 1, and -1, respectively on the diagonal. In such a basis, the antisymmetric bilinear form (Jx, y) is written

(4) $$(Jx, y) = \sum_{i=1}^{n} (y_i x_{i+n} - x_i y_{i+n})$$

We are interested in the boundary-value problem

$$(5)\qquad \begin{cases} \dfrac{dx}{dt} = JH'_x(t, x) \\ x(0) = x(T) \end{cases}$$

Clearly, since H itself is T periodic in time, any solution x to (5) will be extended as a T-periodic solution over $\mathbb{R}$.

Let us first dispel any notion the reader might entertain that just because the right-hand side is T-periodic, the equation $\dot{x} = JH'_x(t, x)$ will always have a T-periodic solution.

LEMMA 1
Set

$$H(t, x) = \sum_{i=1}^{n} \frac{\omega_i}{2}(x_i^2 + x_{i+n}^2) + \sum_{i=1}^{n} f_i(t) x_{i+n}$$

with $\omega_i \in \mathbb{R}$ *and* f_i $(2\pi/\omega)$ *periodic,* $1 \leqslant i \leqslant n$. *If* $\omega \in \mathbb{Z}\omega_k$ *for some* k, *and* $\int_0^T \dot{f}_k(t) \exp(i\omega_k t) dt \neq 0$, *then problem* (5) *has no solution. In all other cases, it has at least one solution.* ▲

Proof. Let us write the equations

$$\begin{aligned} &\textbf{i.} \quad \dot{x}_i = \omega_i x_{i+1} + f_i(t) \\ &\textbf{ii.} \quad \dot{x}_{i+n} = -\omega_i x_i \end{aligned} \qquad 1 \leqslant i \leqslant n$$

which are equivalent to

$$\begin{aligned} &\textbf{i.} \quad \ddot{x}_i + \omega_i^2 x_i = \dot{f}_i(t) \\ &\textbf{ii.} \quad \dot{x}_{i+n} = -\omega_i x_i \end{aligned} \qquad 1 \leqslant i \leqslant n$$

In other words, we are dealing with n uncoupled harmonic oscillators. The result then follows from the standard theory of second-order equations with constant coefficients. If $\omega \notin \mathbb{Z}\omega_i$ for all i, then there is precisely one $2\pi/\omega$-periodic solution (the so-called particular solution). If $\omega \in \mathbb{Z}\omega_k$, we have to expand the right-hand side $\dot{f}_i$ in Fourier series. If the coefficient of $\exp(i\omega_k t)$ does not vanish, we are in the so-called resonant case, and we know that all solutions must contain the nonperiodic term $t \exp(i\omega_k t)$. Hence, the result. ■

In the case of *autonomous* systems, that is, when the Hamiltonian H does not depend on time, there is a further complication. The differential equation

$$\frac{dx}{dt} = JH'(x)$$

may have constant solutions. To be precise, any point $x_0 \in \mathbb{R}^{2n}$ where $H'(x_0)=0$ is called an *equilibrium*, and $x(t) \equiv x_0$ is a solution of the equation. Such constant solutions are T periodic for all T. The question then is whether there are other kinds of periodic solutions.

A classical method for approaching these problems is by means of the least action principle. It is generally attributed to Maupertuis, who was the first to formulate it in modern language, but mathematicians in the seventeenth century, particularly Fermat, were well aware of it.

We shall state the least action principle in the language of the calculus of variations. Recall that an *extremal* of the integral

$$\int L(t, x(t), \dot{x}(t))dt \tag{6}$$

is a solution $\bar{x}$ of the corresponding Euler–Lagrange equation

$$\frac{d}{dt}\frac{\partial L}{\partial \dot{x}_i}=\frac{\partial L}{\partial x_i}, \quad 1 \leqslant i \leqslant 2n \tag{7}$$

PROPOSITION 2 (LEAST ACTION PRINCIPLE)
The solutions of Hamilton's equation

$$\frac{dx}{dt}=JH'_x(t, x) \tag{8}$$

are precisely the extremals of the integral

$$\int [\tfrac{1}{2}(J\dot{x}, x)+H(t, x)]dt \tag{9}$$

▲

Proof. Write the Euler–Lagrange equation

$$\frac{1}{2}\frac{d}{dt} J^*x=\frac{1}{2} J\dot{x}+H'_x(t, x)$$

Since $J^*=J^{-1}=-J$, this is precisely Hamilton's equation ■

$\bar{x}$ is an extremal if the first variation of the integral at $\bar{x}$, among all smooth curves satisfying appropriate boundary conditions, is zero. In other words, $\bar{x}$ is a critical point of the integral on a appropriate subspace of C^1. Note that $\bar{x}$ need not minimize the integral, and in most situations in physics, it does not. It would be more appropriate to speak of a "stationary action" principle.

The least action principle must be tailored to suit the specific problem we are working with. This means taking into account the boundary conditions and choosing the function space we want to work with, preferably a Hilbert space or a reflexive Banach space.

In the case at hand, problem (5), we shall work in the Sobolev space $W^{1,\alpha}(0, T: \mathbb{R}^{2n})$, with $1<\alpha<\infty$. Recall that

$$W^{1,\alpha}(0, T; \mathbb{R}^{2n})=\left\{x \in L^\alpha \,\middle|\, \frac{dx}{dt} \in L^\alpha\right\} \tag{10}$$

Denote by $W^{1,\alpha}_{\text{per}}$ the subspace consisting of T-periodic curves

$$\begin{aligned} W^{1,\alpha}_{\text{per}}(0, T; \mathbb{R}^{2n}) &= \{x \in W^{1,\alpha} | x(0)=x(T)\} \\ &= \left\{x \in W^{1,\alpha} \,\middle|\, \int_0^T \frac{dx}{dt}\, dt=0\right\} \end{aligned} \tag{11}$$

The action functional on $W^{1,\alpha}_{\text{per}}$, given by integral (9), is the sum of two terms

$$\Phi_1(x)=\frac{1}{2}\int_0^T (J\dot{x}, x)dt \tag{12}$$

$$\Phi_2(x)=\int_0^T H(t, x(t))dt \tag{13}$$

The first term, $\Phi_1(x)$, is well defined, because all functions in $W^{1,\alpha}$ are continuous. It is easy to see that the canonical injection $W^{1,\alpha}\to C^0(0, T; \mathbb{R}^{2n})$ is continuous, so Φ_1 is a continuous quadratic form on $W^{1,\alpha}_{\text{per}}$ and hence, indefinitely differentiable.

Recall that the Hamiltonian $H(t, x)$ is assumed at this stage to be C^2. The map

$$x \to \int_0^T H(t, x(t))dt$$

from C^0 to $\mathbb{R}$ then is C^2 also. Composing it on the left with the canonical injection $W^{1,\alpha}\to C^0$, we see that Φ_2 is a C^2 map. The Fréchet derivative at $x \in W^{1,\alpha}$ is given by (see Chapter 1)

$$\Phi_2'(x)y=\int_0^T (H_x'(t, x(t)), y(t))dt$$

We now state the least action principle in this setting

PROPOSITION 3

Define the action functional Φ on $W^{1,\alpha}_{\text{per}}$ by

$$\Phi(x) = \int_0^T [\tfrac{1}{2}(J\dot{x}, x) + H(t, x)]dt$$

Then Φ is C^2, and $\Phi'(x)=0$ if and only if x solves problem (5). ▲

Proof. Let us write $\Phi'(x)=0$. For all $y \in W_{\text{per}}^{1,\alpha}$, we have

$$\int_0^T [\tfrac{1}{2}(J\dot{y}, x) + \tfrac{1}{2}(J\dot{x}, y) + (H'_x(t, x), y)]dt = 0$$

Integrating the second term by parts and taking into account the fact that x and y are T periodic, we obtain

$$\int_0^T (y, H'_x(t, x) + J\dot{x})dt = 0, \qquad \text{for all } y \in W_{\text{per}}^{1,\alpha}$$

Now $H'_x(t, x(t))$ is a continuous function, and $J\dot{x}$ belongs to L^α, so their sum is in L^α. On the other hand, $W_{\text{per}}^{1,\alpha}$ is dense in the dual L^β of L^α, so we have

$$\int_0^T (y, H'_x(t, x) + J\dot{x})dt = 0, \qquad \text{for all } y \in L^\beta, \qquad \alpha^{-1} + \beta^{-1} = 1$$

and hence,

$$H'_x(t, x) + J\dot{x} = 0 \quad \text{in} \quad L^\alpha$$

So $\dot{x}(t) = JH'_x(t, x(t))$ almost everywhere. Since H'_x is C^1 in the (t, x) variables, this implies that x is a classical solution to the differential equation $\dot{x} = JH'_x(t, x)$. Since $x \in W_{\text{per}}^{1,\alpha}$, we also have $x(0) = x(T)$. ■

2. A DUAL ACTION PRINCIPLE

This section is concerned with nonconvex duality. We first prove abstract results and then apply them to the special case of Hamiltonian problems.

We begin with a very simple setting. Let X be a reflexive Banach space, Q a continuous quadratic form on X, and $F: X \to \mathbb{R} \cup \{+\infty\}$ a convex l.s.c. function. We are interested in the function $\Phi: X \to \mathbb{R} \cup \{+\infty\}$ defined by

(1) $$\Phi = Q + F$$

It is essential to note that unless Q is positive, an assumption we do not make, Φ is not a convex function. We shall say that $\bar{v}$ is a *critical point* of Φ if

$$Q'(\bar{v}) + \partial F(\bar{v}) \ni 0$$

If F happens to be Gâteaux differentiable at $\bar{v}$, then so is Φ, and the definition becomes $\Phi'(\bar{v})=0$. In the general case, $Q'(\bar{v})+\partial F(\bar{v})$ is clearly the generalized gradient of the function Φ at $\bar{v}$ (see Chapter 7, Section 3), and the definition can be written $0 \in \partial\Phi(\bar{v})$. It follows from Chapter 7, Section 6 that if $\bar{v}$ minimizes Φ locally, then $\bar{v}$ is a critical point. It is also true that all local maximizers are critical points.

It is well known that there is a unique self-transposed operator $A\colon X \to X^*$, given by

$$\langle Au, v\rangle = [Q(u+v)-Q(u)-Q(v)] \tag{2}$$

$$Q(v)=\tfrac{1}{2}\langle Av, v\rangle \tag{3}$$

Thus we have

$$A=A^* \tag{4}$$

$$Q'(v)=Av \in X^* \tag{5}$$

We now turn to our duality result.

THEOREM 1

Consider the two functionals Φ and Ψ on V

$$\Phi(v)=\tfrac{1}{2}\langle Av, v\rangle + F(v) \tag{6}$$

$$\Psi(v)=\tfrac{1}{2}\langle Av, v\rangle + F^*(-Av) \tag{7}$$

If $\bar{v}$ is a critical point of Φ, it is also a critical point of Ψ. If $\bar{v}$ is a critical point of Ψ and if

$$0 \in \operatorname{Int}(A(X)+\operatorname{Dom} F^*) \tag{8}$$

then there is some $\bar{w} \in \operatorname{Ker} A$ such that $\bar{u}=\bar{v}-\bar{w}$ is a critical point of Φ. ▲

Proof. Let $\bar{v}$ be a critical point of Φ. By definition

$$0 \in A\bar{v}+\partial F(\bar{v})$$

This can be written

$$-A\bar{v} \in \partial F(\bar{v})$$

Using the Legendre reciprocity formula (theorem 4.4.4), we obtain

$$\bar{v} \in \partial F^*(-A\bar{v})$$

Applying A to both sides,

$$A\bar{v} \in A\partial F^*(-A\bar{v})$$

Set $G(v)=F^*(-Av)$. We have

$$A\partial F^*(-A\bar{v}) \subset -\partial G(\bar{v})$$

Hence, $A\bar{v}+\partial G(\bar{v}) \ni 0$, and $\bar{v}$ is a critical point of Ψ.

Let $\bar{v}$ be a critical point of Ψ. Condition (8) implies that (see corollary 4.3.6)

$$A\partial F^*(-A\bar{v}) = -\partial G(\bar{v})$$

so that the equation $A\bar{v}+\partial G(\bar{v}) \ni 0$ can be written

$$A[\bar{v}-\partial F^*(-A\bar{v})] \ni 0$$

This means that there is some $\bar{w} \in V$ such that

$$\bar{v}-\partial F^*(-A\bar{v}) \ni \bar{w} \qquad \text{and} \qquad A\bar{w}=0$$

So $\bar{w} \in \operatorname{Ker} A$, and $\bar{u}=\bar{v}-\bar{w}$ satisfies

$$\bar{u} \in \partial F^*(-A\bar{v})$$

Using the Legendre reciprocity formula (theorem 4.4.4)

$$-A\bar{v} \in \partial F(\bar{u})$$

But $A\bar{v}=A(\bar{v}-\bar{w})=A\bar{u}$ since $A\bar{w}=0$. Hence,

$$A\bar{u}+\partial F(\bar{u}) \ni 0$$

and $\bar{u}$ is a critical point of Φ. ∎

We shall now apply this result to the least action principle for Hamiltonian systems. We first recast the least action principle to fit this framework.

Set $X:=L^{\alpha}(0, T; \mathbb{R}^{2n})$, with $1<\alpha<\infty$. Its dual X^* is $L^{\beta}(0, T; \mathbb{R}^{2n})$, with $\alpha^{-1}+\beta^{-1}=1$. We introduce the closed subspace

$$L_0^{\alpha}=\left\{y \in L^{\alpha}(0, T; \mathbb{R}^{2n}) \middle| \int_0^T y(t)dt=0\right\} \tag{9}$$

The condition on y can be rewritten $\langle z, y\rangle=0$ for all constant functions z.

The constant functions z form a subspace of L^β that we identify with $\mathbb{R}^{2n}$. So,

$$L_0^\alpha = (\mathbb{R}^{2n})^\perp \qquad \text{and} \qquad \mathbb{R}^{2n} = (L_0^\alpha)^\perp \tag{10}$$

For each $y \in L_0^\alpha$, define Πy to be the primitive of y with zero mean

$$\frac{d}{dt}\Pi y = y \qquad \text{and} \qquad \int_0^T \Pi y(t)\,dt = 0 \tag{11}$$

The map

$$x \to \left(\dot{x}, \frac{1}{T}\int_0^T x(t)\,dt\right) \tag{12}$$

is an isomorphism from $W_{\text{per}}^{1,\alpha}$ onto $L_0^\alpha \times \mathbb{R}^{2n}$. Its inverse is the map

$$(y, \xi) \mapsto \Pi y + \xi \tag{13}$$

Replacing $W_{\text{per}}^{1,\alpha}$ by $L_0^\alpha \times \mathbb{R}^{2n}$, we reformulate the action integral as follows:

$$\int_0^T [\tfrac{1}{2}(Jy, \Pi y + \xi) + H(t, \Pi y + \xi)]\,dt$$

$$= \int_0^T [\tfrac{1}{2}(Jy, \Pi y) + H(t, \Pi y + \xi)]\,dt \tag{14}$$

since $y \in (\mathbb{R}^{2n})^\perp$, which leads to proposition 2.

PROPOSITION 2

Assume that $H(t, x)$ is a continuous function of (t, x), convex with respect to x for every fixed t and

$$\phi(|x|) \leqslant H(t, x) \leqslant \psi(|x|) \qquad \textit{for all } (t, x) \tag{15}$$

with $s^{-1}\phi(s) \to +\infty$ and $s^{-\beta}\psi(s)$ bounded when $s \to \infty$.

Then (y, ξ) is a critical point of the functional

$$\Phi(y, \xi) = \frac{1}{2}\int_0^T (Jy, \Pi y)\,dt + \int_0^T H(t, \Pi y + \xi)\,dt \tag{16}$$

on $L_0^\alpha \times \mathbb{R}^{2n}$ if and only if $x = \Pi y + \xi$ solves the boundary value problem

$$\begin{cases} \dot{x} \in J\partial_x H(t, x(t)) & \text{a.e.} \\ x(0) = x(T) \end{cases} \tag{17}$$

▲

Proof. First note that $\Phi = \Phi_1 + \Phi_2$, with Φ_1 quadratic and Φ_2 convex. We have

$$\Phi_1(y, \xi) = \frac{1}{2}\int_0^T (Jy, \Pi y)dt = \frac{1}{2}\int_0^T (-J\Pi y, y)dt$$

On account of the boundary condition, it is easily seen that $J: L_0^\alpha \to L^\beta$ is self-transposed. So $(-J\Pi, 0)$ is the self-transposed operator associated with Φ_1 on $L_0^\alpha \times \mathbb{R}^{2n}$.

On the other hand, the integral

$$I(x) = \int_0^T H(t, x(t))dt \tag{18}$$

defines a convex and finite function on L^β. The growth assumption on H implies that it is bounded from below. Using Fatou's lemma, as in Chapter 1, we see that I is lower semicontinuous. Since it is finite everywhere, it is also continuous and, hence, subdifferentiable everywhere. We have

$$\partial I(x) = \{z \in L^\alpha | z(t) \in \partial_x H(t, x(t)) \quad \text{a.e.}\} \tag{19}$$

The proof of this formula is rather technical and uses the measurable selection theorem (see Ekeland–Temam [1972] lemma 10.4.1).

We are interested in the function $\Phi_2 = I \circ A$ on $L_0^\alpha \times \mathbb{R}^{2n}$, with

$$A(y, \xi) = \Pi y + \xi$$

Here, A sends $L_0^\alpha \times \mathbb{R}^{2n}$ into L^β, so A^* will send L^α into the dual of $L_0^\alpha \times \mathbb{R}^{2n}$, which is $L_0^\beta \times \mathbb{R}^{2n}$. An integration by parts gives

$$A^*(z) = (-\Pi z, \int_0^T z(t)dt)$$

where Πz denotes the primitive of $z - 1/T \int_0^T z(t)dt$ with mean zero.

Since I is continuous, we have

$$\partial\Phi_2(y, \xi) = A^* I(A(y, \xi))$$

$$= \left\{(-\Pi z, \int_0^T z(t)dt) | z \in \partial I(\Pi y + \xi)\right\}$$

Let us now write that (y, ξ) is a critical point of Φ

$$\Phi_1'(y, \xi) + \partial\Phi_2(y, \xi) \ni 0$$

This means that there is some $z \in L^\alpha$ such that

$$z(t) \in \partial_x H(t, \Pi y(t) + \xi)$$

$$-J\Pi y - \Pi z = 0 \quad \text{in} \quad L^\beta$$

$$0 + \int_0^T z(t)dt = 0 \quad \text{in} \quad \mathbb{R}^{2n}$$

Differentiating the second equation gives $Jy + z = 0$, and substituting it into the first, we obtain

$$y(t) \in J\partial_x H(t, \Pi y(t) + \xi)$$

$$\int_0^T y(t)dt = J \int_0^T z(t)dt = 0$$

Setting $x(t) = \Pi y(t) + \xi$, so that $\dot{x} = y$, we obtain precisely problem (17). ■

We now wish to recast this equation into the form of theorem 1. The first step consists in eliminating ξ from the functional $\Phi(y, \xi)$, thereby obtaining a function $\bar{\Phi}(y)$ defined on L_0^α only.

COROLLARY 3

Take $H(t, x)$ as in the preceding equation, and consider the functionals

(20) $$\bar{\Phi}(y) = \frac{1}{2} \int_0^T (Jy, \Pi y)dt + G(\Pi y), \qquad y \in L_0^\alpha$$

(21) $$G(x) = \min_{\xi \in \mathbb{R}^{2n}} \int_0^T H(t, x(t) + \xi)dt, \qquad x \in L^\beta$$

Then G is a convex, continuous function on L^β, and y is a critical point of $\bar{\Phi}$ on L_0^α if and only if there is some $\xi \in \mathbb{R}^{2n}$ such that $x = \Pi y + \xi$ solves problem (17). ▲

Proof. Let $G(x) := \min\{I(x + \xi), \quad \xi \in \mathbb{R}^{2n}\}$.

First, we observe that the infimum in $G(x)$ is achieved. This follows from the fact that the function

$$\xi \to \int_0^T H(t, x(t) + \xi)dt = I(x + \xi)$$

is continuous on $\mathbb{R}^n$ and goes to $+\infty$ when $|\xi| \to \infty$.

We compute $\partial G(x)$. Set $V(x, \xi) = I(x+\xi)$. Since

$$G(x) = \min_{\xi \in \mathbb{R}^{2n}} V(x, \xi) = V(x, \bar{\xi})$$

then proposition 4.6.1 implies that $q \in \partial G(x)$ if and only if $(q, 0)$ belongs to $\partial V(x, \bar{\xi})$. Subdifferential calculus shows that

$$\partial V(x, \bar{\xi}) = \left\{ y, \int_0^T y(t)dt \right\}_{y \in \partial I(x+\xi)}$$

So,

$$q \in \partial G(x) \Leftrightarrow q \in \partial I(x+\bar{\xi}) \text{ and } \int_0^T q(t)dt = 0$$

It follows that y is a critical point of $\bar{\Phi}$ if and only if

$$0 \in J\Pi y + \Pi \partial G(\Pi y) \qquad \text{since} \qquad \Pi^* = -\Pi$$

This means that there is some $\xi \in \mathbb{R}^{2n}$ and $q \in L^\alpha$ such that

$$0 \in J\Pi y + \Pi q$$

$$G(\Pi y) = I(\Pi y + \xi)$$

$$\int_0^T q(t)dt = 0$$

$$q \in \partial I(\Pi y + \xi)$$

In other words, $q \in L_0^\alpha$, $J\Pi y = -\Pi q$, and by formula (19),

$$q(t) \in \partial_x H(t, \Pi y(t) + \xi) \qquad \text{a.e.}$$

Setting $x(t) = \Pi y(t) + \xi$, this can be written

$$\dot{x} \in J\partial_x H(t, x(t)) \qquad \text{a.e.}$$

and $x(0) = x(T)$. This is the desired result. ∎

We now apply theorem 1 to obtain a dual version of the least action principle in Hamiltonian mechanics.

THEOREM 4

Assume that $H(t, x)$ *is a continuous function of* (t, x), *convex with respect to* x *for every fixed* t, *and*

$$\phi(\|x\|) \leq H(t, x) \leq \psi(\|x\|) \qquad \text{for all } (t, x) \tag{22}$$

with $s^{-1}\phi(x) \to +\infty$ *and* $s^{-\beta}\psi(s)$ *bounded when* $s \to \infty$. *Define a functional* Ψ *on* L_0^α *by*

$$\Psi(y) := \int_0^T \left[\tfrac{1}{2}(Jy, \Pi y) + H^*(t, -Jy)\right] dt \tag{23}$$

where $H^*(t, \cdot)$ *is the conjugate of* $H(t, \cdot)$ *with respect to* x.

Then y *is a critical point of* Ψ *if and only if there is some* $\xi \in \mathbb{R}^{2n}$ *such that* $x = \Pi y + \xi$ *solves the boundary value problem*

$$\begin{cases} \dot{x} = JH'_x(t, x) \\ x(0) = x(T) \end{cases} \tag{24}$$

▲

Proof. Apply theorem 1 to Ψ, with $X = L_0^\alpha$, $A = -J\Pi$, and

$$F(y) = \int_0^T H^*(t, -Jy)\, dt \tag{25}$$

Let us compute F^*. We first extend the function F to a function $\bar{F}$ defined on the whole of L^α by the same formula

$$\bar{F}(y) = \int_0^T H^*(t, -Jy)\, dt \tag{26}$$

So F is the restriction of $\bar{F}$ to L_0^α. Since L_0^α is the kernel of the map $\theta: y \to \int_0^T y(t)\,dt$, we can write, using the indicator function of the set $\{0\}$:

$$F(y) = \bar{F}(y) + \psi_{\{0\}}(\theta y)$$

We apply corollary 4.4.12. We have to check that 0 belongs to the interior of $\theta(\text{Dom } F)$. By condition (22), $H^*(t, x)$ is bounded on every bounded subset of $\mathbb{R} \times \mathbb{R}^{2n}$. It follows that $F(\xi)$ is finite for all constant functions ξ. Since the restriction of θ to the constant functions is the multiplication $\xi \mapsto T\xi$, $\theta(\text{Dom } F)$ is the whole space $\mathbb{R}^{2n}$.

Consequently,

$$F^*(q) = \min_{\xi \in \mathbb{R}^{2n}} \bar{F}^*(q + \xi) \tag{27}$$

because the transpose $\theta^*: \mathbb{R}^{2n} \to L^\beta$ is the map associating the constant function $y(t) \equiv \xi$ in L^β with $\xi \in \mathbb{R}^{2n}$.

Define $K: L^\alpha \to \mathbb{R} \cup \{+\infty\}$ by

$$K(y) = \int_0^T H^*(t, y)dt \tag{28}$$

We have

$$K^*(x) = \int_0^T H(t. x)dt \tag{29}$$

Hence, since $\bar{F}(y) = K(-Jy)$ and J is an isomorphism,

$$\bar{F}^*(x) = K^*(-Jx) \tag{30}$$

$$F^*(q) = \min_{\xi \in \mathbb{R}^{2n}} \int_0^T H(t, -Jx + \xi)dt \tag{31}$$

Using the definition of G, formula (21), we obtain

$$F^*(q) = G(-Jx) \tag{32}$$

The dual formula Φ associated with Ψ by formula (7) now reads

$$\begin{aligned} \Phi(y) &= \frac{1}{2}\int_0^T (y, -J\Pi y)dt + F^*(J\Pi y) \\ &= \frac{1}{2}\int_0^T (Jy, \Pi y)dt + G(\Pi y) \end{aligned} \tag{33}$$

This is just what we called $\bar{\Phi}(y)$ in corollary 3. Its critical points correspond to solutions of (17), and the result is proved. ■

In the following sections, we shall find critical points of Ψ by three different methods: the inverse function theorem, global minimization, and the Ambrosetti–Rabinowitz theorem.

3. NONRESONANT PROBLEMS

We shall prove the following result stated in theorem 1.

THEOREM 1
Let $H \in C^2(\mathbb{R}^{2n}, \mathbb{R})$. Assume that two real numbers α and β can be found with

$$\frac{2\pi}{T} m < \alpha \leqslant \beta < \frac{2\pi}{T}(m+1) \qquad \text{for some integer } m \geqslant 0 \tag{1}$$

$$\alpha I \leqslant H''_{xx}(x) \leqslant \beta I \qquad \text{for all } x \in \mathbb{R}^{2n} \tag{2}$$

Then for all $f \in L^1(0, T; \mathbb{R}^{2n})$, the boundary value problem

$$\begin{cases} \dot{x} = JH'_x(x) + f(t) \\ x(0) = x(T) \end{cases} \tag{3}$$

has one solution at least. ▲

Inequality (2) must be understood in the sense of symmetric matrices: $A_1 \leqslant A_2$ if the eigenvalues of $(A_2 - A_1)$ are nonnegative. Together with (1), it is a nonresonance condition ■

We begin by casting the problem into Hamiltonian form. Set

$$m = \frac{1}{T} \int_0^T f(t)dt \tag{4}$$

and denote by F the antiderivative of $f - m$ with zero mean value

$$\frac{d}{dt} F(t) = f(t) - m \qquad \text{and} \qquad \int_0^T F(t)dt = 0 \tag{5}$$

Set

$$z(t) = x(t) - F(t) \tag{6}$$

Note that $z(T) = z(0)$ if and only if $x(T) = x(0)$
We introduce the Hamiltonian

$$K(t, z) = H(z + F(t)) - (Jm, z + F(t)) \tag{7}$$

Problem (3) is equivalent to the following:

$$\begin{cases} \dot{z}(t) \in JK'_z(t, z(t)) \qquad \text{a.e.} \\ z(0) = z(T) \end{cases} \tag{8}$$

The function F is continuous, and inequality (2) sets growth conditions on H

$$\tfrac{1}{2}\alpha\|x\|^2+\alpha'\|x\|+\alpha''\leqslant H(x)\leqslant\tfrac{1}{2}\beta\|x\|^2+\beta'\|x\|+\beta''$$

for adequate constants α', α'', β', and β''. So the new Hamiltonian $K(t, x)$ satisfies all the assumptions of theorem 4 with $\alpha=2=\beta$.

Solving problem (3) is therefore equivalent to finding critical points of the functional

$$\Psi(y)=\int_0^T [\tfrac{1}{2}(Jy, \Pi y)+K^*(t, -Jy)]dt \tag{9}$$

on the space L_0^2. Here $K^*(t, \cdot)$ is the conjugate of $K(t, \cdot)$.

LEMMA 2

For every fixed t, the function $K^*(t, \cdot)$ is C^2 on $\mathbb{R}^{2n}$, and

$$K_{yy}^{*\prime\prime}(t, y)=K_{zz}''(t, z)^{-1}=H_{xx}''(t, x)^{-1} \tag{10}$$

where $y=K_z'(t, z)$, $z=K_y^{*\prime}(t, y)$, $x=z+F(t)$ in $\mathbb{R}^{2n}$. We also have

$$\frac{1}{\beta}I\leqslant K_{yy}^{*\prime\prime}(t, y)\leqslant\frac{1}{\alpha}I \qquad \text{for all } (t, y) \tag{11}$$

▲

Proof. By definition of the conjugate function, we have

$$K^*(t, y)=yz-K(t, z), \qquad \text{with} \qquad y=K_z'(t, z) \tag{12}$$

For fixed t, we can apply the inverse function theorem to y and z, since the derivative

$$K_{zz}''(t, z)=H_{xx}''(z+F(t))$$

is invertible by assumption (2). We have

$$\left(\left(\frac{\partial z_i}{\partial y_j}\right)\right)=K_{zz}''(t, z)^{-1}$$

with obvious notations.

By writing z in terms of y in the right side of equation (12), it follows immediately that $K^*(t, \cdot)$ is C^2. Differentiating the Legendre reciprocity formula,

$$K_y^{*\prime}(t, K_z'(t, z))=z \tag{13}$$

yields equations (10), from which (11) follows. ■

LEMMA 3
The function Ψ *is* C^1 *and twice weakly differentiable Gâteaux*

$$(\psi''(y)z_1, z_2)=\int_0^T [\tfrac{1}{2}(Jz_1, \Pi z_2)+(K_{yy}^{*}{}''(t, -Jy)z_1, z_2)]dt \tag{14}$$ ▲

Proof. The first term in Ψ is quadratic, continuous, and, hence, C^∞. The second term satisfies all the assumptions of example 5 in Section 1.4. Hence, the result. ■

Now remember that $-J\Pi$ is a compact self-adjoint operator on L_0^2, so it has a sequence $\lambda_n \to 0$ of real eigenvalues, and there is an orthonormal basis of eigenvectors. An easy computation from the equation $\lambda y = -J\Pi y$ gives

$$\lambda_k = \frac{1}{k}\frac{T}{2\pi}, \qquad k \in \mathbb{Z}, \qquad k \neq 0 \tag{15}$$

All these eigenvalues have multiplicity $2n$, and the corresponding eigenspaces are

$$E_k = \left\{ y(t) = \exp\left(-2k\pi \frac{Jt}{T}\right)\xi \mid \xi \in \mathbb{R}^{2n} \right\} \tag{16}$$

They are of course orthogonal, and we have the Hilbertian sum

$$L_0^2 = \bigoplus_{k\neq 0} E_k = E' \oplus E'', \qquad \text{with} \tag{17}$$

$$E' = \bigoplus_{k=-m}^{k=-1} E_k \qquad \text{and} \qquad E'' = \left(\bigoplus_{k=-\infty}^{-(m+1)} E_k\right) \oplus \left(\bigoplus_{k=1}^{\infty} E_k\right) \tag{18}$$

Any $z \in L_0^2$ can be written $z = z' + z''$ with $z' \in E'$ and $z'' \in E''$.
By lemma 2, we have

$$(\Psi''(x)z', z') = \sum_{k=-m}^{k=-1} \frac{1}{k}\frac{T}{2\pi} z_k^2 + \int_0^T (K_{yy}^{*}{}''(t, -Jy)z', z')dt \tag{19}$$

$$(\Psi''(x)z'', z'') = \sum_{\substack{k\geq 1 \\ k \leq -m-1}} \frac{1}{k}\frac{T}{2\pi} z_k^2 + \int_0^T (K_{yy}^{*}{}''(t, -Jy)z'', z'')dt \tag{20}$$

Here is where we use assumptions (1) and (2), substituting inequality (11) into the preceding equations, we obtain

$$(\Psi''(x)z', z') \leqslant \sum_{k=-m}^{k=-1} \frac{T}{2k\pi} z_k^2 + \int_0^T \frac{1}{\alpha}(z', z')dt$$

$$= \sum_{k=-m}^{k=-1} \left(\frac{T}{2k\pi} + \frac{1}{\alpha}\right) z_k^2$$

$$\leqslant \left(\frac{-T}{2\pi m} + \frac{1}{\alpha}\right) \|z'\|^2$$

$$= -\left(\frac{T}{2\pi m} - \frac{1}{\alpha}\right) \|z'\|^2 \tag{21}$$

$$(\Psi''(x)z'', z'') \geqslant \sum_{\substack{k \geqslant 1 \\ k \leqslant -m-1}} \frac{T}{2k\pi} z_k^2 + \int_0^T \frac{1}{\beta}(z'', z'')dt$$

$$= \sum_{k \geqslant 1} \left(\frac{T}{2k\pi} + \frac{1}{\beta}\right) z_k^2 + \sum_{k \leqslant -m-1} \left(\frac{T}{2k\pi} + \frac{1}{\beta}\right) z_k^2$$

$$\geqslant \frac{1}{\beta} \sum_{k \geqslant 1} z_k^2 + \sum_{k \leqslant -m-1} \left(\frac{1}{\beta} - \frac{T}{2(m+1)\pi}\right) z_k^2$$

$$\geqslant \left(\frac{1}{\beta} - \frac{T}{2(m+1)\pi}\right) \|z''\|^2 \tag{22}$$

Set $a = (T/2\pi m - 1/\alpha)$ and $b = (1/\beta - T/2(m+1)\pi)$. By assumption, $a > 0$ and $b > 0$, and we have

$$\forall z' \in E', \qquad (\Psi''(x)z', z') \leqslant -a\|z'\|^2 \tag{23}$$

$$\forall z'' \in E'', \qquad (\Psi''(x)z'', z'') \geqslant b\|z''\|^2 \tag{24}$$

It follows that $\Psi''(x)$ is nondegenerate.

LEMMA 4

There is some $k > 0$ such that

$$\forall z \in E, \qquad \|\Psi''(x)z\| \geqslant k\|z\| \tag{25}$$ ▲

Proof. Assume otherwise. Then there is a sequence $z_n = z_n' + z_n''$ such that:

$$\|z_n\|^2 = \|z_n'\|^2 + \|z_n''\|^2 = 1 \tag{26}$$

$$\Psi''(x)z_n \to 0 \quad \text{in} \quad L_0^2 \tag{27}$$

We also have

$$(z'_n, \Psi''(x)z_n) = (z'_n, \Psi''(x)z'_n) + (z'_n, \Psi''(x)z''_n)$$

$$\leqslant -a\|z'_n\|^2 + (z'_n, \Psi''(x)z''_n) \tag{28}$$

$$(z''_n, \Psi''(x)z_n) = (z''_n, \Psi''(x)z'_n) + (z''_n, \Psi''(x)z''_n)$$

$$\geqslant (z''_n, \Psi''(x)z'_n) + b\|z''_n\|^2 \tag{29}$$

Letting $n \to \infty$, this yields

$$\liminf a\|z'_n\|^2 \leqslant \liminf (z'_n, \Psi''(x)z''_n) \tag{30}$$

$$\limsup (z''_n, \Psi''(x)z'_n) \leqslant \limsup -b\|z''_n\|^2 \tag{31}$$

But $(z'_n, \Psi''(x)z''_n) = (z''_n, \Psi''(x)z'_n)$, so that

$$\lim (z'_n, \Psi''(x)z''_n) = 0$$

Inequalities (30) and (31) now give us

$$\limsup a\|z'_n\|^2 \leqslant 0$$

$$\limsup b\|z''_n\|^2 \leqslant 0$$

which is impossible, since $\|z'_n\|^2 + \|z''_n\|^2 = 1$. Hence the result. ■

We now have the situation in example 8, section 5.5, so Ψ has at least one critical point y, and by theorem 2.4, there is some $\xi \in \mathbb{R}^{2n}$ such that $x = \Pi y + \xi$ solves problem (3).

4. RESONANT PROBLEMS

We shall prove the following result of Clarke and Ekeland.

THEOREM 1

Let $H \in C^2(\mathbb{R}^{2n}, \mathbb{R})$. Assume that it is convex and

$$\phi(\|x\|) \leqslant H(x) \leqslant \psi(\|x\|) \qquad \textit{for all } x \tag{1}$$

with $s^{-1}\phi(s) \to +\infty$ and $s^{-2}\psi(s) \to k/2 > 0$ when $s \to \infty$.

Assume $kT < 2\pi$. Then for all $f \in L^1(0, T; \mathbb{R}^{2n})$ the boundary value problem

$$(2) \qquad \begin{cases} \dot{x} = JH'_x(x) + f(t) \\ x(0) = x(T) \end{cases}$$

has one solution at least. ▲

Note that if we take $m=0$ in theorem 3.1, we have $0<\alpha\leqslant\beta<2\pi T^{-1}$ and $\alpha I\leqslant H''_{xx}(x)\leqslant\beta I$. Setting $\alpha=0$ is not allowed in theorem 3.1, but the situation we obtain falls within the range of theorem 1. In other words, theorem 1 allows us to "touch" the $m=0$ resonance.

The proof is quite straightforward: We shall prove that the functional Ψ of theorem 2.4 has a global minimum on L_0^α. The point where it is attained is the critical point we were looking for.

We introduce the same Hamiltonian $K(t, x)$ as in the preceding section and the dual action function Ψ defined by

$$(3) \qquad \Psi(y) = \int_0^T \left[\tfrac{1}{2}(Jy, \Pi y) + K^*(t, -Jy)\right] dt$$

We begin with some estimates.

LEMMA 1

$$\|\Pi y\| \leqslant \frac{T}{2\pi} \|y\|$$ ▲

Proof. Expand y in Fourier series

$$y = \sum_{k \in \mathbb{Z}} y_k \left(\frac{2ik\pi t}{T}\right)$$

Note $y_0 = 0$ since $y \in L_0^2$. Now integrate termwise

$$\Pi y = \sum_{k \neq 0} \frac{T}{2ik\pi} y_k \exp\left(\frac{2ik\pi t}{T}\right)$$

We have

$$\|\Pi y\|^2 = \sum_{k>0} \frac{T}{4k^2\pi^2} \|y_k\|^2 T$$

$$\leqslant \frac{T^2}{4\pi^2} \sum_{k>0} \|y_k\|^2 = \frac{T^2}{4\pi^2} \|y\|^2$$

LEMMA 2
For any choice of $k' > k$, there is some constant c such that

$$K^*(t, x) \geqslant \frac{1}{2k'} \|x\|^2 - c \qquad \text{for all } (t, x) \tag{4}$$ ▲

Proof. Recall that

$$K(t, z) = H(z + F(t)) - (Jm, z + F(t))$$

Since $F(t)$ is continuous, it is bounded on $[0, T]$ by some constant M, and we have

$$\sup_t K(t, z) \leqslant \psi(\|z\| + M) - \|Jm\|(\|z\| - M)$$

So,

$$\limsup_{\|z\| \to \infty} \|z\|^{-2} \sup_t K(t, z) \leqslant \frac{k}{2}$$

This means that for any choice of $k' > k$, there will be some constant c such that

$$K(t, z) \leqslant \frac{k'}{2} \|z\|^2 + c$$

Taking conjugates, we obtain the desired inequality. ■

LEMMA 3
Ψ attains its minimum on L_0^2. ▲

Proof. Since $k < 2\pi T^{-1}$, we can choose k' so that

$$k < k' < 2\pi T^{-1}$$

It follows that

$$\begin{aligned}
\Psi(y) &= \frac{1}{2} \int_0^T (Jy, \Pi y)dt + \int_0^T K^*(t, -Jy)dt \\
&\geqslant -\tfrac{1}{2}\|Jy\| \, \|\Pi y\| + \int_0^T \left[\frac{1}{2k'} \|Jy\|^2 - c\right] dt \\
&\geqslant -\frac{T}{4\pi} \|y\|^2 + \frac{1}{2k'} \|y\|^2 - cT \\
&= \frac{1}{2}\left(\frac{1}{k'} - \frac{T}{2\pi}\right) \|y\|^2 - cT
\end{aligned}$$

Now let y_n be a minimizing sequence for Ψ

$$\Psi(y_n) \to \inf\{\psi(y) \| y \in L_0^2\}$$

The sequence $\Psi(y_n)$ is bounded from above by some constant c'. Substituting this into the preceding inequality

$$c' \geqslant \Psi(y_n) \geqslant \frac{1}{2}\left(\frac{1}{k'} - \frac{T}{2\pi}\right) \|y_n\|^2 - cT$$

$$\frac{1}{2}\left(\frac{1}{k'} - \frac{T}{2\pi}\right) \|y_n\|^2 \leqslant cT + c'$$

Since $k' < 2\pi T^{-1}$, the coefficient on the left is positive, from which it follows that

$$\|y_n\|^2 \leqslant 2(cT + c')\left(\frac{1}{k'} - \frac{T}{2\pi}\right)^{-1} \tag{6}$$

We now extract from the bounded sequence y_n a weakly convergent subsequence, which we still denote by y_n. Let y be its weak limit

$$y_n \to y \qquad \text{weakly}$$

The second term in Ψ is convex and continuous (see Section 2) and, hence, weakly lower semicontinuous

$$\liminf \int_0^T K^*(t, -Jy_n)dt \leqslant \int_0^T K^*(t, -Jy)dt$$

The first term is weakly continuous. To be precise, we have

$$\Pi y_n \to \Pi y$$

$$(\Pi y_n, Jy_n) - (\Pi y, Jy) = (\Pi y_n - \Pi y, Jy_n) + (\Pi y, Jy_n - Jy)$$

The first term on the right converges to zero because $\|Jy_n\|$ stays bounded (Banach–Steinhaus theorem), and the second term converges to zero, since $(Jy_n - Jy) \to 0$.

Finally,

$$\lim \Psi(y_n) \geqslant \Psi(y)$$

So y must be a minimizer.

Theorem 1 now follows from theorem 2.4. ■

We now turn to autonomous problems, that is, we set $f=0$. Of course, theorem 1 still applies, but the solution we obtain might be the trivial one (as pointed out in Section 1). To exclude this solution, we need one more assumption on H.

THEOREM 4

Let $H \in C^2(\mathbb{R}^{2n}, \mathbb{R})$ be convex and $H(x) \geqslant H(0)=0$. Set

$$\liminf_{r\to 0} \frac{1}{r^2}\min\{H(x) \mid \|x\|=r\}=\frac{K}{2} \tag{7}$$

$$\limsup_{R\to\infty} \frac{1}{R^2}\max\{H(x) \mid \|x\|=R\}=\frac{k}{2} \tag{8}$$

Assume $0<k<K\leqslant\infty$. Then for all T in the open interval $(2\pi K^{-1}, 2\pi k^{-1})$, the equation

$$\dot{x}=JH'_x(x)$$

has at least one periodic solution with smallest period T. ▲

If x is T periodic, it is also $2T$ periodic, $3T$ periodic, and so on. The solution that theorem 4 gives us will not be $T/2$ periodic or T/k for any integer $k>1$. In particular, it cannot be constant.

To prove theorem 4, we first apply theorem 1. We have found a minimizer y for Ψ, and $x=\Pi y+\xi$ is a T periodic solution.

Assume that x is actually T/k periodic for some integer $k>1$. Then so is y. Set

$$y_k=y\left(\frac{t}{k}\right)$$

We have $y_k \in L_0^2$. We claim that $\Psi(y_k)<\Psi(y)$, which contradicts the fact that y minimizes Ψ and concludes the proof.

Indeed,

$$\Psi(y_k)=\int_0^T [\tfrac{1}{2}(\Pi y_k, Jy_k)+H^*(-Jy)]dt$$

$$=\int_0^T \frac{1}{2}\left[\left(k\Pi y\left(\frac{t}{k}\right), Jy\left(\frac{t}{k}\right)\right)+H^*\left(-Jy\left(\frac{t}{k}\right)\right)\right]dt$$

$$=\int_0^{T/k}\left[\frac{k}{2}(\Pi y(s), Jy(s))+H^*(-Jy(s))\right]kds$$

$$= \int_0^T \frac{k}{2}(\Pi y(t), Jy(t)) + H^*(-Jy(t))dt$$

$$= k\Psi(y) + (1-k)\int_0^T H^*(-Jy)dt$$

Now $H(x) \geq H(0) = 0$, so $H^*(z) \geq H^*(0) = 0$. Hence,

$$\Psi(y_k) \leq k\Psi(y) \tag{9}$$

At this point in the proof, we need an intermediate result.

LEMMA 5

$$\Psi(y) = \min \Psi < 0$$

▲

Proof. Clearly, $\Psi(y) \leq \Psi(0) = 0$, but this is not sharp enough. Fix $\xi \in \mathbb{R}^{2n}$ with $\|\xi\| \neq 0$. With every $s > 0$ associate the path

$$y_s(t) = s \exp\left(-\frac{2\pi Jt}{T}\right)\xi$$

We have $y_s \in L_0^2$ and $\|y_s\| = s\|\xi\| T^{1/2}$. Moreover,

$$\int_0^T \frac{1}{2}(Jy_s, \Pi y_s)dt = \frac{s^2}{2}\int_0^T \left(J\exp\left(\frac{-2\pi Jt}{T}\right)\xi, \frac{T}{2\pi} J\exp\left(\frac{-2\pi Jt}{T}\right)\xi\right)dt$$

$$= -\frac{s^2}{2}\frac{T}{2\pi}\|\xi\|^2 T$$

Now it follows from our assumptions that

$$\limsup_{r\to 0} \frac{1}{r^2}\{H^*(x) | \|x\| = r\} = \frac{1}{2K}$$

Pick some K' with $2\pi T^{-1} < K' < K$, and choose $|s|$ so small that

$$\int_0^T H^*(-Jy_s)dt \leq \int_0^T \frac{1}{2K'}\|Jy_s\|^2 dt$$

$$\leq \frac{1}{2K'}\alpha^2\|\xi\|^2 T$$

We have

$$\Psi(y_s) = \int_0^T \frac{1}{2}(Jy_s, \Pi y_s)dt + \int_0^T H^*(-Jy_s)dt$$

$$\leqslant -\frac{s^2}{2}\left(\frac{T}{2\pi} - \frac{1}{k'}\right)\|\xi\|^2 T < 0$$

Hence, $\Psi(y) \leqslant \Psi(y_s) < 0$ as desired. ■

Since $\Psi(y) < 0$, we have $k\Psi(y) < \Psi(y)$ whenever $k > 1$ and hence the desired contradiction that concludes the proof of theorem 4

(10) $$\Psi(y_k) < \Psi(y)$$

COROLLARY 6
Assume H is convex and for some α with $1 < \alpha < 2$,

(11) $$\liminf_{r \to 0} \; r^{-\alpha} \min H(x) | \, \|x\| = r\} > 0$$

(12) $$\limsup_{R \to \infty} R^{-\alpha} \max H(x) | \, \|x\| = R\} < +\infty$$

Then for all $T > 0$, the equation $\dot{x} = JH'_x(x)$ has a periodic solution with minimal period T. ▲

5. TRANSRESONANT PROBLEMS

We shall now deal with the case when $H''(x)$ ranges from zero to $+\infty$. This is a more difficult situation to handle, and the nonautonomous case, for instance, is not fully understood. We shall be content to give a simple result.

THEOREM 1
Let $H \in C(\mathbb{R}^{2n}, \mathbb{R})$. Assume that

(1) *H is strictly convex*

(2) $$H(x) \geqslant H(0) = 0$$

Assume moreover that for some $\beta > 2$, we have

(3) $$(x, H'(x)) \geqslant \beta H(x)$$

Then for any $T>0$, the equation

$$\dot{x}=JH'(x) \tag{4}$$

has at least one nonconstant T-periodic solution. ▲

Assumption (3) can be put in a more readily accessible form

$$H(\lambda x)\geqslant\lambda^{\beta}H(x) \qquad \text{for all } \lambda>1, \qquad x\in\mathbb{R}^{2n} \tag{5}$$

Since $\beta>2$, we obtain (note $\lambda>1$ and not $\lambda>0$),

$$\|x\|^{-2}H(x)\rightarrow+\infty \qquad \text{when} \qquad \|x\|\rightarrow\infty \tag{6}$$

$$\|x\|^{-2}H(x)\rightarrow 0 \qquad \text{when} \qquad \|x\|\rightarrow 0 \tag{7}$$

We shall first prove the theorem under the added assumption that for some constant $k>0$

$$H(x)\leqslant\frac{k^{\beta}}{\beta}\|x\|^{\beta}, \qquad \text{for all } x\in\mathbb{R}^{2n} \tag{8}$$

The general case will follow later.

LEMMA 2

Set $c^{\beta}=\beta\min\{H(x)|\ \|x\|=1\}$. It is strictly positive, and we have

$$\forall x, \qquad H(x)\geqslant\frac{c^{\beta}}{\beta}(\|x\|^{\beta}-1) \tag{9}$$

$$\|x\|\leqslant 1\Rightarrow\|H'(x)\|\leqslant\frac{1}{\beta}(k^{\beta}2^{\beta}-c^{\beta})\|x\|^{\beta-1} \tag{10}$$

▲

Proof. The first inequality follows immediately from assumption (3). For the second, start with the convexity inequality

$$(H'(x), z-x)\leqslant H(z)-H(x)$$

Take the supremum over all z such that $\|z-x\|=\|x\|$. Using assumption (3), we obtain

$$\|x\|\ \|H'(x)\|\leqslant\frac{k^{\beta}}{\beta}\|2x\|^{\beta}-\frac{c^{\beta}}{\beta}\|x\|^{\beta}$$

which is the desired result.

LEMMA 3
H^* is everywhere finite and C^1. ▲

Proof. Taking conjugate functions on both sides of inequality (9), we obtain

$$H^*(y) \leqslant \frac{c^{-\alpha}}{\alpha} \|y\|^\alpha + \frac{c^\beta}{\beta} \tag{11}$$

where α is the conjugate exponent of β, so $\alpha^{-1} + \beta^{-1} = 1$.

The C^1 property follows from the fact that H'' is positive definite, as in lemma 3.3.1. ■

LEMMA 4
H^* satisfies all the following:

$$H^*(y) \geqslant \frac{k^{-\alpha}}{\alpha} \|y\|^\alpha \tag{12}$$

$$\alpha H^*(y) \geqslant (y, H^{*\prime}(y)) \tag{13}$$

$$\|y\| \geqslant 1 \Rightarrow \|H^{*\prime}(y)\| \leqslant \frac{1}{\alpha} (2^\alpha c^{-\alpha} - k^{-\alpha}) \|y\|^{\alpha - 1} \tag{14}$$

▲

Proof. The first inequality follows from (8) by taking conjugates of both sides. The second follows from assumption (8) and the Legendre formula

$$\begin{aligned} H^*(y) &= (x, H'(x)) - H(x) \\ &\geqslant (1 - \beta^{-1})(H'(x), x) \\ &= \alpha^{-1}(H'(x), x) \\ &= \alpha^{-1}(y, H^{*\prime}(y)) \end{aligned}$$

with $x = H^{*\prime}(y)$ and $y = H'(x)$.

The third inequality proceeds from (11) and (12) as in lemma 1. ■

We can now proceed to the heart of the proof. By theorem 2.4, we shall be seeking a critical point of the dual action functional on L_0^α

$$\Psi(y) = \int_0^T [\tfrac{1}{2}(Jy, \Pi y) + H^*(-Jy)]dt \tag{15}$$

The origin $y = 0$ is an obvious critical point. It does not interest us, since the corresponding solution $x = \Pi y + \xi$ is constant, that is, the equilibrium. We are looking for another critical point, which we shall find by applying the Ambrosetti–Rabinowitz theorem 5.5.5.

LEMMA 5

Constants $\gamma > 0$ and $r > 0$ can be found such that

$$\|y\| = r \Rightarrow \Psi(y) \geq \gamma \tag{16}$$

$$0 < \|y\| \leq r \Rightarrow \Psi(y) > \Psi(0) = 0 \tag{17}$$

▲

Proof. Clearly, $\Psi(0) = 0$. By condition (12), we have

$$\Psi(y) \geq \int_0^T \frac{1}{2}(Jy, \Pi y) + \frac{k^{-\alpha}}{\alpha} |Jy|^\alpha]dt$$

Using Cauchy–Schwarz on the first term, we obtain

$$\Psi(y) \geq \frac{k^{-\alpha}}{\alpha} \|y\|_\alpha^\alpha - \tfrac{1}{2}\|y\|_\alpha \|\Pi y\|_\beta$$

Now Π sends L_0^α into C^0, which injects into L^β, and $\|y\|_\beta \leq b\|y\|_\alpha$ for some constant b. Hence,

$$\Psi(y) \geq \frac{k^{-\alpha}}{\alpha} \|y\|_\alpha^\alpha - \frac{b}{2}\|y\|_\alpha^2$$

Since $\alpha < 2$, the first term outweighs the other near the origin, and the result follows. ■

LEMMA 6

There is some $y \in L^\alpha$ where $\psi(y) < 0$. ▲

Proof. Define y_s as in lemma 4.5, so that

$$\int_0^T \frac{1}{2}(Jy_s, \Pi y_s)dt = -\frac{s^2}{2}\frac{T}{2\pi}\|\xi\|^2 T$$

On the other hand, by inequality (11), we have

$$\int_0^T H^*(-Jy_s)dt \leq \frac{c^{-\alpha}}{\alpha}\|y_s\|_\alpha^\alpha + \frac{c^\beta}{\beta} T$$

Finally,

$$\Psi(y_s) \leq \frac{c^{-\alpha}}{\alpha}\|\xi\|^\alpha T s^\alpha - \frac{T^2}{4\pi}\|\xi\|^2 s^2 + \frac{c^\beta}{\beta} T$$

Since $\alpha<2$, we have $\Psi(y_s)<0$ for large s. Hence, the result. ■

We now have the situation in examples 6 and 7 in Section 5.5, so that there exists at least one nonzero critical point $\bar{y}$ and the result is proved.

We can also locate this solution in the phase space $\mathbb{R}^{2n}$ by giving an upper bound for its energy level h.

We first relate h to the critical value $v=\Psi(\bar{y})$ we have just found, starting from

$$v=\int_0^T [\tfrac{1}{2}(J\bar{y},\ \Pi\bar{y})+H^*(-J\bar{y})]dt \tag{18}$$

and recalling that $x=\Pi\bar{y}+\xi\in\partial H^*(-J\bar{y})$, since $\bar{y}$ is a critical point. Hence,

$$\begin{aligned}
v&=\int_0^T \tfrac{1}{2}(J\bar{y},\ \Pi\bar{y})+(-J\bar{y},\ \Pi\bar{y}+\xi)-H(\Pi\bar{y}+\xi)]dt\\
&=-\int_0^T [\tfrac{1}{2}(J\dot{x},\ x)+H(x)]dt\\
&=\int_0^T [\tfrac{1}{2}(H'(x),\ x)-H(x)]dt\\
&\geqslant\int_0^T \left(\frac{\beta}{2}-1\right)H(x)dt
\end{aligned} \tag{19}$$

But $H(x(t))=h$ along the trajectory. Hence, $v\geqslant h(\beta/2-1)T$.

We then estimate v; remember that it is found by applying the Ambrosetti–Rabinowitz theorem 3.5.5, which defines v as follows:

$$v=\inf_{\gamma\in\Gamma}\max_{0\leqslant s\leqslant 1}\psi(\gamma(s)) \tag{20}$$

where Γ is the set of all paths γ: $[0,1]\to L_0^\alpha$ such that $\gamma(0)=0$ and $\gamma(1)=y$, the latter being a fixed point where $\psi(y)<0$.

Taking the path γ: $s\to y_s$ described in lemma 5 gives

$$\begin{aligned}
v&\leqslant\max_{0\leqslant s\leqslant 1}\Psi(y_s)\\
&\leqslant\max_{0\leqslant s\leqslant 1}\left\{\frac{c^{-\alpha}}{\alpha}\|\xi\|^\alpha Ts^\alpha-\frac{T^2}{4\pi}\|\xi\|^2s^2\right\}+\frac{c^\beta}{\beta}T\\
&=c^{-2\alpha/(2-\alpha)}T^{2(1-\alpha)/(2-\alpha)}(2\pi)^{\alpha/(2-\alpha)}\left(\frac{1}{\alpha}-\frac{1}{2}\right)+\frac{c^\beta}{\beta}T\\
&=c^{-2\beta/(\beta-2)}T^{-2/(\beta-2)}(2\pi)^{\beta/(\beta-2)}\left(\frac{1}{2}-\frac{1}{\beta}\right)+\frac{c^\beta}{\beta}T
\end{aligned} \tag{21}$$

Finally, we obtain an a priori estimate for the energy level

$$h \leq \frac{1}{\beta} c^{-2/(\beta-2)} T^{-\beta/(\beta-2)} (2\pi)^{\beta/(\beta-2)} + \frac{2c^{\beta}}{\beta(\beta-2)} \tag{22}$$

This allows us to prove theorem 1 for the general case, that is, when the additional assumption (8) is not satisfied.

Indeed, note that the constant k in (8) plays no role in the energy estimate we have just found.

Given any H in $C^1(\mathbb{R}^{2n}, \mathbb{R})$, satisfying conditions (1), (2), and (3) but not (8), we compute the right-hand side of (22) and call it h_0 [recall that $\beta^{-1}c^{\beta}$ is the minimum of $H(x)$ over the unit sphere $|x|=1$].

With any constant $r>0$, associate the convex function

$$G_r(x) = \sup\{(x, y) - H^*(y) | \, \|y\| \leq r\} \tag{23}$$

to the set

$$K_r = \{x | \, \|H'(x)\| \leq r\} \tag{24}$$

and the numbers

$$M_r = \max\{H(x) | x \in K_r\}, \qquad m_r = \min\{H(x) | x \notin K_r\} \tag{25}$$

We have

$$G_r(x) = H(x), \qquad \text{for all } x \in K_r$$

$$G_r(x) \leq r\|x\| + M_r, \qquad \text{for all } x \in \mathbb{R}^{2n}$$

Define ϕ_r: $[0, \infty) \to \mathbb{R}$ as follows:

$$\begin{cases} \phi_r(t) = t & \text{if } 0 \leq t \leq m_r \\ \phi_r(t) = at^{\beta} - b & \text{if } t \geq m_r \end{cases} \tag{26}$$

the constants $a>0$ and $b>0$ being adjusted so that ϕ_r is C^1 at $t=m_r$ and hence everywhere. Since $\beta>1$ and $m_r>1$ (for r large enough), ϕ_r will be convex and increasing.

Now consider the Hamiltonian

$$H_r(x) = \phi_r(G_r(x)) \tag{27}$$

It is C^1 and strictly convex. We have

$$H(x) \leq m_r \Rightarrow H(x) = H_r(x) \tag{28}$$

$$H(x) \geqslant m_r \Rightarrow H_r(x) \leqslant a(r\|x\| + M_r)^\beta - b \tag{29}$$

It follows from condition (3) or (5) that

$$\|x\| \leqslant 1 \Rightarrow H(x) \leqslant \|x\|^\beta \max\{H(x) | \|x\| = 1\} \tag{30}$$

Conditions (28), (29), and (30) together imply that for any r, we can choose some k_r large enough so that

$$H_r(x) \leqslant \frac{1}{\beta} k_r^\beta \|x\|^\beta \tag{31}$$

So condition (8) will hold for H_r.

The proof of theorem 1 for H now runs as follows. Choose $r > 0$ so large that $m_r \geqslant h_0$. Apply theorem 1 to H_r, and find a T-periodic solution of $\dot{x} = JH_r'(x)$ with energy level $h \leqslant h_0$. It follows that, for all t

$$H_r(x(t)) \leqslant h_0 \leqslant m_r$$

By (27) and (28), this implies that

$$H_r(x(t)) = H(x(t))$$

So in fact we have solved the equation $\dot{x} = JH'(x)$ as desired. ■

TABLE 1 Tangent Cones and Support Functions

Operations	Tangent cone	Contingent cone	Tangent cone to a convex subset	Support functions of closed convex subsets
$x \in K \subset L$		$T_K(x) \subset T_L(x)$	$T_K(x) \subset T_L(x)$	$\sigma_K(p) \leqslant \sigma_L(p)$
$K = \bigcup_{i=1}^{n} K_i$		$T_K(x) = \bigcup_{i=1}^{n} T_{K_i}(x)$		$\sigma_K(p) = \sup_{i=1,\ldots,n} \sigma_{K_i}(p)$
$K = K_1 \cap K_2$	If $C_{K_1}(x) - C_{K_2}(x) = X$, $C_{K_1}(x) \cap C_{K_2}(x) \subset C_K(x)$	$T_{K_1}(x) \cap T_{K_2}(x) \supset T_K(x)$	If $0 \in \operatorname{int}(K_1 - K_2)$, $T_{K_1}(x) \cap T_{K_2}(x) = T_K(x)$	If $0 \in \operatorname{Int}(K_1 - K_2)$, $\exists \bar{q}$ such that $\sigma_K(p) = \sigma_{K_1}(p - \bar{q}) + \sigma_2(\bar{q})$ $= \inf_q (\sigma_{K_1}(p - q) + \sigma_2(q))$
$K = \prod_{i=1}^{n} K_i$	$\prod_{i=1}^{n} C_{K_i}(x_i) = C_K(x)$	$\prod_{i=1}^{n} T_{K_i}(x_i) \supset T_K(x)$	$\prod_{i=1}^{n} T_{K_i}(x_i) = T_K(x)$	$\sigma_K(p) = \sum_{i=1}^{n} \sigma_{K_i}(p_i)$
A is a C^1 map from X to Y and $K \subset X$		$\nabla A(x) T_K(x) \subset T_{A(K)}(Ax)$	If $A \in \mathscr{L}(X, Y)$, $(A T_K(x)) = T_{A(K)}(Ax)$	If $A \in \mathscr{L}(X, Y)$, $\sigma_{A(K)}(p) = \sigma_K(A^* p)$
A is a C^1 map from X to Y and $L \subset Y$	If $\operatorname{Im} \nabla A(x) - C_L(Ax) = Y$, $\nabla A(x)^{-1} C_L(Ax) \subset C_{A^{-1}(L)}(x)$	$\nabla A(x)^{-1} T_L(Ax) \supset T_{A^{-1}(L)}(x)$	If $A \in \mathscr{L}(X, Y)$ and $0 \in \operatorname{Int}(\operatorname{Im} A - L)$, $A^{-1} T_{L(Ax)} = T_{A^{-1}(L)}(x)$	If $A \in \mathscr{L}(X, Y)$ and $0 \in \operatorname{Int}(\operatorname{Im} A - L)$, $\exists \bar{q} \in A^{*-1}(p)$ such that $\sigma_{A^{-1}(L)}(p) = \sigma_L(\bar{q})$ $= \inf \{\sigma_L(q) \mid A^* q = p\}$

TABLE 2 Derivative of a Set-Valued Map F from X to Y at $(x_0, y_0) \in \text{Graph}(F)$

Operations	Derivative	Contingent derivative	Derivative of F with closed convex graph
B is a differentiable map from Y to Z. $\boxed{BF}$		$D(BF)(x_0, By_0) \supset \nabla B(y_0) DF(x_0, y_0)$	If $B \in \mathscr{L}(Y, Z)$, $B \cdot DF(x_0, y_0) = D(BF)(x_0, By_0)$
A is a continuously differentiable map from X_0 to X. $\boxed{FA}$	If $X = \text{Im}\, \nabla A(x_0) - \text{Dom}\, CF(Ax_0, y_0)$, then $C(FA)(x_0, y_0) \supset CF(Ax_0, y_0) \nabla A(x_0)$ and $C(FA)(x_0, y_0)^* \subset \nabla A(x_0)^* CF(Ax_0, y_0)^*$	$D(FA)(x_0, y_0) \subset DF(Ax_0, y_0) \nabla A(x_0)$	If $A \in \mathscr{L}(X, Y)$ and $0 \in \text{Int}(\text{Im}\, A - \text{Dom}\, F)$, $D(FA)(x_0, y_0) = DF(Ax_0, y_0) A$ and $D(FA)(x_0, y_0)^* = A^* DF(Ax_0, y_0)$
Let $K \subset X$ be a closed subset and $F\|_K$ denote the restriction of F to K. $\boxed{F\|_K}$	If $X = \text{Dom}\, CF(x_0, y_0) - C_K(x_0)$ then $CF(x_0, y_0)\|_{C_K(x_0)} \subset C(F\|_K)(x_0, y_0)$ and $C(F\|_K)(x_0, y_0)^*(\cdot) \subset CF(x_0, y_0)^*(\cdot) + N_K(x_0)$	$DF(x_0, y_0)\|_{T_K(x_0)} \supset D(F\|_K)(x_0, y_0)$	If $0 \in \text{Int}(\text{Dom}\, F - K)$ and K is convex, $DF(x_0, y_0)\|_{T_K(x_0)} = D(F\|_K)(x_0, y_0)$ and $D(F\|_K)(x_0, y_0)^*(0)$ $= DF(x_0, y_0)^*(\cdot) + N_K(x_0)$

TABLE 3 Epiderivative of a Proper Real-Valued Function $V: X \to \mathbb{R} \bigcup \{+\infty\}$ at $x_0 \in \text{Dom } V$

Operations	Epi derivative of a lower semicontinuous function when $\dim(X) < +\infty$	Epicontingent derivative	Epiderivative of a lower semicontinuous convex function
Let A be a continuously differentiable map from X_0 to X. $\boxed{VA}$	If $X = \text{Im } \nabla A(x_0) - \text{Dom } C_+V(Ax_0)$, then $C_+V(Ax_0)\nabla A(x_0) \geqslant C_+(VA)(x_0)$ and $\partial(VA)(x_0) \subset \nabla A(x_0)^* \partial V(Ax_0)$	$D_+V(Ax_0)\nabla A(x_0) \leqslant D_+(VA)(x_0)$	If $A \in \mathscr{L}(X_0, X)$ and $0 \in \text{Int}(\text{Im } A - \text{Dom } V)$, then $D_+V(Ax_0)(\nabla A(x_0)) = D_+(VA)(x_0)$ and $\partial(VA)(x_0) = A^* \partial V(Ax_0)$
Let K be a closed subset of X. $\boxed{V\|_K}$	If $X = C_K(x_0) - \text{Dom } C_+V(x_0)$, then $C_+(V\|_K)(x_0) \leqslant C_+V(x_0)\|_{C_K(x_0)}$ and $\partial(V\|_K)(x_0) \subset \partial V(x_0) + N_K(x_0)$	$D_+(V\|_K)(x_0) \geqslant D_+V(x_0)\|_{T_K(x_0)}$	$D_+(V\|_K)(x_0) = D_+V(x_0)\|_{T_K(x_0)}$ and $\partial(V\|_K)(x_0) = \partial V(x_0) + N_K(x_0)$ when K is convex and $0 \in \text{Int}(K - \text{Dom } V)$

Let $V_i: X \to \mathbb{R} \bigcup \{+\infty\}$ $\boxed{\max V_i}$ If $x_0 \in \bigcap_{i=1}^{n} \text{Int Dom } V_i$, we set $J(x_0) = \left\{ i \,\middle|\, \max_i V_i(x_0) = V_j(x_0) \right\}$	If the functions are locally Lipschitz, $C_+(\max V_i)(x_0) \leqslant \max_{j \in J(o)} C_+ V_j(x_0)$ and $\partial(\max V_i)(x_0) \subset \bar{co} \bigcup_{j \in J(x_0)} \partial V_j(x_0)$	$D_+(\max V_i)(x_0) \geqslant \max_{j \in J(x_0)} D_+ V_j(x_0)$	$D_+(\max V_i)(x_0) = \max_{j \in J(x_0)} D_+ V_j(x_0)$ and $\partial(\max V_i)(x_0) = \overline{co} \bigcup_{j \in J(x_0)} \partial V_j(x_0)$
V and W are proper functions and $x_0 \in \text{Dom } V \cap \text{Dom } W$. $\boxed{V + W}$	If V is Lipschitz around x_0, then $C_+(V+W)(x_0) \leqslant C_+ V(x_0) + C_+ W(x_0)$ and $\partial(V+W)(x_0) \subset \partial V(x_0) + \partial W(x_0)$	$D_+(V+W)(x_0) \geqslant D_+ V(x_0) + D_+ W(x_0)$	If $0 \in \text{Int}(\text{Dom } V - \text{Dom } W)$, then $D_+(V+W)(x_0) = D_+ V(x_0) + D_+ W(x_0)$ and $\partial(V+W)(x_0) = \partial V(x_0) + \partial W(x_0)$

Comments

CHAPTER 2: SMOOTH ANALYSIS

Section 1

The inverse function theorem is so classical that we don't even know who started it. In modern times, the need has arisen for an inverse function theorem that would cover situations where the underlying space is C^∞ or C^ω (which are not Banach spaces) and the linear inverse loses derivatives (sends C^p onto C^{p-1}, say). The answer is the Nash–Moser inverse function theorem, where Newton's method, because of its very rapid convergence, plays an essential role in the proof. It falls outside the scope of this book; see Moser (1973) and Nirenberg (1974) for an exposition; Hamilton (1982) for a survey; Ray (1983) for an original proof; and Arnold (1978) for applications to classical mechanics (the Kolmogorov–Arnold–Moser theorem).

Section 2

Milnor's proof is in Milnor (1978). Other proofs use either a combinatorial lemma of Knaster–Kuratowski–Mazurkiewicz type (1926) or homology theory under more or less clever disguises (Brouwer, 1910).

Section 3

Theorem 1 is due to Whitney. Our proof of the infinite–dimensional Morse lemma (theorem 8) follows Lang. In finite dimensions, the Morse lemma is but the first step of singularity theory, a good introduction to the subject being Bröcker's book. For a proof of the Morse lemma in the C^2 case on a Hilbert space, see Cambini (1973).

Section 4

This follows Cerf's exposition in $\Gamma_4 = 0$.

Section 5

The results in this section are taken from Crandall and Rabinowitz (1971, 1973) and so are the proofs. Nirenberg's exposition (1974) follows the same pattern. Theorem 1 will also be found in Prodi and Ambrosetti (1973). See the books by Chow and Mallet-Paret (1982), Hassard et al. (1981), and Iooss and Joseph (1980) for further information on bifurcation theory.

Section 6 and 7

Transversality theory is a creation of René Thom. See Abraham and Robbin (1967) for a more complete exposition, and a proof of Sard's theorem. We have followed this excellent reference in our presentation.

Smale's method was given in (1976c). Here we have chosen different boundary conditions. See Milnor (1975) for the impact of the Sard–Brown theorem on topology. Smale's method itself came out as a smooth version of Scarf's algorithm (1967b) for computing fixed points. Continuation methods such as Smale's are very popular nowadays because they are simple, flexible, and often very efficient in solving numerically a large variety of nonlinear problems see Robinson (1980) and the references therein; Garcia and Gould (1978, 1980), and Lasry and Siconolfi (1983).

CHAPTER 3: SET-VALUED MAPS

Definitions of continuity for set-valued maps and most results of the first section are now classical. For more details, we refer to the forthcoming book by Rockafellar and Wets. Some related material appears in the book by Berge (1954). The extension to set-valued maps with closed convex graphs of the closed graph theorem and the open mapping principle was found by Robinson (1976a) and Ursescu (1975). The concept of transpose of closed convex processes was used by Rockafellar (1967b) for applications to economic models (1974b). (See also Makarov and Rubinov, 1970, 1973).

The economic model studied in the fourth section is due to von Neumann (1937) and has been thoroughly studied since (see Gale, 1956; Nikaïdo, 1968, etc.). The extension to the nonlinear case is due to Ky Fan (1958). Frobenius's theorem goes back to 1908. This section is taken from Aubin (1978c). The book by Castaing and Valadier (1971) gives a thorough study of measurable set-valued maps.

A study of continuous selections of set-valued maps can be found in the first chapter of the book by Aubin and Cellina (1983).

CHAPTER 4: CONVEX ANALYSIS AND OPTIMIZATION

Convex analysis and optimization are by now very established. It started with the fundamental work of Fenchel (1949, 1951), followed by the works of Moreau and Rockafellar, and summarized in Moreau (1967) and Rockafellar (1970a, 1974a,, 1976).

We refer to these books (among others) for further comments. We mention only that the duality theory for constrained minimization problems was first established in the special case of linear programming, thanks to the pionneering work of Dantzig (1963) and Kuhn and Tucker (1951), following earlier ideas of J. von Neumann (unpublished).

The concepts of derivatives and co-differentials was introduced in Aubin (1981) and Pchenitchny (1980). The computation of tangent cones to $L \cap A^{-1}(M)$ is due to Aubin (1979c). Section 3.7 on the regularity of the set of solutions and Lagrange multipliers of convex minimization problems is taken from Aubin (1982a).

CHAPTER 5: A GENERAL VARIATIONAL PRINCIPLE

Section 1

The most important results of this section are theorems 7 (the Banach contraction principle) and 14 (the Caristi fixed-point theorem). The first one is classical and has already been used in Chapter 1 to prove the inverse function theorem. The second one is recent (Caristi, 1976; Kirk and Caristi, 1975) and came as a surprise to experts: it was the first fixed-point theorem that did not require the self-map to be continous. Since then, this vein has been exploited, with further interesting results, by Kirk, Ray, and their students.

The connection between Caristi's fixed-point theorem and Ekeland's ε-variational principle was first noted by F. Browder. The approach we use here, through dissipative dynamical systems, is due to J. P. Aubin and J. Siegel (1980). Our exposition follows theirs.

Section 2

Theorem 1 is due to F. Browder (1965a). See also related work by Browder (1965b), Edelstein (1963, 1965), and Kirk (1965). Theorem 7 is a nonlinear version of the mean ergodic theorem (von Neumann, 1932). It is due to Baillon (1978). The proof is due to Pazy (1979).

Section 3

The ε-variational principle (theorem 1) is due to Ekeland (1972, 1973, 1974). Many consequences, including some we describe in this book, are given in the survey paper Ekeland (1979a).

Example 9 is classical in the Russian literature on the calculus of variations (see the excellent book by Ioffe and Tihomirov, 1974). Much more can be said about this problem and related ones: it would take us into relaxation theory, which is described in the books of Ioffe and Tihomirov and Ekeland and Temam.

Section 4

Theorem 3 is due to Brøndsted and Rockafellar. Proposition 4 is an abstract version of Temam's results for the Plateau problem (see Temam, 1971 or Chapter 5 of the book by Ekeland and Temam), which we describe in the remainder of the section.

Section 5

Palais and Smale introduced condition (C) in a series of papers where they extended Morse theory and the Liusternik–Schnirelman theory to infinite-dimensional manifolds.

The condition (weak C) appears here for the first time. Proposition 4 and theorem 5 strengthen existing results. Theorem 5, with the assuption that U be C^1, is a celebrated result of Ambrosetti and Rabinowitz (1973). We found a simpler proof, relying on the ε-variational principle instead of the so-called deformation lemma, and Brézis showed us how to extend it to the case when U is not C^1.

Example 8 is found in Ekeland (1979a).

Section 6

The results of this section are due to Ekeland and Lebourg (1976) (see also Ekeland, 1979a).

Theorem 6 solves half of the so-called Asplund conjecture (see Asplund, 1968) for the first steps in this direction, and the book by Diestel and Uhl for connections with the Radon–Nikodym property). The other half remains open (if every continuous convex function on a Banach space is Fréchet differentiable on a dense G_δ, must the space have an equivalent Fréchet differentiable norm?), as do all related questions when we replace Fréchet differentiability by Gâteaux differentiability.

Section 7

Proposition 12 is the main result of Ekeland and Lebourg's paper (1976). The approach we use here, however, is different, and due to Lebourg. Proposition 13 is due to Edelstein (1968). Similar results are known when one seeks to maximize the distance to a given point in a closed subset (instead of minimizing it). They will be found in Edelstein (1966) and also follow from the results of this section.

CHAPTER 6: SOLVING INCLUSIONS

The main concepts of game theory, which we present in the first section, were defined by early economists. Cournot, for instance, introduced the duopoly model in his book (1838). Even if we don't give everyone his due, we should at least quote the very influential books of Walras (1874) and Pareto (1909).

Their work, however, never achieved great levels of formal sophistication. The first one to formulate a truly mathematical theory was, as in so many other instances, J. von Neumann (1944). His very early minimax theorem (1928), which he deduced from Brouwer's fixed point theorem (1910), has set standards of rigor for the whole field and has received many extensions.

Ky Fan's inequality (theorem 3.5) was proved in 1972. The extension to monotone functions (theorem 3.9) is due to Brezis, Nirenberg, and Stampacchia (1972). The existence of noncooperative equilibria in game theory is due to Nash (1950a).

The first fixed-point theorem for set-valued maps (with compact convex values) was proved by Kakutani (1941) for the needs of game theory. The particular needs of mathematical economics led to successive extensions of Kakutani's theorem, culminating in the Gale–Nikaïdo–Debreu result (theorem 4.4). The inner trend of pure mathematics also led to fixed-point theorems for set-valued maps, such as theorem 4.13, which is essentially due to F. Browder (1968) and Ky Fan (1972), and theorem 4. 15, which is due to Rogalski (1972). Haddad and Lasry (1983) have also given fixed-point theorems for maps with nonconvex values.

Leray and Schauder proved their famous theorem in their 1934 paper, and applied it to nonlinear partial differential equations. The extension to set-valued maps follows a method given in Granas (1976). Theorem 4.21 on quasi-variational inequalities originates in Arrow and Debreu (1954) and was extended by Joly and Mosco (1974).

The KKM lemma was found by Knaster, Kuratowski, and Mazurkiewicz (1926) to give a combinatorial proof of Brouwer's fixed-point theorem. It was generalized by Shapley (1973) (theorem 4.24), who used his result to provide a proof that the core of a balanced game is nonempty, a fact first stated by Scarf (1967a). The proof we give of the KKMS lemma is due to Ichiishi (1981a, b), and further extended by Ky Fan.

The concept of Walras equilibrium was introduced by Walras (1874) and since then many partial proofs were offered, until the basic one due to Arrow and Debreu (1954). The alternative modelization of an equilibrium was proposed by Aubin (see Aubin and Cellina, 1983).

Monotone maps were introduced by Zarantonello (1960). For a detailed account of monotone maps and variational inequalities, see Brezis (1968, 1973), J. L. Lions (1969), and F. Browder (1976).

Maximal monotone maps were introduced and characterized by Minty (1965).

The proof of the theorem on the sum of two maximal monotone maps is due to Attouch (1981), extending a result due to Brezis, Crandall, and Pazy (1970) and Rockafellar (1974a).

We refer to the paper by Brezis and Haraux (1976) for the study of the range of the sum of two maximal monotone maps, and applications to Hammerstein equations (see also Brezis and Browder, 1975 and F. Browder, 1975). The existence and uniqueness of a solution to a differential inclusion for maximal monotone maps is due to Crandall and Pazy (1969).

The literature on fixed-point theory and monotone maps is quite large. Further references are listed in the bibliography.

CHAPTER 7: NONSMOOTH ANALYSIS

Nonsmooth analysis started at the end of the 1960s when the need to extend the successful subdifferential calculus to nonconvex and nonsmooth functions or to use convenient "tangent cones" for expressing the necessary conditions became pressing. Let us quote, among many other works, Dubovitskii and Miljutin (1971), Ioffe and Tihomirov (1972), Laurent (1972), and Neustadt (1976).

The concept of generalized gradient and normal cone introduced by Clarke (1975) gave a new impetus in the field and was at the origin of a considerable amount of work. Other attempts for defining other concepts of generalized gradients were made by Russian mathematicians (see, e.g., Demianov and Vassiliev, 1981; Demianov and Rubinov, 1983, and Pchenitchny, 1980. The part of this chapter dealing with generalized gradient and normal cones is based on the works of Clarke (1975, 1976d, 1977b, 1981a) regrouped in Clarke (1983), and the works of Rockafellar (1979a, b, c, 1980). The importance of the role of the Bouligand tangent cone (Bouligand, 1930, 1932) in viability theory for differential inclusion was recognized by Haddad (1981a), following many papers (Brézis, 1970; Clarke, 1975; Crandall, 1970; Gautier and Penot, 1979; Ladde and Lakshmikantham, 1974; Redheffer, 1972; Yorke, 1967, 1969). See the book by Aubin and Cellina (1983) for further comments. The fact that the tangent cone is the Kuratowski liminf of the contingent cone was discovered by Cornet (1981a), Penot (1981), and Rockafellar and Wets (unpublished). Many works were denoted to tangent cones and derivatives (Auslender, 1978a, b; Crouzeix, 1977, for quasiconvex functions; Gauvin, 1979; Gollan, 1981; Halkin, 1976; Hiriart-Urruty, 1978, 1979a, b, c; Hiriart-Urruty and Thibault, 1980; Hogan, 1973; Ioffe, 1981a; Janin, 1982; Lebourg, 1975, 1979; Lemarechal, 1975, Lempio-Maurer, 1980; Penot, 1974, 1978a, b, c; Shi Shu Chung, 1980; Thibault, 1979; Warga, 1976, 1978a).

In Frankowska (to appear), one can find an intermediate cone lying between $C_K(x)$ and $T_K(x)$, namely,

$$P_K(x) := \liminf_{h \to 0} \frac{1}{h}(K - x)$$

which coincides with the tangent space of a differentiable manifold. She also introduces the asymptotic cone T^∞ of a nonconvex cone T, defined by

$$T^\infty := \{v \ \varepsilon T | v + T \subset T\}$$

which is a convex subcone of T. The asymptotic cone $P_K^\infty(x)$ of $P_K(x)$, which is a closed convex cone larger than $C_K(x)$, plays quite an important role.

The associated generalized gradient $\partial_\infty V(x)$, the set of p such that $(p, -1) \in P^\infty{}_{p(V)}(x, V(x))^-$, is smaller than the generalized gradient, and is reduced to the usual gradient when V is only Fréchet differentiable (instead of being of class C^1 as for Clarke's generalized gradient).

Epi-contingent derivatives are quite useful for the theory of Hamilton–Jacobi equations (see Aubin, 1981 and Aubin and Cellina, 1983) and are related to the concept of generalized solutions introduced in Crandall and Lions (1981) and P.-L. Lions (1981a, 1982).

Many concepts of generalized derivatives of vector-valued maps have been proposed and studied. Let us mention the fans, introduced by Ioffe (1979, 1982). See also Aubin (1982b), Ioffe (1981c), and the papers of Clarke (1976e), Hiriart-Urruty (to appear), Kutateladze (1977), McLeod (1965), Sweester (1977), and Thibault (1982),

The concept of contingent derivative of set-valued map was introduced in Aubin (1981) and the concept of derivative in Aubin (1982a).

Other concepts of derivatives of set-valued maps were proposed by Banks and Jacobs (1970), de Blasi (1976), Boudourides and Shinas (1981), Gautier (unpublished), Mirica (1980), Nurminski (1978), Petcherskaja (1980), Shinas and Boudourides (1981), and Spingarn (1981).

By using the asymptotic tangent cone $P_K^\infty(x)$ she introduced in 1983, H. Frankowska has defined asymptotic derivatives and asymptotic co-differentials of set-valued maps and has used them for proving necessary conditions for optimal trajectories of differential inclusions.

The inverse function theorem is taken from Aubin (1982a). See also the papers of F. Clarke (1976c), Halkin (1976), Ioffe (1981c), and Warga (1978). Corollary 6 subsumes many earlier results of "normal solvability theory": see the papers by F. Browder (1976) and Kirk (1975).

CHAPTER 8: HAMILTONIAN SYSTEM

There is a vast literature on the subject of periodic solutions for Hamiltonian systems, beginning with the classical work of Poincaré, "Les méthodes nouvelles de la mécanique céleste" (1892–1899). We refer to the book by Moser (1973) for a survey and a bibliography, and to Jorna (1978) and Lichtenberg and Lieberman (1983) for a practicioner's view on the subject.

Most of this literature deals with systems depending on a small parameter, by the various methods of perturbation theory. The interest in global, topological methods was rekindled by the Rabinowitz paper (1978). See the surveys by Berestycki (1983) and Ambrosetti (1983) for a bibliography of recent developments.

The results in Section 2 can be traced to F. Clarke (1978, 1980a). Theorem 2.1 is found in Ekeland and Lasry (1980a, 1983). Theorem 2.4 is found in Clarke and Ekeland (1978, 1980). There is also related work by Aubin and Ekeland (1980) and Brézis, Coron and Nirenberg (1980).

Theorem 3.1 (the nonresonant case) is well known: see, for instance, Mahwin (1976) and the book by Fucik (1983). The proof we give is of course new.

Theorem 4.1 (the resonant case) is due to Clarke and Ekeland (1978, 1980). See also Clarke and Ekeland (1982), Ekeland (1981a, 1981b), and Willem (to appear) for a study of nonautonomous systems by this method.

Theorem 5.1 (the trans-resonant case) is a particular case of the paper by Rabinowitz (1978), where convexity is replaced by a weaker assumption. The use of the Ambrosetti–Rabinowitz theorem in this context was initiated by Ekeland (1979b) and extended by Brézis, Coron, and Nirenberg (1980) to a nonlinear wave equation. See Ambrosetti and Mancini (1981) for a different proof, and progress on the question of minimality: does the solution found in theorem 5.1 have minimal period T?

The nonautonomous trans-resonant case has been solved by Bahri and Berestycki (1981, 1983), who have found infinitely many periodic solutions.

The question of finding periodic solutions with prescribed energy (instead of prescribed period) has also given rise to interesting developments; see Weinstein (1973), Moser (1976), Ekeland and Lasry (1980), and Ekeland (1984).

Finally, note that direct variational methods, without the help of the duality theory of Section 2, have also proved quite useful; see Rabinowitz (1978), Benci and Rabinowitz (1979), and Bahri and Berestycki (1981, 1983).

Bibliography

Abraham, R., and Robbin, J. (1967). *Transversal mappings and flows*. Benjamin, New York.

Amann, H., and Zehnder, E. T. (1980). Nontrivial solutions for a class of nonresonance problems and application to nonlinear differential equations. *Ann. Sc. N. Sup. Pisa.*, **7**, 539–603.

Ambrosetti, A., and Mancini, G. (1981). Solutions of minimal period for a class of convex Hamiltonian systems. *Math. Ann.*, **255**, 405–421.

Ambrosetti, A., and Mancini, G. (1981). On a theorem of Ekeland and Lasry concerning the number of periodic Hamiltonian trajectories. *J. Diff. Eq.*, **43**, 1–6.

Ambrosetti, A., and Prodi, G. (1973). On the inversion of some differential mappings with singularities between Banach spaces. *Ann. Mat. Pura Appl.*, **93**, 291–247.

Ambrosetti, A., and Rabinowitz, P. (1973). Dual variational methods in critical point theory and applications. *J. Funct. Anal.*, **14**, 349–381.

Antosiewicz, H. A., and Cellina, A. (1975). Continuous selections and differential relations. *J. Diff. Eq.*, **19**, 386–398.

Antosiewicz, H. A., and Cellina, A. (1977). Continuous extension of multifunctions. *Ann. Polonisi Mat.*, **34**, 107–111.

Arnold, V. (1974). *Methodes mathèmatiques de la mécanique classique* (French translation) MIR, Moscow, 1976.

Arnold, V. (1978). *Chapîtres Supplèmentaires de la Théorie des Équations Différentielles Ordinaires.* (French translation). MIR, Moscow, 1980.

Arrow, K. J., and Debreu, G. (1954). Existence of an equillibrium for a competitive economy. *Econometrica*, **22**, 265–290.

Arrow, K. J., and Hahn, F. M. (1971). *General Competitive Analysis*. Holden-Day, San Francisco.

Artstein, Z. (1974). On the calculus of closed set-valued functions. *Indiana Univ. Math. J.*, **24**, 433–441.

Asplund, E. (1966). Farthest points in reflexive locally uniformly rotund Banach spaces. *Isr. J. Math.*, **4**, 213–216.

Asplund, E. (1968). Fréchet differentiability of convex functions. *Acta Math.*, **121**, 31–47.

Asplund, E., and Rockafellar, R. T. (1969). Gradients of convex functions. *Trans. Am. Math. Soc.*, **139**, 443–467.

Attouch, H. (1979). Famille d'opérateurs maximaux monotones et mesurabilité. *Ann. Mat. Pura Appl.*, **120**, 35–111.

Attouch, H. (1981). On the maximality of the sum of two maximal monotone operators. *Nonlinear Anal. TAM*, **5**, 143–147.

Attouch, H. (To apppear). *Variational Convergence for Functions and Operators*, Research Notes in Mathematics. Pitman, London.

Attouch, H., and Damlamian, A. (1972). On multivalued evolution equations in Hilbert spaces. *Isr. J. Math.*, **12**, 373–390.

Attouch, H., and Damlamian, A. (1975). Problèmes d'évolution dans les Hilbert et applications. *J. Math. Pure Appl.*, **54**, 53–74.

Attouch, H., and Wets, R. (1983). A convergence for bivariate functions aimed at the convergence of saddle values. In *Mathematical Theory of Optimization*, P. Cecconi and T. Zolezzi (Eds.), Springer-Verlag, Berlin.

Attouch, H., and Wets, R. (1983). Convergence de points min/sup et de points fixes. *CRAS*, **296**, 657–660.

Attouch, H., and Wets, R. (1983). A convergence theory for saddle functions. *Trans. Am. Math. Soc.*, in press.

Aubin, J. P. (1963). Un théorème de compacité. *CRAS*, **265**, 5042–5045.

Aubin, J. P. (1970). Abstract boundary-value operators and their adjoints. *Rend. Sem. Padova*, **43**, 1–33.

Aubin, J. P. (1972). Théorème du minimax pour une classe de fonctions. *CRAS*, **274**, 455–458.

Aubin, J. P. (1974). Règles de décision optimales en théorie des jeux à deux personnes. *CRAS*, **279**, 173–176.

Aubin, J. P. (1977). Evolution monotone d'allocations de biens disponibles. *CRAS*, **285**, 293–296.

Aubin, J. R., (1977). *Applied Abstract Analysis*. Wiley-Interscience, New York.

Aubin, J. P. (1978). Propriété de Perron-Frobenius pour des correspondances. *CRAS*, **286**, 911–914.

Aubin, J. P. (1978). Analyse fonctionnelle nonlinéaire et applications à l'équilibre économique. *Ann. Sc. Math. Quebec*, **2**, 5–47.

Aubin, J. P. (1978). Gradients généralisés de Clarke. *Ann. Sc. Math. Québec*, **2**, 197–252.

Aubin, J. P. (1979). Cônes tangents à un sous-ensemble convexe fermé. *Ann. Sc. Math. Québec*, **3**, 63–80.

Aubin, J. P. (1979). *Mathematical Methods of Game and Economic Theory*. North-Holland, Amsterdam.

Aubin, J. P. (1979). *Applied functional analysis*. Wiley-Interscience, New York.

Aubin, J. P. (1980). Further properties of Lagrange multipliers in nonsmooth optimization. *Appl. Math. Opt.*, **6**, 79–90.

Aubin, J. P. (1981). Contingent derivatives of set-valued maps and existence of solutions to nonlinear inclusions and differential inclusions. *In Advances in Mathematics Supplementary Studies*, L. Nachbin (Ed.), Academic Press, New York, pp. 160–232.

Aubin, J. P. (1982). Ioffe's fans and generalized derivatives of vector-valued maps. In *Convex Analysis and Optimization*, J. P. Aubin and R. Vinter (Ed.), Pitman, London.

Aubin, J. P. (1982). Comportement lipschitzien des solutions de problèmes de minimisation convexes. *CRAS*, **295**, 235–238.

Aubin, J. P., and Cellina, A. (1984). *Differential Inclusions*. Springer-Verlag, Berlin.

Aubin, J. P., and Clarke, F. M. (1977). Monotone invariant solutions to differential inclusions. *J. London Math. Soc.*, **16**, 357–366.

Aubin J. P., and Clarke, F. M. (1977). Multiplicateurs de Lagrange en optimisation nonconvexe et applications. *J. London Math. Soc.*, **16**, 357–366.

Aubin J. P., and Clarke F. M. (1979). Shadow prices and duality for a class of optimal control problems. *SIAM J. Opt. Control*, **17**, 567–586.

Aubin, J. P., and Cornet, B. (1976). Règles de décision en théorie des jeux et théorèmes de point fixe. *CRAS*, **283**, 11–14.

Aubin, J. P., Ekeland, I. (1976). Estimation of the duality gap in nonconvex optimization. *Math. Op. Res.*, **1**, 1–14.

Aubin, J. P., Ekeland, I. (1980). Second order evolution equation with convex Hamiltonian. *Can. Math. Bull.*, **23**, 81–94.

Aubin J. P., Siegel, J. (1980). Fixed points and stationary points of dissipative multivalued maps. *Proc. Am. Math. Soc.*, **78**, 391–398.

Aubin, J. P., Vinter, R. (1982). *Convex Analysis and Optimization*. Pitman, Boston.

Aumann, R. J. (1965). Integrals of set-valued functions. *J. Math. Anal. Appl.*, **12**, 1–12.

Auslender, A. (1976). *Optimisation: Méthodes Numériques*. Masson, Paris.

Auslender, A. (1977). Minimisation sans contraintes de fonctions localement lipschitziennes. *CRAS*, **284**, 959–961.

Auslender, A. (1978). Stabilité différentiable en programmation nonconvexe nondifférentiable. *C. R. Acad. Sci. Paris*, **286**, 575–577.

Auslender, A. (1978). Differentiable stability in nonconvex and nondifferentiable programming. In *Mathematical Programming Study*, No. 10, P. Huard (Ed.).

Averbuh, V. I., and Smolyanov, O. G. (1968). Different definitions of derivatives in linear topological spaces. *Usp. Mat. Nauk*, **23**, 67–116.

Bahri, A., and Berestycki, H. (1981). A perturbation method in critical point theory. *Trans. Am. Math. Soc.*, **267**, 1–32.

Bahri, A., and Berestycki, H. (to appear). Existence of forced oscillations for some nonlinear differential equations. *Comm. Pure. Appl. Math.*

Bahri, A., Berestycki, H. (to appear). Forced vibrations of superquadratic Hamiltonian systems. *Acta Mathem.*

Baillon, J. B. (1978). Générateurs de semi-groupes dans les espaces de Banach uniformément lisses. *J. Funct. Anal.*, **29**, 199–212.

Baillon, J. B., and Brézis, H. (1976). Une remarque sur le comportement asymptotique des semigroupes nonlinéaires. *Houston J. Math.*, **2**, 5–7.

Baillon, J. B., and Haddad, G. (1977). Quelques propriétés des opérateurs angle-bornés et n-cycliquement monotones. *Isr. J. Math.*, **26**, 137–150.

Baiocchi, C., and Capelo, A. (1978). *Disequazioni variazionali e quasivariazionali. Applicazioni a problemi di frontiera libera.* Pitagora Editrice, Bologna.

Banks, H. T., and Jacobs, M. Q. (1970). A differential calculus for multi-functions. *J. Math. An. Appl.*, **29**, 246–272.

Barbu, V. (1976). *Nonlinear Semi-Groups and Differential Equations in Banach Spaces*. Noordhoff, Leiden, Netherlands.

Barbu, V. (1981). Necessary conditions for nonconvex distributed control problems. *J. Math. Anal. Appl.*, **80**, 566–597.

Barbu, V., and Cellina, A. (1970). On the surjectivity of multivalued dissipative mappings. *Boll. Un. Mat. Ital.*, **3**, 817–826.

Beer, G. (to appear). On functions that approximate relations. *Proc. Am. Math. Soc.*

Begle, E. (1950). A fixed-point theorem. *Ann. Math.*, **51**, 544–550.

Belluce, L. P., and Kirk, W. A. (1969). Some fixed point theorems in metric and Banach spaces. *Can. Math. Bull.*, **127**, 481–489.

Belluce, L. P., and Kirk, W. A. (1969). Fixed point theorems for families of contraction mappings. *Proc. Am. Math. Soc.*, **20**, 141–146.

Benci, V. (1981). A geometrical index for the group S^1 and some applications to the study of periodic solutions to ordinary differential equations. *Comm. Pure Appl. Math.*, **34**, 393–432.

Benci, V., and Rabinowitz, P. (1979). Critical point theorems for indefinite functionals. *Inv. Math.*, **52**, 241–273.

Bensoussan, A. (1982). *Stochastic Control by Functional Analysis Methods*. North-Holland, Amsterdam.

Bensoussan, A., and Lions, J. L. (1978). *Applications des inéquations variationnelles et contrôle stochastique*. Dunod, Paris.

Bensoussan, A., and Lions, J. L. (1982). *Contrôle Impulsionnel et Inéquations Quasi-Variationnelles*. Dunod, Paris.

Berestycki, H. (1983). Solutions périodiques de systèmes Hamiltoniens, *Séminaire Bourbaki* No. 603.

Berestycki, H., Lasry, J. M., Mancini, G., and Ruf, B. (to appear). Existence of multiple periodic solutions on star-shaped Hamiltonian surfaces. *Comm. Pure Appl. Math.*

Berge, C. (1959). *Espaces Topologiques et Fonctions Multivoques*. Dunod, Paris.

Bishop, E., and Phelps, R. (1963). The support functional of a convex set. *Proc. Symp. Pure Math.*, **7**, 27–35.

de Blasi, F. S. (1976). On differentiability of multifunctions. *Pac. J. Math.*, **66**, 67–81.

Bliss, G. A. (1946). *Lectures on the Calculus of Variations*. University of Chicago Press, Chicago.

Bondareva, O. N. (1962). Theory of the core in n-person games, (in Russian). *Vestn. Leningrad*, **13**, 141–142.

Bony, J. M. (1969). Principe du maximum, inegalité de Harnack, et unicité du problèm de Cauchy pour des operateurs elliptiques dégériéiés *Ann. Inst. Fourier*, **19**, 277–304.

Borwein, J. (1978). Weak tangent cones and optimization in Banach spaces. *SIAM J. Cont. Opt.*, **16**, 512–522.

Boudourides, M., and Shinas, J. (1981). The mean value theorem for multifunctions, *Bull. Math Roum.*, **25**, 128–141.

Bouligand, G. (1930). Sur les surface dépourvues de points hyperlimites. *Ann. Soc. Polon. Math.*, **9**, 32–41.

Bouligand, G. (1932). *Introduction à la gêométrie Infinitésimale Directe*. Gauthier-Villars, Paris.

Bourbaki, N. (1981). *Espaces Vectoriels Topologiques*. Masson, Paris, Chaps. 1–5.

Bourbaki, N. (1982). *Variétés Différentielles et Analytiques*. Masson, Paris, Fasc. 1–5.

Bourbaki, N. (1982). *Topologie Générale*. Masson, Paris, Chap. 1–4.

Brézis, H. (1968). Equations et inéquations nonlinéaires dans les espaces vectoriels en dualité. *Ann. Inst. Fourier*, **18**, 115–175.

Brézis, H. (1970). On a characterization of flow invariant sets. *Comm. Pure Appl. Math.*, **23**, 261–263.

Brézis, H. (1973). *Opérateurs Maximaux Monotones et Semi-Groupes de Contractions dans les Espaces de Hilbert*. North-Holland, Amsterdam.

Brézis, H. (1983). *Analyse Fonctionnelle et Applications*. Masson, Paris.

Brézis, H. (1983). Periodic solutions of nonlinear vibrating strings and duality principles. *Bull. Am. Math. Soc.* **8**, 409–426.

Brézis, H., Coron, J. M., and Nirenberg, L. (1980). Free vibrations for a nonlinear wave equations and a theorem of P. Rabinovitz. *Comm. Pure App. Math.*, **33**, 667–689.

Brézis, H., and Nirenberg, L. (1978). Characterization of the range of some nonlinear operators and applications to boundary valve problems *Ann. Scuola Norm. Sup. Pisa*, **5**, 225–326.

Brézis, H., and Browder, F. (1974). Some new results about Hammerstein equations, *Bull. Am. Math. Soc.*, **80**, 567–572.

Brézis, H., and Browder, F. (1975). Nonlinear integral equations and systems of Hammerstein type. *Adv. Math.*, **18**, 115–147.

Brézis, H., and Browder, F. (1976). Nonlinear egodic theorems. *Bull. Am. Math. Soc.*, **82**, 959–961.

Brézis, H., and Browder, F. (1976). A general principle on ordered sets in nonlinear functional analysis. *Adv. Math.*, **21**, 355–369.

Brézis, H., and Browder, F. (1977). Remarks on nonlinear ergodic theory. *Adv. Math.*, **25**, 165–177.

Brézis, H., Crandall, M., and Pazy, A. (1970). Perturbations of nonlinear maximal monotone sets in Banach spaces. *Comm. Pure Appl. Math.*, **23**, 123–144.

Brézis, H., Crandall, M., and Pazy, A. (1976). Un principe variationnel associé à certaines équations paraboliques. *CRAS*, **282**, 971–974.

Brézis H., and Haraux, A. (1976). Image d'une somme d'opérateurs monotones et applications. *Isr. J. Math.*, **23**, 165–186.

Brézis, H., and Lions, P. L. (1978). Produits infinis de résolvantes. *Isr. J. Math.*, **29**, 329–345.

Brézis, H., Nirenberg, L., and Stampacchia, G. (1972). A remark on Ky Fan's minimax principle. *Boll. Un. Mat. Ital.*, **6**, 293–300.

Bröcker, T. (1972). *Differenzierbare Abbildungen* Der Regensburger Trichter. (English translation: Cambridge University Press, Cambridge, U.K.).

Brondsted, A., and Rockafellar, R. T. (1965). On the subdifferentiability of convex functions. *Proc. Am. Math. Soc.*, **16**, 605–611.

Brouwer, L. (1910). Uber eindeutige stetige Transformationen von Flächen in sich. *Math. Ann.*, **67**, 176–180.

Brouwer, L. (1911). Beweis der Invarianz der Dimensionzahl. *Math. Ann.*, **70**.

Brouwer, L. E. J. (1912). Uber Abbildungen von Manningfaltigkeiten. *Math. Ann.*, **71**, 97–115.

Browder, F. (1965). Fixed point theorems for non-compact mappings in Hilbert spaces. *Proc. Natl. Acas. Sci. USA*, **53**, 1272–1276.

Browder, F. (1965). Non expansive nonlinear operators in a Banach space. *Proc. Natl. Acad. Sci. USA*, **54**, 1041–1044.

Browder, F. (1968). The fixed point theory of multivalued mappings in topological vector spaces. *Math. Ann.*, **177**, 183–301.

Browder, F. (1976). *Nonlinear operators and nonlinear equations of evolution in Banach spaces. Am. Math. Soc. Proc. Symp. Pure Math.*, **18**, 2.

Browder, F. (1976). Normal solvability for nonlinear mappings into Banach spaces. *Bull. Am. Math. Soc.*, **79**, 328–350.

Browder, F. (1983). Fixed point theory and nonlinear problems. *Bull. Am. Math. Soc.*, **9**, (1), 1–39.

Bruckner, A. J., Leonard, J. L. (1966). 'Derivatives' in the Slaught memorial papers. *Am. Math. Mon.*, **73**, 24–56.

Cambini, A. (1973). Sul lemma di Morse. *Boll. Un. Mat. Ital.*, **7**, 87–93.

Carathéodory, C. (1967). *Calculus of Variations and Partial Differential Equations of the First Order*. Holden-Day, San Francisco.

Caristi, J. (1976). Fixed point theorems for mappings satisfying inwardness conditions. *Trans. Am. Math. Soc.*, **215**, 241–251.

Castaing, C., Valadier, M. (1977). *Convex analysis and measurable multifunctions. Lecture Notes in Mathematics.*, Volume 580. Springer-Verlag, Berlin.

Castro, A., and Lazer, A. (1979). Critical point theory and the number of solutions of a nonlinear Dirichlet prob. *Ann. Mat. Pura. Appl.*, **120**, 113–137.

Céa, J. (1971). *Optimisation: Théorie et Algorithmes*. Dunod, Paris.

Céa, J. Glowinski, R., and Nedelec, J. C. (1971). Minimisation de fonctionelles non différentiables. *Lecture Notes in Mathematics*, 228, Morris (Ed.), Springer-Verlag, Berlin.

Cellina, A. (1969). Approximation of set valued functions and fixed point theorems. *Ann. Mat. Pura Appl.*, **82**, 17–24.

Cellina, A. (1969). A theorem on the approximation of compact multivalued mappings. *Rend. Accad. Naz. Lincei*, **47**, 434–440.

Cellina, A. (1969). Multivalued Functions and Multivalued Flows. University of Maryland Tech. Note BN 615.

Cellina, A. (1970). A further result on the approximation of set valued mappings. *Rend. Accad. Naz. Lincei*, **48**, 230–234.

Cellina, A. (1970). Multivalued differential equations and ordinary differential equations. *SIAM J. Appl. Math.*, **18**, 533–538.

Cellina, A. (1971). The role of approximation in the theory of multivalued mappings. In *Differential Games and Related Topics*, H. W. Kuhn and G. P. Szego (Eds.), North-Holland, Amsterdam.

Cellina, A. (1971). On mappings defined by differential equations *Zesz. Nauk. Uniw. Jagiellon. Pr. Mat.*, **15**, 17–19.

Cellina, A. (1976). A selection theorem. *Rend. Sem. Univ. Padova*, **55**, 143–149.

Cellina, A. (1980). On the differential inclusion $x' \in (-1, +1)$. *Rend. Acc. Naz. Lincei*, **69**, 1–6.

Cellina, A., and Lasota, A. (1969). A new approach to the definition of topological degree for multivalued mappings. *Rend. Accad. Naz. Lincei*, **47**, 434–440.

Cellina, A., and Marchi, M. V. (1982). Nonconvex perturbations of maximal monotone differential inclusions.

Cerf, J. (1966). $\Gamma^4 = 0$. *Lecture Notes in Mathematics*, Volume 25. Springer-Verlag, Berlin.

Chaney, R. W. (1982). Second order sufficiency conditions for non differentiable programming problems. *SIAM J. Contrôl Opt.*, **20**, 20–33.

Chow, S. Mallet-Paret, J. (1982). *Methods of bifurcation theory*, Springer.

Clark, C. W., Clarke, F., and Munro, G. R. (1979). The optimal exploitation of renewable resource stocks. *Economics*, **47**, 25–47.

Clarke, F. H. (1975). The Euler-Lagrange differential inclusion. *J. Diff. Eq.*, **19**, 80–90.

Clarke, F. H. (1975). Generalized gradients and applications. *Trans. Am. Math. Soc.*, **205**, 247–262.

Clarke, F. H. (1976). On the inverse function theorem. *Pac. J. Math.*, **64**, 97–102.

Clarke, F. H. (1976). The maximum principle under minimal hypotheses. *SIAM J. Contrôl Opt.*, **14**, 1078–1091.

Clarke, F. H. (1976). The generalized problem of Bolze. *SIAM J. Contrôl Opt.*, **14**, 682–699.

Clarke, F. H. (1976). A new approach to Lagrange multipliers. *Math. Op. Res.*, **1**, 165–174.

Clarke, F. H. (1976). Optimal solutions to differential inclusions. *J. Opt. Th. Appl.*, **19**, 469–478.

Clarke, F. H. (1977). External arcs and extended Hamiltonian systems. *Trans. Am. Math. Soc.*, **231**, 349–367.

Clarke, F. H. (1977). Multiple integrals of Lipschitz functions in the calculus of variations. *Proc. Am. Math. Soc.*, **64**, 260–264.

Clarke, F. H. (1978). Pointwise contraction criteria for the existence of fixed points. *Bull. Can. Math. Soc.*, **21**, 7–11.

Clarke, F. H. (1978). Nonsmooth analysis and optimization. *Proceedings of the International Congress of Mathematicians*, Helsinki, 1978.

Clarke, F. H. (1978). Solutions périodiques des équations Hamiltoniennes. *CRAS*, **287**, 951–952.

Clarke, F. H. (1979). Optimal control and the true Hamiltonian. *SIAM Rev.*, **21**, 157–166.

Clarke, F. H. (1980). The Erdmann condition and Hamiltonian inclusions in optimal control and calculus of variations *Can. J. Math.*, **32**, 494–509.

Clarke, F. H. (1981). Generalized gradients of Lipschitz functionals. *Adv. Math.*, **40**, 52–67.

Clarke, F. H. (1981). Periodic solutions of Hamiltonian inclusions. *J. Diff. Eq.*, **40**, 1–6.

Clarke, F. H. (1982). On Hamiltonian flows and symplectic transformations. *SIAM J. Control Opt.*, **20**, 355–359.

Clarke, F. H. (1983). Optimization and nonsmooth analysis. Wiley-Interscience, New York.

Clarke F. H., and Ekeland, I. (1978). Solutions périodiques, de période donnée, des équations Hamiltoniennes. *CRAS*, **287**, 1031–1015.

Clarke, F. H., and Ekeland, I. (1980). Hamiltonian trajectories having prescribed minimal period. *Comm. Pure Appl. Math.*, **33**, 103–116.

Clarke, F. H., and Ekeland, I. (1982). Nonlinear oscillations and boundary-value problems for Hamiltonian systems. *Arc. Ration. Mech. Anal.*, **78**, 315–333.

Conley, C., and Zehnder, E. (1983). The Birkhoff-Lewis fixed point theorem and a conjecture of V. I. Arnold. *I. Math.*, **73**, 33–49.

Conley, C., and Zehnder, E. (to appear). Morse type index theory for flows and periodic solutions for Hamiltonian equations. *Comm. Pure Appl. Math.*

Cornet, B. (1977). Accessibilité des optima de Pareto par des processus monotones. *CRAS, Sér. A*, 641–644.

Cornet, B. (1977). An abstract theorem for planning procedures. *Lec. Notes Econ. Math. Syst.*, **144**, 53–59.

Cornet, B. (1981). Regularity properties of tangent and normal cones. *Cah. Math. Décision, Univ. Paris Dauphine*, 81–30.

Cornet, B. (1981). Contributions à la théorie mathematique des mécanismes dynamiques d'allocation des ressources. Thèse de doctorat d'Etat, Université de Paris-Dauphine.

Cornet, B., and Lasry, J. M. (1976). Un théorème de surjectivité pour une procédure de planification. *CRAS*, **282**, 1375–1378.

Costa, D., and Willem, M. (1983). Multiple critical points of invariant functionals and applications. MRC Technical summary report 2532.

Cournot, A. A. (1838). *Recherches sur les Principes Mathéorie des Richesses*. H. Rivière and Co., Paris.

Crandall, M. G. (1970). Differential equations on convex sets. *J. Math. Soc. J.*, **22**, 443–455.

Crandall, M. G. (1972). A generalization of Peano's existence theorem and flow invariance. *Proc. Am. Math. Soc.*, **36**, 151–155.

Crandall, M. G. (1977). An introduction to constructive aspects of bifurcation and the implicit funct. th. Applications of bifurcation theory, P. Rabinowitz (Ed.), pp. 1–35. Academic Press, New York.

Crandall, M. G. (1979). *Nonlinear Evolution Equations*. Academic Press, New York.

Crandall, M. G., and Lions, P. L. (1981). Conditions d'unicité pour les solutions généralisées des equations d'Hamilton-Jacobi. *CRAS*, **292**, 183–186.

Crandall, M. G., and Pazy, A. (1969). Semi-groups of nonlinear contractions and dissipative sets. *J. Funct. Anal.*, **3**, 376–418.

Crandall, M. G., and Pazy, A. (1970). On accretive sets in Banach spaces. *J. Funct. Anal.*, **5**, 204–217.

Crandall, M. G., and Rabinowitz, P. (1971). Bifurcation from simple eigenvalues. *J. Funct. Anal.*, **8**, 321–340.

Crandall, M. G., and Rabinowitz, P. (1973). Bifurcation, perturbation of simple eigenvalues and linearized stability. *Arch. Rat. Mech. An.*, **52**, 161–180.

Cronin, J. (1964). Fixed points and topological degree in nonlinear analysis. *Am. Math. Soc. Math. Surveys*, **11**.

Crouzeix, J. P. (1977). Contribution à l'étude des fonctions quasi-convexes. Ph D Thesis, Universite de Clermont.

Crouzeiz, J. P. (1980). On second order conditions for quasi-convexity. *Math. Prog.*, **18**, 349–352.

Danes, J. (1979). Two fixed point theorems in topological and metric spaces. *Bull. Aust. Math. Soc.*, **14**, 259–265.

Dantzig, G. B. (1963). *Linear Programming and Extensions*. Princeton University Press.

Debreu, G. (1959). *Theory of Value*. Wiley, New York.

Demianov, V. F. (1982). *Non Smooth Analysis* (in Russian). Leningrad University Press.

Demianov, V. F., and Malosemov, V. N. (1972). *Introduction to Minimax* (in Russian). Nauka, Moscow.

Demianov, V. F., and Rubinov, A. M. (1980). On quasidifferentiable functionals (in Russian). *Dokl. Acad. Nauk SSSR Sov. Math. Dokl.*, **21**, 14–17.

Demianov, V. F., and Vassiliev, L. V. (1981). *Non differentiable optimization* (in Russian). Nauka, Moscow.

Desolneux-Moulis, N. (1979). Orbites périodiques des systèmes Hamiltoniens autonômes. *Séminaires Bourbaki*, **32**, (552).

Diestel, J. (1975). Geometry of Banach spaces. In *Lecture Notes in Mathematics*, Vol. 485, Springer-Verlag, Berlin.

Diestel, J. (1977). Vector measures. *Am. Math. Soc., Math Surveys*, **15**.

Dolecki, Z., Salinetti, G., and Wets, R. (1983). Convergence of functions: equi-semicontinuity. *Trans. Am. Math. Soc.*, **276**, 409–429.

Downing, D., and Kirk, W. A. (1977). A generalization of Caristi's theorem and application to nonlinear mapping theory. *Pac. J. Math.*, **69**, 339–347.

Downing, D., and Kirk, W. A. (1977). Fixed point theorems for set-valued mappings in metric and Banach spaces. *Math. Jpn.*, **22**, 99–112.

Downing, D., and Ray, W. (to appear). Renorming and the theory of phi-accretive set-valued mappings. *Pac. J. Math.*

Dubovicki, A. I., and Milyutin, A. M. (1971). *Necessary Conditions for a Weak Extremum in the General Problem of Optimal Control* (in Russian). Nauka, Moscow.

Dugundji, J. (1951). An extension of Tietze's theorem. *Pac. J. Math.*, **1**, 353–367.

Dugundji, J. (1954). *Topology*. Allyn and Bacon, Boston.

Dunford, N., and Schwartz, J. T. (1958). *Linear Operators*. Wiley-Interscience, New York.

Duvaut, G., and Lions, J. L. (1972). *Les Inéquations en Mécanique et en Physique*. Dunod, Paris.

Eaves, B. C. (1972). Homotopies for computation of fixed points. *Math. Program.*, **3**, 1–22.

Edelstein, H. (1963). A theorem on fixed points under isometries. *Am. Math. Mon.*, **70**, 298–300.

Edelstein, H. (1965). On nonexpansive mappings of Banach spaces. *Proc. Cambridge Phil. Soc.*, **60**, 439–447.

Edelstein, H. (1966). Farthest points of set uniformity convex Banach spaces. *Isr. J. Math.*, **4**, 171–176.

Edelstein, H. (1968). On nearest points of sets in uniformly convex Banach spaces. *J. London Math. Soc.*, **43**, 375–377.

Edelstein, H. (1972). The construction of an asymptotic center with a fixed point property. *Bull. Am. Math. Soc.*, **78**, 206–208.

Edelstein, H. (1974). Fixed point theorems in uniformly convex spaces. *Proc. Am. Math. Soc.*, **44**, 369–374.

Edelstein, H. (1975). On some aspects of fixed point theory in Banach spaces. In *The Geometry of Metric and Linear Spaces, Lecture Notes*, Vol. 490, Springer-Verlag, Berlin.

Eggleston, H. C. (1958). Convexity. *Cambridge Tracts in Mathematics*, Volume 47. Cambridge University Press, London.

Eilenberg, S., and Montgomery, D. (1966). Fixed point theorem for multivalued transformations. *Am. J. Math.*, **58**, 214–222.

Ekeland, I. (1972). Remarques sur les problèmes variationnels 1. *CRAS*, **275**, 1057–1059.

Ekeland, I. (1973). Remarques sur les problèmes variationnels 2. *CRAS*, **276**, 1347–1348.

Ekeland, I. (1974). On the variational principle. *J. Math. Anal. Appl.*, **47**, 324–353.

Ekeland, I. (1979). Periodic solutions of Hamilton's equations and a theorem of P. Rabinowitz. *J. Diff. Eq.*, **34**, 523–534.

Ekeland, I. (1979). Nonconvex minimization problems. *Bull. Am. Math. Soc.*, **1**, 443–474.

Ekeland, I. (1981). Oscillations forcées de systèmes Hamiltoniens nonlinéaires. *Bull. SMF*, **109**, 297–330.

Ekeland, I. (1981). Forced oscillations for nonlinear Hamiltonian systems. In *Advances in Mathematics.*, L. Nachbin (Ed.), Academic Press, New York.

Ekeland, I. (1982). Dualité et stabilité des systèmes Hamiltoniens. *CRAS*, **294**, 673–676.

Ekeland, I. (1983). A perturbation theory near convex Hamiltonian systems. *J. Diff. Eq.*, **50**, 407–440.

Ekeland, I. (1984). Une théorie de Morse pour les systèmes Hamiltoniens convexes. *Ann. IHP, Nonlinear Anal.* **1**, 19–78.

Ekeland, I., and Lasry, J. M. (1980). Problèmes variationnels non convexes en dualité. *CRAS*, **291**, 493–495.

Ekeland, I., and Lasry, J. M. (1980). On the number of closed trajectories for a Hamiltonian flow on a convex energy surface. *Ann. Math.*, **112**, 283–319.

Ekeland, I., and Lebourg, G. (1976). Generic Frechet-differentiability and perturbed optimization problems in Banach spaces. *Trans. Am. Math. Soc.* **224**, 193–216.

Ekeland, I., and Temam, R. (1976). *Convex Analysis and Variational Problems*. Elsevier North-Holland, Amsterdam.

Fadell, E. (1978). Recent results in the fixed point theory of continuous maps. *Bull. Am. Math. Soc.*, **76**, 10–29.

Fadell, E., and Rabinowitz, P. (1978). Generalized cohomological index theories for group actions with an application to bifurcation questions for Hamiltonian systems *Inv. Math.*, **45**, 139–174.

Fan, Ky (1952). Fixed point and minimax theorems in locally convex topological linear spaces. *Proc. Natl. Acad. Sci. USA*, **38**, 121–126.

Fan, Ky. (1953). Minimax theorems. *Proc. Natl. Acad. Sci. USA*, **39**, 42–47.

Fan, Ky. (1956). On systems of linear inequalities. *Ann. Math. Stud.* **38**, 99–156.

Fan, Ky. (1957). Existence theorems and extreme solutions for inequalities concerning convex functions or linear translations. *Math. Z.*, **68**, 205–216.

Fan, Ky. (1958). On the equilibrium value of a system of convex and concave functions. *Math. Z.*, **70**, 271–280.

Fan, Ky. (1961). A generalization of Tychonoff's fixed point theorem. *Math. Ann.*, **142**, 305–310.

Fan, Ky. (1963). On the Krein-Milman theorem. *Proc. Symp. Pure Math. Am. Math. Soc.* **7**, 211–219.

Fan, Ky. (1964). Sur un théorème minimax. *C.R. Acad. Sci.*, **259**, 3925–3928.

Fan, Ky. (1965). A generalization of the Alaoglu-Bourbaki theorem and its applications. *Math. Z.*, **88**, 48–60.

Fan, Ky. (1966). Applications of a theorem concerning sets with convex sections. *Math. Ann.*, **163**, 189–203.

Fan, Ky. (1970). A conbinatorial property of pseudomanifolds and covering properties of simplexes. *J. Math. Anal. Appl.*, **31**, 68–80.

Fan, Ky. (1972). A minimax inequalities and applications. In *Inequalities*, Volume 3, Shisha (Ed.). Academic Press, New York, pp. 103–113.

Fan, Ky. (1979). Fixed point theorems and related theorems for non-compact convex sets. In *Game Theory and Related Topics*, North-Holland, Amsterdam, pp. 151–156.

Fenchel, W. (1949). On conjugate convex functions. *Can. J. Math.*, **1**, 73–77.

Fenchel, W. (1951). *Convex Cones, Sets and Functions*, Mimeographed Lecture Notes. Princeton University.

Fenchel, W. (1952). A remark on convex sets and polarity. *Medd. Lunds Univ. Mat. Sem.* (Supplement Band), 82–89.

Fenchel, W. (1956). Uber konvexe Functionen mit vorgeschriebenen Niveaumannigfaltigkeiten. *Math. Z.*, **63**, 496–506.

Figueiredo, D. (1967). Topics in nonlinear analysis. Lecture Notes 48, University of Maryland.

Frankowska, H. (1983). Inclusions adjointes associées aux trajectoires minimales d'inclusions diff. *CRAS*, (in press).

Frankowska, H. (to appear). First order necessary conditions for nonsmooth variational control problems. *SIAM. J. Opt.*

Frankowska, H. (to appear). On the single-valuedness of Hamilton–Jacobi operators. *J. Nonlinear Anal. TAM.*

Frobenius, G. (1908). Uber Matrizen aus positiven elementen. *Sitzungsber Pr. Akad. Wiss.*, 471–476.

Fucik, S. (1983). *Nonlinear Problems.* Reidel, Dordrecht, Netherlands.

Furi, M., Martelli, M., and Vignoli, A. (1970). On minimum problems for families of functionals. *Ann. Mat. Pura Appl.*, **86**, 181–187.

Gale, D. (1956). The closed linear model of production. *Ann. Math. Stud.*, **38**, 285–303.

Garcia, C. B., and Gould, F. J. (1978). A theorem on homotopy paths. *Math. Op. Res.*, **3**, 282–289.

Garcia, C. B. and Gould, F. J. (1980). Relations between several path following algorithms and local and global Newton methods. *SIAM Rev.*, **22**, 263–274.

Gautier, S. (1973). Différentiabillité des multiapplications. *Publ. Math. Pau.*

Gautier, S., and Penot, J. P. (1973). Fermés invariants par un système dynamique. *CRAS*, **276**, 1457–1460.

Gauvin, J. (1979). The generalized gradient of a marginal function in mathematical programming. *Math. Opt. Res.*, **4**, 458–463.

Gollan, B. (1981). Higher order necessary conditions for an abstract optimization problem. *Math. Program. Study*, **14**, 69–76.

Gollan, B. (to appear). A general perturbation theory for abstract optimization problems. *J. Opt. Th. Appl.*

Granas, A. (1962). Sur la multiplication cohomotopique dans les espaces de Banach. *CRAS*, **254**, 56–57.

Granas, A. (1976). Sur la méthode de continuité de Poincaré. *CRAS*, **282**, 983–986.

Haddad, G. (1981). Monotone viable trajectories for functional differential inclusions. *J. Diff. Eq.*, **42**, 1–24.

Haddad, G. (1981). Monotone trajectories of differential inclusions with memory. *Isr. J. Math.*, **39**, 83–100.

Haddad, G. (1981). Topological properties of the set of solutions for functional differential inclusions. *Nonlinear Anal. Theory, Meth. Appl.*, **5**, 1349–1366.

Haddad, G., and Lasry, J. M. (to appear). Periodic solutions of functional differential inclusions and fixed points of σ-selectionable correspondences. *J. Math. Anal. Appl.*

Hale, J. (1977). *Theory of Functional Differential Equations*. Springer-Verlag, Berlin.

Halkin, H. (1972). Extremal properties of biconvex contingent equations. In Ordinary Differentiable Equations (NRL-MRC Conference), Academic Press, New York.

Halkin, H. (1976). Interior mapping theorem with set-valued derivative. *J. Anal. Math.*, **30**, 200–207.

Halkin, H. (1976). Mathematical programming without differentiability. In *Calculus of Variations and Control Theory*, D. L. Russell (Ed.), Academic Press, New York, 279–288.

Halpern, B. R. (1968). A general fixed point theorem. *Proc. Symp. Nonlinear Funct. Anal. Amer. Math. Soc.*

Halpern, B. R., and Berginan, G. M. (1968). A fixed point theorem for inward and outward maps. *Trans. Am. Math. Soc.*, **130**, 353–358.

Hamilton, R. (1982). The inverse function theorem of Nash and Moser. *Bull. Am. Math. Soc.*, **7**, 1–64.

Hassard, B., Kazarinoff, N., and Wan, Y. (1981). Theory and applications of Hopf bifurcation. London Mathematical Society Lecture Notes 41, Cambridge University Press, London.

Hildenbrand, W. (1974). *Core and Equilibria of a Large Economy*. Princeton University Press, Princeton.

Himmelberg, C. J. (1972). Fixed points of compact multifunctions. *Indiana Univ. Math. J.*, **22**, 719–729.

Himmelberg, C. J. (1975). Measurable relations. *Fund. Math.*, **87**, 53–72.

Himmelberg, C. J., Jacobs M. Q., and Van Vleck, F. S. (1969). Measurable multifunctions, selectors and Filippov's implicit function lemma. *J. Math. Anal. Appl.*, **25**, 276–284.

Himmelberg, C. J., and Van Vleck, F. S. (1971). Selection and implicit function theorems for multifunctions with Souslin graph. *Bull. Acad. Pol. Sc.*, **19**, 911–916.

Himmelberg, C. J., Van Vleck, F. S. (1972). Lipschitzian generalized differential equations. *Rend. Sem. Mat. Padova*, **48**, 159–169.

Himmelberg, C. J., and Van Vleck, F. S. (1972). Fixed points of semi-condensing multifunctions. *Boll. Un. Mat. Ital.*, **5**, 187–194.

Hiriart-Urruty, J. B. (1978). Gradients généralisés de fonctions marginales. *SIAM J. Cont. Opt.*, **16**, 301–316.

Hiriart-Urruty, J. B. (1979). Refinements of necessary optimality conditions in nondifferntiable programming 1. *Appl. Math. Opt.*, 63–82.

Hiriart-Urruty, J. B. (1979). New concepts in nondifferentiable programming. *Bull. Soc. Math. F. Mém.*, **60**, 57–85.

Hiriart-Urruty, J. B. (1979). Tangent cones, generalized gradients and mathematical programming in Banach spaces. *Math. Oper. Res.*, **4**, 79–97.

Hiriart-Urruty, J. B. (1980). Mean value theorems in nonsmooth analysis. *Numer. Funct. Anal. Opt.*, **2**(1), 1–30.

Hiriart-Urruty, J. B. (1981). Optimality conditions for discrete nonlinear norm-approximation problems. In *Optimization and Optimal Control*, Springer-Verlag, Berlin, pp. 29–41.

Hiriart-Urruty, J. B. (to appear). Refinements of necessary optimality conditions in nondifferentiable programming 2. *Math. Programming Study*.

Hiriart-Urruty, J. B., and Thibault, L. (1980). Existence et caractérisation de différentielles généralisées. *CRAS*, **290**, 1091–1094.

Hirsch, M. W. (1976). *Differentiable Topology*. Springer-Verlag, Berlin.

Hirsch, M. W., and Smale, S. (1974). *Differential Equations, Dynamical Systems and Linear Algebra*. Academic Press, New York.

Hirsch, M. W., and Smale, S. (1978). On algorithms for solving $f(x)=0$. *Not. Am. Math. Soc.*, **25**, 501–544.

Hogan, W. (1973). Directional derivative for extremal value functions with applications to the completely convex case. *Oper. Res.*, **21**, 188–209.

Holmes, R. D. (1976). Fixed point for local radial contractions. *Proc Sympos Fixed Point Theory and Its Appl*, Dalhousie University Academic Press, New York, pp. 79–89.

Hopf, H. (1928). Eine Verallgemeinerung der Euler-Poincaréschen Formel. *Nachr. Ges. Wiss. Goettingen*, 127–136.

Hopf, H. (1929). Uber die algebraische Anzahl Fixpunkten. *Math. Z.*, **29**, 493–524.

Huard, P. (1975). Optimization algorithms and point to set maps. *Math Program.*, **8**, 308–331.

Ichiishi, T. (1981). A social coalitional equilibrium existence lemma. *Econometrica*, **49**.

Ichiishi, T. (1981). On the Knaster-Kuratowski-Mazurkiewicz-Shapley theorem. *J. Math. Anal. Appl.*

Ioffe, A. D. (1976). An existence theorem for a general Bolza problem. *SIAM J. Cont. Opt.*, **14**, **458**–466.

Ioffe, A. D. (1977). On lower semicontinuity of integral functionals 1. *SIAM J. Cont. Opt.*, **15**, 521–538.

Ioffe, A. D. (1978). Survey of measurable selection theorems: Russian literature supplement. *SIAM. J. Cont. Opt.*, **16**, 728–723.

Ioffe, A. D. (1979). Differentielles généralisées d'applications localement lipschitziennies d'un espace de Banach dans un autre. *CRAS*, **289**, 637–639.

Ioffe, A. D. (1981). A new proof of the equivalence of the Hahn-Banach extension. *Proc. Am. Math. Soc.*, **82**, 385–390.

Ioffe, A. D. (1981). Nonsmooth analysis: differential calculus of nondifferentiable mappings. *Trans. Am. Math. Soc.*, **266**, 1–56.

Ioffe, A. D. (1981). Sous-différentielles approchées de fonctions numériques. *CRAS*, **292**, 675–678.

Ioffe, A. D. (1982). Nonsmooth analysis and the theory of fans. *Convex Analysis and Optimization*, J. P. Aubin and R. Vinter (Eds.), Pitman, Boston, 93–117.

Ioffe, A. D. (to appear). Approximate subdifferential of nonconvex functions. *Trans. Am. Math. Soc.*

Ioffe, A. D., and Levin, V. L. (1972). Subdifferentials of convex functions. *Trans. Moscow Math. Soc.*, **26**, 1–72.

Ioffe, A. D., and Tihomirov, V. M. (1974). *The Theory of Extremal Problems.* Nauka, Moscow (English translation, North-Holland, Amsterdam, 1979).

Ioss, G., and Joseph, D. (1980). *Elementary Stability and Bifurcation Theory*. Springer-Verlag, Berlin

Istratescu, V. I. (1981). *Fixed point theory.* Reidel, Dordrecht.

Itoh, S., and Takahashi, W. (1977). Single valued mappings, multivalued mappings and fixed-point theorems. *J. Math. Anal. Appl.*, **59**, 514–521.

James, R. C. (1964). Weakly compact sets. *Trans. Amer. Math. Soc.*, **113**, 129–140.

Janin, R. (1982). Sur des multiapplications qui sont des gradients généralisés. *CRAS*, **294**, 115–117.

Jorna, S. (1978). Topics in nonlinear dynamics, Conference Proceedings, 46. American Institute of Physics, New York.

Kakutani, S. (1941). A generalization of Brouwer's fixed point theorem. *Duke Math. J.*, **8**, 457–459.

Kato, T. (1967). Nonlinear semi-groups and evolution equations. *J. Math. Soc. J.*, **19**, 508–520.

Kirk, W. A. (1965). A fixed point theorem for mappings which do not increase distance. *Am. Math. Mon.*, **72**, 1004–1006.

Kirk, W. A. (1976). Caristi's fixed point theorem and metric convexity. *Colloq. Math.*, **81**–86.

Kirk, W. A., and Caristi, J. (1975). Mapping theorems in metric and Banach spaces. *Bull. Acad. Polon. Sci.*, **23**, 391–394.

Knaster, B., Kuratowski, C., and Mazurkiewicz, C. (1926). Ein Beweis des Fixpunktsatzes fur n dimensionale Simplexe. *Fund. Math.*, **14**, 132–137.

Krasnoselski, M. A. (1963). *Topological Methods in the Theory of Nonlinear Integral Equations.* Pergamon Press, New York.

Krasnoselski, M. A. (1964). *Positive Solutions of Operator Equations*. Noordhoff, Groningen.

Krasnoselski, M. A., and Rutickii, Y. B. (1961). *Convex Functions and Orlicz Spaces*. Noordhoff, Groningen.

Krasnoselski, M. A., Zabreiko, P. P. (1975). *Geometrical methods of nonlinear analysis* (in Russian). Nauka, Moscow.

Kuhn, H. W., and Tucker, A. W. (1951). *Nonlinear Programming. Proceedings of the 2d Berkeley Symposium on Mathematical Statistics and Probability*, University of California Press, Berkeley. pp. 481–492.

Kuratowski, K. (1958). *Topologie*, Vols. 1 and 2 4th. Ed. corrected. Panstowowe Wyd Nauk, Warszawa. (Academic Press, New York, 1966.

Kutateladze, S. S. (1977). Subdifferentials of convex operators. *J. Math. Sibirsk*, **13**, 1057–1064.

Ladde, G., and Lakshmikantham, V. (1974). On flow invariant sets. *Pac. J. Math.*, **51**, 215–220.

Landesman, E., and Lazer, A. (1970). Nonlinear perturbations of linear elliptic boundary value problems at resonance. *J. Math. Mech.*, **19**, 609–623.

Lang, S. (1962). *Introduction to differentiable Manifolds*. Wiley-Interscience, New York.

Lasota, A., and Olech, C. (1966). On the closedness of the set of trajectories of a control system. *Bull. Acad. Pol. Sc.*, **14**, 615–621.

Lasota, A., and Olech, C. (1968). On Cesari's semicontinuity condition for set valued mappings. *Bull. Acad. Pol. Sc.*, **16**, 711–716.

Lasota, A., and Opial, Z. (1965). An application for the Kakutani-Ky Fan theorem in the theory of ordinary differential equations. *Bull. Acad. Pol. Sc.*, **13**, 781–786.

Lasota, A., Opial, Z. (1968). Fixed point theorems for multivalued mappings and optimal control problems. *Bull. Acad. Pol. Sc.*, **16**, 645–649.

Laurent, P. J. (1972). *Approximation et Optimisation*. Hermann, Paris.

Lazer, A. (1972). Application of a lemma on bilinear forms to nonlinear oscillations. *Proc. Am. Math. Soc.*, **33**, 89–94.

Lazer, A., and Sanchez, P. (1969). On periodically perturbed conservative systems. *Michigan Math. J.*, **16**, 193–200.

Leach, E. B. (1961). A note on inverse function theorems. *Proc. Am. Math. Soc.*, **12**, 694–697.

Lebourg, G. (1975). Valeur moyenne pour un gradient généralisé. *CRAS*, **281**, 795–797.

Lebourg, G. (1978). Problèmes d'optimalisation dépendant d'un paramètre à valeurs dans in espace de Banach. Séminaire d'Analyse Fonctionnelle, 1978–1979, Ecole Polytechnique, 11.

Lebourg, G. (1979). Generic differentiability of Lipschitzian functions. *Trans. Am. Math. Soc.*, **256**, 125–144.

Leibniz, G. (1684). *Nova methoda pro maximis et minimis.*

Lemaréchal, C. (1975). An extension of Davidon methods to nondifferentiable problems. Math programming study. *Nondifferentiability Optimization*, North-Holland, Amsterdam, 95–109.

Lempis, F., and Maurer, H. (1980). Differential stability in infinite dimensional nonlinear programming. *Appl. Math. Opt.*, **6**, 139–152.

Leray, J. (1946). Sur les équations et les transformations. *J. Math. Pure Appl.*, **24**, 201–248.

Leray, J. (1952). La théorie des points fixes et ses applications en analyse. *Proc. Int. Cong. Math.* (Cambridge 1950) *Am. Math. Soc.*, **2**, 202–208.

Leray, J. (1959). Théorie des points fixes, indice total et nombre de Lefschetz. *Bull. Soc. Math. Fr.*, **87**, 221–233.

Leray, J., and Lions, J. L. (1965). Quelques résultats de Visik sur les problèmes elliptiques nonlinéaires. *Bull. Soc. Math. Fr.*, **93**, 97–107.

Leray, J., Schauder, J. (1934). Topologie et équations fonctionnelles. *Ann. Sci. ENS*, **51**, 45–78.

Levitin, E. S. (1974). On differential properties in the optimum value in parametric problems of mathematices programming. *Dokl. Acad. Nauk SSSR*, **215**, (Sov. Math. Dokl. 1974, **15**, 603–608).

Lichtenberg, A., and Liebermann, M. (1983) *Regular and stochastic motion*, Springer-Verlag, Heidelberg.

Lions, J. L. (1968). *Contrôle Optimal de Systèmes Gouvernés par des Équations aux Dérivées Partielles*. Dunod, Paris.

Lions, J. L. (1969). *Quelques Méthodes de Résolution de Problèmes Nonlinéaires*. Dunod, Paris.

Lions, J. L. (1976). *Sur Quelques Questions d'Analyse, de Mécanique et de Contrôle Optimal*. Presses de l'Université de Montréal, Montréal.

Lions, P. L. (1977). Approximation de point fixe de contractions. *CRAS*, **284**, 1357–1359.

Lions, P. L. (1978). Produits infinis de résolvantes. *Isr. J. Math.*, **29**, 329–345.

Lions, P. L. (1978). Une méthode itérative de résolution d'une équation variationnelle. *Isr. J. Math.*, **31**, 204–208.

Lions, P. L. (1980). Résolution des équations de Hamilton-Jacobi-Bellman pour des opérateurs uniformément elliptiques. *CRAS*, **290**, 1049–1052.

Lions, P. L. (1981). Minimization problems in L^1. *J. Funct. Anal.*, **41**, 236–275.

Lions, P. L. (1981). Solutions généralisées des équations de Hamilton-Jacobi du ler ordre. *CRAS*, **292**, 953–956.

Lions, P. L. (1982). *Generalized Solutions of Hamilton-Jacobi Equations*. Pitman, Boston.

Lions, J. L., and Magenes, E. (1968). *Problèmes aux Limites Non Homogènes*. (3 vol.) Dunod-Gauthier-Villars, Paris.

Lions, J. L., and Stampacchia, G. (1965). Inéquations variationnelles noncoercives. *CRAS*, **261**, 25–27.

Lions, J. L., and Stampacchia, G. (1967). Variational inequalities. *Comm. Pure Appl. Math.*, **20**, 493–519.

Lipschitz, R. (1877). *Lehrbuch der Analyse*, Bonn.

Liusternik, L. A. (1947). *The Topology of the Calculus of Variations in the Large*. (English trans. Vol. 16, American Mathematical Society, 1966).

Liusternik, L. A., and Schnirelman, L. G. (1930). *Méthodes Topologiques dans les Problèmes Variationnels*. Hermann, Paris.

Makarov, V. L., and Rubinov, A. M. (1970). Superlinear point to set mappings of economic dynamics (in Russian). *Uspehi Mat. Nauk*, **25**, 125–169.

Makarov, V. L., and Rubinov, A. M. (1973). Mathematical theory of economical dynamics and equilibria (in Russian). Nauka, Moscow.

Mangasarian, O. L. (1966). Sufficient conditions for the optimal control of nonlinear systems. *SIAM J. Contr. Opt.*, **4**, 139–152.

Mangasarian, O. L. (1969). *Nonlinear Programming*. McGraw-Hill, New York.

Mangasarian, O. L., and Fromovitz, S. (1967). The Fritz John necessary optimality conditions in presence of equality and inequality constraints. *J. Math. Anal. Appl.*, **17**, 37–47.

Markin, J. T. (1973). Continuous dependence of fixed point sets. *Proc. Am. Math. Soc.*, **38**, 545–547.

Martelli, M., and Vignoli, A. (1974). On differentiability of multivalued maps. *Boll. Un. Mat. Stat.*, **10**, 701–712.

Martin, R. M. (1973). Differential equations on closed subsets of a Banach space. *Trans. Am. Math. Soc.*, **179**, 399–414.

Martin, R. M. (1976). *Nonlinear Operators and Differential Equations in Banach Spaces*. Wiley-Interscience, New York.

Maschler, M., and Peleg, B. (1976). Stable sets and stable points of set-valued dynamical systems. *Siam J. Cont. Opt.*, **14**, 985–995.

Maurer, H. (1979). First order sensitivity of the optional value function in mathematical programming and optimal control. Proc. Symp. Math. Prog. with Data Perturbations. Washington, D.C.

Maurer, H. (1979). Differential stability in optimal control problems. *Appl. Math. Opt.*, **5**, 283–295.

McCormick, G. P. (1975). Optimality criteria in nonlinear programming. In R. W. Cottle (ed.), Proceedings of the Symposium on Applied Mathematics, Vol. 9.

Mc Kenzie, L. (1959). On the existence of general equilibrium for a competitive market. *Econometrica*, **27**, 54–71.

McLeod, R. M. (1965). Mean value theorems for vector-valued functions. *Proc. Edinburgh Math. Soc.* **14**(2), 197–209.

McShane, E. J. (1939). On multipliers for Lagrange problems. *Am. J. Math.*, **61**, 809–819.

Michael, E. (1956). Continuous selections 1. *Ann. Math.*, **63**, 361–381.

Michael, E. (1956). Continuous selections 2. *Ann. Math.*, **64**, 562–580.

Michael, E. (1957). Continuous selections 3. *Ann. Math.*, **65**, 375–390.

Michael, E. (1959). A theorem on semicontinuous set valued functions. *Duke Math. J.*, **26**, 647–651.

Michel, P. (1974). Problèmes d'optimisation définispar des fonctions qui sont sommes de fonctions convexes et dérivables. *J. Math. Pure. Appl.*, **53**, 321–330.

Milnor, J. (1963). *Morse Theory*. Princeton University Press.

Milnor, J. (1965). *Topology from the Differentiable Viewpoint*. University of Virginia, Charlottesville.

Milnor, J. (1978). Analytic proof of the hairy ball theorem and the Brouwer fixed point theorem. *Am. Math. Mon.*, **85**, 521–524.

Minty, G. (1961). On the maximal domain of a monotone function. *Mich. Math. J.*, **8**, 135–137.

Minty, G. (1962). Monotone (nonlinear) operators in a Hilbert space. *Duke Math. J.*, **29**, 341–348.

Minty, G. (1964). On the monotonicity of the gradient of a convex function. *Pac. J. Math.*, **14**, 243–247.

Minty, G. (1965). A theorem on maximal monotone sets in Hilbert space. *J. Math. Anal. Appl.*, **11**, 434–439.

Minty, G. (1967). On the generalizing of the direct method of the calculus of variations. *Bull. Am. Math. Soc.*, **73**, 315–321.

Minty, G. (1974). A finite-dimensional tool-theorem in monotone operator theory. *Adv. Math.*, **12**, 1–7.

Mirica, S. (1980). A note on the generalized differentiability of mappings. *Trans. Am. Math. Soc.*, **4**, 567–575.

Moreau, J. J. (1965). Proximité et dualité dans un espace Hilbertien. *Bull. Soc. Math. Fr.*, **93**, 273–299.

Moreau, J. J. (1967). *Fonctionnelles Convexes*. Collège de France, Paris.

Moreau, J. J. (1974). La convexité en statistique. *Analyse Convexe et ses Applications*, J. P. Aubin (Ed.), Springer-Verlag, Berlin.

Moreau, J. J. (1978). Un cas de convergence des itérés d'une contraction d'un espace Hilbertien. *CRAS*, **286**, 143–144.

Mosco, U. (1969). Convergence of convex sets and of solutions of variational inequalities. *Adv. Math.*, **3**, 510–585.

Mosco, U. (1970). Perturbations of variational inequalities. *Proc. Symp. Pure Math., Am. Math. Soc.*, **18**, 182–194.

Mosco, U. (1976). *Implicit Variational Problems and Quasi-Variational Inequalities*. Lecture Notes 543, Springer-Verlag, Berlin.

Moser, J. (1966). A rapidly converging iteration method and nonlinear partial differential equations. *Ann. Scuola Norm. Pisa*, 265–315, 499–535.

Moser, J. (1973). *Stable and Random motion in dynamical systems*. Princeton University Press, Princeton.

Moser, J. (1976). Periodic orbits near an equilibrium and a theorem by A. Weinstein. *Comm. PAM*, **29**, 727–747.

Moulin, H. (1980). *Théorie des Jeux pour l'Économie et la Politique*. Hermann, Paris.

Nadler, S. B. (1969). Multivalued contraction mappings. *Pac. J. Math.*, **30**, 475–488.

Nash, J. (1950). Equilibrium points in n-person games. *Proc. Natl. Acad. Sci. USA*, **36**, 48–49.

Nash, J. (1950). The bargaining problem. *Econ.*, **18**, 128–160.

Nash, J. (1956). The imbedding problem for Riemannian manifolds. *Ann. Math.*, **63**, 20–63.

von Neumann, J. (1929). Zur allgemeinen Theorie des Masses. *Fund. Math.*, **13**, 73–116.

von Neumann, J. (1932). Zur Operatorenmethode in der klassischen Mechanik. *Ann. Math.*, **33**, 587–643.

von Neumann, (1937). Uber ein Okonomische Gleichungssystem und eine Verallgemeinerung des Brouwersschen Fixpunktsatzes. *Ergebnisse eines Math. Collo.*, **8**, 73–83.

von Neumann, J., and Morgenstern, O. (1944). *Theory of Games and Economic Behaviour*. Princeton University Press, Princeton.

Neustadt, L. (1976). *Optimization*. Princeton University Press.

Nikaido, H. (1968). *Convex Structure and Economic Theory*. Academic Press, New York.

Nirenberg. L. (1972). An abstract form of the Cauchy-Kowalewski theorem. *J. Diff. Geom.*, **6**, 561–576.

Nirenberg, L. (1974). *Topics in Nonlinear Functional Analysis*. Lecture Notes, New York University, New York.

Nirenberg, L. (1981). Variational and topological methods in nonlinear problems. *Bull. Am. Math. Soc.*, **4**, 267–301.

Numinskii, E. A. (1978). On differentiability of multifunctions. *Kibernetika*, **6**, 46–48.

Olech, C. (1965). A note concerning extremal points of a convex set. *Bull. Acad. Pol. Sc.*, **13**, 347–351.

Olech, C. (1965). A note concerning set-valued measurable functions. *Bull. Acad. Pol. Sc.*, **13**, 317–321.

Olech, C. (1967). Lexicographical order, range of integral and bang-bang principle. *Mathematical Theory of Control*, Academic Press, New York, 35–45.

Olech, C. (1968). Approximation of set-valued functions by continuous functions. *Colloq. Math.*, **19**, 285–293.

Olech, C. (1969). Existence theorems for optimal control problems involving multiple integrals. *J. Diff. Eq.*, **6**, 512–526.

Olech, C. (1975). Existence of solutions of nonconvex orientor fields. *Boll. Un. Mat. Ital.*, **11**(4), 189–197.

Opial, Z. (1967). Weak convergence of the sequence of successive approximations for nonexpansive mappings. *Bull. Am. Math. Soc.*, **73**, 591–597.

Palais, R. S. (1963). Morse theory on Hilbert manifolds. *Topology*, **2**, 299–340.

Palais, R. S. (1966). Ljusternik-Schnirelman theory on Banach manifolds. *Topology*, **5**, 115–132.

Palais, R. S. (1970). Critical point theory and the minimax principle. *Proc. Symp. Pure Math. Am. Math. Soc.*, **15**, 185–212.

Palais, R. S., and Smale, S. (1964). A generalized Morse theory. *Bull. Am. Math. Soc.*, **70**, 165–171.

Panagiotopoulos, P. D. (1980). Superpotentials in the sense of Clarke and in the sense of Warga and applications. *ZAMM*, **62**.

Panagiotopoulos, P. D. (to appear). Optimal control and parametrized identification of structures with convex or nonconvex strain energy density. *Solid Mech. Archives*.

Papageorgiou, N. S. (to appear). Nonsmooth analysis on partially ordered spaces, Part 2: nonconvex case, Clarke's theory. *Pac. J. Math*.

Pareto, V. (1909). *Manuel d'Économie Politique*. Girard & Brière, Lausanne.

Pascali, D., and Sburlan, S. (1979). *Nonlinear Mappings of Monotone Type*. Noordhoff, Leiden, Netherlands.

Pazy, A. (1977). On the asymptotic behaviour of iterates of nonexpensive mappings in Hibert spaces. *Isr. J. Math.*, **26**, 197–204.

Pazy, A. (1978). On the asymptotic behaviour of semigroups of nonlinear contractions in Hilbert spaces. *J. Funct. Anal.*, **27**, 293–307.

Pazy, A. (1979). Remarks on nonlinear egodic theory in Hilbert spaces. *Nonlinear Anal. TMA*, **3**, 863–871.

Pazy, A. (1983). *Semi-groups of Linear Operators and Applications*. Springer-Verlag, Berlin.

Pchenitchny, B. W. (1971). *Necessary Conditions for an Extremum*. Dekker, New York.